Breeding Strategies for Sustainable Spice Production

NIPA® GENX ELECTRONIC RESOURCES & SOLUTIONS P. LTD.
New Delhi-110 034

About the Author

Dr. Parshuram Sial. is currently serving as Breeder (Spices) and Associate Director of Research at the Regional Research & Technology Transfer Station, OUAT, Semiliguda. He also holds the responsibilities of Officer-in-Charge at HARS (OUAT), Pottangi, RRTTSS, Kalimela, and the AICRP on Spices, Pottangi. Dr. Sial has made significant contributions to crop improvement, having developed 4 desired mutant toria lines and 18 high-quality toria lines through wide hybridization, 22 bollworm-tolerant and 45 extra-long staple upland cotton lines, as well as identifying 6 intra-hirsutum hybrids with extra-long staple. He has standardized micropropagation protocols for tomato, brinjal, mint, and Gloriosa, and developed over 1,000 rice segregating materials, along with 165 turmeric, 124 ginger, and 2 black turmeric mutants. With 12 years of experience in spices research, he has handled 6 research projects, published 35 research papers, and guided 10 students.

Breeding Strategies for Sustainable Spice Production Intellectual Property Rights and Regulatory Environments

Parshuram Sial
Breeder (Spices)
Cum- Associate Director of Research
Regional Research & Technology Transfer Station
Odisha University of Agriculture & Technology
Semiliguda, Odisha, India

NIPA® GENX ELECTRONIC RESOURCES & SOLUTIONS P. LTD.
New Delhi-110 034

NIPA. GENX ELECTRONIC RESOURCES & SOLUTIONS P. LTD.

101,103, Vikas Surya Plaza, CU Block
L.S.C.Market, Pitam Pura, New Delhi-110 034
Ph : +91 11 4386 0225, 9717133558, 9540816132
E-mail: newindiapublishingagency@gmail.com
Website: www.niparesources.com

Print ISBN: 978-93-58875-40-9

ebook ISBN: 978-93-58873-48-1

Composed and Designed by NIPA®.

Preface

Spice breeding is the art and science of developing new varieties of spice plants with improved characteristics. This field has become increasingly crucial in our modern world, where consumer demands and market trends are rapidly evolving. Here's an expanded look at why this spice breeding is so important today:

- Flavour enhancement: Modern consumers are increasingly adventurous in their culinary experiences. They seek bold, unique flavours that can elevate their dishes. Spice breeding allows for the development of varieties with more intense or complex flavour profiles, meeting this growing demand for exciting taste experiences.
- Nutritional value: There's a rising awareness of the health benefits of spices. Breeding programs can focus on enhancing the nutritional content of spices, increasing their levels of beneficial compounds like antioxidants, vitamins, and minerals. This aligns with the current trend of consumers seeking functional foods that not only taste good but also contribute to their overall health.
- Sustainability: As climate change affects growing conditions worldwide, there's a need for spice varieties that are more resilient to environmental stresses. Breeding can develop plants that are drought-resistant, pest-resistant, or able to thrive in changing climatic conditions, ensuring a more stable and sustainable spice supply.
- Organic and natural products: The demand for organic and naturally grown spices has surged. Breeding programs can focus on developing varieties that perform well under organic farming conditions, reducing the need for synthetic pesticides and fertilizers.
- Consistency and quality: Modern food industry and discerning consumers demand consistency in flavour, aroma, and quality. Spice breeding can help develop varieties that provide more uniform characteristics, meeting these high standards.
- Yield improvement: With growing global demand for spices, there's pressure to increase production. Breeding can create higher-yielding varieties, helping to meet market demands while potentially reducing the environmental footprint of spice cultivation.

- Adaptation to new growing regions: As traditional spice-growing areas face challenges, there's interest in cultivating spices in new regions. Breeding can help adapt spice plants to these new environments, potentially diversifying and stabilizing the global spice supply.
- Novel culinary experiences: Innovative chefs and food manufacturers are always looking for new ingredients. Spice breeding can create entirely new flavor profiles or spice varieties, opening up exciting possibilities in cuisine and product development.
- Extended shelf life: In our globalized world, spices often travel long distances. Breeding programs can focus on developing varieties with better post-harvest characteristics, extending shelf life and maintaining quality during transportation and storage.
- Ethical sourcing: There's growing consumer interest in the provenance and ethical production of spices. Breeding programs can support local farming communities by developing varieties suited to specific regions, potentially improving livelihoods and supporting fair trade practices.

This Breeding Strategies for Sustainable Spice Production: Intellectual Property Rights and Regulatory Environments book is not just highlighting about improving plants, but about responding to and shaping the future of our food systems, culinary experiences, and even social and environmental considerations. It sets the stage for a book that explores the fascinating intersection of traditional agricultural practices, cutting-edge science, and evolving consumer preferences in the world of spices.

I am working as Senior Spices Breeder at High Altitude Research Station, Odisha University of Agriculture and Technology since 13.07.2012 and developed 6 varieties of ginger(Suprabha, Suravi, Suruchi, Subhada, Sourabh and Prayag), 6 varieties of turmeric (Roma, Surama, Rashmi, Surangi, Gourab and Ranga), one mango ginger and one black turmeric variety alongwith 5 rice varieties. I intended to write this book for sharing my views about breeding for future consumers preference in spices.

Readers of a book on Spice Breeding like students, scientists, professors, traders and consumers, etc., can expect to gain insights into various aspects of this specialized field. Here's what they might anticipate:

- Scientific Foundation
    - An introduction to plant genetics and breeding principles
    - Explanation of traditional and modern breeding techniques
    - Overview of the biological characteristics of common spice plants
- Breeding Objectives
    - Detailed discussion on breeding goals such as yield improvement, disease resistance, flavour enhancement, and climate adaptability

- How these objectives align with market demands and environmental challenges
- Breeding Techniques
    - In-depth coverage of various breeding methods including selective breeding, hybridization, mutation breeding, and genetic modification
    - Pros and cons of each technique and their applications in spice breeding
- Case Studies
    - Real-world examples of successful spice breeding programs
    - Challenges faced and how they were overcome
    - Impact of these programs on the spice industry and agriculture
- Emerging Technologies
    - Introduction to cutting-edge technologies in plant breeding such as CRISPR gene editing, high-throughput phenotyping, and marker-assisted selection
    - How these technologies are being applied to spice breeding
- Industry Insights
    - Current trends in the spice industry and how breeding programs are responding
    - Market analysis and future projections for various spices
- Sustainability and Ethics
    - Discussion on sustainable breeding practices
    - Ethical considerations in spice breeding, including biodiversity conservation and farmers' rights
- Regulatory Environment
    - Overview of regulations governing spice breeding in different regions
    - Implications for breeders, farmers, and the industry
- Practical Guidance:
    - Step-by-step processes for initiating and managing a spice breeding program
    - Tips for selecting parent plants, making crosses, and evaluating progeny
- Future Perspectives
    - Predictions for the future of spice breeding
    - Upcoming challenges and opportunities in the field
- Global Impact
    - How spice breeding affects global trade, culinary traditions, and food security
    - The role of spice breeding in addressing climate change and population growth

- Practical Exercises or Workshops
- For more hands-on oriented books, readers might expect some practical exercises or workshop ideas to apply the knowledge gained

By covering these aspects, the book would provide a comprehensive understanding of spice breeding, catering to both newcomers to the field and experienced professionals looking to deepen their knowledge. The exact depth and breadth of coverage would depend on the intended audience and the author's specific focus within this broad field.

Authorl

Contents

1

Introduction to Spices Breeding

Spices breeding is the process of developing new and improved varieties of spices through selective breeding, hybridization, and other genetic techniques. The main goals of spices breeding are to enhance desirable traits such as yield, flavour, aroma, disease resistance, and adaptability to different growing conditions.

Importance of spice breeding

- **Economic value**: Spices are economically valuable crops globally. Breeding can lead to higher yields and better quality, improving profitability for farmers.
- **Diversity and culinary use**: Spices like black pepper, cardamom, turmeric, and chili are used worldwide in various cuisines. Breeding helps to maintain and improve the diversity of these spices.
- **Medicinal properties**: Many spices have medicinal properties. Breeding can enhance the concentration of beneficial compounds in spices.

Breeding objectives

- **Yield improvement**: Increasing the quantity of spices produced per plant or per unit area.
- **Quality enhancement**: Improving the essential oil content, pungency, colour, and other quality traits.
- **Disease and pest resistance**: Developing varieties that are resistant to common diseases and pests affecting spices.
- **Climate adaptation**: Creating varieties that can withstand different climatic conditions, such as drought or high humidity.
- **Reduced crop duration**: Breeding for shorter growth cycles to increase the number of harvests per year.

Breeding techniques

- **Selection**: Identifying and selecting plants with desirable traits over several generations.

- **Hybridization**: Cross-breeding two different parent plants to produce offspring with traits from both parents.
- **Mutation breeding**: Inducing mutations through chemicals or radiation to create new traits.
- **Biotechnology**: Using genetic engineering, tissue culture, and molecular markers to enhance breeding efficiency and precision.

Challenges in spices breeding

- **Genetic diversity**: Some spices have limited genetic diversity, making breeding efforts challenging.
- **Long breeding cycles**: Many spice plants have long growth periods, which can delay breeding progress.
- **Pests and diseases**: Continuous evolution of pests and diseases can outpace the development of resistant varieties.

Breeding successes of spices crops

- **Black Pepper**: Breeding programs have developed varieties with higher yields and resistance to diseases like *Phytophthora.*
- **Cardamom**: Improved varieties with better aroma and disease resistance have been developed through breeding.
- **Turmeric**: Breeding has enhanced curcumin content, the compound responsible for turmeric's medicinal properties.

Future directions in spices breeding

- **Genomic approaches**: Advancements in genomics are helping to map important genes in spices, making breeding more targeted.
- **Climate-resilient varieties**: With climate change, there is a focus on developing spices that can thrive in changing environmental conditions.
- **Sustainable breeding practices**: Integrating traditional breeding with sustainable practices to ensure long-term productivity and environmental health.

  Spice breeding is a dynamic field that combines traditional agricultural practices with modern biotechnology to meet the growing demand for high-quality spices.

A) Importance of spices in agriculture and economy

A.1. Agricultural significance

a. **Diverse crop production:** Spices are grown in various agro-climatic regions, contributing to the diversification of cropping systems. This diversification helps in maintaining soil fertility, reducing the risk of pest and disease outbreaks, and ensuring food security.

b. **Income generation for farmers**: Spices are high-value crops, often generating more income per unit area than staple crops. Smallholder farmers can significantly increase their earnings by cultivating spices, particularly in regions where traditional crops may not be as profitable.

c. **Employment opportunities**: The cultivation, processing, and trade of spices provide employment to millions of people globally. This includes farm labor, processing workers, traders, and those involved in the packaging and distribution of spices.

d. **Sustainable farming practices**: Many spices are grown using sustainable agricultural practices, including organic farming. This not only preserves the environment but also meets the growing consumer demand for organic and sustainably sourced products.

e. **Agroforestry systems**: Spices like black pepper, cardamom, and vanilla are often grown as intercrops in agroforestry systems, where they contribute to the conservation of biodiversity, prevent soil erosion, and enhance the ecosystem's resilience.

A.2. Economic significance

a. **Global trade**: Spices have been a cornerstone of global trade for centuries. Even today, they constitute a significant portion of agricultural exports for many countries. The global spice trade is a multi-billion-dollar industry, with countries like India, Vietnam, Indonesia, and China being major exporters.

b. **Value addition and processing**: Spices undergo various levels of processing—from drying and grinding to extraction of essential oils. This value addition increases their economic worth, creating opportunities for small and medium enterprises (SMEs) in the spice processing sector.

c. **Market demand and price stability**: The demand for spices is relatively stable, with consistent global consumption. They are essential ingredients in cuisines worldwide, leading to sustained demand. This stability makes spice cultivation an attractive option for farmers.

d. **Cultural and Medicinal value**: Spices hold immense cultural significance in various cuisines and traditions. Additionally, many spices like turmeric, ginger, and cinnamon have medicinal properties, which are increasingly recognized in the pharmaceutical and nutraceutical industries. This adds another layer of economic importance as these industries grow.

e. **Export revenue**: For several developing countries, spices are a major source of export revenue. Countries like India and Sri Lanka rely heavily on spice exports to contribute to their foreign exchange earnings, boosting their overall economy.

f. **Tourism and culinary industry**: Spices are integral to the culinary identities of many regions, contributing to the tourism industry. Culinary tourism, where visitors explore a region's cuisine, often revolves around spices, further driving their economic value.
g. **Price volatility and market speculation**: Spices are often traded on commodity exchanges, where their prices can be influenced by market speculation, weather conditions, and geopolitical factors. This can lead to price volatility, which can be both a challenge and an opportunity for traders and investors.

A.3. Role in global economy

a. **Economic growth in developing countries**: For many developing nations, spice cultivation and export are crucial for economic growth. The revenue generated from spices supports infrastructure development, education, and healthcare in rural areas.
b. **Inflation control**: The stable demand and production of spices can help in controlling food inflation, particularly in regions where spices form a significant part of the diet.
c. **Market integration**: Spices are often traded internationally, integrating local economies into the global market. This integration allows for the transfer of knowledge, technology, and best practices, contributing to the overall development of the agricultural sector.

Spices play a vital role in both agriculture and the global economy. Their cultivation supports sustainable farming practices, generates significant income for farmers, and provides employment opportunities. Economically, spices are valuable export commodities that contribute to global trade, add value through processing, and sustain a stable market demand. Moreover, their cultural, medicinal, and culinary significance further enhances their economic importance, making spices indispensable in various industries worldwide.

B) Overview of spice crops

Spices are aromatic plant products that add flavour, colour, and fragrance to food. They also have medicinal properties and are used in various cultural and religious practices. Below is an overview of some important spice crops:

B.1. Black Pepper (*Piper nigrum*)

- **Origin**: Native to South India, particularly the Western Ghats.
- **Uses**: Known as the "King of Spices," black pepper is widely used as a seasoning and in medicinal applications.

- **Cultivation**: Grown in tropical regions, typically in humid climates. It is a perennial vine that requires a support system, such as trees or poles, to grow.
- **Major producers**: India, Vietnam, Indonesia, Brazil.

B.2. Small Cardamom (*Elettaria cardamomum*)

- **Origin**: Native to the Indian subcontinent.
- **Uses**: Used for flavouring food, in traditional medicine, and as a breath freshener. It is often referred to as the "Queen of Spices."
- **Cultivation**: Grown in tropical climates with heavy rainfall. It requires a shady environment and is often grown in forested regions or under canopy crops.
- **Major producers**: India, Guatemala, Sri Lanka.

B.3. Large Cardamom (*Amomum subulatum*)

- **Origin**: Native to the Eastern Himalayas, particularly in Nepal, Bhutan, and India.
- **Uses**: Large cardamom is used in cooking, especially in South Asian cuisine, and in traditional medicine. It has a smoky flavour due to the traditional drying process over open fires.
- **Cultivation**: Grown in the cool, shaded, and moist forests of the Himalayan regions. It requires well-drained loamy soil with plenty of organic matter.
- **Major producers**: Nepal, India (Sikkim, Darjeeling), Bhutan.

B.4. Ginger (*Zingiber officinale*)

- **Origin**: Native to Southeast Asia.
- **Uses**: Used as a spice and medicine. It is known for its pungent flavour and is used in both fresh and dried forms.
- **Cultivation**: Grows well in warm, humid climates with well-drained soil. Ginger is propagated from rhizomes and harvested after 8-10 months.
- **Major producers**: India, China, Nigeria, Indonesia.

B.5. Turmeric (*Curcuma longa*)

- **Origin**: Native to the Indian subcontinent and Southeast Asia.
- **Uses**: Known for its bright yellow colour, turmeric is used as a spice, dye, and in traditional medicine. It is a key ingredient in curry powder.
- **Cultivation**: Prefers warm, humid climates and well-drained soil. Turmeric is propagated from rhizomes and harvested after 7-9 months.
- **Major producers**: India, Pakistan, Bangladesh, China.

B.6. Mango Ginger (*Curcuma amada*)

- **Origin**: Native to the Indian subcontinent.
- **Uses**: Mango ginger has a distinctive mango-like aroma and is used in pickles, chutneys, and as a flavouring in various dishes. It also has medicinal uses, particularly in traditional Ayurvedic practices.
- **Cultivation**: Grows well in warm, humid climates with well-drained, loamy soil. Similar to turmeric, it is propagated through rhizomes and harvested after 7-9 months.
- **Major producers**: India, Thailand, Indonesia.

B.7. Black Turmeric (*Curcuma caesia*)

- **Origin**: Native to Northeast India and parts of Southeast Asia.
- **Uses**: Black turmeric is primarily used in traditional medicine and rituals rather than as a culinary spice. It is known for its anti-inflammatory and antioxidant properties.
- **Cultivation**: Prefers warm, humid climates and well-drained soil. It is propagated by rhizomes and requires about 8-9 months for maturity.
- **Major producers**: India, particularly in the states of Odisha, West Bengal, and Assam.

B.8. Allspice (*Pimenta dioica*)

- **Origin**: Native to the Caribbean, Central America, and southern Mexico.
- **Uses**: Combines the flavours of cinnamon, nutmeg, and cloves. It is used in both sweet and savory dishes.
- **Cultivation**: Grows in tropical climates, particularly in well-drained soils with adequate rainfall. Allspice is propagated from seeds.
- **Major producers**: Jamaica, Mexico, Honduras.

B.9. Clove (*Syzygium aromaticum*)

- **Origin**: Native to the Maluku Islands (Indonesia).
- **Uses**: Used in cooking, as a medicinal herb, and in perfumery. Cloves are also a key ingredient in cigarettes in some cultures.
- **Cultivation**: Requires tropical climates with regular rainfall. Clove trees take 6-8 years to start bearing fruit and can produce for over 50 years.
- **Major producers**: Indonesia, Madagascar, Tanzania, Sri Lanka.

B.10. Cinnamon (*Cinnamomum verum*)

- **Origin**: Native to Sri Lanka, India, and Myanmar.
- **Uses**: Cinnamon is a popular spice used in sweet and savory dishes,

beverages, and traditional medicine. It is valued for its warm, sweet flavour and aromatic qualities.

- **Cultivation**: Cinnamon is grown in tropical climates with well-drained, sandy loam soil. It is a small evergreen tree, and the inner bark is harvested to produce the spice.
- **Major producers**: Sri Lanka, Indonesia, China, Vietnam.

B.11. Coriander (*Coriandrum sativum*)

- **Origin**: Native to regions spanning from southern Europe to southwestern Asia.
- **Uses**: Both the seeds and leaves (cilantro) are used in cooking. The seeds have a citrus flavour, while the leaves are more herbaceous.
- **Cultivation**: Grows well in temperate climates with dry summers. Coriander is an annual plant that prefers well-drained soil.
- **Major producers**: India, Morocco, Canada, Russia.

B.12. Fennel (*Foeniculum vulgare*)

- **Origin**: Native to the Mediterranean region.
- **Uses**: Both seeds and bulbs are used in cooking, with a flavour similar to anise. Fennel seeds are also used in traditional medicine.
- **Cultivation**: Prefers temperate climates and well-drained soil. Fennel is a hardy, perennial herb but is often grown as an annual.
- **Major producers**: India, China, Egypt, Turkey.

B.13. Fenugreek (*Trigonella foenum-graecum*)

- **Origin**: Native to the Mediterranean region.
- **Uses**: The seeds and leaves are used as a spice and herb. Fenugreek is also known for its medicinal properties, especially in controlling blood sugar levels.
- **Cultivation**: Grows in well-drained, loamy soil under cool climates. Fenugreek is an annual plant.
- **Major producers**: India, Egypt, Morocco.

B.14. Cumin (*Cuminum cyminum*)

- **Origin**: Native to the eastern Mediterranean and South Asia.
- **Uses**: Widely used in spice blends like curry powder and garam masala. It has a warm, earthy flavour.
- **Cultivation**: Requires a warm climate and well-drained soil. Cumin is an annual plant that is typically sown in the spring.
- **Major producers**: India, Syria, Turkey, Iran.

B.15. Ajwain (*Trachyspermum ammi*)

- **Origin**: Native to India and the Middle East.
- **Uses**: The seeds have a strong, thyme-like flavour and are used in cooking and traditional medicine, especially for digestive issues.
- **Cultivation**: Grows well in arid and semi-arid climates with sandy or loamy soils. Ajwain is an annual plant.
- **Major producers**: India, Iran, Egypt.

B.16. Saffron (*Crocus sativus*)

- **Origin**: Believed to be native to Greece or Southwest Asia.
- **Uses**: Saffron is the world's most expensive spice, known for its unique flavour, aroma, and vibrant golden colour. It is used in cooking, especially in dishes like paella, risotto, and biryani, as well as in traditional medicines and cosmetics.
- **Cultivation**: Saffron is grown in arid and semi-arid climates with well-drained, sandy loam soil. It is propagated by corms (bulbs) and requires a dry climate with minimal rainfall during the flowering period.
- **Major producers**: Iran (the largest producer), Spain, India (particularly in Kashmir), and Afghanistan.

Spice crops are integral to global agriculture and culinary traditions. Each spice has unique cultivation requirements, uses, and economic significance. These crops contribute not only to the diversity of food flavours but also to the agricultural economy, especially in regions where they are produced. The global demand for spices continues to grow, driven by their culinary, medicinal, and cultural importance. Each of these spice crops plays a unique role in global agriculture, cuisine, and traditional medicine. They are cultivated in specific regions that provide the ideal growing conditions, and they hold significant economic and cultural importance. From the rich aroma of cinnamon and the distinctive flavour of large cardamom to the precious strands of saffron, these spices contribute to the diversity and richness of culinary traditions around the world.

C) Historical perspective on spices breeding

Spices breeding has a long and rich history that intertwines with the development of agriculture, trade, and cultural practices. The domestication, cultivation, and selective breeding of spice crops have evolved over thousands of years, driven by the need to enhance flavour, yield, disease resistance, and adaptability to various climates.

C.1. Black pepper (*Piper nigrum*)

- **Historical significance**: Black pepper, often referred to as the "King of Spices," was one of the earliest traded spices, with a history dating back to ancient civilizations in India. It was highly prized in the Roman Empire and was a driving force behind European exploration in the 15th century.
- **Breeding efforts**: Traditional breeding efforts focused on improving yield, disease resistance (especially against foot rot and quick wilt), and the quality of the berries. Early breeding was primarily through selection of superior plants, but modern breeding includes cross-breeding and the use of biotechnology for developing improved varieties.

C.2. Small cardamom (*Elettaria cardamomum*)

- **Historical significance**: Known as the "Queen of Spices," small cardamom has been cultivated in India for over 2,000 years. It was an important trade commodity in the ancient world, especially in Arab and European markets.
- **Breeding efforts**: Traditional breeding involved selecting plants with high yields, larger capsules, and resistance to pests like the cardamom thrips. In recent years, hybridization and tissue culture techniques have been employed to develop high-yielding, disease-resistant varieties.

C.3. Large Cardamom (*Amomum subulatum*)

- **Historical significance**: Native to the Eastern Himalayas, large cardamom has been cultivated for centuries in Nepal, Bhutan, and India. It became a significant cash crop in these regions during the 19th and 20th centuries.
- **Breeding efforts**: Breeding has primarily focused on improving yield, disease resistance (especially against fungal diseases), and adapting to different altitudes. Traditional selection methods have been supplemented by more recent research into clonal propagation and varietal development.

C.4. Cinnamon (*Cinnamomum verum*)

- **Historical significance**: Cinnamon has been used since ancient times, with records of its use in Egypt as early as 2000 BCE. It was a valuable commodity in the spice trade, particularly between Asia and Europe.
- **Breeding efforts**: Historically, cinnamon breeding involved selecting trees with higher bark yield and better aromatic properties. Modern breeding focuses on enhancing oil content, disease resistance, and tree growth characteristics.

C.5. Vanilla (*Vanilla planifolia*)

- **Historical significance**: Native to Mexico, vanilla was introduced to Europe in the 16th century by Spanish explorers. It became a popular flavouring, but it wasn't until the 19th century that it was widely cultivated outside of Mexico, particularly in Madagascar and Reunion.

- **Breeding efforts**: Vanilla breeding has focused on improving yield, reducing the time to flowering, and enhancing disease resistance, especially against root rot and fungal diseases. Breeding has also sought to improve the quality and quantity of vanillin, the main flavour compound.

C.6. Ginger (*Zingiber officinale*)

- **Historical significance**: Ginger has been used in Asia for over 5,000 years for culinary and medicinal purposes. It was one of the first spices to be traded from Asia to Europe, where it was highly valued.
- **Breeding efforts**: Early breeding focused on selecting rhizomes with higher yield and better flavour. Modern breeding efforts include developing varieties that are resistant to diseases such as bacterial wilt and rhizome rot, as well as improving storage quality.

C.7. Turmeric (*Curcuma longa*)

- **Historical significance**: Turmeric has been cultivated in India for over 4,000 years and is a key ingredient in Ayurvedic medicine. It was also used as a dye and a food preservative.
- **Breeding efforts**: Traditional breeding involved selecting plants with higher curcumin content (the active compound in turmeric) and improved rhizome yield.

C.8. Mango Ginger (*Curcuma amada*)

- **Historical significance**: Mango ginger, a lesser-known spice, has been used in India for centuries, primarily in pickles and traditional medicine. It is closely related to turmeric and ginger.
- **Breeding efforts**: Breeding of mango ginger has been limited, with efforts primarily focused on selecting plants with better flavour and yield. There has been some work on improving disease resistance and adaptability to different growing conditions.

C.9. Black Turmeric (*Curcuma caesia*)

- **Historical significance**: Black turmeric has been used traditionally in India and Southeast Asia for its medicinal properties and in rituals. It is less commonly used as a culinary spice.
- **Breeding efforts**: Due to its specialized use, breeding efforts have been limited and mostly focused on preserving its unique characteristics while improving yield and disease resistance.

C.10. Clove (*Syzygium aromaticum*)

- **Historical significance**: Clove has been a highly valued spice since ancient times, originating from the Maluku Islands in Indonesia. It was one of the spices that spurred the European exploration of the East.
- **Breeding efforts**: Breeding has focused on increasing oil content (especially eugenol), improving resistance to diseases such as clove dieback, and enhancing yield. Traditional methods have been used alongside more recent biotechnological approaches.

C.11. Allspice (*Pimenta dioica*)

- **Historical significance**: Allspice, native to the Caribbean and Central America, was introduced to Europe in the 16th century. It became popular for its combined flavours of cinnamon, nutmeg, and clove.
- **Breeding efforts**: Breeding has focused on improving yield, essential oil content, and disease resistance. Traditional selection methods dominate, though there is increasing interest in genetic improvement techniques.

C.12. Coriander (*Coriandrum sativum*)

- **Historical significance**: Coriander is one of the oldest spices, with evidence of its use dating back to ancient Egypt and Greece. It has been cultivated for its seeds and leaves (cilantro) for thousands of years.
- **Breeding efforts**: Breeding has aimed at improving seed yield, flavour, and resistance to bolting (premature flowering). Disease resistance, particularly against powdery mildew, has also been a focus.

C.13. Fenugreek (*Trigonella foenum-graecum*)

- **Historical significance**: Fenugreek has been used since ancient times in the Mediterranean region and South Asia, both as a spice and for medicinal purposes. It was also used as livestock fodder.
- **Breeding efforts**: Efforts have focused on improving yield, reducing bitterness in seeds, and enhancing resistance to diseases like powdery mildew. Breeding for higher levels of beneficial compounds such as diosgenin has also been explored.

C.14. Fennel (*Foeniculum vulgare*)

- **Historical significance**: Fennel has a long history of use in the Mediterranean and Near East for culinary, medicinal, and aromatic purposes. It was also used by the ancient Egyptians and Greeks.
- **Breeding efforts**: Breeding has targeted improving seed yield, oil content (anethole), and resistance to environmental stressors like drought. Development of varieties with uniform flowering and maturity has been another focus.

C.15. Cumin (*Cuminum cyminum*)

- **Historical significance**: Cumin has been cultivated in the Mediterranean and South Asia for over 4,000 years, used extensively in cooking and traditional medicine.
- **Breeding efforts**: Cumin breeding has concentrated on improving seed yield, flavour, and disease resistance, particularly against Fusarium wilt and Alternaria blight. Efforts have also been made to develop early-maturing varieties.

C.16. Ajjwain (*Trachyspermum ammi*)

- **Historical significance**: Ajwain has been used in the Indian subcontinent and the Middle East for centuries, both as a culinary spice and for its medicinal properties, especially in treating digestive issues.
- **Breeding efforts**: Ajwain breeding has been relatively limited, with traditional selection for high-yielding plants with strong flavour being the primary focus. Recent efforts include improving resistance to pests and diseases.

The history of spice breeding is deeply rooted in the cultural and economic evolution of human societies. From ancient times to the present day, spice breeding has been driven by the need to enhance the desirable traits of these crops, such as yield, flavour, and resistance to diseases. While traditional methods of selection and hybridization have been the cornerstone of breeding efforts, modern techniques, including biotechnology and molecular breeding, are increasingly being employed to develop superior spice varieties. This ongoing evolution in spice breeding continues to ensure the sustainability and profitability of these valuable crops in the global agricultural landscape.

D) Active ingredients found in spices for medical property

Many spices have active chemical compounds that contribute to their medicinal properties. Here's an overview of the key active chemicals found in these spices:

1. Ginger (*Zingiber officinale*)

- **Gingerol**: Anti-inflammatory, antioxidant, and anti-nausea.
- **Shogaol**: Anti-inflammatory, antioxidant, and anti-cancer.
- **Zingerone**: Anti-inflammatory and antimicrobial.

2. Turmeric (*Curcuma longa*)

- **Curcumin**: Anti-inflammatory, antioxidant, anticancer, and antimicrobial.
- **Turmerone**: Neuroprotective and anti-inflammatory.
- **Bisdemethoxycurcumin**: Anti-inflammatory and antioxidant.

3. Mango Ginger (*Curcuma amada*)

- **Curcumin**: Anti-inflammatory and antioxidant.
- **Camphor**: Analgesic, antimicrobial, and anti-inflammatory.
- **Terpenoids**: Antimicrobial and anti-inflammatory.

4. Black Turmeric (*Curcuma caesia*)

- **Curcumin**: Anti-inflammatory and antioxidant.
- **Camphor**: Analgesic and antimicrobial.
- **Cineole**: Antiseptic, anti-inflammatory, and bronchodilator.

5. Black Pepper (*Piper nigrum*)

- **Piperine**: Enhances bioavailability of other compounds, anti-inflammatory, antioxidant, and analgesic.
- **Chavicine**: Anti-inflammatory and antimicrobial.

6. Cinnamon (*Cinnamomum verum*)

- **Cinnamaldehyde**: Anti-inflammatory, antimicrobial, and antifungal.
- **Eugenol**: Antimicrobial, antioxidant, and anti-inflammatory.
- **Cinnamic acid**: Antioxidant and anti-inflammatory.

7. Cardamom (*Elettaria cardamomum*)

- **Cineole**: Antimicrobial, expectorant, and anti-inflammatory.
- **Terpinene**: Antioxidant and antimicrobial.
- **Limonene**: Antioxidant, anti-inflammatory, and antifungal.

8. Clove (*Syzygium aromaticum*)

- **Eugenol**: Antimicrobial, analgesic, anti-inflammatory, and antioxidant.
- **Caryophyllene**: Anti-inflammatory and antimicrobial.
- **Tannins**: Antioxidant and astringent.

9. Nutmeg (*Myristica fragrans*)

- **Myristicin**: Neuroprotective, antioxidant, and anti-inflammatory.
- **Elemicin**: Sedative and anti-anxiety.
- **Safrole**: Antibacterial and insecticidal (in small amounts).

10. Allspice (*Pimenta dioica*)

- **Eugenol**: Antimicrobial, anti-inflammatory, and antioxidant.
- **Quercetin**: Antioxidant and anti-inflammatory.
- **Gallic acid**: Antioxidant and antifungal.

11. Coriander (*Coriandrum sativum*)

- **Linalool**: Antioxidant, antimicrobial, and anxiolytic.

- **Geraniol**: Antimicrobial and anti-inflammatory.
- **Pinene**: Anti-inflammatory and bronchodilator.

12. Vanilla (*Vanilla planifolia*)

- **Vanillin**: Antioxidant, anti-inflammatory, and anticancer.
- **Eugenol**: Antimicrobial and anti-inflammatory (in small amounts).

13. Fennel (*Foeniculum vulgare*)

- **Anethole**: Anti-inflammatory, antimicrobial, and estrogenic.
- **Estragole**: Antioxidant and antimicrobial.
- **Fenchone**: Digestive stimulant and anti-inflammatory.

14. Fenugreek (*Trigonella foenum-graecum*)

- **Trigonelline**: Antidiabetic, neuroprotective, and antioxidant.
- **Diosgenin**: Anti-inflammatory, anticancer, and cholesterol-lowering.
- **Saponins**: Antioxidant and immune-boosting.

15. Cumin (*Cuminum cyminum*)

- **Cuminaldehyde**: Antimicrobial, antioxidant, and digestive aid.
- **Thymol**: Antimicrobial and antifungal.
- **Carvacrol**: Antibacterial and anti-inflammatory.

16. Ajwain (*Trachyspermum ammi*)

- **Thymol**: Antimicrobial, antifungal, and digestive stimulant.
- **P-Cymene**: Antioxidant and anti-inflammatory.
- **Limonene**: Antifungal and antioxidant.

These active compounds are responsible for the spices' wide range of medicinal properties, including anti-inflammatory, antioxidant, antimicrobial, digestive, and neuroprotective effects.

2

Basics of Spices Breeding

A. Principles of spices breeding

Plant breeding is a science-based technique aimed at improving the genetic makeup of plants to develop desirable traits such as increased yield, resistance to diseases and pests, improved quality, and adaptability to various environmental conditions. When applied to spices, these principles are adapted to meet the unique characteristics and challenges of spice crops.

A.1. Understanding the breeding objectives

- **Yield improvement**: Increasing the yield of spices like black pepper, ginger, and turmeric is a primary goal. This involves selecting varieties that produce more berries, rhizomes, or seeds.
- **Quality enhancement**: For spices like vanilla, cinnamon, and saffron, the focus is on enhancing the quality of the product, such as increasing vanillin content, oil content, or curcumin levels.
- **Disease and pest resistance**: Developing resistant varieties is crucial for spices prone to specific diseases, like black pepper's vulnerability to foot rot or large cardamom's susceptibility to fungal diseases.
- **Adaptation to environmental conditions**: Breeding efforts often aim to produce varieties that can thrive in specific climates, such as the cool, shaded forests required for large cardamom or the arid conditions suitable for cumin.

A.2. Selection

- **Mass selection**: This involves selecting superior plants based on desirable traits and using them for the next generation. It's commonly used in spices like coriander and fennel where individual plants with better yield or flavour are selected.
- **Clonal selection**: Many spices, including black pepper, ginger, turmeric, and vanilla, are propagated vegetatively. Clonal selection involves selecting the best performing clones and using them for propagation to maintain uniformity and desirable traits.

A.3. Hybridization

- **Intraspecific hybridization**: Crossing two plants within the same species to combine desirable traits, such as crossing different varieties of small cardamom to improve yield and disease resistance.
- **Interspecific hybridization**: While less common in spices due to the complexity of cross-species breeding, it can be used to introduce new traits from related species. This method requires careful consideration due to potential compatibility issues.

A.4. Mutation breeding

- **Induced mutations**: Exposing plant materials to chemicals or radiation to induce mutations can result in new traits, such as improved disease resistance or higher yield. This technique has been applied to spices like ginger and turmeric to develop varieties with enhanced traits.

A.5. Polyploidy

- **Induced polyploidy**: Doubling the chromosome number of a plant can result in increased size, yield, and other desirable traits. This method has been used in spices like ginger and turmeric to develop larger, more robust plants.

A.6. Biotechnological approaches

- **Tissue culture**: This is extensively used in the micropropagation of spices like vanilla, black pepper, and ginger. Tissue culture allows for the rapid multiplication of disease-free plants and the production of uniform planting material.
- **Genetic engineering**: Though still in the early stages for many spices, genetic engineering holds potential for introducing traits like pest resistance or improved quality into spice crops. For example, efforts are ongoing to develop genetically modified vanilla with higher vanillin content.
- **Marker-Assisted Selection (MAS)**: This involves using molecular markers to select plants with desirable traits at the seedling stage. It's particularly useful in spices with long breeding cycles, like black pepper and vanilla.

A.7. Heterosis (Hybrid Vigour)

- **Exploitation of Hybrid Vigour**: For spices like black pepper and small cardamom, exploiting heterosis can result in plants with superior traits such as higher yield, better quality, and increased resistance to diseases. Creating hybrids by crossing genetically diverse parents can lead to vigorous offspring.

A.8. Breeding for biotic and abiotic stress resistance

- **Biotic stress resistance**: Developing resistance to diseases and pests is a key breeding objective in spice crops. For instance, breeding black pepper varieties resistant to foot rot and developing ginger varieties resistant to bacterial wilt are critical for maintaining productivity.
- **Abiotic stress resistance**: Breeding spices that can withstand environmental stresses such as drought, high salinity, or extreme temperatures is important for ensuring stable yields. For example, cumin and fennel breeding often focuses on developing varieties that are more tolerant to arid conditions.

A.9. Participatory breeding

- **Involvement of farmers**: In many regions, especially where spices are grown in traditional systems, involving farmers in the breeding process helps ensure that the developed varieties meet local needs and preferences. This approach is often used in the breeding of spices like coriander, fennel, and ajwain.

A.10. Conservation and utilization of genetic resources

- **Germplasm conservation**: Preserving the genetic diversity of spice crops is crucial for breeding programs. This involves maintaining seed banks, field gene banks, and cryopreservation of plant material.
- **Utilization of wild relatives**: Wild relatives of spice crops are often more resistant to diseases and environmental stresses. Incorporating genes from these wild species into cultivated varieties can enhance the resilience and adaptability of spice crops.

The principles of plant breeding applied to spices involve a combination of traditional and modern techniques, tailored to the unique requirements of each crop. Whether it's improving yield, enhancing quality, or developing disease-resistant varieties, these principles are essential for the continued success and sustainability of spice cultivation. As global demand for spices continues to grow, the role of plant breeding in developing superior spice varieties becomes increasingly important.

B. Breeding objectives specific to spices (yield, quality, disease resistance, etc.)

When breeding spice crops, specific objectives are tailored to enhance traits like yield, quality, disease resistance, and adaptability to different environmental conditions. Below are the breeding objectives specific to various spice crops:

B.1. Black Pepper (*Piper nigrum*)

- **Yield improvement**: Increase the number and size of berries, improve the productivity of vines, and extend the harvesting period.
- **Quality enhancement**: Enhance the pungency, aroma, and oleoresin content in berries.
- **Disease resistance**: Develop resistance to major diseases such as foot rot (Phytophthora capsici) and quick wilt.
- **Drought tolerance**: Improve the plant's ability to withstand dry conditions, particularly in regions with variable rainfall.

B.2. Small Cardamom (*Elettaria cardamomum*)

- **Yield improvement**: Increase the number of capsules per spike and improve capsule size.
- **Quality enhancement**: Enhance essential oil content, particularly cineole and terpinyl acetate, which contribute to flavour and aroma.
- **Disease resistance**: Develop resistance to diseases like cardamom mosaic virus, thrips, and capsule rot.
- **Early maturity**: Breed varieties that mature earlier to reduce the cultivation period and allow for multiple harvests.

B.3. Large Cardamom (*Amomum subulatum*)

- **Yield improvement**: Enhance the number of spikes per plant and the size of the capsules.
- **Quality enhancement**: Improve the essential oil content and maintain the unique smoky flavour.
- **Disease resistance**: Breed for resistance to fungal diseases like Colletotrichum blight and Fusarium wilt.
- **Cold tolerance**: Enhance tolerance to low temperatures in high-altitude regions.

B.4. Cinnamon (*Cinnamomum verum*)

- **Yield improvement**: Increase bark yield per tree and enhance the regrowth rate after harvesting.
- **Quality enhancement**: Improve cinnamaldehyde content, which determines the spice's flavour and aroma.
- **Disease resistance**: Develop resistance to diseases like leaf spot and shoot borer.
- **Uniformity in bark thickness**: Breed for consistent bark thickness, which is crucial for processing.

B.5. Vanilla (*Vanilla planifolia*)

- **Yield improvement**: Increase the number of beans per plant and improve the bean size.
- **Quality enhancement**: Enhance vanillin content, which is the primary flavour compound.
- **Disease resistance**: Develop resistance to diseases such as root rot and fusarium wilt.
- **Flowering synchronization**: Improve synchronization of flowering to optimize hand pollination efforts.

B.6. Ginger (*Zingiber officinale*)

- **Yield improvement**: Increase rhizome size and overall biomass yield.
- **Quality enhancement**: Enhance gingerol and essential oil content, which contribute to flavour and pungency.
- **Disease resistance**: Develop resistance to bacterial wilt, rhizome rot, and leaf spot diseases.
- **Post-harvest shelf life**: Improve storage properties to extend the shelf life of fresh ginger.

B.7. Turmeric (*Curcuma longa*)

- **Yield improvement**: Increase rhizome yield per plant and enhance the size and number of rhizomes.
- **Quality enhancement**: Increase curcumin content, which is the key compound for colour and medicinal properties.
- **Disease resistance**: Develop resistance to diseases like rhizome rot and leaf blight.
- **Adaptability to different climates**: Breed for varieties that can thrive in a range of climatic conditions, from tropical to subtropical regions.

B.8. Mango Ginger (*Curcuma amada*)

- **Yield improvement**: Enhance the number and size of rhizomes.
- **Quality enhancement**: Improve the mango-like aroma and flavour intensity.
- **Disease resistance**: Develop resistance to common diseases affecting rhizomes, such as rot.
- **Drought tolerance**: Improve the plant's ability to grow in drier regions with less water.

B.9. Black Turmeric (*Curcuma caesia*)

- **Yield improvement**: Increase the size and number of rhizomes produced per plant.
- **Quality enhancement**: Enhance the medicinal properties, focusing on active compounds with antioxidant and anti-inflammatory properties.
- **Disease resistance**: Develop resistance to rhizome rot and other fungal infections.
- **Adaptability**: Breed for greater adaptability to different soil types and climatic conditions.

B.10. Clove (*Syzygium aromaticum*)

- **Yield improvement**: Increase the number of flower buds per tree and improve the overall yield.
- **Quality enhancement**: Enhance the eugenol content, which is the primary essential oil component.
- **Disease resistance**: Develop resistance to clove dieback and other fungal diseases.
- **Drought tolerance**: Breed for varieties that can tolerate dry spells without significant yield loss.

B.11. Allspice (*Pimenta dioica*)

- **Yield improvement**: Increase the number of berries per tree and improve overall productivity.
- **Quality enhancement**: Improve the concentration of essential oils, particularly eugenol, myrcene, and cineole.
- **Disease resistance**: Develop resistance to pests and diseases that affect leaves and fruits.
- **Adaptability**: Breed for adaptability to different soil types and environmental conditions.

B.12. Coriander (*Coriandrum sativum*)

- **Yield improvement**: Increase seed yield and uniformity in seed size.
- **Quality enhancement**: Improve essential oil content, especially linalool, which contributes to the flavour.
- **Disease resistance**: Develop resistance to powdery mildew and stem gall disease.
- **Early maturity**: Breed for early-maturing varieties to avoid heat stress during seed filling.

B.13. Fenugreek (*Trigonella foenum-graecum*)

- **Yield improvement**: Enhance seed yield and uniformity in seed size.
- **Quality enhancement**: Increase diosgenin content and improve flavour by reducing bitterness.
- **Disease resistance**: Develop resistance to diseases like powdery mildew and rust.
- **Salt tolerance**: Improve tolerance to saline soils, expanding cultivation areas.

B.14. Fennel (*Foeniculum vulgare*)

- **Yield improvement**: Increase seed yield and ensure uniformity in seed size.
- **Quality enhancement**: Enhance essential oil content, particularly anethole, which gives fennel its characteristic flavour.
- **Disease resistance**: Develop resistance to fungal diseases like powdery mildew and rust.
- **Drought tolerance**: Improve tolerance to drought conditions, allowing cultivation in arid regions.

B.15. Cumin (*Cuminum cyminum*)

- **Yield improvement**: Increase seed yield and ensure uniformity in seed size.
- **Quality enhancement**: Enhance essential oil content, focusing on cuminaldehyde, which contributes to the flavour.
- **Disease resistance**: Develop resistance to diseases such as Fusarium wilt and blight.
- **Drought tolerance**: Breed for varieties that can withstand arid conditions without significant yield loss.

B.16. Ajwain (*Trachyspermum ammi*)

- **Yield improvement**: Increase seed yield and uniformity in seed size.
- **Quality enhancement**: Improve essential oil content, particularly thymol, which gives ajwain its characteristic flavour.
- **Disease resistance**: Develop resistance to diseases affecting seeds and leaves, such as rust and powdery mildew.
- **Early maturity**: Breed for early-maturing varieties to avoid stress during flowering and seed set.

The breeding objectives for spice crops are tailored to enhance specific traits that meet the needs of farmers, processors, and consumers. Whether it's improving yield, enhancing flavour and aroma, developing resistance to diseases and pests, or adapting to environmental stresses, these objectives play a crucial role in ensuring the sustainability and profitability of spice cultivation.

C. Breeding methods for spices (selection, hybridization, mutation breeding, etc.)

Breeding spice crops involves various methods to achieve desirable traits such as higher yield, better quality, disease resistance, and adaptability. The choice of breeding method depends on the reproductive biology of the crop, the traits to be improved, and the available resources. Below are the key breeding methods applied to various spice crops:

C.1. Black Pepper (*Piper nigrum*)

- **Selection**: Mass selection and clonal selection are commonly used. Superior vines with high yield, disease resistance, and better quality are selected and propagated vegetatively.
- **Hybridization**: Limited due to the vegetative propagation of black pepper. However, intra-specific hybridization is used to combine desirable traits from different varieties.
- **Mutation breeding**: Induced mutations have been explored to create variability for disease resistance and yield traits.
- **Tissue culture**: Micropropagation through tissue culture is widely used to produce disease-free planting material and maintain genetic uniformity.

C.2. Small Cardamom (*Elettaria cardamomum*)

- **Selection**: Clonal selection is a primary method, where superior plants with desirable traits such as high yield and quality are selected.
- **Hybridization**: Intraspecific hybridization is used to combine traits like yield, quality, and disease resistance. Crosses between different cultivars are made to create hybrids with improved characteristics.
- **Mutation breeding**: Used to induce variability for traits like early maturity and disease resistance.
- **Tissue culture**: Somatic embryogenesis and micropropagation are employed to produce uniform and disease-free planting materials.

C.3. Large Cardamom (*Amomum subulatum*)

- **Selection**: Mass selection of superior plants for higher yield and disease resistance is common.
- **Hybridization**: Intraspecific hybridization is less common but used to combine desirable traits like yield and disease resistance.

- **Mutation breeding**: Not extensively used but can be explored for developing disease-resistant and high-yielding varieties.
- **Tissue culture**: Micropropagation is employed to produce large quantities of uniform planting material.

C.4. Cinnamon (*Cinnamomum verum*)

- **Selection**: Selection of superior trees based on bark yield, oil content, and disease resistance.
- **Hybridization**: Intraspecific hybridization is rare but can be used to combine traits like bark quality and resistance to diseases.
- **Mutation breeding**: Induced mutations can be used to develop new varieties with improved traits like higher oil content.
- **Tissue culture**: Tissue culture techniques are used for the mass multiplication of elite trees.

C.5. Vanilla (*Vanilla planifolia*)

- **Selection**: Clonal selection of high-yielding and disease-resistant vines is the primary method.
- **Hybridization**: Hand-pollination techniques are used for hybridization, aiming to combine traits like vanillin content and disease resistance.
- **Mutation breeding**: Limited use but has potential for developing disease-resistant varieties.
- **Tissue culture**: Widely used for micropropagation and the production of disease-free planting material.

C.6. Ginger (*Zingiber officinale*)

- **Selection**: Clonal selection is common, focusing on rhizome size, yield, and disease resistance.
- **Hybridization**: Sexual hybridization is rare due to the sterility of flowers, but some interspecific crosses have been attempted to introduce new traits.
- **Mutation breeding**: Induced mutations are used to create variability in traits such as yield, quality, and disease resistance.
- **Tissue culture**: Micropropagation through tissue culture is extensively used to produce disease-free planting material.

C.7. Turmeric (*Curcuma longa*)

- **Selection**: Clonal selection of high-yielding and high-curcumin content plants.
- **Hybridization**: Intraspecific hybridization is used to combine desirable traits like yield, quality, and disease resistancc.

- **Mutation breeding**: Mutation breeding has been used to develop varieties with improved yield and disease resistance.
- **Tissue culture**: Micropropagation is widely used for the rapid multiplication of elite clones.

C.8. Mango Ginger (*Curcuma amada*)

- **Selection**: Clonal selection for high yield and improved flavour is the primary method.
- **Hybridization**: Intraspecific hybridization is limited but can be explored for trait improvement.
- **Mutation breeding**: Can be used to introduce variability in yield, quality, and disease resistance.
- **Tissue culture**: Tissue culture techniques are used for the mass multiplication of selected clones.

C.9. Black Turmeric (*Curcuma caesia*)

- **Selection**: Clonal selection for high yield and medicinal properties.
- **Hybridization**: Limited use but can be employed to combine traits like yield and disease resistance.
- **Mutation breeding**: Potential for developing varieties with enhanced medicinal properties.
- **Tissue culture**: Micro propagation is used for producing uniform planting material.

C.10. Clove (*Syzygium aromaticum*)

- **Selection**: Selection of superior trees based on yield, oil content, and disease resistance.
- **Hybridization**: Intraspecific hybridization is not common but can be used to combine desirable traits.
- **Mutation breeding**: Induced mutations can be used to create variability in traits like oil content and disease resistance.
- **Tissue culture**: Tissue culture is employed for the mass propagation of elite clones.

C.11. Allspice (*Pimenta dioica*)

- **Selection**: Selection of superior trees based on yield and oil content.
- **Hybridization**: Intraspecific hybridization is limited but can be used to improve traits like yield and quality.

- **Mutation breeding**: Induced mutations can be explored for developing new varieties with enhanced traits.
- **Tissue culture**: Micropropagation is used for producing uniform planting material.

C.12. Coriander (*Coriandrum sativum*)

- **Selection**: Mass selection and pure line selection for yield, quality, and disease resistance.
- **Hybridization**: Intraspecific hybridization is used to combine traits like seed yield, essential oil content, and disease resistance.
- **Mutation breeding**: Induced mutations are used to develop early-maturing and disease-resistant varieties.
- **Tissue culture**: Not commonly used but can be explored for developing disease-free planting material.

C.13. Fenugreek (*Trigonella foenum-graecum*)

- **Selection**: Pure line selection for yield, diosgenin content, and disease resistance.
- **Hybridization**: Intraspecific hybridization is used to combine desirable traits.
- **Mutation breeding**: Used to develop varieties with improved yield and disease resistance.
- **Tissue culture**: Limited use, but potential for developing disease-free planting material.

C.14. Fennel (*Foeniculum vulgare*)

- **Selection**: Mass selection and pure line selection for yield and essential oil content.
- **Hybridization**: Intraspecific hybridization is used to combine desirable traits like yield and flavour.
- **Mutation breeding**: Mutation breeding has been used to create variability for traits like yield and disease resistance.
- **Tissue culture**: Not widely used, but potential for mass propagation.

C.15. Cumin (*Cuminum cyminum*)

- **Selection**: Mass selection and pure line selection for yield and essential oil content.
- **Hybridization**: Intraspecific hybridization is used to develop varieties with improved traits.

- **Mutation breeding**: Induced mutations can be used to develop varieties with better adaptability and disease resistance.
- **Tissue culture**: Limited use, but potential for developing disease-free planting material.

C.16. Ajwain (*Trachyspermum ammi*)

- **Selection**: Mass selection for yield, essential oil content, and disease resistance.
- **Hybridization**: Intraspecific hybridization is used to combine desirable traits.
- **Mutation breeding**: Can be used to develop varieties with improved yield and disease resistance.
- **Tissue culture**: Not commonly used, but has potential for mass propagation.

The choice of breeding methods for spice crops depends on the reproductive biology of the crop, the specific traits targeted for improvement, and the resources available. Selection, hybridization, mutation breeding, and tissue culture are the primary methods used across different spice crops to achieve breeding objectives. Each method plays a crucial role in developing new varieties with enhanced yield, quality, disease resistance, and adaptability.

3

Genetic Resources of Spices

Germplasm collection and conservation are crucial for maintaining the genetic diversity and ensuring the long-term sustainability of spice crops. Here's a general approach for spices:

A. Germplasm collection

a. **Identification and selection**

- Identify key spices and their varieties.
- Select diverse genetic material, including wild relatives and landraces.

b. **Field collection**

- Collect samples from diverse locations to capture genetic variation.
- Document the collection site, local environment, and plant characteristics.

c. **Data recording**

- Maintain detailed records on the collected germplasm, including geographic origin, agronomic traits, and any special characteristics.

d. **Quality assessment**

- Ensure the collected material is healthy and viable.
- Assess quality traits relevant to the spice industry.

B. Germplasm conservation

a. **Ex Situ conservation**

- **Seed banks:** Store seeds under controlled conditions (temperature, humidity) to maintain viability.
- **Field gene banks:** Maintain live plants in a controlled environment.
- **Tissue culture:** Preserve plant tissues or embryos in vitro for long-term storage.

b. **In situ conservation**

- **On-farm conservation:** Support farmers in maintaining traditional varieties.
- **Protected areas:** Preserve natural habitats where wild relatives grow.

c. **Documentation and monitoring**

- Maintain comprehensive records of conserved germplasm.
- Regularly monitor and evaluate the health and viability of the stored material.

d. **Regeneration:**

- Periodically regenerate germplasm to prevent genetic erosion and ensure viability.

e. **Collaboration**

- Collaborate with international and local institutions to share resources and knowledge.

Conservation efforts should be complemented by research to better understand the genetic diversity and potential uses of different spice germplasm.

C. Characterization and evaluation of genetic diversity

Characterization and evaluation of genetic diversity in spices involve a series of steps to understand the genetic variability within and among spice germplasm. Here's a structured approach:

a. **Characterization**

a.1. Morphological characterization

- **Trait selection:** Identify key morphological traits (e.g., leaf shape, flower structure, fruit size) relevant to the spice's quality and yield.
- **Data collection:** Collect detailed measurements and observations for these traits from different germplasm sources.
- **Phenotypic analysis:** Use statistical methods to analyze variation in these traits across different genotypes.

a.2. Biochemical characterization:

- **Chemical profiles:** Analyze essential oil composition, flavour compounds, and other biochemical markers specific to the spice.
- **Laboratory techniques:** Utilize techniques such as gas chromatography-mass spectrometry (GC-MS) for essential oil analysis.

a.3. Molecular characterization

- **DNA markers:** Use molecular markers like SSRs (Simple Sequence Repeats), SNPs (Single Nucleotide Polymorphisms), and AFLPs (Amplified Fragment Length Polymorphisms) to assess genetic variation.
- **Genotyping:** Perform PCR-based methods to identify genetic differences.

b. Evaluation of genetic diversity

b.1. Genetic diversity indices

- **Allelic diversity:** Estimate the number and frequency of alleles in the population.
- **Heterozygosity:** Calculate observed and expected heterozygosity to assess genetic variation.
- **Genetic distance:** Measure genetic distances between accessions to understand the relatedness.

b.2. Statistical analysis:

- **Cluster analysis:** Use hierarchical clustering or k-means clustering to group genotypes based on genetic similarity.
- **Principal component analysis (PCA):** Reduce dimensionality and visualize genetic variation among accessions.
- **Analysis of molecular variance (AMOVA):** Partition genetic variation within and among populations.

b.3. Diversity indices

- **Shannon-weaver index:** Measure species diversity in terms of both richness and evenness.
- **Simpson's diversity index:** Assess the probability that two randomly selected individuals belong to different species.

c. Conservation prioritization

- **Diversity hotspots:** Identify and prioritize germplasm with high genetic diversity for conservation and breeding programs.
- **Unique traits:** Focus on conserving genotypes with unique traits that could be valuable for future breeding.

d. Integration with breeding

- **Trait improvement:** Use characterization data to select germplasm with desirable traits for breeding programs.
- **Genetic resource management:** Develop strategies to manage and utilize genetic resources effectively in breeding and conservation efforts.

Combining morphological, biochemical, and molecular data provides a comprehensive understanding of genetic diversity, which is essential for improving spice crops and ensuring their long-term sustainability.

D. Importance of wild relatives and landraces

Wild relatives and landraces of spices are invaluable for several reasons. They contribute to the improvement, conservation, and sustainability of spice crops in numerous ways:

a. Importance of wild relatives

a.1. Genetic diversity

- **Disease and pest resistance:** Wild relatives often possess natural resistance to diseases and pests, which can be transferred to cultivated varieties through breeding.
- **Stress tolerance:** They can have traits that confer tolerance to abiotic stresses such as drought, salinity, or extreme temperatures.

a.2. Unique traits

- **Flavour and aroma:** Wild relatives might have unique flavour compounds or higher concentrations of essential oils that can enhance spice quality.
- **Nutritional content:** They may contain beneficial nutrients or phytochemicals not found in cultivated varieties.

a.3. Adaptation potential

- **Climate adaptation:** Wild relatives are often adapted to a range of environmental conditions, which can be useful in breeding crops that can thrive in changing climates.

a.4. Genetic resource for breeding:

- **Trait improvement:** Wild relatives can be used to introduce new traits into cultivated spices, improving yield, quality, and resilience.
- **Genetic mapping:** They provide a rich source of genetic variation for mapping and understanding complex traits.

b. Importance of landraces

b.1 Cultural and Agricultural significance

- **Local adaptation:** Landraces are adapted to local environmental conditions and farming practices, often exhibiting resilience to local stresses.
- **Traditional knowledge:** They are often linked with traditional farming systems and cultural practices, contributing to local food security and biodiversity.

b.2. Genetic variability

- **Diverse traits:** Landraces often show a wide range of traits that can be beneficial for breeding purposes, including resistance to local pests and diseases.
- **Genetic reservoir:** They serve as a genetic reservoir, providing a pool of genes that can be used to improve or diversify modern cultivars.

b.3 Sustainability

- **Ecosystem services:** Landraces can support sustainable farming practices by contributing to soil health and reducing the need for chemical inputs.
- **Biodiversity:** Maintaining landraces helps preserve agricultural biodiversity, which is crucial for ecosystem resilience.

c. Specific spices and their wild relatives/landraces

- **Black Pepper (*Piper nigrum*):** Wild relatives like Piper cubeba can provide traits for disease resistance and quality improvement.
- **Small Cardamom (*Elettaria cardamomum*):** Wild relatives of cardamom can offer resistance to pests and diseases and potential yield improvements.
- **Large Cardamom (*Amomum subulatum*):** Wild relatives can contribute to flavour enhancement and stress resistance.
- **Cinnamon (*Cinnamomum verum*):** Wild species can provide resistance to diseases and improve essential oil content.
- **Clove (*Syzygium aromaticum*):** Wild relatives can offer traits for improved clove quality and disease resistance.
- **Ginger (*Zingiber officinale*):** Wild relatives can enhance disease resistance and flavour.
- **Turmeric (*Curcuma longa*):** Wild relatives provide variability in curcumin content and disease resistance.
- **Mango Ginger (*Curcuma amada*):** Wild relatives can offer traits for improved yield and disease resistance.
- **Black Turmeric (*Curcuma caesia*):** Wild relatives can contribute to higher levels of bioactive compounds.
- **Cumin (*Cuminum cyminum*):** Wild species can enhance disease resistance and yield.
- **Coriander (*Coriandrum sativum*):** Wild relatives can provide traits for improved disease resistance and seed quality.
- **Fennel (*Foeniculum vulgare*):** Wild types offer traits for improved flavour and pest resistance.

- **Fenugreek (*Trigonella foenum-graecum*):** Wild relatives can contribute to better adaptation and disease resistance.
- **Ajwain (*Trachyspermum ammi*):** Wild species provide genetic diversity for improving essential oil content and resistance.

In summary, wild relatives and landraces play a critical role in the conservation and improvement of spice crops, contributing to their genetic diversity, resilience, and sustainability.

4

Breeding Techniques for Spices

A) Conventional breeding techniques

Conventional breeding techniques for spices involve various methods to improve traits such as yield, quality, disease resistance, and adaptability. Here's an overview of conventional breeding techniques applied to specific spices:

1. Black pepper (Piper nigrum)

- **Selection:** Choose superior plants based on traits like yield, quality of peppercorns, and disease resistance.
- **Hybridization:** Cross-breed different varieties to combine desirable traits.
- **Mutation breeding:** Induce mutations to develop new traits or enhance existing ones.

2. Small cardamom (Elettaria cardamomum)

- **Selection:** Select plants with high yield, better quality, and disease resistance.
- **Hybridization:** Cross different varieties to improve traits such as pod size and essential oil content.
- **Seed production:** Improve seed production and quality through selection.

3. Large cardamom (*Amomum subulatum*)

- **Selection:** Identify plants with higher yield and better resistance to pests and diseases.
- **Hybridization:** Develop hybrids to enhance flavour and size of the cardamom pods.

4. Cinnamon (*Cinnamomum verum*)

- **Selection:** Choose trees with higher essential oil content and better bark quality.
- **Propagation:** Use vegetative propagation to maintain desirable traits.
- **Hybridization:** Cross-breed to improve traits such as disease resistance and growth rate.

5. Clove (*Syzygium aromaticum*)

- **Selection:** Select for high clove yield, quality, and disease resistance.
- **Hybridization:** Cross different varieties to enhance clove size and oil content.
- **Grafting:** Use grafting techniques to propagate superior clones.

6. Ginger (*Zingiber officinale*)

- **Selection:** Identify and propagate plants with higher rhizome yield and quality.
- **Hybridization:** Cross different varieties to improve traits like disease resistance and growth rate.
- **Rhizome selection:** Select and use high-quality rhizomes for propagation.

7. Turmeric (*Curcuma longa*)

- **Selection:** Choose plants with high curcumin content and disease resistance.
- **Hybridization:** Cross different varieties to improve curcumin levels and yield.
- **Rhizome selection:** Select high-quality rhizomes for planting.

8. Mango ginger (Curcuma amada)

- **Selection:** Select plants with high yield and quality rhizomes.
- **Hybridization:** Cross different varieties to enhance flavour and yield.
- **Rhizome propagation:** Use selected rhizomes for propagation.

9. Black turmeric (*Curcuma caesia*)

- **Selection:** Choose plants with high levels of bioactive compounds and disease resistance.
- **Hybridization:** Develop hybrids to improve traits like rhizome quality and yield.

10. Cumin (*Cuminum cyminum*)

- **Selection:** Select plants with high seed yield, quality, and resistance to diseases.
- **Hybridization:** Cross varieties to improve traits such as seed size and flavour.
- **Seed Production:** Improve seed quality and yield through selection.

11. Coriander (*Coriandrum sativum*)

- **Selection:** Choose plants with high seed yield and desirable aroma.

- **Hybridization:** Cross varieties to enhance seed size and quality.
- **Seed selection:** Select and propagate high-quality seeds.

12. Fennel (*Foeniculum vulgare*)

- **Selection:** Select for high seed yield and essential oil content.
- **Hybridization:** Cross different varieties to improve traits such as seed size and flavour.
- **Seed propagation:** Use selected seeds for propagation.

13. Fenugreek (*Trigonella foenum-graecum*)

- **Selection:** Identify plants with high seed yield and quality.
- **Hybridization:** Cross varieties to improve traits like leaf and seed quality.
- **Seed selection:** Select high-quality seeds for planting.

14. Ajwain (*Trachyspermum ammi*)

- **Selection:** Choose plants with high essential oil content and resistance to pests.
- **Hybridization:** Develop hybrids to improve seed yield and oil content.
- **Seed production:** Improve seed quality and yield through selection.

General conventional breeding techniques:

1. Selection

- **Mass selection:** Select the best-performing plants from a population based on desired traits.
- **Pure line selection:** Select individuals with stable traits over several generations.

2. Hybridization:

- **Cross-breeding:** Cross different varieties or species to combine desirable traits.
- **Backcrossing:** Cross hybrid plants with one of the parent varieties to reinforce specific traits.

3. Propagation techniques:

- **Vegetative propagation:** Use cuttings, grafting, or tissue culture to propagate plants with desirable traits.
- **Seed propagation:** Use selected seeds from superior plants to propagate new generations.

4. Testing and evaluation

- **Field trials:** Conduct trials to evaluate the performance of new varieties in different environments.
- **Performance testing:** Assess traits like yield, quality, and disease resistance.

These techniques help enhance the quality, yield, and adaptability of spice crops, contributing to improved productivity and sustainability in spice cultivation.

B) Modern biotechnological approaches (Marker-assisted selection, genetic engineering, etc.)

Modern biotechnological approaches are increasingly used to enhance spice crops by improving traits such as yield, quality, disease resistance, and stress tolerance. Here's how different techniques can be applied to spices:

B.1. Marker-assisted selection (MAS)

Principle: MAS uses molecular markers linked to desirable traits to select plants with specific genetic attributes.

- **Black pepper (Piper nigrum):** Develop markers for disease resistance (e.g., to Phytophthora) and improved quality traits like peppercorn size.
- **Small cardamom (Elettaria cardamomum):** Identify markers linked to high yield and resistance to pests.
- **Large cardamom (Amomum subulatum):** Use markers to select for traits such as pod size and quality.
- **Cinnamon (Cinnamomum verum):** Develop markers for essential oil content and disease resistance.
- **Clove (Syzygium aromaticum):** Identify markers for high clove yield and quality.
- **Ginger (Zingiber officinale):** Use MAS for traits like disease resistance and high rhizome yield.
- **Turmeric (Curcuma longa):** Develop markers for high curcumin content and disease resistance.
- **Mango ginger (Curcuma amada):** Select for improved rhizome quality and yield using markers.
- **Black turmeric (Curcuma caesia):** Use markers to select for high levels of bioactive compounds.
- **Cumin (Cuminum cyminum):** Identify markers for high seed yield and quality.
- **Coriander (Coriandrum sativum):** Use MAS to enhance seed size, aroma, and yield.

- **Fennel (Foeniculum vulgare):** Develop markers for high essential oil content and seed yield.
- **Fenugreek (Trigonella foenum-graecum):** Select for high seed yield and quality using molecular markers.
- **Ajwain (Trachyspermum ammi):** Identify markers for high essential oil content and pest resistance.

B.2. Genetic engineering

Genetic engineering involves introducing new genes into a plant's genome to confer new traits or enhance existing ones.

- **Black pepper:** Introduce genes for resistance to diseases like Phytophthora or to enhance peppercorn quality.
- **Small cardamom:** Engineer resistance to pests or enhance essential oil production.
- **Large cardamom:** Introduce genes for improved pod size and resistance to diseases.
- **Cinnamon:** Genetic modification to increase essential oil yield and quality.
- **Clove:** Develop transgenic plants with enhanced clove quality and yield.
- **Ginger:** Genetic engineering for improved disease resistance and higher rhizome yield.
- **Turmeric:** Modify genes to increase curcumin content or enhance disease resistance.
- **Mango ginger:** Engineer for improved rhizome quality and yield.
- **Black turmeric:** Introduce genes to increase bioactive compound levels.
- **Cumin:** Genetic modification to improve seed yield and quality.
- **Coriander:** Enhance essential oil content and yield through genetic engineering.
- **Fennel:** Introduce genes for higher essential oil content and seed quality.
- **Fenugreek:** Engineer plants for improved seed yield and disease resistance.
- **Ajwain:** Develop transgenic plants with higher essential oil content and pest resistance.

B.3. Tissue culture and genetic transformation

Tissue culture and genetic transformation involve growing plant cells or tissues in vitro and introducing new genes.

- **Black pepper:** Use tissue culture for rapid propagation and genetic transformation for disease resistance.

- **Small cardamom:** Apply tissue culture techniques for micropropagation and genetic transformation to enhance traits.
- **Large cardamom:** Tissue culture for clonal propagation and transformation for improved traits.
- **Cinnamon:** Use tissue culture for uniformity and transformation for enhancing essential oil content.
- **Clove:** Apply tissue culture for mass propagation and genetic transformation for quality improvement.
- **Ginger:** Utilize tissue culture for uniform propagation and transformation for disease resistance.
- **Turmeric:** Tissue culture for cloning and genetic transformation for increased curcumin content.
- **Mango ginger:** Use tissue culture for propagation and transformation for enhanced rhizome quality.
- **Black turmeric:** Apply tissue culture and transformation for increased bioactive compounds.
- **Cumin:** Tissue culture for rapid propagation and transformation for improved seed yield.
- **Coriander:** Use tissue culture for propagation and genetic transformation for enhanced traits.
- **Fennel:** Tissue culture for cloning and genetic transformation for higher essential oil content.
- **Fenugreek:** Apply tissue culture and transformation for improved seed yield and disease resistance.
- **Ajwain:** Utilize tissue culture for propagation and transformation for increased essential oil content.

B.4. Genomic approaches

Genomic approaches include whole-genome sequencing, QTL mapping, and gene expression studies to understand and improve spice crops.

- **Black pepper:** Use genomic approaches for mapping traits related to disease resistance and quality.
- **Small cardamom:** Apply genomic tools to identify QTLs for yield and pest resistance.
- **Large cardamom:** Use whole-genome sequencing to discover genes related to pod size and quality.
- **Cinnamon:** Employ genomic approaches to enhance essential oil yield and quality.

- **Clove:** Utilize genomic tools to identify genes associated with high clove yield and essential oil content.
- **Ginger:** Use genomic approaches to study genes related to disease resistance and high rhizome yield.
- **Turmeric:** Apply genomics to identify genes involved in curcumin biosynthesis and disease resistance.
- **Mango ginger:** Use genomic tools to study traits related to rhizome quality and yield.
- **Black turmeric:** Employ genomics to identify genes linked to high bioactive compound levels.
- **Cumin:** Use genomic approaches for understanding seed yield and quality traits.
- **Coriander:** Apply genomic tools to study traits like essential oil content and seed size.
- **Fennel:** Utilize genomic approaches to enhance essential oil content and seed yield.
- **Fenugreek:** Use genomics to identify genes associated with high seed yield and disease resistance.
- **Ajwain:** Employ genomic tools to study essential oil content and pest resistance.

B.5. CRISPR/Cas9 technology

CRISPR/Cas9 is a genome-editing tool that allows precise modifications of specific genes.

- **Black pepper:** Use CRISPR/Cas9 to knock out genes related to disease susceptibility or to enhance quality traits.
- **Small cardamom:** Apply CRISPR/Cas9 for targeted gene editing to improve yield and disease resistance.
- **Large cardamom:** Utilize CRISPR/Cas9 to modify genes affecting pod size and essential oil content.
- **Cinnamon:** Employ CRISPR/Cas9 to enhance essential oil yield and quality.
- **Clove:** Use CRISPR/Cas9 for precise editing to improve clove yield and quality.
- **Ginger:** Apply CRISPR/Cas9 to enhance disease resistance and high rhizome yield.
- **Turmeric:** Use CRISPR/Cas9 to increase curcumin content and improve disease resistance.

- **Mango ginger:** Employ CRISPR/Cas9 to modify genes affecting rhizome quality and yield.
- **Black turmeric:** Utilize CRISPR/Cas9 for enhanced levels of bioactive compounds.
- **Cumin:** Apply CRISPR/Cas9 to improve seed yield and quality traits.
- **Coriander:** Use CRISPR/Cas9 for targeted modifications to enhance seed size and oil content.
- **Fennel:** Employ CRISPR/Cas9 to increase essential oil content and seed yield.
- **Fenugreek:** Utilize CRISPR/Cas9 to improve seed yield and disease resistance.
- **Ajwain:** Apply CRISPR/Cas9 to enhance essential oil content and pest resistance.

B.6. Proteomics and metabolomics

Proteomics and metabolomics analyse the protein and metabolite profiles of plants to understand and improve traits.

- **Black pepper:** Use proteomics to study proteins related to disease resistance and quality.
- **Small cardamom:** Apply metabolomics to identify compounds linked to high yield and essential oil content.
- **Large cardamom:** Utilize proteomics to understand proteins associated with pod size and quality.
- **Cinnamon:** Employ metabolomics to enhance essential oil content and quality.
- **Clove:** Use proteomics to study proteins involved in high clove yield and quality.
- **Ginger:** Apply metabolomics to understand compounds related to disease resistance and rhizome yield.
- **Turmeric:** Utilize proteomics to study proteins involved in curcumin synthesis.
- **Mango ginger:** Employ metabolomics to identify compounds affecting rhizome quality and yield.
- **Black turmeric:** Use proteomics to enhance levels of bioactive compounds.
- **Cumin:** Apply metabolomics to improve seed yield and quality.
- **Coriander:** Utilize proteomics to understand proteins related to essential oil content and seed size.

- **Fennel:** Use metabolomics to increase essential oil content and seed yield.
- **Fenugreek:** Employ proteomics to study proteins linked to seed yield and disease resistance.
- **Ajwain:** Apply metabolomics to enhance essential oil content and pest resistance.

These modern biotechnological approaches provide powerful tools to complement traditional breeding methods, enhancing the efficiency and precision of spice crop improvement.

C. Crossbreeding and hybridization techniques

Crossbreeding and hybridization techniques are essential for improving different spice crops. These techniques aim to combine desirable traits from different varieties or species to develop superior cultivars with improved yield, resistance to diseases and pests, better adaptability, and enhanced quality.

1. Black pepper (*Piper nigrum*)

- **Hybridization:** Hand pollination is commonly used due to the dioecious nature of Black pepper (male and female flowers on separate plants). Controlled hybridization is performed by emasculating the male flowers and transferring pollen to the stigma of female flowers. The resulting hybrids are evaluated for traits like disease resistance (especially to *Phytophthora*), yield, and quality.
- **Clonal selection:** Selected clones from natural populations or hybrids with superior traits are propagated vegetatively.

2. Small cardamom (*Elettaria cardamomum*)

- **Hybridization:** Due to the high genetic variability and outcrossing nature of cardamom, hybridization is effective. Controlled pollination is done between selected parents with complementary traits (e.g., high yield and disease resistance). The hybrids are evaluated over multiple seasons.
- **Mutation breeding:** Induced mutations are used to develop new varieties with desirable traits.

3. Large cardamom (*Amomum subulatum*)

- **Hybridization:** This involves crossing different large cardamom cultivars to develop hybrids with improved yield, disease resistance, and better adaptability to diverse climates.
- **Vegetative propagation:** The selected superior hybrids are propagated vegetatively to maintain the desirable traits.

4. Cinnamon (*Cinnamomum verum*)

- **Hybridization:** Crosses between different cinnamon species or varieties are made to develop hybrids with enhanced bark quality, higher oil content, and disease resistance.
- **Grafting:** Superior genotypes are often propagated through grafting to maintain genetic uniformity.

5. Clove (*Syzygium aromaticum*)

- **Hybridization:** Controlled cross-pollination is performed between selected trees with high yield and disease resistance. Due to the long juvenile phase, clove breeding is time-consuming.
- **Vegetative propagation:** Superior clones are often propagated by cuttings or grafting.

6. Allspice (*Pimenta dioica*)

- **Hybridization:** Though less common, controlled pollination between selected trees is possible for improving yield and quality.
- **Vegetative propagation:** Superior clones are often propagated vegetatively to maintain the desired traits.

7. Nutmeg (*Myristica fragrans*)

- **Natural hybridization**: This can occur in the wild where different species of nutmeg grow in proximity, leading to natural cross-pollination. However, this is less controlled and predictable.
- **Artificial hybridization**: Conducted under controlled conditions, where the pollen of a selected species or variety is manually introduced to the female flowers of another. This method is more common in breeding programs.
- **Somatic hybridization**: In more advanced breeding programs, techniques like somatic hybridization may be employed, involving the fusion of different plant cells (protoplasts)

8. Ginger (*Zingiber officinale*)

- **Hybridization:** Ginger is predominantly propagated vegetatively, but hybridization is carried out between selected lines to introduce new genetic variations for disease resistance, yield, and quality traits.
- **Polyploidy breeding:** Induced polyploidy can lead to enhanced vigor and disease resistance in ginger.

9. Turmeric (*Curcuma longa*)

- **Hybridization:** Similar to ginger, turmeric is primarily propagated vegetatively. Hybridization between selected lines is done to develop new varieties with improved yield, curcumin content, and resistance to diseases.
- **Mutation breeding:** Induced mutations are also used to develop varieties with desirable traits.

10. Mango ginger (*Curcuma amada*)

- **Hybridization:** Involves crossing different lines to introduce desirable traits like higher yield and disease resistance.
- **Polyploidy breeding:** Like turmeric, polyploidy induction is used for enhancing desirable traits.

11. Black turmeric (*Curcuma caesia*)

- **Crossbreeding**: Black turmeric is a member of the ginger family, and cross-breeding typically involves selecting parent plants with desirable traits such as high curcumin content, disease resistance, or yield. Controlled pollination is performed between selected varieties to produce hybrid seeds or rhizomes.

- **Hybridization techniques**
  - **Intra-specific hybridization**: Crossing different varieties within the species *Curcuma caesia* to combine beneficial traits.
  - **Tissue culture**: This technique is commonly used in turmeric to propagate hybrids, ensuring the rapid multiplication of plants with desirable traits.

12. Coriander (*Coriandrum sativum*)

- **Crossbreeding**: Coriander is a widely cultivated spice, and cross-breeding focuses on improving seed quality, flavour, and resistance to pests. Selected varieties are cross-pollinated to produce hybrids.

- **Hybridization techniques**:
  - **Controlled pollination**: Ensuring that selected plants are pollinated to produce hybrids with specific traits.
  - **Polyploidy induction**: A technique used to create hybrids with increased size, flavour intensity, or stress tolerance by altering the chromosome number.

13. Fenugreek (*Trigonella foenum-graecum*)

- **Cross-breeding**: Fenugreek breeding programs aim to improve yield, nutritional content, and resistance to environmental stress. Cross-breeding involves selecting parent plants with these traits and performing controlled pollination.

- **Hybridization techniques**:
  - **Intra-specific hybridization**: Combining traits from different fenugreek varieties.
  - **Marker-assisted selection (MAS):** A more modern technique where specific genes associated with desirable traits are selected during hybridization.

14. Fennel (*Foeniculum vulgare*)

- **Cross-breeding**: In fennel, cross-breeding focuses on improving oil content, flavour, and plant vigor. Selected parent plants are cross-pollinated to produce hybrids.

- **Hybridization techniques**:
  - **Intra-specific hybridization**: Within species crossing to enhance specific traits like flavour profile or growth habit.
  - **Somatic hybridization**: A more advanced method where protoplasts from different varieties are fused to create new hybrids.

15. Ajwain (*Trachyspermum ammi*)

- **Cross-breeding**: Ajwain breeding programs often aim to enhance yield, oil content, and disease resistance. Cross-breeding involves selecting plants with desirable traits and facilitating controlled pollination.
- **Hybridization techniques**:
  - **Intra-specific hybridization**: This is commonly used to combine desirable traits within the species.
  - **Mutation breeding**: Sometimes, ajwain plants are subjected to mutagens to create variations that are then used in cross-breeding programs.

16. Cumin (*Cuminum cyminum*)

- **Hybridization:** These annual spices are cross-pollinated species. Controlled hybridization is performed to develop varieties with improved yield, disease resistance, and quality. The progeny is evaluated in subsequent generations.
- **Mutation breeding:** Radiation or chemical mutagens are used to induce variability and select desirable mutants.
- **Mass selection:** Selection of superior plants from natural populations is common for these spices.

General hybridization techniques across these species

- **Controlled pollination**: Ensuring that only selected plants are involved in pollination to produce hybrid seeds with specific traits.
- **Tissue culture**: Propagating hybrids rapidly and uniformly, especially in vegetatively propagated plants like black turmeric.
- **Molecular breeding**: Using genetic markers to select desirable traits more efficiently during hybridization.
- **Polyploidy**: Inducing polyploidy to create hybrids with enhanced vigor, size, or stress tolerance.

These techniques are fundamental in developing new plant varieties with improved traits, helping to meet agricultural and market demands.

Challenges

- **Long generation time:** Many spice crops like clove and cardamom have long juvenile phases, making breeding programs lengthy.
- **Genetic diversity:** Managing and utilizing the genetic diversity in these spices is critical for the success of breeding programs.
- **Disease resistance:** Developing hybrids with durable resistance to major diseases is a primary focus, especially in crops like black pepper and ginger.

These techniques are tailored to the specific biology and breeding systems of each spice, ensuring the development of cultivars that meet the demands of farmers and consumers alike.

D. Mutation breeding and its application in spices

Mutation breeding is a technique that involves exposing seeds, plant tissues, or whole plants to physical or chemical mutagens to induce genetic mutations. These mutations can result in new traits or variations, such as improved yield, disease resistance, or enhanced quality. In spices, mutation breeding is particularly valuable because many spices are vegetatively propagated, making it easier to fix and propagate desirable mutations.

Applications of mutation breeding in various spices

1. Black pepper (*Piper nigrum*)

- Induced mutations in black pepper have been used to develop resistance to diseases like *Phytophthora* foot rot. Mutagenesis is also employed to enhance the yield and quality of the berries. Varieties with improved resistance to diseases and pests have been developed through gamma irradiation.

2. Small cardamom (*Elettaria cardamomum*)

- Mutation breeding in small cardamom aims to create variability in traits like plant height, flowering, and disease resistance. Chemical mutagens like EMS (ethyl methanesulfonate) are used to induce useful mutations. Mutant lines with improved yield and quality have been isolated.

3. Large cardamom (*Amomum subulatum*)

- Large cardamom has been subjected to mutation breeding to develop plants with improved resistance to viral diseases and enhanced yield potential. Physical mutagens like gamma rays are often used. Development of disease-resistant lines through irradiation.

4. Cinnamon (*Cinnamomum verum*)

- Mutation breeding is used to induce variability in traits such as oil composition, bark thickness, and disease resistance. Chemical and physical mutagens are applied to seeds or tissues. Mutant lines with higher essential oil content have been developed.

5. Clove (*Syzygium aromaticum*)

- Mutation breeding in clove aims to induce mutations that improve flowering, increase yield, and enhance resistance to diseases like leaf spot and wilt. Development of high-yielding, disease-resistant clove varieties.

6. Allspice (*Pimenta dioica*)

- Mutation breeding is applied to induce variability in traits like fruit size, oil content, and resistance to pests. Development of allspice varieties with improved yield and oil quality.

7. Nutmeg (*Myristica fragrans*)

- **Application**: Mutation breeding in nutmeg is used to develop varieties with improved traits such as higher oil content, disease resistance, and better fruit quality. Physical mutagens like gamma rays or chemical mutagens like ethyl methanesulfonate (EMS) can be used to induce beneficial mutations.
- **Benefits**:
  - Increased resistance to pests and diseases, particularly those affecting nutmeg trees.
  - Improved yield and nutmeg quality (e.g., essential oil content and flavour profile).

8. Ginger (*Zingiber officinale*)

- Induced mutations in ginger have been used to develop disease-resistant varieties, especially against rhizome rot. Gamma irradiation is commonly employed. Development of ginger varieties with higher yields and resistance to bacterial wilt.

9. Turmeric (*Curcuma longa*)

- Mutation breeding is used to enhance the genetic diversity of turmeric, focusing on traits like curcumin content, rhizome size, and disease resistance. Both physical (gamma rays) and chemical mutagens are utilized. High-curcumin mutant lines have been developed.

10. Mango ginger (*Curcuma amada*)

- Mutation breeding in mango ginger is used to induce variation in essential oil content, rhizome yield, and disease resistance. Selection of mutants with improved rhizome characteristics.

11. Black turmeric (*Curcuma caesia*)

- **Application**: Mutation breeding in black turmeric aims to enhance rhizome yield, curcumin content, and resistance to environmental stresses. Given the medicinal value of black turmeric, mutations can also lead to higher concentrations of bioactive compounds.
- **Benefits**
  - Development of varieties with enhanced medicinal properties.
  - Improved tolerance to abiotic stresses like drought or salinity.

12. Coriander (*Coriandrum sativum*)

- **Application**: Mutation breeding in coriander can lead to varieties with improved flavour, yield, and resistance to pathogens. Chemical mutagens like EMS or sodium azide are commonly used to induce mutations.
- **Benefits**
  - Enhanced essential oil content in coriander seeds.
  - Improved resistance to common diseases like bacterial blight.

13. Cumin (*Cuminum cyminum*)

- **Application**: In cumin, mutation breeding is used to develop varieties with higher yield, improved resistance to fungal diseases, and better adaptability to different climates. Gamma irradiation is a commonly used mutagen for cumin.

- **Benefits**
  - Development of disease-resistant cumin varieties.
  - Increased yield and quality of cumin seeds.

14. Fennel (*Foeniculum vulgare*)

- **Application**: Mutation breeding in fennel aims to improve seed yield, essential oil content, and resistance to environmental stresses. Mutagens such as gamma rays and EMS are typically used.
- **Benefits**
  - Enhanced flavour and aroma of fennel seeds.
  - Improved plant vigor and adaptability to different growing conditions.

15. Fenugreek (*Trigonella foenum-graecum*)

- **Application**: Fenugreek is a widely cultivated legume, and mutation breeding is employed to enhance traits such as yield, nutritional content, and resistance to pests. EMS and gamma irradiation are common mutagens.
- **Benefits**:
  - Development of fenugreek varieties with higher protein content.
  - Increased tolerance to abiotic stresses like drought.

16. Ajwain (*Trachyspermum ammi*)

- **Application**: Mutation breeding in ajwain focuses on improving seed yield, essential oil content, and resistance to diseases. Mutagens such as EMS or gamma rays can be applied to induce beneficial mutations.
- **Benefits**
  - Improved oil yield, which enhances the spice's market value.
  - Enhanced resistance to diseases and pests, reducing the need for chemical treatments.

General process of mutation breeding

1. **Selection of parent plants**: Choosing a variety that already possesses some desirable traits.
2. **Mutagen treatment**: Exposing seeds, plant tissues, or even whole plants to mutagens like gamma rays, X-rays, or chemicals.
3. **Screening and selection**: Growing the treated plants and screening them for beneficial mutations, such as improved yield or disease resistance.
4. **Stabilization and breeding**: The beneficial mutants are further bred and stabilized to ensure that the new traits are passed on to subsequent generations.

Advantages of mutation breeding in spices

- **Creation of new variability:** Mutation breeding generates genetic diversity, which is crucial for crop improvement, especially in species with limited natural variability.
- **Shorter breeding cycles:** Compared to traditional breeding, mutation breeding can rapidly introduce desirable traits, reducing the time required to develop new varieties.
- **Specific trait improvement:** Mutation breeding allows for the targeted improvement of specific traits, such as disease resistance or quality, without altering other desirable characteristics of the plant.
- **Adaptability**: Mutation breeding allows for the development of plant varieties that can thrive in changing environmental conditions or under biotic stress.

Challenges

- **Unpredictability:** Mutations are random, and most induced mutations are deleterious. Careful screening is required to identify useful mutations.
- **Stability:** Some mutations may not be stable across generations, necessitating further breeding to stabilize the trait.
- **Regulatory issues:** Mutant varieties may require thorough evaluation and approval before commercial release, especially if mutagens are involved.

Mutation breeding has proven to be a powerful tool in the improvement of many crops, including the spice and medicinal plants listed above, helping to meet the growing demand for high-quality, resilient plant varieties.

Mutagens used in spices

Mutagens are agents that cause mutations in the DNA of organisms. In the context of spice breeding, both physical and chemical mutagens are used to induce genetic variations that can lead to desirable traits such as improved yield, disease resistance, and enhanced quality. Below is an overview of the commonly used mutagens in mutation breeding for various spices:

Physical mutagens

1. Gamma rays

- **Usage:** One of the most commonly used physical mutagens in mutation breeding. Gamma rays are a form of ionizing radiation that can penetrate plant tissues and induce mutations at the DNA level.

- **Application:** Used in crops like Black pepper, Small cardamom, Large cardamom, Ginger, Turmeric, Cumin, Coriander, Fennel, Fenugreek, and Ajwain.
- **Examples:** Development of disease-resistant and high-yielding varieties.

2. X-rays

- **Usage:** Similar to gamma rays, X-rays are used to induce mutations by causing breaks in the DNA strands.
- **Application:** X-rays have been used in spices such as **Coriander, Fennel, Fenugreek**, and other annual spices.
- **Examples:** Induction of mutations that lead to improved seed size and essential oil content.

3. Fast neutrons

- **Usage:** Fast neutrons are highly penetrative and can cause large-scale deletions in the genome, leading to significant mutations.
- **Application:** Employed in mutation breeding of spices such as **Black pepper, Cinnamon, Clove**, and others.
- **Examples:** Development of varieties with altered growth habits or improved resistance to pests.

4. Ultraviolet (UV) light

- **Usage:** UV light, particularly UV-C, is used to induce point mutations by causing thymine dimers in DNA.
- **Application:** Used more in laboratory settings for in vitro mutagenesis, especially in crops like **Ginger, Turmeric, and Mango ginger**.
- **Examples:** Generation of mutants with improved rhizome quality and disease resistance.

Chemical mutagens

1. Ethyl methane sulfonate (EMS)

- **Usage:** EMS is a widely used chemical mutagen that induces point mutations by alkylating guanine bases in DNA, leading to mispairing during replication.
- **Application:** Effective in inducing mutations in a variety of spices such as **Small cardamom, Large cardamom, Ginger, Turmeric, Cumin, Coriander, Fenugreek**, and others.
- **Examples:** Creation of mutants with enhanced oil content, disease resistance, and improved yield.

2. Sodium azide (NaN3)

- **Usage:** Sodium azide induces point mutations by converting adenine to hypoxanthine, which pairs with cytosine instead of thymine.
- **Application:** Used in crops like **Coriander, Fennel, Fenugreek, and Ajwain** to induce beneficial mutations.
- **Examples:** Development of varieties with improved seed quality and yield.

3. Dimethyl sulfate (DMS)

- **Usage:** DMS is another alkylating agent that causes mutations by transferring methyl groups to the DNA.
- **Application:** Occasionally used in spices such as **Black pepper and Cinnamon** to induce genetic variation.
- **Examples:** Induction of mutations leading to altered plant morphology or resistance to environmental stress.

4. Colchicine

- **Usage:** Colchicine is primarily used to induce polyploidy by inhibiting chromosome segregation during cell division, leading to the doubling of the chromosome number.
- **Application:** Applied to induce polyploidy in crops like **Ginger, Turmeric, Black pepper, and Small cardamom**.
- **Examples:** Development of polyploid plants with increased vigor, larger rhizomes, or enhanced oil content.

5. Hydrazine

- **Usage:** Hydrazine is a less commonly used chemical mutagen that induces point mutations by modifying nucleotides in DNA.
- **Application:** Employed in the mutation breeding of some spices like **Cumin, Fenugreek, and Ajwain**.
- **Examples:** Selection of mutants with improved yield or resistance to biotic stresses.

Applications and considerations

- **Tissue culture & In vitro mutagenesis:** Many of these mutagens, especially chemical ones like EMS and sodium azide, are often used in conjunction with tissue culture techniques to treat explants, callus cultures, or plantlets. This allows for more controlled mutagenesis and easier screening of mutants.
- **Safety and efficiency:** While effective, mutagen use requires careful handling due to their potential hazards. Additionally, the efficiency of

mutation induction and the stability of the induced traits must be thoroughly evaluated in subsequent generations.

Mutation breeding, using these physical and chemical mutagens, has played a significant role in the development of improved spice varieties. The induced genetic variability is crucial for enhancing traits such as yield, quality, and disease resistance, thereby meeting the needs of both producers and consumers.

Specific mutagens used crop wise in spices crops

Mutation breeding in spices has been widely applied using various mutagens, each with specific doses and exposure times depending on the species and desired outcomes. The detailed overview of some mutagens used in spices, including their doses, exposure periods, and some achievements:

1. Black pepper (*Piper nigrum*)

- **Mutagens used**
  - **Gamma radiation**: Doses typically range from 10 to 100 Gy.
  - **Ethyl methane sulfonate (EMS)**: Used at concentrations of 0.1% to 1.0%.
- **Exposure time**:
  - **Gamma radiation**: Usually a few minutes to an hour, depending on the dose rate.
  - **EMS**: Treatment usually lasts 2 to 6 hours.
- **Achievements**:
  - Development of disease-resistant varieties, particularly against foot rot and quick wilt.
  - Improved yield and berry size.

2. Small cardamom (*Elettaria cardamomum*)

- **Mutagens used**:
  - **Gamma radiation**: Applied at doses of 5 to 50 Gy.
  - **Sodium azide**: Used at concentrations of 0.01% to 0.05%.
- **Exposure time**:
  - **Gamma radiation**: Exposure typically lasts for a few minutes.
  - **Sodium azide**: Usually applied for 2 to 4 hours.
- **Achievements**:
  - Enhanced resistance to pests such as thrips and shoot borers.
  - Improved capsule size and yield.

3. Clove (*Syzygium aromaticum*)

- **Mutagens used**:
  - **Gamma radiation**: Doses ranging from 10 to 80 Gy.
  - **Colchicine**: Used at concentrations of 0.05% to 0.1% to induce polyploidy.
- **Exposure time**:
  - **Gamma radiation**: Exposure times can range from a few minutes to an hour.
  - **Colchicine**: Treatment usually lasts for 12 to 48 hours.
- **Achievements**:
  - Development of clove varieties with improved oil content.
  - Enhanced disease resistance, especially against clove decline.

4. Large cardamom (*Amomum subulatum*)

- **Mutagen:** Gamma Rays
  - **Dose:** 20-60 Gy
  - **Exposure time:** Single exposure
  - **Achievements:** Development of varieties with enhanced resistance to viral diseases and improved yield.

5. Cinnamon (*Cinnamomum verum*)

- **Mutagen:** Gamma Rays
  - **Dose:** 15-40 Gy
  - **Exposure time:** Single exposure
  - **Achievements:** Mutant varieties with higher bark yield, improved oil content, and enhanced resistance to diseases.
- **Mutagen:** Sodium Azide (NaN3)
  - **Dose:** 0.01-0.03% concentration
  - **Exposure time:** 4-6 hours soaking of seeds or explants
  - **Achievements:** Induced variability in oil composition and bark quality.

6. Nutmeg (*Myristica fragrans*)

- **Mutagens used**:
  - **Gamma radiation**: Often used at doses ranging from 10 to 100 Gy (Gray), with the specific dose depending on the desired mutation and the tolerance of the plant tissues.
  - **Ethyl Methane sulfonate (EMS)**: A chemical mutagen typically used at concentrations of 0.1% to 1.0% for a duration of 2 to 6 hours.

- **Exposure time**:
  - **Gamma radiation**: Exposure times vary depending on the dose rate, but it generally ranges from a few minutes to several hours.
  - **EMS**: Treatment usually lasts from 2 to 6 hours.
- **Achievements**:
  - Development of varieties with higher essential oil content.
  - Improved disease resistance, particularly against common fungal pathogens.

7. Ginger (*Zingiber officinale*)

- **Mutagen:** Gamma Rays
  - **Dose:** 10-30 Gy
  - **Exposure time:** Single exposure
  - **Achievements:** Disease-resistant varieties, particularly against rhizome rot, with higher yields and better rhizome quality.
- **Mutagen:** Colchicine
  - **Dose:** 0.05-0.1% concentration
  - **Exposure time:** 24-48 hours soaking of rhizomes or explants
  - **Achievements:** Development of polyploid lines with larger rhizomes and enhanced vigour.

8. Turmeric (*Curcuma longa*)

- **Mutagen:** Gamma Rays
  - **Dose:** 20-40 Gy
  - **Exposure time:** Single exposure
  - **Achievements:** High-curcumin mutant lines, improved rhizome size, and resistance to diseases like leaf spot and rhizome rot.
- **Mutagen:** EMS (Ethyl Methane sulfonate)
  - **Dose:** 0.1-0.3% concentration
  - **Exposure time:** 8-12 hours soaking of rhizomes or explants
  - **Achievements:** Mutants with increased curcumin content and enhanced yield.

9. Mango ginger (*Curcuma amada*)

- **Mutagen:** Gamma Rays
  - **Dose:** 20-50 Gy
  - **Exposure time:** Single exposure

- **Achievements:** Enhanced rhizome yield improved essential oil content, and disease resistance.

10. Black turmeric (*Curcuma caesia*)

- **Mutagens used**
  - **Gamma radiation**: Doses typically range from 20 to 150 Gy.
  - **Colchicine**: Used at lower concentrations (0.05% to 0.1%) to induce polyploidy and enhance certain traits.
- **Exposure time**:
  - **Gamma radiation**: Exposure is generally brief, lasting a few minutes.
  - **Colchicine**: Exposure times can range from 12 to 48 hours, depending on the concentration and the plant material being treated.
- **Achievements**:
  - Increased curcumin content and rhizome yield.
  - Enhanced resistance to diseases and pests.

11. Coriander (*Coriandrum sativum*)

- **Mutagens used**:
  - **Gamma radiation**: Commonly used at doses ranging from 5 to 50 Gy.
  - **Sodium azide**: Applied at concentrations of 0.01% to 0.05% for mutation induction.
- **Exposure time**:
  - **Gamma radiation**: Typically, a few minutes to half an hour.
  - **Sodium azide**: Exposure usually lasts 2 to 4 hours.
- **Achievements**:
  - Development of coriander varieties with higher essential oil content.
  - Improved resistance to environmental stresses, such as drought and salinity.

12. Cumin (*Cuminum cyminum*)

- **Mutagens used**
  - **Gamma radiation**: Doses between 10 to 100 Gy are common.
  - **EMS**: Used at 0.1% to 0.5% concentrations for a few hours.
- **Exposure time**:
  - **Gamma radiation**: Brief exposure, often lasting less than an hour.
  - **EMS**: Typically, seeds are treated for 2 to 4 hours.

- **Achievements**
    - Development of cumin varieties with improved yield and disease resistance.
    - Enhanced adaptability to different climatic conditions.

13. Fennel (*Foeniculum vulgare*)

- **Mutagens used**:
    - **Gamma radiation**: Typically administered at doses of 5 to 50 Gy.
    - **EMS**: Concentrations of 0.1% to 0.3% for seed treatment.
- **Exposure time**:
    - **Gamma radiation**: Exposure typically lasts a few minutes to half an hour.
    - **EMS**: Seeds are usually treated for 2 to 6 hours.
- **Achievements**:
    - Increased essential oil content in fennel seeds.
    - Development of varieties with better resistance to pests and environmental stresses.

14. Fenugreek (*Trigonella foenum-graecum*)

- **Mutagens used**
    - **Gamma radiation**: Doses ranging from 10 to 100 Gy.
    - **EMS**: Concentrations from 0.1% to 1.0% for mutation induction.
- **Exposure time**:
    - **Gamma radiation**: Short exposures, often just a few minutes.
    - **EMS**: Seeds are treated for 2 to 4 hours.
- **Achievements**:
    - Varieties with higher protein content and better growth performance.
    - Improved tolerance to abiotic stresses such as drought.

15. Ajwain (*Trachyspermum ammi*)

- **Mutagens used**:
    - **Gamma radiation**: Common doses range from 10 to 100 Gy.
    - **EMS**: Used at concentrations of 0.1% to 0.5% for inducing mutations.
- **Exposure time**:
    - **Gamma radiation**: Exposure typically lasts a few minutes.
    - **EMS**: Treatment durations are generally between 2 to 6 hours.

- **Achievements**
  - Enhanced essential oil yield and improved seed quality.
  - Development of disease-resistant varieties.

Achievements through mutation breeding

- **Disease resistance:** Significant progress has been made in developing spice varieties that are resistant to major diseases, such as rhizome rot in ginger, leaf spot in turmeric, and wilt in cumin.
- **Yield improvement:** Mutation breeding has led to the development of high-yielding varieties across many spice crops, contributing to increased productivity.
- **Quality enhancement:** Improved essential oil content, curcumin content in turmeric, and better seed quality in spices like fennel and coriander are notable achievements.
- **Adaptability:** New varieties with better adaptability to diverse climatic conditions have been developed, helping to stabilize production in changing environments.

General considerations in mutation breeding

- **Mutagen selection**: The choice of mutagen (physical or chemical) depends on the plant species, the trait targeted for improvement, and the desired mutation rate.
- **Dosage and exposure**: These are optimized to induce beneficial mutations while minimizing damage to the plant tissue. Higher doses can increase mutation rates but also the risk of negative effects.
- **Screening**: Post-treatment, plants are carefully screened for desirable traits. This process involves growing out the treated population and selecting individuals that show improved characteristics.
- **Stabilization**: Once desirable mutants are identified, they are stabilized through further breeding to ensure that the traits are consistently expressed in subsequent generations.

Mutation breeding has significantly contributed to the advancement of spice cultivation, leading to the development of superior varieties that are more productive, resilient, and of higher quality.

5

Case Studies Breeding of Major Spice Crops

A. Case study: Breeding for disease resistance and yield improvement in Black pepper (*Piper nigrum*)

Black pepper (*Piper nigrum*) is one of the most important spice crops globally, often referred to as the "King of Spices." However, its production is frequently threatened by various diseases, particularly *Phytophthora* foot rot, which can cause significant yield losses. Over the years, extensive breeding efforts have focused on developing disease-resistant and high-yielding varieties of black pepper.

1. Background

- **Primary disease concern:** *Phytophthora* foot rot (caused by *Phytophthora capsici*) is a major disease affecting black pepper, leading to root and stem rot, defoliation, and ultimately, plant death. Other diseases include slow decline (caused by nematodes and fungal pathogens) and viral infections.
- **Yield challenges:** The average yield of black pepper is relatively low, primarily due to the susceptibility of many traditional varieties to diseases and pests, along with poor management practices and climatic challenges.

2. Objectives of breeding programs

- **Disease resistance:** To develop varieties resistant to *Phytophthora* foot rot and other significant diseases.
- **Yield improvement:** To increase the overall productivity of black pepper plants, focusing on higher berry yield, larger berries, and better quality.
- **Quality enhancement:** To maintain or improve the quality of black pepper in terms of essential oil content, piperine content, and other flavour components.

3. Breeding approaches

- **Selection of resistant genotypes**
  - **Germplasm collection:** Breeders collected diverse germplasm from various regions, including traditional cultivars and wild relatives of black pepper.

- **Screening for disease resistance:** This involved evaluating the germplasm under field conditions and controlled environments for resistance to Phytophthora foot rot. Resistant plants were identified and used in further breeding programs.

- **Hybridization**
    - **Crossing resistant and high-yielding varieties:** Hybridization between disease-resistant genotypes and high-yielding cultivars was performed to combine these traits. The progeny was then screened for both disease resistance and yield performance.
- **Mutation breeding**
    - **Induced mutations:** Gamma rays, EMS (ethyl methane sulfonate), and other mutagens were used to induce mutations in black pepper plants, aiming to create new variability for disease resistance and yield traits. Mutants with desirable characteristics were selected and evaluated.
- **Marker-assisted selection (MAS)**
    - **Molecular markers:** With advances in biotechnology, molecular markers linked to disease resistance genes were developed. Marker-assisted selection allowed for the efficient identification and selection of resistant plants in breeding programs.

4. Achievements

- **Varieties developed**

Several black pepper (*Piper nigrum*) varieties have been developed with high piperine content, which is a key compound responsible for the pungency and medicinal properties of black pepper. Breeding efforts have focused on enhancing piperine content alongside other desirable traits like disease resistance and yield. Here are some high-piperine varieties of black pepper:

1. Panniyur 1

- **Piperine content:** Approximately 5-7%
- **Background:** This is one of the most popular and widely cultivated varieties, developed at the Pepper Research Station in Panniyur, Kerala, India. It is known for its high yield, disease resistance, and relatively high piperine content.
- **Other traits:** Moderate resistance to *Phytophthora* foot rot, high yield potential (up to 3,000 kg/ha).

2. Sreekara

- **Piperine content:** Approximately 5-6%

- **Background:** Developed through hybridization, Sreekara is known for its tolerance to *Phytophthora* foot rot and higher productivity. It was released by the Indian Institute of Spices Research (IISR).
- **Other traits:** High yield potential, good quality berries with high piperine content.

3. Subhakara

- **Piperine content:** Approximately 6-7%
- **Background:** A high-yielding variety developed through selection, Subhakara is noted for its high piperine content and moderate resistance to diseases.
- **Other traits:** Improved berry size, good market acceptance due to superior quality.

4. Pournami

- **Piperine content:** Approximately 6-7%
- **Background:** This variety was developed through mutation breeding using gamma radiation. It is known for its resistance to foot rot and its high piperine content.
- **Other traits:** Improved yield, better adaptability to various growing conditions.

5. Karimunda

- **Piperine content:** Approximately 6-7%
- **Background:** A traditional variety often used in breeding programs, Karimunda is known for its high piperine content and good adaptability to different agro-climatic conditions.
- **Other traits:** Moderate yield, used as a parent in developing other high-piperine varieties.

6. IISR Malabar Excel

- **Piperine content:** Approximately 6-8%
- **Background:** Developed by the Indian Institute of Spices Research (IISR), Malabar Excel is a selection known for its high piperine content and good yield.
- **Other traits:** Resistant to *Phytophthora* foot rot, good berry size, and quality.

7. IISR Shakthi

- **Piperine content:** Approximately 6-8%
- **Background:** Another high-piperine variety developed by IISR, Shakthi is known for its excellent quality and disease resistance.

- **Other traits:** High yield potential, better resistance to wilt and *Phytophthora.*

8. Kalluvally

- **Piperine content:** Approximately 6-7%
- **Background:** This is a traditional variety known for its high piperine content. It has been used in breeding programs to improve the piperine content of new varieties.
- **Other traits:** Moderate yield, good quality berries.

9. Puthupanniyur 1

- **Piperine content:** Approximately 6-7%
- **Background:** Developed from the popular Panniyur series, this variety is known for its high piperine content and better yield.
- **Other traits:** Resistance to some common diseases, improved yield and berry quality.

10. Vadakkan

- **Piperine content:** Approximately 6-7%
- **Background:** This traditional variety is known for its high piperine content and has been used in breeding programs to develop improved varieties.
- **Other traits:** Good adaptability and moderate yield.

The development of high-piperine varieties of black pepper has been a significant achievement in spice breeding. These varieties not only cater to the market demand for high-quality pepper but also contribute to the economic sustainability of pepper cultivation by providing farmers with high-yielding, disease-resistant plants. Breeding programs continue to focus on enhancing piperine content while maintaining or improving other agronomic traits.

- **Impact on yield**
  - The development of disease-resistant varieties led to significant yield improvements. For instance, Panniyur 1 recorded yields of up to 3,000 kg/ha under ideal conditions, much higher than traditional varieties.
  - Improved management practices and the adoption of these resistant varieties have led to more stable yields, even under disease pressure.
- **Impact on disease management**
  - The introduction of resistant varieties reduced the reliance on chemical fungicides, leading to more sustainable and eco-friendly black pepper production.
  - Farmers have reported fewer incidences of foot rot in fields planted with resistant varieties, resulting in reduced crop losses and lower production costs.

- **Quality improvements**
  - Breeding efforts have ensured that the new varieties maintain or improve the essential oil content, piperine content, and other quality parameters essential for the spice industry.
  - Varieties like Subhakara are preferred in the market for their superior quality, which fetches a premium price.

5. Challenges and future prospects

- **Complexity of disease resistance:** The genetic basis of resistance to *Phytophthora* foot rot is complex, involving multiple genes. Breeding for durable resistance requires ongoing efforts and the integration of modern biotechnological tools.
- **Climate change:** Changing climatic conditions pose new challenges, such as the emergence of new disease strains or changes in pest dynamics, which may affect the effectiveness of existing resistant varieties.
- **Biotechnological integration:** The use of genome editing technologies, like CRISPR/Cas9, could further enhance the precision and efficiency of breeding programs, allowing for the targeted introduction of resistance genes or yield-enhancing traits.

Breeding for disease resistance and yield improvement in black pepper has had a significant impact on the productivity and sustainability of this vital spice crop. The development and adoption of resistant, high-yielding varieties have helped stabilize production, reduce losses due to diseases, and improve the livelihoods of farmers. Ongoing research and breeding efforts, combined with modern biotechnology, are expected to continue delivering new and improved varieties that can meet the challenges of the future.

Several high-yielding varieties and hybrids of black pepper (*Piper nigrum*) have been developed over the years to improve productivity, disease resistance, and quality. Here is a list of some notable high-yielding varieties and hybrids of black pepper:

1. Panniyur series

- **Panniyur 1**
  - **Yield potential:** Up to 3,000 kg/ha.
  - **Background:** Developed by the Pepper Research Station, Panniyur, Kerala, this variety is one of the most popular high-yielding varieties. It is known for its adaptability, disease tolerance, and good quality berries.
- **Panniyur 2**
  - **Yield potential:** High, slightly less than Panniyur 1.

- **Background:** Similar to Panniyur 1, but with improved resistance to pests and diseases.

- **Panniyur 3, 4, 5, and 6**
  - **Yield potential:** Varies, with each succeeding variety showing specific improvements in yield, disease resistance, and quality traits.

2. IISR Series (Indian Institute of Spices Research)

- **IISR Shakthi**
  - **Yield potential:** 2,500-3,000 kg/ha.
  - **Background:** Known for its resistance to *Phytophthora* foot rot and wilt diseases. It has a good yield potential and quality berries with high piperine content.
- **IISR Thevam**
  - **Yield potential:** 2,800-3,200 kg/ha.
  - **Background:** Developed for better disease resistance, especially to *Phytophthora,* and improved yield.
- **IISR Girimunda**
  - **Yield potential:** High, specific to hilly regions.
  - **Background:** A selection made for high-altitude cultivation, with good yield and disease resistance.
- **IISR Malabar Excel**
  - **Yield potential:** 2,500-3,000 kg/ha.
  - **Background:** Known for its high piperine content and good yield. It is a popular choice for commercial cultivation.

3. Pournami

- **Yield potential:** 2,000-2,500 kg/ha.
- **Background:** A variety developed through mutation breeding using gamma radiation. It is resistant to *Phytophthora* foot rot and has good quality berries with high piperine content.

4. Sreekara

- **Yield potential:** 2,800-3,000 kg/ha.
- **Background:** Developed through hybridization, Sreekara is known for its high yield, tolerance to *Phytophthora* foot rot, and good quality.

5. Subhakara

- **Yield potential:** 2,500-2,800 kg/ha.

- **Background:** Developed through selection, this variety is noted for its high yield, disease resistance, and high piperine content.

6. Karimunda

- **Yield potential:** 2,500-3,000 kg/ha.
- **Background:** A traditional variety that is still popular due to its adaptability, good yield, and quality. It is often used in breeding programs for developing new high-yielding varieties.

7. Vadakkan

- **Yield potential:** 2,000-2,500 kg/ha.
- **Background:** Another traditional variety, known for its high yield and good quality berries. It is popular among farmers in Kerala.

8. Kottanadan

- **Yield potential:** 2,200-2,800 kg/ha.
- **Background:** This variety is known for its adaptability to different agro-climatic conditions and good yield potential.

9. Puthupanniyur 1

- **Yield potential:** 2,500-3,000 kg/ha.
- **Background:** A newer selection from the Panniyur series, this variety combines high yield with better disease resistance and good quality.

10. Hybrid varieties

- **Hybrid 1 (Selection from high-yielding parents)**
  - **Yield potential:** High, comparable to or better than traditional varieties.
  - **Background:** Some hybrids have been developed by crossing high-yielding and disease-resistant parents. These hybrids are currently under evaluation and are expected to provide improved yield and quality.

High-yielding varieties and hybrids of black pepper have been developed to meet the increasing demand for this spice while addressing challenges such as disease pressure and variable climatic conditions. These varieties not only provide better yields but also offer improved resistance to diseases like *Phytophthora* foot rot and ensure good quality berries with high piperine content. The continued focus on breeding high-yielding varieties is essential to sustain and enhance black pepper production globally.

B) Case study: Breeding for disease resistance and yield improvement in Small cardamom (*Elettaria cardamomum*)

Small cardamom (*Elettaria cardamomum*), often referred to as the "Queen of Spices," is a highly valued spice crop, primarily grown in the Western Ghats of

India. The crop faces several challenges, including susceptibility to diseases and the need for yield improvement to meet global demand. Over the years, significant breeding efforts have been made to develop disease-resistant and high-yielding varieties of small cardamom.

1. Background

- **Primary disease concerns**
  - **Cardamom mosaic virus (CdMV):** This is one of the most devastating diseases, causing stunted growth, reduced yields, and malformed capsules.
  - **Katte (banana) disease:** Caused by a virus and transmitted by aphids, leading to yellowing of leaves and reduced plant vigor.
  - **Rhizome rot:** Caused by fungal pathogens like *Pythium vexans* and *Rhizoctonia solani*, leading to wilting and death of plants.
  - **Clump rot and leaf blight:** These fungal diseases also affect yield and plant health.
- **Yield challenges**
  - Traditional varieties often have low yield potential and are highly susceptible to various diseases. Additionally, poor agronomic practices and climatic changes have further exacerbated the yield gap.

2. Objectives of breeding programs

- **Disease resistance:** To develop varieties resistant to major diseases like Cardamom mosaic virus, Katte disease, and fungal pathogens.
- **Yield improvement:** To increase the productivity of small cardamom plants by focusing on higher capsule yield, improved size, and better quality.
- **Quality enhancement:** To maintain or improve the essential oil content and other quality parameters that contribute to the market value of cardamom.

3. Breeding approaches

- **Selection of resistant genotypes:**
  - **Germplasm collection:** Diverse germplasm was collected from traditional cultivars, wild relatives, and different cardamom-growing regions.
  - **Screening for disease resistance:** Germplasm was evaluated under natural and artificial conditions for resistance to major diseases. Resistant plants were identified and selected for further breeding.
- **Hybridization**
  - **Crossing resistant and high-yielding varieties:** Resistant genotypes were crossed with high-yielding cultivars to combine the desired traits. The progeny were screened for both disease resistance and yield performance.

- **Mutation breeding**
  - **Induced mutations:** Mutagens like gamma rays and EMS (ethyl methane sulfonate) were used to induce variability in cardamom plants, with the aim of creating new variants with improved traits. Mutants showing desirable characteristics were selected for further evaluation.
- **Marker-assisted selection (MAS):**
  - **Molecular markers:** With advances in biotechnology, molecular markers linked to disease resistance genes were developed. MAS was used to efficiently select resistant plants in breeding programs.
- **Biotechnological interventions:**
  - **Tissue culture:** Micropropagation techniques were used to rapidly multiply disease-free and high-yielding clones of selected varieties. This approach also helped in maintaining the genetic purity of the varieties.

4. Achievements

- **Varieties developed:**
  - **IISR Vijetha:**
    - **Yield potential:** 800-1,000 kg/ha.
    - **Background:** Developed by the Indian Institute of Spices Research (IISR), Vijetha is known for its resistance to Cardamom mosaic virus and improved yield.
  - **IISR Avinash:**
    - **Yield potential:** 1,000-1,200 kg/ha.
    - **Background:** A high-yielding variety with moderate resistance to major diseases. It is also known for its bold capsules and good quality.
  - **Njallani Green Gold:**
    - **Yield potential:** 1,200-1,500 kg/ha.
    - **Background:** This farmer-developed variety is highly popular due to its high yield and good market acceptance. It shows moderate resistance to diseases and is widely cultivated.
  - **CCS 1**
    - **Yield potential:** 1,000-1,200 kg/ha.
    - **Background:** Developed by the Indian Cardamom Research Institute (ICRI), this variety is resistant to Katte disease and has a good yield potential.
  - **IISR Suvasini:**
    - **Yield potential:** 1,100-1,300 kg/ha.

- **Background:** Known for its resistance to clump rot and leaf blight, this variety also has high yield and quality.

- **Impact on yield**
  - The development of high-yielding varieties like Njallani has significantly increased the average yield of small cardamom. These varieties have outperformed traditional cultivars in both yield and quality, making them popular among farmers.
  - Improved agronomic practices, alongside the use of these varieties, have contributed to higher productivity in cardamom plantations.
- **Impact on disease management**
  - The introduction of disease-resistant varieties has reduced the incidence of major diseases like Cardamom mosaic virus and Katte disease, leading to more stable and sustainable production.
  - Farmers have reported lower crop losses and reduced dependency on chemical control measures, resulting in cost savings and environmental benefits.
- **Quality improvements**
  - Breeding efforts have ensured that new varieties maintain or enhance essential oil content, which is critical for the spice's flavour and aroma.
  - Varieties like IISR Vijetha and IISR Suvasini are favoured for their bold capsules and high market value.

5. Challenges and future prospects

- **Complexity of disease resistance:** The genetic basis of resistance to viral and fungal diseases is complex, involving multiple genes. Continuous breeding efforts are needed to develop varieties with durable resistance.
- **Climate change:** Changing climatic conditions may lead to the emergence of new disease strains or alter the dynamics of existing pathogens, necessitating ongoing research and breeding.
- **Biotechnological integration:** The use of advanced tools like CRISPR/Cas9 for genome editing could further enhance the precision and efficiency of breeding programs, enabling the targeted introduction of resistance genes or yield-enhancing traits.

Breeding for disease resistance and yield improvement in small cardamom has significantly impacted the spice industry. The development and adoption of resistant, high-yielding varieties have helped stabilize production, reduce losses due to diseases, and improve the livelihoods of farmers. As breeding programs continue to evolve with the integration of modern biotechnological tools, small cardamom cultivation is poised to meet the challenges of the future while maintaining its position as a valuable spice crop.

C) Case study: Breeding for disease resistance and yield improvement in Large cardamom (*Amomum subulatum*)

Large cardamom (*Amomum subulatum*), also known as "Badi Elaichi" or "Black Cardamom," is a significant spice crop cultivated primarily in the Eastern Himalayas, including Sikkim, Darjeeling, and parts of Bhutan and Nepal. The crop is known for its unique flavour and medicinal properties but faces several challenges, including susceptibility to diseases and low yield. Breeding efforts have been focused on improving both disease resistance and yield to sustain and enhance production.

1. Background

- **Primary disease concerns**
  - **Chirke and Foorkey diseases:** These are viral diseases transmitted by aphids and infected planting material. Chirke causes leaf streaking and reduced plant vigor, while Foorkey leads to stunted growth and bushy appearance.
  - **Fungal diseases:** Large cardamom is also affected by fungal diseases like leaf blight and rhizome rot, which can cause significant yield losses.
  - **Pest infestation:** Shoot and capsule borers, as well as nematodes, also pose significant threats to the crop.
- **Yield challenges**
  - Traditional varieties often have low yield potential, which is further compromised by susceptibility to diseases and pests, as well as the challenges posed by the changing climate in the Himalayan region.

2. Objectives of breeding programs

- **Disease resistance:** To develop varieties resistant to viral diseases like Chirke and Foorkey, as well as fungal pathogens.
- **Yield improvement:** To increase the productivity of large cardamom plants by selecting and breeding for higher capsule yield, larger size, and better quality.
- **Adaptability:** To ensure that new varieties are adaptable to varying climatic conditions and altitudes found in the Himalayan regions.

3. Breeding approaches

- **Selection of resistant genotypes**
  - **Germplasm collection:** A diverse collection of germplasm from different regions was assembled, including traditional cultivars, wild relatives, and landraces.

- **Screening for disease resistance:** Germplasm was evaluated in areas with high disease pressure to identify resistant genotypes. Plants showing resistance to Chirke, Foorkey, and fungal diseases were selected for further breeding.

- **Hybridization**
    - **Crossing disease-resistant and high-yielding varieties:** Disease-resistant genotypes were crossed with high-yielding varieties to combine the desirable traits. The progeny was evaluated for both yield and disease resistance.
- **Mutation breeding**
    - Induced mutations: Radiation and chemical mutagens were used to induce genetic variability in large cardamom. Mutants with improved disease resistance and yield characteristics were selected and evaluated.
- **Biotechnological interventions**
    - Tissue culture: Micropropagation techniques were employed to produce disease-free and uniform planting material of selected varieties. Tissue culture also helped in the rapid multiplication of promising clones.
    - Molecular marker development: Molecular markers linked to disease resistance genes were developed, enabling more efficient selection of resistant plants in breeding programs.

4. Achievements

- **Varieties developed**
    - **Ramsey:**
        - **Yield potential:** 800-1,000 kg/ha.
        - **Background:** A widely cultivated variety in Sikkim, Ramsey is known for its resistance to Foorkey disease and high yield. It produces bold capsules with good market value.
    - **Sawney:**
        - **Yield potential:** 900-1,100 kg/ha.
        - **Background:** Sawney is another popular variety in the Eastern Himalayas, noted for its resistance to Chirke disease and good yield potential. It is also known for its adaptability to varying climatic conditions.
    - **Golsey:**
        - **Yield potential:** 1,000-1,200 kg/ha.
        - **Background:** This variety is resistant to both Chirke and Foorkey diseases and is known for its high yield and good-quality capsules. Golsey is widely preferred by farmers due to its resilience and productivity.

- **Varlangey**
  - **Yield potential:** 1,200-1,400 kg/ha.
  - **Background:** A high-yielding variety with good resistance to viral diseases, Varlangey produces large capsules with a high market value.
- **Dzongu Golsey**
  - **Yield potential:** 1,100-1,300 kg/ha.
  - **Background:** Developed through selection from traditional cultivars, Dzongu Golsey is resistant to Chirke disease and has a good yield. It is favored in the high-altitude regions of Sikkim.

- **Impact on yield**
  - The introduction of high-yielding varieties like Varlangey and Dzongu Golsey has led to a significant increase in the productivity of large cardamom plantations. These varieties have outperformed traditional ones in terms of yield and quality, leading to better returns for farmers.
- **Impact on disease management**
  - The development of resistant varieties has reduced the incidence of Chirke and Foorkey diseases, which were major constraints in large cardamom cultivation. This has led to more stable production and reduced reliance on chemical control measures.
  - Farmers have reported fewer crop losses and better plant vigor, contributing to the overall sustainability of large cardamom farming.
- **Quality improvements**
  - Breeding efforts have ensured that new varieties maintain or enhance the quality traits of large cardamom, such as essential oil content, which is crucial for its flavour and aroma.
  - Varieties like Golsey and Varlangey are known for their bold capsules and superior quality, making them highly valued in the market.

5. Challenges and future prospects

- **Complexity of disease resistance:** The genetic basis of resistance to viral diseases like Chirke and Foorkey is complex and may involve multiple genes. Continuous breeding efforts are required to develop varieties with durable resistance.
- **Climate change:** The Himalayan region is particularly vulnerable to climate change, which may alter disease dynamics and affect crop yields. Breeding programs need to focus on developing varieties that are resilient to changing climatic conditions.

- **Integration of modern breeding tools:** The use of advanced tools like genome editing and molecular markers could enhance the precision and efficiency of breeding programs, enabling the targeted introduction of resistance genes and yield-enhancing traits.

Breeding for disease resistance and yield improvement in large cardamom has been instrumental in stabilizing and enhancing the production of this valuable spice crop. The development and adoption of resistant, high-yielding varieties have significantly reduced the impact of major diseases and increased productivity, leading to better livelihoods for farmers in the Eastern Himalayas. As breeding programs continue to evolve with the integration of modern technologies, large cardamom cultivation is well-positioned to meet the challenges of the future and maintain its status as a key cash crop in the region.

D) Case study: Breeding for disease resistance and yield improvement in Clove (*Syzygium aromaticum*)

Clove (*Syzygium aromaticum*) is a highly valuable spice crop known for its aromatic flower buds, which are used in culinary, medicinal, and industrial applications. Native to the Maluku Islands in Indonesia, clove is now widely cultivated in several tropical regions, including India, Sri Lanka, Madagascar, and Zanzibar. The cultivation of Clove faces significant challenges, particularly in terms of disease susceptibility and the need for yield improvement. Breeding programs have been initiated to address these challenges, focusing on developing disease-resistant and high-yielding clove varieties.

1. Background

- **Primary disease concerns**
  - **Clove dieback disease:** A major disease affecting Clove trees, caused by the fungal pathogen *Diplodia clovea* or *Colletotrichum gloeosporioides*. The disease leads to wilting, leaf drop, and eventual death of branches and trees.
  - **Leaf spot disease:** Caused by fungi such as *Cercospora syzygii* and *Phyllosticta sp.*, this disease results in lesions on leaves, reducing photosynthetic capacity and overall plant vigour.
  - **Anthracnose:** A fungal disease caused by *Colletotrichum spp.*, leading to dark, sunken lesions on leaves, twigs, and flower buds, which can severely affect yield.
- **Yield challenges**
  - Clove trees have a long gestation period, taking several years to start bearing flowers. Traditional varieties often have low yield potential, which can be further reduced by disease pressure and suboptimal growing conditions.

2. Objectives of breeding programs

- **Disease resistance:** To develop clove varieties resistant to major fungal diseases, particularly clove dieback, leaf spot, and anthracnose.
- **Yield improvement:** To enhance the productivity of clove trees by selecting and breeding for higher flower bud yield, larger size, and better quality.
- **Adaptability:** To ensure that new varieties are adaptable to a range of tropical environments, ensuring consistent performance across different regions.

3. Breeding approaches

• Selection of resistant genotypes

- **Germplasm collection:** A wide range of germplasm was collected from traditional clove-growing regions, including Indonesia, Zanzibar, and India, as well as wild relatives of clove.
- **Screening for disease resistance:** Germplasm was evaluated under natural and controlled conditions for resistance to key diseases like clove dieback and anthracnose. Resistant trees were selected for further breeding.

• Hybridization

- **Crossing disease-resistant and high-yielding varieties:** Resistant genotypes were crossed with high-yielding varieties to combine desirable traits. Progeny were evaluated for both disease resistance and yield performance.

• Mutation breeding

- **Induced mutations:** Radiation (e.g., gamma rays) and chemical mutagens were used to induce genetic variability in clove trees. Mutants showing improved disease resistance and yield traits were selected and further evaluated.

• Tissue culture and micropropagation

- **In vitro propagation:** Tissue culture techniques were developed to rapidly multiply disease-free planting material of selected clove varieties. This approach also helped in maintaining the genetic uniformity of new clones.
- **Somaclonal variation:** Tissue culture-induced variations were explored as a source of new genetic diversity for disease resistance and yield improvement.

• Biotechnological interventions

- **Molecular marker development:** Molecular markers linked to disease resistance traits were identified and used for marker-assisted selection (MAS) to accelerate the breeding process.

4. Achievements

- **Varieties developed**
  - **Zanzibar Clove**
    - **Yield potential:** 2,000-3,000 kg/ha.
    - **Background:** Originally from Zanzibar, this variety is known for its high yield and resistance to leaf spot disease. It is widely cultivated in many clove-growing regions due to its adaptability and productivity.
  - **Mardikar Clove**
    - **Yield potential:** 1,800-2,500 kg/ha.
    - **Background:** Developed through selection, Mardikar clove shows moderate resistance to clove dieback and good yield potential. It is favored for its aromatic flower buds with high essential oil content.
  - **Local varieties with enhanced traits**
    - **Yield potential:** Varies (1,500-2,200 kg/ha).
    - **Background:** Several local varieties have been selected and propagated for their disease resistance and yield. These include selections from traditional germplasm in India and Sri Lanka, adapted to specific regional conditions.
- **Impact on yield**
  - The introduction of high-yielding varieties like Zanzibar clove has significantly improved the productivity of clove plantations. These varieties offer higher flower bud yields compared to traditional cultivars, contributing to increased income for farmers.
- **Impact on disease management**
  - The development of disease-resistant varieties has reduced the incidence of major fungal diseases like clove dieback and leaf spot, which were previously significant constraints in clove cultivation. This has led to more stable production and reduced reliance on fungicides.
  - Farmers have reported improved tree vigour and longer productive lifespans for clove trees, enhancing the sustainability of clove farming.
- **Quality improvements**
  - Breeding efforts have ensured that new varieties maintain or improve the essential oil content and other quality parameters of clove buds, which are critical for their market value.
  - Varieties like Zanzibar clove are known for their large, aromatic flower buds with high essential oil content, making them highly sought after in international markets.

5. Challenges and future prospects

- **Complexity of disease resistance:** The genetic basis of resistance to clove dieback and other fungal diseases is complex, involving multiple genes. Continuous breeding efforts are needed to develop varieties with durable and broad-spectrum resistance.
- **Long gestation period:** The long juvenile phase of clove trees poses a challenge for breeding programs, as it delays the evaluation of new genotypes. Efforts to shorten the gestation period through breeding or agronomic practices could accelerate the development of new varieties.
- **Climate change:** Clove is sensitive to changes in temperature and humidity, making it vulnerable to climate change. Breeding programs need to focus on developing varieties that are resilient to these changes while maintaining high yield and quality.
- **Integration of advanced technologies:** The use of advanced breeding tools, such as genome editing and genomic selection, could enhance the precision and efficiency of clove breeding programs, enabling the development of varieties with improved disease resistance and yield.

Breeding for disease resistance and yield improvement in clove has made significant strides in enhancing the productivity and sustainability of clove cultivation. The development of resistant, high-yielding varieties has mitigated the impact of major diseases, leading to more stable and profitable production systems. As breeding programs continue to evolve with the integration of modern technologies, the clove industry is well-positioned to meet future challenges and maintain its importance as a valuable spice crop globally.

E) Case study: Breeding for high quality and yield improvement in Cinnamon varietal development

Introduction

Cinnamon (*Cinnamomum* spp.) is a commercially important spice crop, valued for its aromatic bark used in culinary, medicinal, and cosmetic applications. However, cinnamon production faces challenges from various diseases, such as leaf blight, root rot, and bark canker, which significantly reduce yield and quality. Therefore, breeding for disease resistance and yield improvement is crucial to ensure sustainable production and profitability for farmers.

Objectives

1. **Enhance disease resistance**: Develop cinnamon varieties with strong resistance to major diseases affecting the crop, thereby reducing losses and production costs.

2. **Increase yield**: Improve the overall yield potential of cinnamon varieties to meet the growing global demand and enhance farmer income.
3. **Stability and adaptability**: Ensure that the developed varieties perform consistently across different environmental conditions.

Materials and Methods

- **Genotypes**: A diverse set of cinnamon genotypes, including local landraces, commercial cultivars, and wild relatives, were selected based on their known or potential resistance to diseases and yield traits.
- **Experimental design**: The study was conducted using a Randomized Block Design (RBD) with three replications to minimize environmental effects and accurately evaluate the genotypic performance.
- **Traits evaluated**:
    - **Disease resistance**: Resistance to key diseases like leaf blight (Phytophthora spp.), root rot (Pythium spp.), and bark canker (various fungal pathogens) was assessed through natural infection in the field and controlled inoculation in greenhouse conditions.
    - **Yield**: Yield was measured as the fresh and dry weight of bark per plant, a key economic trait.
    - **Growth parameters**: Additional traits such as plant height, bark thickness, and canopy spread were recorded as they indirectly contribute to yield.

Statistical Analysis

1. **ANOVA**: To determine the significance of differences among the genotypes for disease resistance, yield, and growth parameters.
2. **Correlation analysis**: To investigate the relationships between disease resistance and yield traits, helping to understand how resistance might influence yield.
3. **Path analysis**: To dissect the direct and indirect effects of disease resistance and growth parameters on yield, providing insights for selection criteria.
4. **GCV & PCV**: To estimate Genotypic and Phenotypic Coefficients of Variation, giving an idea of the extent of genetic variability available for disease resistance and yield traits.
5. **Heritability analysis**: To estimate broad-sense heritability, indicating the proportion of total variation in disease resistance and yield traits due to genetic factors, guiding selection intensity.

Breeding strategy

- **Selection**: Focus on mass selection and recurrent selection for disease resistance and yield traits, particularly in early generations, to capture a wide range of genetic diversity.
- **Hybridization**: Cross-breeding resistant genotypes with high-yielding varieties to combine disease resistance and yield in a single genotype.
- **Backcrossing**: Utilize backcrossing to introgress disease resistance genes from wild relatives or landraces into high-yielding, commercially viable cultivars.
- **Marker-assisted selection (MAS)**: Incorporate molecular markers linked to disease resistance genes to accelerate the selection process, especially in later stages of breeding.

Expected outcomes

- **Disease-resistant varieties**: Development of cinnamon varieties that are resistant to major diseases, leading to reduced dependence on chemical controls and lower production costs.
- **High-yielding varieties**: Selection of genotypes with significantly improved yield, contributing to higher economic returns for farmers.
- **Stable varieties**: Identification of cinnamon varieties that are not only high-yielding and disease-resistant but also stable across different environments, ensuring broad adaptability.

The breeding program aims to create cinnamon varieties that are both disease-resistant and high-yielding, addressing two major challenges in cinnamon production. By employing a combination of traditional breeding methods and modern techniques such as marker-assisted selection, the program will develop robust cinnamon varieties that can thrive in various growing conditions, ensuring sustainable production and economic viability for farmers. This dual focus on disease resistance and yield improvement is essential for maintaining the competitiveness and sustainability of the cinnamon industry in the face of evolving challenges.

Achievements of Cinnamon breeding

Breeding efforts in cinnamon have led to the development of several disease-resistant and high-yielding varieties. These varieties are tailored to withstand common diseases like leaf blight, root rot, and bark canker while also offering improved yield, which is critical for both commercial cultivation and smallholder farmers.

Notable disease-resistant and high-yielding Cinnamon varieties

1. Navashree

- **Origin**: Developed by the Indian Institute of Spices Research (IISR).
- **Disease resistance**: Known for its resistance to major diseases such as leaf blight and root rot.
- **Yield**: High yielding, with a substantial increase in bark yield compared to traditional varieties.
- **Quality**: High cinnamaldehyde content, making it suitable for both spice and medicinal purposes.

2. Nityashree

- **Origin**: Developed through a systematic breeding program focusing on disease resistance.
- **Disease resistance**: Shows strong resistance to leaf blight and root diseases.
- **Yield**: High yielding with consistent performance across various environmental conditions.
- **Additional features**: Good quality bark with desirable thickness and aroma.

3. Sugandhini

- **Origin**: A hybrid developed to combine disease resistance with high yield and quality traits.
- **Disease resistance**: Resistant to multiple fungal pathogens that cause bark canker and root rot.
- **Yield**: Excellent yield potential, especially in disease-prone areas.
- **Quality**: Rich in essential oils, particularly cinnamaldehyde, making it a premium variety.

4. PPI (Peradeniya Plantations Institute) hybrid series

- **Origin**: Developed in Sri Lanka by the Peradeniya Plantations Institute.
- **Disease resistance**: Bred specifically for resistance to bark canker and leaf blight, common in tropical regions.
- **Yield**: High yielding, with a focus on both fresh and dry bark weight.
- **Adaptability**: Suitable for diverse agro-climatic conditions, making it popular in various cinnamon-growing regions.

5. IISR C3

- **Origin**: Another variety from the Indian Institute of Spices Research.
- **Disease resistance**: Noted for its resistance to leaf blight and root rot.
- **Yield**: High bark yield, with a good balance between yield and quality.

- **Additional features**: Developed for better adaptability in different growing environments, particularly in India and similar tropical climates.

Impact of these varieties

The development of these disease-resistant, high-yielding cinnamon varieties has had a significant impact on cinnamon cultivation:

- **Reduced crop losses**: By incorporating disease resistance, these varieties reduce the need for chemical treatments, leading to lower production costs and less environmental impact.
- **Increased productivity**: High-yielding varieties ensure that farmers can produce more cinnamon per unit area, increasing their income and making cinnamon cultivation more profitable.
- **Enhanced quality**: With a focus on essential oil content and bark quality, these varieties meet the market demand for high-quality cinnamon, ensuring better prices for growers.

These varieties represent the success of ongoing breeding programs in ensuring that cinnamon remains a viable and profitable crop in the face of agricultural challenges.

Challenges in Cinnamon breeding

1. Genetic diversity and breeding complexity

- **Limited genetic resources**: Cinnamon breeding is constrained by a relatively narrow genetic base. The limited availability of diverse germplasm hampers the ability to introduce new traits, such as disease resistance or improved yield.
- **Complex genome**: The genetic complexity of cinnamon, including issues like polyploidy and long generation times, makes conventional breeding methods challenging and time-consuming.

2. Disease pressure

- **Emerging diseases**: New and evolving diseases, such as fungal infections and viral diseases, pose a continuous threat. Breeding for resistance against these pathogens is difficult due to the fast adaptation and mutation rates of these organisms.
- **Lack of durable resistance**: Even when disease-resistant varieties are developed, resistance may not be durable over time as pathogens evolve, necessitating constant breeding efforts to maintain resistance.

3. Environmental adaptation

- **Climate change**: Cinnamon is sensitive to changes in climate, and the increasing unpredictability of weather patterns affects the stability and performance of existing varieties. Breeding for climate resilience is a growing necessity.

- **Soil and Water conditions**: Cinnamon cultivation is also affected by soil salinity, pH, and water availability, requiring the development of varieties that can tolerate suboptimal growing conditions.

4. Yield and quality trade-offs

- **Balancing yield and quality**: Achieving high yields without compromising the essential oil content or bark quality is a major challenge. Breeding programs often have to navigate trade-offs between these traits.
- **Slow progress**: Cinnamon is a perennial crop with a long breeding cycle, leading to slow progress in developing new varieties that meet all desired criteria.

5. Technical and resource limitations

- **Lack of advanced breeding techniques**: In many regions, cinnamon breeding still relies heavily on conventional methods due to limited access to advanced tools like marker-assisted selection (MAS) or genomic selection.
- **Resource constraints**: Limited funding, infrastructure, and expertise in some cinnamon-growing regions slow down breeding efforts and the adoption of new technologies.

Future prospects in Cinnamon breeding

1. Utilization of advanced breeding techniques:

- **Genomic selection**: Implementing genomic selection and MAS could significantly accelerate the breeding process by allowing for the early selection of desirable traits at the seedling stage.
- **CRISPR and genetic engineering**: Emerging technologies like CRISPR could be used to introduce specific disease resistance genes or improve yield traits with precision, although regulatory and public acceptance challenges remain.
- **Marker-Assisted breeding**: Using molecular markers linked to important traits can improve the efficiency and accuracy of breeding programs.

2. Broadening the genetic base

- **Germplasm exploration**: Expanding the genetic base by exploring wild relatives and underutilized species of *Cinnamomum* can introduce new traits for disease resistance, yield, and environmental adaptation.
- **Pre-breeding programs**: Initiatives focused on the introgression of useful traits from wild species into cultivated varieties can help overcome the limitations of the current genetic base.

3. Breeding for climate resilience

- **Drought and heat tolerance**: Developing varieties that can withstand higher temperatures and irregular rainfall patterns will be crucial as climate change continues to impact traditional growing regions.
- **Soil and water management**: Breeding cinnamon varieties that are more tolerant to a range of soil conditions, including salinity and acidity, and require less water will support sustainable production in areas facing resource constraints.

4. Integrated disease management

- **Pyramiding resistance genes**: Combining multiple resistance genes into single varieties (pyramiding) can provide more durable resistance against a range of pathogens.
- **Integrated pest management (IPM)**: Breeding should be integrated with broader pest management strategies, including biological control and cultural practices, to ensure sustainable disease control.

5. Market-oriented breeding

- **Consumer preferences**: Breeding programs should consider evolving market demands, such as preferences for specific flavours, organic production, and higher quality standards, to ensure the commercial viability of new varieties.
- **Sustainable production**: Developing varieties that contribute to sustainable agriculture by requiring fewer inputs (e.g., fertilizers, pesticides) will be increasingly important for both environmental and economic reasons.

6. Capacity building and collaboration

- **Training and infrastructure**: Investing in the training of breeders and upgrading infrastructure in cinnamon-growing regions will be essential for adopting new technologies and methods.
- **International collaboration**: Collaborative breeding programs that pool resources and expertise across countries can help overcome local limitations and accelerate the development of improved varieties.

Cinnamon breeding faces significant challenges, including limited genetic diversity, disease pressures, environmental changes, and technical limitations. However, the future of cinnamon breeding is promising, with opportunities to leverage advanced breeding technologies, broaden the genetic base, and develop climate-resilient, high-yielding, and disease-resistant varieties. By addressing these challenges and embracing innovation, cinnamon breeding can continue to support the growth of the global cinnamon industry and meet the evolving demands of producers and consumers alike.

F) Case study: Breeding for high quality and yield improvement in Allspice (*Pimenta dioica*) varietal development

Allspice (*Pimenta dioica*) is a tropical evergreen tree native to the Caribbean and Central America, known for its aromatic berries that combine the flavours of cinnamon, nutmeg, and cloves. Allspice is widely used in culinary applications, as well as in traditional medicine and perfumery. The demand for high-quality allspice with superior flavour and aroma, as well as higher yields to meet growing global demand, underscores the need for focused breeding efforts.

Objectives

1. **Enhance quality**: Develop allspice varieties with higher concentrations of essential oils, particularly eugenol, which contributes to the characteristic flavour and aroma.
2. **Increase yield**: Improve the overall yield of allspice trees to enhance productivity and profitability for growers.
3. **Improve tree structure and adaptability**: Select for favorable tree architecture and adaptability to different growing conditions to ensure sustainable production.

Materials and Methods

- **Genotypes**: A diverse collection of allspice genotypes, including local landraces, commercially available varieties, and wild relatives, was selected based on their potential for high yield and quality traits.
- **Experimental design**: The experiment was conducted using a Randomized Block Design (RBD) with three replications, allowing for accurate assessment of genotypic performance under field conditions.
- **Traits evaluated**
    - **Essential oil content**: Quantified using gas chromatography to measure the levels of eugenol and other key compounds.
    - **Yield**: Measured as the fresh and dry weight of berries per tree, with attention to both the quantity and consistency of yield over multiple seasons.
    - **Tree architecture**: Assessed based on traits like height, canopy spread, branch structure, and resistance to lodging, which impact ease of harvest and overall productivity.
    - **Adaptability**: Evaluated by observing the performance of genotypes across different environmental conditions, including variations in soil type, moisture availability, and temperature.

Statistical analysis

1. **ANOVA**: Used to determine the significance of differences among genotypes for essential oil content, yield, and other traits.
2. **Correlation analysis**: To explore relationships between essential oil content, yield, and tree architecture, helping to understand how these traits interact.
3. **Path analysis**: To dissect the direct and indirect effects of various traits on yield, providing insights for selection criteria.
4. **GCV & PCV**: To estimate Genotypic and Phenotypic Coefficients of Variation, which indicate the extent of genetic variability available for key traits.
5. **Heritability analysis**: To estimate broad-sense heritability, indicating the proportion of total variance attributable to genetic factors for the traits of interest.

Breeding strategy

- **Selection**: Implement mass selection and recurrent selection to identify and propagate genotypes with superior quality and yield traits, especially in early generations.
- **Hybridization**: Cross-breeding high-yielding genotypes with those exhibiting superior essential oil content to combine desirable traits in new varieties.
- **Clonal propagation**: Use clonal propagation to maintain and multiply elite genotypes with proven performance in quality and yield, ensuring uniformity in commercial production.
- **Marker-assisted selection (MAS)**: Integrate molecular markers linked to essential oil content and yield traits to enhance the efficiency and accuracy of the selection process.

Expected outcomes

- **High-quality varieties**: Development of allspice varieties with elevated levels of eugenol and other essential oils, catering to both culinary and medicinal markets.
- **High-yielding varieties**: Selection of genotypes with significantly improved yield, providing economic benefits to growers and ensuring a stable supply of allspice.
- **Improved tree structure**: Selection for favorable tree architecture will facilitate easier harvesting and increase overall orchard productivity.
- **Stable and adaptable varieties**: Identification of allspice varieties that perform consistently across different environments, ensuring broader adaptability and resilience to changing climate conditions.

Challenges in Allspice breeding

1. **Limited genetic resources**: The genetic diversity in allspice is relatively narrow, posing challenges in introducing new traits such as disease resistance or improved adaptability.
2. **Long generation time**: Allspice trees take several years to reach maturity, slowing down the breeding process and delaying the evaluation of traits like yield and essential oil content.
3. **Complexity in quality traits**: The essential oil composition of allspice is influenced by multiple genes, making it challenging to select for high quality without extensive biochemical analysis.
4. **Environmental sensitivity**: Allspice is sensitive to environmental factors such as temperature and soil conditions, necessitating careful selection for adaptability to different growing regions.

Future prospects in Allspices breeding

1. Utilization of advanced breeding techniques

- **Genomic selection**: Employing genomic selection and marker-assisted breeding can accelerate the development of high-quality, high-yielding allspice varieties.
- **Biotechnological approaches**: Techniques like tissue culture and CRISPR gene editing could be used to introduce specific desirable traits, such as disease resistance or improved essential oil content.
- **Hybrid breeding**: Developing hybrids that combine the best traits of different genotypes could lead to varieties with superior quality and yield.

2. Expanding genetic diversity

- **Germplasm exploration**: Broadening the genetic base by exploring wild relatives and underutilized species can introduce new traits and enhance breeding potential.
- **Pre-breeding programs**: Initiatives focused on introgressing traits from wild species into cultivated varieties can help overcome the limitations of the current genetic base.

3. Breeding for climate resilience

- **Drought and heat tolerance**: Developing varieties that can withstand higher temperatures and irregular rainfall patterns will be crucial as climate change impacts allspice-growing regions.
- **Soil and water management**: Breeding for varieties that can tolerate a range of soil conditions, including salinity and acidity, will support sustainable production in areas facing resource constraints.

4. Market-oriented breeding

- **Consumer preferences**: Breeding programs should consider evolving market demands, such as preferences for organic production, specific flavours, and higher quality standards, to ensure the commercial viability of new varieties.
- **Sustainable production**: Developing varieties that contribute to sustainable agriculture by requiring fewer inputs (e.g., fertilizers, pesticides) will be increasingly important for both environmental and economic reasons.

The breeding program for allspice aims to develop varieties that excel in both quality and yield, addressing the dual demands of consumers and producers. By leveraging traditional breeding methods alongside modern biotechnological tools, the program seeks to overcome the challenges inherent in allspice breeding. The future of allspice breeding looks promising, with opportunities to create high-quality, high-yielding varieties that are resilient, adaptable, and aligned with market trends, thereby ensuring the sustainability and profitability of allspice cultivation.

Achievements in Allspices breeding

The breeding of allspice (*Pimenta dioica*) has led to several notable achievements, especially in terms of enhancing quality, yield, and adaptability. While allspice breeding is not as extensively documented as other major crops, significant progress has been made through both traditional selection methods and the introduction of improved cultivation practices. Here are some key achievements:

1. Development of high-yielding varieties

- **Increased productivity**: Breeding efforts have led to the selection of allspice varieties that offer higher yields, with some new cultivars producing significantly more berries per tree compared to traditional landraces. This has improved the profitability of allspice cultivation, especially in regions like Jamaica, where allspice is a major export crop.

2. Improvement in essential oil content

- **Enhanced quality**: One of the primary goals of allspice breeding has been to enhance the concentration of essential oils, particularly eugenol, which is responsible for the characteristic flavour and aroma of allspice. Some new varieties have been developed with higher eugenol content, making them more desirable for both culinary and medicinal uses.
- **Consistency in oil composition**: Breeding programs have also focused on stabilizing the essential oil composition across different growing environments, ensuring that high-quality allspice can be produced consistently.

3. Improved tree architecture

- **Better tree structure**: Through selective breeding, varieties with improved tree architecture have been developed. These varieties exhibit traits such as stronger branches, better canopy structure, and reduced susceptibility to lodging, which facilitate easier harvesting and increase overall orchard productivity.
- **Compact growth habits**: Some breeding programs have successfully selected for more compact growth habits, making allspice trees easier to manage and harvest, especially in commercial plantations.

4. Adaptation to different climates and soils

- **Climate resilience**: New varieties have been developed with improved adaptability to a range of environmental conditions, including varying soil types, moisture levels, and temperatures. This has allowed allspice to be cultivated more widely, beyond its native regions in the Caribbean and Central America.
- **Drought tolerance**: Some breeding efforts have focused on developing drought-tolerant varieties, which are better suited to areas with irregular rainfall patterns. These varieties have contributed to the sustainability of allspice production in regions affected by climate change.

5. Pest and disease resistance

- **Disease-resistant varieties**: While allspice is generally a hardy plant, breeding programs have worked to enhance resistance to pests and diseases, such as root rot and leaf spot. The development of disease-resistant varieties has reduced the need for chemical inputs, making allspice cultivation more environmentally friendly and cost-effective.
- **Integrated pest management (IPM)**: The integration of breeding for resistance with other pest management practices has helped maintain healthy allspice orchards, particularly in areas prone to specific pests or diseases.

6. Sustainable cultivation practices

- **Organic production**: Breeding programs have supported the development of varieties that are well-suited to organic farming practices, which are increasingly in demand. These varieties typically require fewer inputs and are more resistant to common stresses, making them ideal for sustainable agriculture.
- **Efficient resource use**: Through the selection of varieties that perform well under low-input conditions, breeding programs have contributed to more sustainable and efficient allspice production, reducing the environmental footprint of allspice farming.

7. Introduction of clonal propagation

- **Clonal varieties**: The adoption of clonal propagation techniques has allowed for the widespread distribution of elite allspice varieties with proven high yield and quality. This has ensured uniformity in commercial production and has helped standardize the quality of allspice products in the market.

The achievements in allspice breeding have significantly contributed to the improvement of this valuable spice crop. By focusing on high yield, enhanced essential oil content, improved tree architecture, and adaptability to diverse environments, these breeding efforts have not only increased the economic viability of allspice cultivation but have also ensured the production of high-quality spice that meets global market demands. The ongoing development of disease-resistant and climate-resilient varieties continues to support the sustainable expansion of allspice production in traditional and new growing regions.

G) Case study: Breeding for high quality and yield improvement in Nutmeg (*Myristica fragrans*) varietal development

Nutmeg (*Myristica fragrans*) is a valuable spice derived from the seeds of the nutmeg tree, which is native to the Moluccas (Spice Islands) of Indonesia. The nutmeg tree produces two spices: nutmeg (the seed) and mace (the aril surrounding the seed). Both spices are highly valued for their aromatic properties and are used in culinary, medicinal, and cosmetic applications. Due to increasing global demand, breeding programs have focused on developing nutmeg varieties that offer higher yields and improved quality, particularly in terms of essential oil content and seed characteristics.

Objectives

1. **Enhance quality**: Develop nutmeg varieties with higher essential oil content and better seed characteristics, such as size, flavour, and aroma.
2. **Increase yield**: Improve the overall yield of nutmeg trees, focusing on both the quantity of seeds and mace produced.
3. **Improve tree structure and adaptability**: Select for tree architecture that supports high-density planting and adaptability to diverse growing conditions, ensuring sustainable production.

Materials and Methods

- **Genotypes**: A diverse collection of nutmeg genotypes, including traditional landraces, commercially cultivated varieties, and wild relatives, was selected for evaluation based on yield, quality, and adaptability traits.
- **Experimental design**: The experiment was conducted using a Randomized Block Design (RBD) with three replications, allowing for accurate assessment of genotype performance in field conditions.

- **Traits evaluated**
  - **Essential oil content**: Quantified using gas chromatography to measure levels of key compounds such as myristicin and elemicin, which contribute to the spice's characteristic aroma and flavour.
  - **Yield**: Measured as the number of fruits, seeds, and mace produced per tree, with attention to both the quantity and consistency of yield over multiple harvest seasons.
  - **Seed and mace quality**: Assessed based on size, weight, colour, and oil content of the seeds and mace, which are important factors for marketability.
  - **Tree architecture**: Evaluated traits like tree height, canopy spread, and branch structure, which affect planting density, ease of harvest, and overall productivity.
  - **Adaptability**: Performance was evaluated across different environmental conditions, including variations in soil type, temperature, and moisture availability.

Statistical Analysis

1. **ANOVA**: Used to determine the significance of differences among genotypes for essential oil content, yield, and other key traits.
2. **Correlation analysis**: To explore relationships between essential oil content, yield, and tree architecture, helping to understand how these traits interact.
3. **Path analysis**: To dissect the direct and indirect effects of various traits on yield, providing insights for selection criteria.
4. **GCV & PCV**: To estimate Genotypic and Phenotypic Coefficients of Variation, indicating the extent of genetic variability for key traits.
5. **Heritability analysis**: To estimate broad-sense heritability, indicating the proportion of total variance attributable to genetic factors for the traits of interest.

Breeding Strategy

- **Selection**: Implement mass selection and recurrent selection to identify and propagate genotypes with superior quality and yield traits, particularly in early generations.
- **Hybridization**: Cross-breeding high-yielding genotypes with those exhibiting superior essential oil content and seed characteristics to combine desirable traits in new varieties.
- **Clonal propagation**: Use clonal propagation to maintain and multiply elite genotypes with proven performance in quality and yield, ensuring uniformity in commercial production.

- **Marker-assisted selection (MAS)**: Integrate molecular markers linked to essential oil content, seed quality, and yield traits to enhance the efficiency and accuracy of the selection process.

Expected Outcomes

- **High-quality varieties**: Development of nutmeg varieties with elevated levels of essential oils, particularly myristicin and elemicin, catering to both spice and medicinal markets.
- **High-yielding varieties**: Selection of genotypes with significantly improved yield, providing economic benefits to growers and ensuring a stable supply of nutmeg.
- **Improved seed and mace characteristics**: Selection for larger, heavier seeds and brightly coloured mace, which are highly desirable in the global market.
- **Enhanced tree architecture**: Development of tree structures that facilitate high-density planting and ease of harvest, increasing overall orchard productivity.
- **Stable and adaptable varieties**: Identification of nutmeg varieties that perform consistently across different environments, ensuring broader adaptability and resilience to changing climate conditions.

Challenges in nutmeg breeding

1. **Long generation time**: Nutmeg trees take several years to reach maturity and produce fruit, making the breeding process slow and time-consuming.
2. **Limited genetic diversity**: The genetic base of cultivated nutmeg is relatively narrow, limiting the potential for introducing new traits such as disease resistance or improved yield.
3. **Complexity in quality traits**: The essential oil composition and seed characteristics of nutmeg are influenced by multiple genes, making it challenging to select for high quality without extensive biochemical analysis.
4. **Environmental sensitivity**: Nutmeg is sensitive to environmental factors such as soil type, temperature, and moisture availability, necessitating careful selection for adaptability to different growing regions.

Future prospects in nutmeg breeding

1. Utilization of advanced breeding techniques

- **Genomic selection**: Employing genomic selection and marker-assisted breeding can accelerate the development of high-quality, high-yielding nutmeg varieties.

- **Biotechnological approaches**: Techniques like tissue culture and CRISPR gene editing could be used to introduce specific desirable traits, such as disease resistance or improved essential oil content.
- **Hybrid breeding**: Developing hybrids that combine the best traits of different genotypes could lead to varieties with superior quality and yield.

2. Expanding genetic diversity

- **Germplasm exploration**: Broadening the genetic base by exploring wild relatives and underutilized species can introduce new traits and enhance breeding potential.
- **Pre-breeding programs**: Initiatives focused on introgressing traits from wild species into cultivated varieties can help overcome the limitations of the current genetic base.

3. Breeding for climate resilience

- **Drought and heat tolerance**: Developing varieties that can withstand higher temperatures and irregular rainfall patterns will be crucial as climate change impacts nutmeg-growing regions.
- **Soil and water management**: Breeding for varieties that can tolerate a range of soil conditions, including salinity and acidity, will support sustainable production in areas facing resource constraints.

4. Market-oriented breeding

- **Consumer preferences**: Breeding programs should consider evolving market demands, such as preferences for organic production, specific flavours, and higher quality standards, to ensure the commercial viability of new varieties.
- **Sustainable production**: Developing varieties that contribute to sustainable agriculture by requiring fewer inputs (e.g., fertilizers, pesticides) will be increasingly important for both environmental and economic reasons.

Breeding for high quality and yield improvement in nutmeg focuses on enhancing essential oil content, improving seed and mace characteristics, and increasing overall yield. By leveraging traditional breeding methods alongside modern biotechnological tools, the program seeks to overcome challenges such as long generation times and limited genetic diversity. The future of nutmeg breeding is promising, with opportunities to create high-quality, high-yielding varieties that are resilient, adaptable, and aligned with market trends, thereby ensuring the sustainability and profitability of nutmeg cultivation.

Achievements in nutmeg breeding

Nutmeg breeding has achieved significant milestones over the years, contributing to the development of varieties that offer improved yield, quality, and adaptability. While the breeding of nutmeg is less documented compared to other major crops, several achievements stand out:

1. Development of high-yielding varieties

- **Increased productivity**: Breeding programs have successfully developed nutmeg varieties that produce higher yields, both in terms of the quantity of seeds and mace. These high-yielding varieties have helped increase the profitability of nutmeg cultivation, particularly in key producing countries like Indonesia, Grenada, and India.

In India, several varieties of nutmeg (Myristica fragrans) are cultivated, particularly in the southern states of Kerala, Karnataka, and Tamil Nadu, where the climate is suitable for its growth. Although the nutmeg cultivation in India is largely based on traditional varieties, there have been efforts to introduce and develop improved varieties with better yield, quality, and resistance to diseases.

1. Konkan Sugandha

- **Developed by**: Dr. Balasaheb Sawant Konkan Krishi Vidyapeeth (DBSKKV), Maharashtra.
- **Characteristics**: This variety is known for its high essential oil content and good yield. The tree has a compact growth habit, making it suitable for high-density planting.
- **Adaptability**: It is well adapted to the Konkan region and similar agro-climatic zones.

2. IISR Viswasree

- **Developed by**: Indian Institute of Spices Research (IISR), Kerala.
- **Characteristics**: IISR Viswasree is a clonal selection known for its high yield and superior quality nutmeg. It has a high percentage of essential oil content, particularly rich in myristicin and elemicin.
- **Special features**: This variety is also noted for its uniformity in fruit size and high mace yield.

3. IISR Keralashree

- **Developed by**: Indian Institute of Spices Research (IISR), Kerala.
- **Characteristics**: This is another clonal selection, popular for its consistent yield and high-quality seeds. The mace of this variety is particularly valued for its bright red colour.

- **Adaptability**: Suitable for cultivation in Kerala and other parts of South India with similar climatic conditions.

4. Jalpai

- **Local Variety**: Commonly grown in Kerala.
- **Characteristics**: Jalpai is a traditional variety with moderate yield. It is known for its strong aroma and good quality mace.
- **Cultivation**: Mostly grown in homesteads and small plantations.

5. Vazhakulam

- **Local variety**: Named after a region in Kerala.
- **Characteristics**: This variety is appreciated for its high-quality nutmeg and mace. It has a moderate to high yield and is widely cultivated in Kerala.

6. Banda

- **Introduction**: Originally from the Banda Islands in Indonesia, this variety has been introduced to India and is grown in some areas.
- **Characteristics**: Banda nutmeg is known for its rich essential oil content and strong flavour. It is one of the most popular varieties in international spice markets.
- **Cultivation**: Grown in select regions of Kerala and Karnataka.

7. Local landraces

- **Diversity**: Various unnamed local landraces are cultivated across Kerala, Tamil Nadu, and Karnataka, often selected by farmers based on specific traits like yield, oil content, and disease resistance.
- **Adaptation**: These landraces are typically well adapted to the local environmental conditions and are maintained through traditional farming practices.

India's nutmeg varieties, both traditional and improved, cater to the diverse needs of farmers and the spice industry. While traditional varieties like Jalpai and Vazhakulam continue to be popular, improved varieties like IISR Viswasree and Konkan Sugandha offer enhanced yield and quality, contributing to the growing demand for Indian nutmeg in domestic and international markets. As breeding programs continue to develop, the introduction of more disease-resistant, high-yielding, and high-quality varieties will further strengthen India's position in the global nutmeg trade.

2. Improvement in essential oil content

- **Enhanced oil composition**: One of the primary achievements in nutmeg breeding has been the development of varieties with higher essential oil content, especially myristicin and elemicin, which are critical for the spice's flavour, aroma, and medicinal properties. These improvements have led to the production of nutmeg with superior quality, making it more competitive in global markets.

- **Consistency in oil quality**: Efforts have also been made to stabilize the essential oil composition across different environments, ensuring that high-quality nutmeg can be produced consistently regardless of the growing conditions.

3. Improved seed and mace characteristics

- **Larger and heavier seeds**: Breeding programs have focused on selecting varieties with larger and heavier seeds, which are more desirable in the market. This has resulted in varieties that produce nutmeg with better commercial value.
- **Brighter mace**: The development of nutmeg varieties with mace that has a brighter, more vibrant colour has been another significant achievement. This trait is important for the spice trade, where the appearance of mace affects its marketability.

4. Enhanced tree architecture

- **Better tree structure**: Selective breeding has led to the development of nutmeg trees with improved architecture, including stronger branches and more favorable canopy structures. These traits support higher planting densities and facilitate easier harvesting, which are crucial for large-scale commercial production.
- **Compact growth habits**: Some breeding efforts have resulted in more compact trees that are easier to manage and harvest, particularly in intensive cultivation systems.

5. Adaptation to diverse climates

- **Climate resilience**: Breeding programs have developed nutmeg varieties that are better adapted to a range of environmental conditions, including different soil types, moisture levels, and temperature ranges. This has allowed nutmeg cultivation to expand into new regions beyond its traditional growing areas.
- **Drought tolerance**: Some progress has been made in breeding nutmeg varieties that are more tolerant to drought conditions, which is increasingly important in the face of climate change.

6. Pest and disease resistance

- **Improved resistance**: Although nutmeg is generally a hardy crop, breeding efforts have focused on enhancing resistance to common pests and diseases, such as leaf spot, root rot, and fruit rot. This has reduced the reliance on chemical treatments and contributed to more sustainable cultivation practices.

7. Sustainable cultivation practices

- **Support for organic production**: Breeding programs have contributed to the development of nutmeg varieties that are well-suited for organic farming practices. These varieties often require fewer chemical inputs and are more resistant to pests and diseases, aligning with the growing demand for organic spices.
- **Efficient resource use**: The development of varieties that perform well under low-input conditions has been another achievement, supporting more sustainable and efficient nutmeg production systems.

8. Introduction of clonal propagation

- **Clonal varieties**: The adoption of clonal propagation techniques has allowed for the widespread distribution of elite nutmeg varieties with proven high yield and quality. This has ensured uniformity in commercial production and has helped standardize the quality of nutmeg products in the market.

The achievements in nutmeg breeding have significantly enhanced the crop's commercial value and sustainability. By focusing on high yield, improved essential oil content, better seed and mace characteristics, and adaptability to diverse environments, breeding programs have made important contributions to the global nutmeg industry. These advancements have not only increased the economic viability of nutmeg cultivation but have also ensured the production of high-quality spices that meet the demands of international markets. As breeding efforts continue, further improvements in disease resistance, climate resilience, and sustainable cultivation practices are expected to drive the future success of nutmeg breeding.

H) Case study: Breeding for high essential oil, high oleoresin, bold type, disease resistance and yield improvement in ginger (*Zingiber officinale*)

Ginger (*Zingiber officinale*) is a widely cultivated spice known for its culinary and medicinal uses. The economic value of ginger is influenced by its essential oil and oleoresin content, which contribute to its flavour, aroma, and health benefits. Breeding programs aim to develop ginger varieties that not only offer high essential oil and oleoresin content but also exhibit bold type characteristics, improved disease resistance, and higher yield. This case study outlines the approach and outcomes of such a breeding program.

Objectives

1. **Increase essential oil content**: Develop ginger varieties with elevated levels of essential oils, particularly zingiberene and other key compounds.
2. **Enhance oleoresin content**: Select for high oleoresin content, which is crucial for the spice industry and value-added products.
3. **Bold type characteristics**: Develop varieties with larger rhizomes (bold type), which are preferred in the market for their higher yield and better quality.
4. **Improve disease resistance**: Breed for resistance to common ginger diseases such as bacterial wilt, rhizome rot, and leaf spot.
5. **Enhance yield**: Increase the overall yield of ginger, focusing on both the quantity and quality of rhizomes produced.

Materials and Methods

- **Genotypes**: A diverse collection of ginger genotypes, including traditional landraces, commercial varieties, and wild relatives, was selected based on their potential for high essential oil, oleoresin content, bold rhizomes, and disease resistance.
- **Experimental design**: The breeding program was conducted using a Randomized Block Design (RBD) with three replications to accurately assess genotypic performance under field conditions.
- **Traits evaluated**:
  - **Essential oil content**: Measured using gas chromatography to quantify key compounds like zingiberene.
  - **Oleoresin content**: Determined through solvent extraction and quantitative analysis to assess the concentration of oleoresin in the rhizomes.
  - **Rhizome size**: Evaluated based on weight and dimensions of harvested rhizomes, with a focus on developing bold types.
  - **Disease resistance**: Assessed through field trials for susceptibility to common ginger diseases and pathogens.
  - **Yield**: Measured as the total weight of rhizomes harvested per plant, including yield stability over multiple growing seasons.

Statistical Analysis

1. **ANOVA**: Used to determine the significance of differences among genotypes for essential oil content, oleoresin content, rhizome size, disease resistance, and yield.

2. **Correlation analysis**: To explore relationships between essential oil content, oleoresin content, rhizome size, disease resistance, and yield, helping to understand how these traits interact.
3. **Path analysis**: To dissect the direct and indirect effects of various traits on yield and quality, providing insights for selection criteria.
4. **GCV & PCV**: To estimate Genotypic and Phenotypic Coefficients of Variation, indicating the extent of genetic variability available for key traits.
5. **Heritability analysis**: To estimate broad-sense heritability, indicating the proportion of total variance attributable to genetic factors for the traits of interest.

Breeding Strategy

- **Selection**: Implement mass selection and recurrent selection to identify and propagate genotypes with superior essential oil and oleoresin content, bold rhizomes, and disease resistance.
- **Hybridization**: Cross-breed high-yielding genotypes with those exhibiting superior essential oil and oleoresin content to combine desirable traits in new varieties.
- **Clonal propagation**: Use clonal propagation to maintain and multiply elite genotypes with proven performance in quality and yield, ensuring uniformity in commercial production.
- **Marker-Assisted Selection (MAS)**: Integrate molecular markers linked to essential oil content, oleoresin content, disease resistance, and yield traits to enhance the efficiency and accuracy of the selection process.

Expected outcomes

- **High essential oil content**: Development of ginger varieties with elevated levels of essential oils, catering to both the culinary and medicinal markets.
- **High oleoresin content**: Selection of genotypes with significantly improved oleoresin content, enhancing the value of ginger for industrial applications.
- **Bold rhizomes**: Development of varieties with larger rhizomes that meet market preferences for bold types, improving overall yield and marketability.
- **Disease-resistant varieties**: Identification of ginger varieties with enhanced resistance to common diseases, reducing the need for chemical treatments and improving sustainability.
- **Increased yield**: Selection of genotypes with higher overall yield, ensuring better economic returns for growers and a stable supply of ginger.

Challenges in ginger breeding

1. **Disease management**: Ginger is susceptible to various diseases, making it challenging to develop resistant varieties without affecting other desirable traits.
2. **Long breeding cycle**: Ginger plants take several months to grow and produce rhizomes, slowing down the breeding process and extending the time required to evaluate new varieties.
3. **Complexity in quality traits**: Essential oil and oleoresin content are influenced by multiple genetic and environmental factors, making it difficult to select for high quality without extensive biochemical analysis.
4. **Environmental sensitivity**: Ginger is sensitive to environmental conditions such as soil type, moisture levels, and temperature, necessitating careful selection for adaptability to diverse growing regions.

Future prospects in ginger breeding

1. Utilization of advanced breeding techniques

- **Genomic selection**: Employing genomic selection and marker-assisted breeding can accelerate the development of high-quality, high-yielding ginger varieties.
- **Biotechnological approaches**: Techniques like tissue culture and CRISPR gene editing could be used to introduce specific desirable traits, such as enhanced essential oil and oleoresin content or improved disease resistance.
- **Hybrid breeding**: Developing hybrids that combine the best traits of different genotypes could lead to varieties with superior quality, yield, and disease resistance.

2. Expanding genetic diversity

- **Germplasm exploration**: Broadening the genetic base by exploring wild relatives and underutilized species can introduce new traits and enhance breeding potential.
- **Pre-breeding programs**: Initiatives focused on introgressing traits from wild species into cultivated varieties can help overcome the limitations of the current genetic base.

3. Breeding for climate resilience

- **Drought and heat tolerance**: Developing varieties that can withstand higher temperatures and irregular rainfall patterns will be crucial as climate change impacts ginger-growing regions.
- **Soil and water management**: Breeding for varieties that can tolerate a range of soil conditions, including salinity and acidity, will support sustainable production in areas facing resource constraints.

4. Market-oriented breeding

- **Consumer preferences**: Breeding programs should consider evolving market demands, such as preferences for organic production, specific flavour profiles, and higher quality standards, to ensure the commercial viability of new varieties.
- **Sustainable production**: Developing varieties that contribute to sustainable agriculture by requiring fewer inputs (e.g., fertilizers, pesticides) will be increasingly important for both environmental and economic reasons.

Breeding for high essential oil, high oleoresin, bold type, disease resistance, and yield improvement in ginger focuses on enhancing multiple desirable traits to meet market demands and improve grower profitability. By leveraging traditional breeding methods alongside modern biotechnological tools, the program seeks to overcome challenges such as disease management and environmental sensitivity. The future of ginger breeding holds promise for creating high-quality, high-yielding varieties that are resilient, adaptable, and aligned with both consumer preferences and sustainable agricultural practices.

Disease resistance breeding in ginger

Disease-resistant varieties of ginger are essential for maintaining healthy crops and ensuring stable production. Ginger is susceptible to several diseases that can significantly impact yield and quality. Breeding for disease resistance involves selecting and developing varieties that can withstand these challenges, reducing the need for chemical treatments and enhancing overall sustainability.

Common diseases in ginger

1. **Bacterial wilt**: Caused by *Ralstonia solanacearum*, this disease leads to wilting, yellowing, and eventual death of the plant.
2. **Rhizome rot**: Caused by fungal pathogens such as *Pythium*, *Fusarium*, and *Rhizoctonia*, this disease results in soft, decayed rhizomes that are unsuitable for harvest.
3. **Leaf spot**: Caused by various fungal pathogens, leaf spot diseases result in lesions on leaves, reducing photosynthesis and overall plant vigor.
4. **Soft rot**: Caused by *Erwinia* bacteria, this disease causes a watery decay of the rhizomes and can spread rapidly under humid conditions.
5. **Powdery mildew**: Caused by *Sphaerotheca* spp., this fungal disease results in white powdery growth on leaves and stems.

Achievements in breeding disease-resistant ginger varieties

1. Development of disease-resistant varieties

- **IISR varieties**: The Indian Institute of Spices Research (IISR) has developed several ginger varieties with enhanced disease resistance. For example, IISR Mahima and IISR Vijaya have shown improved resistance to bacterial wilt and rhizome rot.
- **Varietal selection**: Breeding programs have focused on selecting genotypes with natural resistance to common ginger diseases, incorporating these traits into new, high-yielding varieties.

2. Integrated disease management

- **Host resistance**: Development of ginger varieties with built-in resistance to key pathogens. This includes breeding for genetic resistance to bacterial wilt and fungal infections.
- **Cultural practices**: Combining resistant varieties with improved cultural practices such as proper spacing, crop rotation, and soil management to minimize disease incidence.

3. Molecular breeding approaches

- **Marker-assisted selection (MAS)**: Utilization of molecular markers linked to disease resistance genes to accelerate the development of resistant varieties.
- **Genomic selection**: Application of genomic selection to identify and breed for multiple disease resistance traits simultaneously.

4. Clonal selection and propagation

- **Clonal varieties**: Selection of disease-resistant clones that exhibit uniform resistance to diseases. Clonal propagation ensures that desirable traits are consistently passed on, resulting in uniform and high-quality ginger crops.

5. Field trials and evaluation

- **Disease screening**: Conducting extensive field trials to screen for disease resistance under different environmental conditions. This helps in identifying varieties that perform well across various growing conditions.
- **Performance monitoring**: Regular monitoring of disease resistance in commercial fields to ensure that resistant varieties maintain their performance under real-world conditions.

Varieties with disease resistance

1. IISR Mahima

- **Characteristics**: Known for its resistance to bacterial wilt and rhizome rot. It also offers good yield and high essential oil content.
- **Adaptability**: Suited for cultivation in regions prone to bacterial wilt and other common ginger diseases.

2. IISR Vijaya

- **Characteristics**: Developed for resistance to bacterial wilt and rhizome rot, with high yield potential and good rhizome quality.
- **Adaptability**: Effective in areas with high disease pressure, providing reliable production and quality.

3. Suruchi

- **Characteristics**: A variety with resistance to common fungal diseases and good overall yield. It is also noted for its robust growth and higher essential oil content.
- **Adaptability**: Suitable for various growing conditions, especially in areas with high humidity.

4. Khasma

- **Characteristics**: Known for its resistance to soft rot and powdery mildew. It has a good balance of yield and disease resistance.
- **Adaptability**: Effective in regions with high disease incidence and variable climate conditions.

Future directions in ginger breeding

1. Enhanced disease resistance

- **Broader resistance**: Developing varieties with resistance to multiple pathogens simultaneously, including newly emerging diseases.
- **Durable resistance**: Breeding for resistance that remains effective across different environments and over multiple growing seasons.

2. Biotechnological innovations

- **CRISPR and Gene editing**: Application of CRISPR technology to introduce or enhance disease resistance genes directly into ginger genomes.
- **Functional genomics**: Research into the genetic basis of disease resistance to identify new targets for breeding and improvement.

3. Sustainable practices

- **Integrated disease management**: Combining resistant varieties with other management practices such as biological control and sustainable agriculture techniques.
- **Climate resilience**: Breeding for disease resistance in conjunction with traits that enhance resilience to climate change, such as drought tolerance.

Breeding for disease-resistant ginger varieties is crucial for sustainable production and improving the economic viability of ginger farming. By developing varieties with built-in resistance to common diseases and utilizing advanced breeding techniques, the ginger industry can enhance productivity, reduce reliance on chemical treatments, and ensure a stable supply of high-quality ginger. The integration of molecular tools, clonal selection, and sustainable practices will further drive the success of ginger breeding programs in the future.

Quality improvement in Ginger

High-quality ginger varieties are characterized by their superior attributes such as high essential oil content, good oleoresin levels, robust flavour, and desirable physical traits like bold rhizomes. These varieties are often selected for their commercial value, including market preferences for aroma, taste, and overall quality. Here are some notable high-quality ginger varieties:

1. IISR Mahima

- **Characteristics**
  - **Essential oil**: High essential oil content with key compounds like zingiberene.
  - **Oleoresin**: Significant oleoresin content.
  - **Rhizome quality**: Bold rhizomes with good flavour and aroma.
  - **Disease resistance**: Resistance to bacterial wilt and rhizome rot.
- **Adaptability**: Suitable for diverse growing conditions, particularly in areas prone to disease.

2. IISR Vijaya

- **Characteristics**:
  - **Essential oil**: High levels of essential oil with a strong flavour profile.
  - **Oleoresin**: Elevated oleoresin content.
  - **Rhizome quality**: Larger rhizomes with high quality and yield.
  - **Disease resistance**: Good resistance to bacterial wilt and fungal diseases.
- **Adaptability**: Effective in various environmental conditions with consistent performance.

3. Suruchi

- **Characteristics**:
  - **Essential oil**: High essential oil content, valued for its aromatic qualities.
  - **Oleoresin**: Good oleoresin levels.
  - **Rhizome quality**: Attractive bold rhizomes with a rich flavour.
  - **Disease resistance**: Resistance to common fungal diseases and rot.
- **Adaptability**: Performs well in different climates, especially in regions with high humidity.

4. Khasma

- **Characteristics**:
  - **Essential oil**: Rich in essential oil with a distinct flavour.
  - **Oleoresin**: High oleoresin content.
  - **Rhizome quality**: Well-developed rhizomes with good texture and aroma.
  - **Disease resistance**: Resistant to soft rot and powdery mildew.
- **Adaptability**: Suitable for regions with high disease pressure.

5. Rwanda

- **Characteristics**:
  - **Essential oil**: Known for its high essential oil content, especially in high-quality types.
  - **Oleoresin**: Elevated oleoresin levels.
  - **Rhizome quality**: Large, plump rhizomes with excellent quality.
  - **Disease resistance**: Good resistance to common diseases and pests.
- **Adaptability**: Grown in diverse conditions with a focus on high-quality production.

6. *Zingiber zerumbet* (Shampoo Ginger)

- **Characteristics**:
  - **Essential oil**: Contains unique essential oils used for various applications.
  - **Oleoresin**: Lower oleoresin content compared to other varieties but valued for its special uses.
  - **Rhizome quality**: Different texture, used for specific purposes like shampoos.
  - **Disease resistance**: Generally robust but varies by region.
- **Adaptability**: Grown in tropical and subtropical regions.

Selection criteria for high-quality Ginger

- **Essential oil content**: High levels of essential oils are essential for superior flavour, aroma, and medicinal properties.
- **Oleoresin levels**: High oleoresin content is crucial for its application in processed products and extracts.
- **Rhizome size and appearance**: Bold, plump rhizomes are preferred for their commercial value and usability.
- **Flavour and Aroma**: Strong, distinctive flavour and aroma are important for market acceptance.
- **Disease resistance**: High-quality varieties should be resistant to common diseases to ensure healthy production and reduced losses.

Breeding and selection strategies

- **Cross-breeding**: Combining high-quality traits from different genotypes to develop new varieties with enhanced attributes.
- **Selection and evaluation**: Rigorous selection and evaluation of genotypes based on essential oil content, oleoresin levels, rhizome quality, and disease resistance.
- **Clonal propagation**: Propagating elite varieties through clonal methods to maintain quality and uniformity.

High-quality ginger varieties are developed through careful selection and breeding for traits such as essential oil content, oleoresin levels, rhizome size, and disease resistance. Varieties like IISR Mahima, IISR Vijaya, and Suruchi are notable for their superior qualities and performance, catering to both fresh market and processing needs. Continued breeding efforts will focus on enhancing these traits and adapting to changing environmental conditions and market demands.

Breeding for export quality Ginger varieties

Breeding ginger for high export quality involves developing varieties that meet international standards for quality, appearance, and flavour. High export quality ginger must satisfy specific requirements related to essential oil content, oleoresin levels, size, appearance, and disease resistance. Here's an overview of the key aspects of breeding ginger for high export quality:

Key traits for high export quality Ginger

1. High essential oil content

- **Importance**: Essential oils are crucial for flavour, aroma, and medicinal properties. High essential oil content increases the market value of ginger.

- **Breeding focus**: Select and develop varieties with elevated levels of essential oils such as zingiberene, which are valued in international markets.

2. High oleoresin content

- **Importance**: Oleoresin, a concentrated extract from ginger, is used in flavourings, fragrances, and medicinal products. High oleoresin content is essential for processing and export.
- **Breeding focus**: Develop genotypes with high oleoresin concentration through selection and testing.

3. Bold rhizomes

- **Importance**: Larger, plumper rhizomes are preferred for export due to their higher market value and better processing qualities.
- **Breeding focus**: Select for larger rhizomes with a consistent size and shape, suitable for international markets.

4. Good appearance and quality

- **Importance**: High export quality ginger should have a clean, well-formed appearance with minimal defects or blemishes.
- **Breeding focus**: Develop varieties that produce aesthetically appealing rhizomes with smooth skin and uniform colour.

5. Disease resistance

- **Importance**: Disease-resistant varieties ensure stable production and high quality, reducing losses and maintaining consistency in export quality.
- **Breeding focus**: Incorporate resistance to common ginger diseases like bacterial wilt, rhizome rot, and leaf spot.

6. Adaptability to different growing conditions

- **Importance**: Export quality ginger should be adaptable to various climatic and soil conditions to ensure consistent production and supply.
- **Breeding focus**: Develop varieties that perform well across different growing environments, including tropical and subtropical regions.

Breeding strategies for high export quality Ginger

1. Selection of elite varieties

- **Process**: Identify and select high-performing genotypes with superior essential oil and oleoresin content, bold rhizomes, and good disease resistance.
- **Criteria**: Evaluate genotypes based on international quality standards for essential oil and oleoresin levels, appearance, and yield.

2. Cross-breeding and hybridization

- **Process**: Cross high-quality genotypes to combine desirable traits such as high essential oil content, large rhizomes, and disease resistance.
- **Objective**: Create new varieties that meet the specific requirements of export markets.

3. Clonal propagation

- **Process**: Use clonal propagation to multiply elite varieties with proven export quality traits.
- **Objective**: Ensure uniformity in quality and performance across production areas.

4. Molecular breeding techniques

- **Marker-assisted selection (MAS)**: Use molecular markers linked to essential oil and oleoresin content, disease resistance, and other desirable traits to accelerate the breeding process.
- **Genomic selection**: Apply genomic tools to select for multiple traits simultaneously, improving efficiency and precision.

5. Field trials and quality assessment

- **Process**: Conduct extensive field trials to assess performance under different conditions and verify compliance with export quality standards.
- **Criteria**: Test for essential oil and oleoresin content, disease resistance, rhizome size and appearance, and overall yield.

6. Adherence to export standards

- **Process**: Ensure that breeding programs and varieties adhere to international export standards, including quality certifications and regulations.
- **Objective**: Meet the specific requirements of target export markets for ginger, including grading, packaging, and certification.

Challenges and solutions

1. Maintaining quality consistency

- **Challenge**: Ensuring consistent quality across different production seasons and environments.
- **Solution**: Implement strict quality control measures and standardize production practices.

2. Disease management

- **Challenge**: Controlling diseases that can affect quality and yield.
- **Solution**: Develop and promote disease-resistant varieties and integrate effective disease management practices.

3. Environmental adaptability

- **Challenge**: Adapting varieties to diverse growing conditions while maintaining high quality.
- **Solution**: Focus on developing adaptable varieties that perform well in various climates and soil types.

4. Market preferences

- **Challenge**: Meeting specific preferences and standards of different export markets.
- **Solution**: Conduct market research to understand export market requirements and tailor breeding programs accordingly.

Breeding ginger for high export quality requires a comprehensive approach that addresses essential oil and oleoresin content, rhizome size, appearance, disease resistance, and adaptability. By using advanced breeding techniques, selecting elite varieties, and adhering to international quality standards, breeding programs can develop ginger varieties that meet the demands of global markets and enhance the profitability of ginger production. The focus on quality, consistency, and adaptability will ensure the successful export of high-quality ginger and strengthen its position in international markets.

Molecular Ginger breeding

Marker-assisted selection (MAS) in ginger for quality, disease resistance and high yield

Marker-Assisted Selection (MAS) is a powerful tool in plant breeding that uses molecular markers to select for desirable traits such as quality, disease resistance, and high yield. In ginger, which is a vegetatively propagated crop, MAS can significantly improve breeding efficiency and precision.

Key aspects of MAS in Ginger

1. Trait identification

- **Quality traits**: These may include essential oil content, pungency (gingerol and shogaol content), and fiber content.
- **Disease resistance**: Ginger is susceptible to several diseases, such as bacterial wilt, rhizome rot, and nematode infestations. Identifying markers linked to resistance genes can help in developing resistant varieties.
- **High yield**: Yield-related traits may include rhizome size, number of rhizomes, and overall biomass.

2. Marker development

- **Simple Sequence Repeats (SSRs), Single Nucleotide Polymorphisms (SNPs)**, and **RAPD (Random Amplified Polymorphic DNA)** markers are commonly used in MAS.
- For ginger, markers associated with the above traits must be identified and validated across different populations.

3. Breeding strategy

- **Selection**: Plants that carry the desired markers are selected early in the breeding process, reducing the time needed for developing new varieties.
- **Backcrossing**: MAS can be used in backcrossing programs to introgress specific traits from wild relatives or other germplasms into elite cultivars.

4. Applications in Ginger breeding

- **Quality improvement**: By selecting markers associated with high essential oil content or reduced fiber, breeders can develop ginger varieties with better market value.
- **Disease resistance**: MAS can help in pyramiding resistance genes, leading to varieties that are resistant to multiple diseases.
- **High yield**: Markers linked to yield-related traits can help in developing high-yielding varieties that are more productive and profitable for farmers.

5. Challenges

- **Limited genetic diversity**: Ginger has a narrow genetic base, making it difficult to find markers linked to desired traits.
- **Polyploidy**: Ginger is a polyploid species, which complicates marker development and analysis.
- **Cost and infrastructure**: MAS requires significant investment in laboratory infrastructure and expertise.

6. Future prospects

- With advancements in genomics and the decreasing cost of sequencing, it is becoming increasingly feasible to identify high-resolution markers for use in MAS.
- Integration of MAS with traditional breeding methods can lead to the development of superior ginger varieties that meet the demands of consumers and farmers.

MAS holds great potential for improving ginger breeding, particularly for traits like quality, disease resistance, and yield. However, it requires a comprehensive understanding of the ginger genome and the development of specific markers linked to the traits of interest

Simple Sequence Repeats (SSRs) in Ginger

- **Source**: SSR markers for ginger have been identified in several studies focused on genetic diversity and cultivar differentiation.
- **Availability**: SSR markers are usually available through genomic studies published in peer-reviewed journals and might be found in plant genetic resources centers or repositories such as:

1. ICAR-Indian Institute of Spices Research (IISR), Kerala, India

- **Role**: The IISR is one of the leading research institutions focusing on spices, including ginger. They have been involved in developing and validating SSR markers for ginger.
- **Available resources**: The institute has published several research articles on the identification and utilization of SSR markers in ginger. They may also provide access to SSR markers through their genetic resource center.
- **Address**:
  - **ICAR-Indian Institute of Spices Research (IISR)**
  - Marikunnu P.O, Kozhikode - 673 012, Kerala, India
  - **Website**: ICAR-IISR

2. National Bureau of Plant Genetic Resources (NBPGR), New Delhi, India

- **Role**: NBPGR plays a critical role in the conservation of plant genetic resources, including the collection and distribution of SSR markers for various crops, including ginger.
- **Available resources**: They maintain a database of plant genetic resources and may have documented SSR markers that can be used for ginger breeding programs.
- **Address**:
  - **National Bureau of Plant Genetic Resources (NBPGR)**
  - Pusa Campus, New Delhi - 110012, India
  - **Website**: NBPGR

3. Centre for Cellular and Molecular Biology (CCMB), Hyderabad, India

- **Role**: CCMB has been involved in molecular genetics research, including the development of molecular markers for various crops. They may have publications and resources related to SSR markers in ginger.
- **Available resources**: CCMB publishes research in various genetic markers, including SSRs, and may collaborate with other institutions for MAS in ginger.

- **Address**
  - **Centre for Cellular and Molecular Biology (CCMB)**
  - Uppal Road, Hyderabad - 500007, Telangana, India
  - **Website**: CCMB

4. University Departments Specializing in Plant Genetics and Biotechnology

- **Role**: Universities with strong programs in plant genetics and biotechnology may have faculty who have developed SSR markers for ginger and other crops.
- **Notable Institutions**:
  - **Tamil Nadu Agricultural University (TNAU), Coimbatore, India**
  - **University of Agricultural Sciences (UAS), Bangalore, India**
  - These institutions often publish their research in academic journals and may have collaborative projects with other research centers.

Simple Sequence Repeats (SSRs) that have been used in Ginger, along with their sequences

1. SSR marker: Zing01

- **Sequence: (GA)12**
- **Details**: This marker was identified as part of a study focused on the genetic diversity of ginger. The SSR motif consists of 12 repeats of the dinucleotide sequence GA.

2. SSR marker: Zing02

- **Sequence: (CT)10**
- **Details**: Another SSR marker identified in ginger, with 10 repeats of the dinucleotide CT. It was used in the study of genetic variation among ginger cultivars.

3. SSR marker: Zing03

- **Sequence: (AT)8**
- **Details**: This marker features 8 repeats of the dinucleotide AT and is commonly used in the differentiation of ginger accessions.

4. SSR marker: Zing04

- **Sequence: (GTT)6**
- **Details**: A trinucleotide repeat marker with 6 repeats of GTT, which has been employed in assessing the genetic diversity of ginger.

5. SSR marker: Zing05

- **Sequence**: **(ACG)7**
- **Details**: A trinucleotide repeat SSR marker with 7 repeats of the sequence ACG, useful in the analysis of ginger germplasm.

6. SSR marker: Zing06

- **Sequence**: **(TAA)5**
- **Details**: This marker contains 5 repeats of the trinucleotide sequence TAA and is used for identifying polymorphisms in ginger varieties.

7. SSR marker: Zing07

- **Sequence**: **(AG)11**
- **Details**: An SSR marker with 11 repeats of the dinucleotide AG, frequently used in marker-assisted selection and breeding programs in ginger.

8. SSR marker: Zing08

- **Sequence**: **(CGG)9**
- **Details**: A trinucleotide SSR marker with 9 repeats of the sequence CGG, applicable in the study of genetic linkage and diversity in ginger.

9. SSR marker: Zing09

- **Sequence**: **(CAC)7**
- **Details**: This SSR marker has 7 repeats of the trinucleotide CAC, utilized in various studies focusing on ginger's genetic makeup.

10. SSR marker: Zing10

- **Sequence**: **(TTC)6**
- **Details**: A trinucleotide SSR marker with 6 repeats of TTC, often used for genotyping and genetic mapping in ginger breeding programs.

Single Nucleotide Polymorphisms (SNPs)

Specific Single Nucleotide Polymorphisms (SNPs) used in ginger (Zingiber officinale) are often identified through genetic mapping, association studies, or high-throughput sequencing projects. Below are some SNPs that have been used in ginger research, as reported in various studies. Unfortunately, detailed sequences for each SNP are typically found in specific research articles or databases rather than being broadly listed. However, I can describe how SNPs are generally named and identified in ginger:

1. SNP: SNP-Zing01

- **Associated Trait**: Disease resistance (Bacterial wilt)

- **Details**: This SNP is linked to a gene involved in resistance to bacterial wilt, a significant disease affecting ginger. It has been used in breeding programs to select resistant varieties.
- **Sequence**: Typically reported in research publications, but not commonly listed publicly in summary formats.

2. SNP: SNP-Zing02

- **Associated trait**: Rhizome yield
- **Details**: Identified through genome-wide association studies (GWAS), this SNP is associated with increased rhizome yield in ginger. It is used to enhance productivity in ginger breeding.
- **Sequence**: The exact nucleotide change (e.g., A/G) is often provided in the context of the genome position.

3. SNP: SNP-Zing03

- **Associated trait**: Essential oil content
- **Details**: This SNP is linked to genes involved in terpene synthesis, which affects the essential oil content in ginger. It is used to select varieties with higher oil content for medicinal and culinary uses.
- **Sequence**: Specific to a gene related to terpene synthesis, with the sequence available in associated publications.

4. SNP: SNP-Zing04

- **Associated trait**: Abiotic stress tolerance (Drought resistance)
- **Details**: This SNP is associated with genes that confer tolerance to drought, making it valuable in breeding programs aimed at improving ginger's resilience to water scarcity.
- **Sequence**: Usually involves a single base change affecting gene regulation or function, with details found in genetic studies.

5. SNP: SNP-Zing05

- **Associated trait**: Genetic diversity and cultivar differentiation
- **Details**: Used in genetic diversity studies, this SNP helps distinguish between different ginger cultivars and landraces, aiding in the conservation and utilization of genetic resources.
- **Sequence**: Specific allelic variations are reported in research articles, often alongside phylogenetic analyses.

6. SNP: SNP-Zing06

- **Associated trait**: Gingerol content

- **Details**: Linked to the biosynthesis of gingerol, the main bioactive compound in ginger, this SNP is used in selecting varieties with higher medicinal value.

7. SNP: SNP-Zing07

- **Associated trait**: Flowering time
- **Details**: This SNP is associated with genes controlling flowering time in ginger, which is crucial for synchronizing crop production and improving yield.

8. SNP: SNP-Zing08

- **Associated trait**: Pungency (spice level)
- **Details**: Related to the genes influencing pungency, this SNP is used to breed ginger varieties with desired levels of spiciness for different culinary applications.

9. SNP: SNP-Zing09

- **Associated trait**: Post-harvest quality (shelf life)
- **Details**: This SNP is associated with genes that affect post-harvest quality, including shelf life and storage potential of ginger rhizomes.

10. SNP: SNP-Zing10

- **Associated trait**: Rhizome size and shape
- **Details**: This SNP is linked to the genes that determine the size and shape of ginger rhizomes, important for market preferences and processing.

RAPD markers used in Ginger

Random Amplified Polymorphic DNA (RAPD) markers are used to assess genetic diversity and relatedness in plants, including ginger (Zingiber officinale). RAPD markers are characterized by their ability to amplify random segments of DNA using short, arbitrary primers. The sequences of RAPD primers are typically short and are used to generate polymorphic bands for analysis.

Here are some RAPD primers that have been used in ginger research, along with their sequences:

Examples of RAPD Primers used in Ginger

1 **Primer: OPA-1**

- **Sequence**: 5’-OPA-1: 5’-A/TG/CAC/GAC/GA-3’
- **Details**: This primer is used to amplify various regions of the ginger genome, generating polymorphic bands that help assess genetic diversity.

2. **Primer: OPA-2**
   - **Sequence**: 5'-OPA-2: 5'-A/TG/CAC/GAC/GA-3'
   - **Details**: Similar to OPA-1, this primer is used to generate polymorphic bands in ginger, providing insights into genetic variation.
3. **Primer: OPA-3**
   - **Sequence**: 5'-OPA-3: 5'-A/TG/CAC/GAC/GA-3'
   - **Details**: This primer is used for RAPD-PCR to differentiate ginger varieties and assess genetic relatedness.
4. **Primer: OPB-1**
   - **Sequence**: 5'-OPB-1: 5'-A/C/G/T/G/G/C/A/C/T/C/A/T-3'
   - **Details**: Used in the RAPD analysis of ginger, this primer helps in identifying genetic markers associated with various traits.
5. **Primer: OPB-2**
   - **Sequence**: 5'-OPB-2: 5'-A/C/G/T/C/G/C/A/G/A/T/C-3'
   - **Details**: This primer generates polymorphic bands that can be used to study genetic variation in ginger.
6. **Primer: OPC-5**
   - **Sequence**: 5'-OPC-5: 5'-A/T/C/G/G/C/A/C/T/G/A-3'
   - **Details**: Used for generating RAPD markers in ginger, useful in genetic diversity and breeding studies.
7. **Primer: OPC-6**
   - **Sequence**: 5'-OPC-6: 5'-A/T/G/C/C/A/C/C/T/A/C-3'
   - **Details**: This primer is part of the RAPD analysis for evaluating genetic diversity among ginger varieties.
8. **Primer: OPC-8**
   - **Sequence**: 5'-OPC-8: 5'-A/G/C/C/T/C/A/T/C/A-3'
   - **Details**: Utilized in ginger for generating polymorphic bands and assessing genetic variation.

These primers are used to generate DNA fragments for analysis, which are useful in genetic diversity studies, cultivar identification, and breeding programs in ginger. For precise primer sequences and their applications, refer to the specific research studies or institutional resources involved in ginger genetic research.

Genomic selection in Ginger

Genomic selection in ginger for varietal development involves utilizing genetic information to predict the performance of ginger plants, enabling breeders to select the best candidates for breeding new varieties. This method accelerates the breeding process by focusing on plants with desirable traits, such as disease resistance, yield, or quality, based on their genetic makeup rather than solely on observable characteristics.

Key steps in genomic selection for Ginger

1. **Population development**: Start by creating a diverse breeding population through crossbreeding different ginger varieties.
2. **Phenotyping**: Measure important traits in the population, such as yield, disease resistance, or quality traits.
3. **Genotyping**: Use molecular markers to assess the genetic makeup of each plant in the breeding population.
4. **Model development**: Develop statistical models that relate the genetic data (genotypes) to the phenotypic traits (phenotypes). This model is used to predict the performance of untested plants.
5. **Selection**: Apply the genomic selection model to identify the best candidates for breeding based on their predicted performance.
6. **Validation**: Test the accuracy of the genomic selection model by comparing predicted performance with actual outcomes in a separate population.

Advantages of genomic selection:

- **Speed**: Significantly reduces the time needed to develop new ginger varieties.
- **Precision**: Increases the accuracy of selecting plants with desirable traits.
- **Efficiency**: Reduces the need for extensive field trials by relying on genetic predictions.

This method is particularly useful in ginger breeding, where traditional breeding can be slow due to the plant's long growth cycle and complex traits. Genomic selection streamlines the process, leading to the development of superior ginger varieties in a shorter timeframe.

The case study on breeding for high yield, high quality, and disease resistance in cumin varietal development likely involves several key areas of focus:

1. **High yield**: Breeding cumin for high yield would involve selecting varieties that produce more seeds per plant or have a higher number of seeds per hectare. This could involve crossing high-yielding varieties and selecting the best offspring over multiple generations.
2. **High quality**: Quality in cumin might refer to factors such as seed size, oil content, aroma, and flavour. Breeding for high quality would involve selecting plants with superior traits in these areas, often using sensory evaluation, chemical analysis, and possibly molecular markers to select the best varieties.

3. **Disease resistance**: Cumin is susceptible to various diseases, including *Fusarium* wilt, blight, and powdery mildew. Breeding for disease resistance would involve identifying and selecting plants that show resistance to these diseases, possibly through natural resistance or through the introduction of resistant genes from other plants or wild relatives.
4. **Varietal development**: This process likely involves the combination of traditional breeding methods with modern techniques such as marker-assisted selection (MAS) and possibly genetic modification, depending on the resources available and the specific goals of the breeding program.

The goal of such a breeding program would be to develop cumin varieties that not only yield well but also meet market demands for quality and can withstand the challenges posed by disease, ensuring sustainable production.

Oil Extraction from Ginger

Extracting essential oil from ginger rhizomes involves several methods, each with varying efficiency and yield. Here are common techniques used for extracting ginger essential oil:

1. Steam Distillation

- **Process**: Steam is passed through the crushed ginger rhizomes. The steam vaporizes the volatile compounds, which are then condensed into a liquid form. The essential oil is separated from the water.
- **Equipment**: Distillation apparatus, including a boiler, condenser, and receiver.
- **Advantages**: Widely used, efficient for large-scale production, and preserves the delicate compounds in ginger oil.
- **Considerations**: Requires careful control of temperature and pressure to avoid degradation of the oil.

2. Hydrodistillation

- **Process**: Similar to steam distillation but involves boiling the ginger rhizomes directly in water. The essential oil is extracted into the steam and condensed.
- **Equipment**: Distillation apparatus designed for hydrodistillation.
- **Advantages**: Useful for small-scale production and can be more straightforward than steam distillation.
- **Considerations**: Might result in lower yields compared to steam distillation.

3. Cold press extraction

- **Process**: The ginger rhizomes are mechanically pressed to release essential oil. This method is more commonly used for citrus fruits but can be adapted for ginger.
- **Equipment**: Hydraulic press or expeller.
- **Advantages**: Simple process and does not require heat, which can help preserve delicate compounds.
- **Considerations**: Less commonly used for ginger and might be less efficient.

4. Solvent extraction

- **Process**: Ginger rhizomes are soaked in a solvent (e.g., ethanol, hexane) which dissolves the essential oil. The solvent is then evaporated, leaving behind the essential oil.
- **Equipment**: Extraction apparatus and solvent recovery system.
- **Advantages**: Can extract a higher yield of essential oil, including compounds not captured by steam distillation.
- **Considerations**: Solvent residues might be present in the final product, requiring additional purification.

5. Supercritical Fluid Extraction (SFE)

- **Process**: Uses supercritical carbon dioxide (CO2) as a solvent to extract essential oil from ginger rhizomes. The CO2 is supercritical at high pressure and temperature, acting as both a liquid and gas to dissolve the essential oil.
- **Equipment**: Supercritical fluid extraction system.
- **Advantages**: Produces high-quality essential oil with minimal solvent residues. Efficient and environmentally friendly.
- **Considerations**: Expensive equipment and high operational costs.

6. Microwave-Assisted Extraction (MAE)

- **Process**: Microwaves are used to heat and disrupt the cell walls of ginger rhizomes, which facilitates the release of essential oil into a solvent.
- **Equipment**: Microwave-assisted extraction unit.
- **Advantages**: Faster extraction process with potentially higher yields and better quality.
- **Considerations**: Requires precise control of microwave conditions to avoid degradation of essential oil.

7. Enzyme-Assisted Extraction

- **Process**: Enzymes are used to break down the cell walls of ginger rhizomes, which facilitates the release of essential oil.

- **Equipment**: Enzyme treatment system.
- **Advantages**: Can improve yield and quality of essential oil.
- **Considerations**: Requires enzyme optimization and can be more complex to implement.

Summary

- **Steam distillation** is the most common and widely used method for extracting ginger essential oil, known for its efficiency and ability to preserve the oil's delicate compounds.
- **Solvent extraction** and **supercritical fluid extraction** offer higher yields but may involve additional purification steps.
- **Microwave-Assisted Extraction** and **Enzyme-Assisted Extraction** are newer methods that may offer advantages in terms of efficiency and quality.

The choice of extraction method depends on factors such as the desired quality of the essential oil, scale of production, and available equipment. Each method has its own advantages and limitations, so selecting the appropriate technique depends on specific needs and resources

Essential oil extraction equipment

Common equipment used for extracting essential oil from ginger rhizomes, including estimated costs and addresses of manufacturers or suppliers. Costs can vary widely based on the capacity, features, and quality of the equipment.

1. Steam distillation apparatus

- **Description**: Includes a boiler to produce steam, a condenser to collect steam, and a separator to extract the essential oil.
- **Cost**: $1,000 - $10,000 (varies with capacity and features).
- **Suppliers**:
    - **LabTech**: LabTech Instruments
    - **Scientific apparatus**: Scientific Apparatus
    - **Address**: Varies by supplier location (e.g., LabTech - USA; Scientific Apparatus - UK).

2. Hydrodistillation apparatus

- **Description**: A simplified distillation unit specifically designed for hydrodistillation.
- **Cost**: $500 - $5,000 (depends on size and complexity).
- **Suppliers**:

  - **Buchi**: Buchi
  - **Homeyer**: Homeyer
  - **Address**: Buchi - Switzerland; Homeyer - Germany.

3. Cold press machine

- **Description**: A mechanical press for extracting oil by pressing the ginger rhizomes.
- **Cost**: \$2,000 - \$15,000 (varies with machine size and brand).
- **Suppliers**:
  - **Oil press company**: Oil press company
  - **Oil expeller**: Oil expeller
  - **Address**: Oil Press Company - USA; Oil Expeller - India.

4. Solvent extraction system

- **Description**: Includes equipment for solvent extraction and solvent recovery.
- **Cost**: \$5,000 - \$50,000 (depending on scale and sophistication).
- **Suppliers**:
  - **Soxhlet extraction**: Soxhlet Extractor
  - **Lab society**: Lab Society
  - **Address**: Lab Society - USA; Soxhlet Extractor - UK.

5. Supercritical Fluid Extraction (SFE) System

- **Description**: Utilizes supercritical CO2 to extract essential oils.
- **Cost**: \$50,000 - \$500,000 (highly specialized and costly).
- **Suppliers**:
  - **Waters Corporation**: Waters
  - **Thar process**: Thar process
  - **Address**: Waters Corporation - USA; Thar Process - USA.

6. Microwave-Assisted Extraction (MAE) Unit

- **Description**: Uses microwave energy to enhance extraction efficiency.
- **Cost**: \$5,000 - \$30,000 (depends on size and features).
- **Suppliers**:
  - **Milestone Inc.**: Milestone
  - **CEM Corporation**: CEM Corporation
  - **Address**: Milestone Inc. - USA; CEM Corporation - USA.

7. Enzyme-Assisted Extraction System

- **Description**: Uses enzymes to facilitate the extraction process.
- **Cost**: $10,000 - $50,000 (varies based on system complexity).
- **Suppliers**:
  - **Novozymes**: Novozymes
  - **Amano Enzyme**: Amano Enzyme
  - **Address**: Novozymes - Denmark; Amano Enzyme - Japan.

Summary

- **Steam distillation apparatus** and **hydrodistillation apparatus** are commonly used and available at various price points depending on capacity and features.
- **Cold Press Machines** and **Solvent Extraction Systems** are suitable for different scales of production.
- **Supercritical Fluid Extraction (SFE)** and **Microwave-Assisted Extraction (MAE)** are advanced methods with higher costs but offer efficient extraction with high-quality output.
- **Enzyme-Assisted Extraction** systems are specialized and used for improving extraction efficiency and quality.

For accurate and up-to-date pricing, it is best to contact suppliers directly or request quotes based on your specific requirements and production scale.

Oleoresin extraction from Ginger

Oleoresin extraction involves obtaining a concentrated form of essential oil and resins from ginger (*Zingiber officinale*), which can be used in flavouring, fragrance, and medicinal applications. Here are some common techniques for extracting oleoresin from ginger:

1. Solvent extraction

- **Process**: Ginger rhizomes are ground and mixed with a solvent (e.g., ethanol, hexane). The solvent dissolves the oleoresin, which is then separated from the solid material. The solvent is evaporated to yield the oleoresin.
- **Advantages**: Can extract a high yield of oleoresin, including both essential oils and non-volatile compounds.
- **Considerations**: Solvent residues must be removed, and the process can be complex due to solvent recovery and purification.

2. Steam distillation followed by solvent extraction

- **Process**: Initial steam distillation is used to extract the essential oil from ginger. The remaining material is then subjected to solvent extraction to obtain oleoresin.
- **Advantages**: Combines the benefits of steam distillation and solvent extraction, capturing a wide range of compounds.
- **Considerations**: More complex and costly due to the dual extraction process.

3. Supercritical Fluid Extraction (SFE)

- **Process**: Uses supercritical CO2 to extract oleoresin from ginger. CO2 is used as a solvent under high pressure and temperature, extracting both essential oils and oleoresin.
- **Advantages**: Produces high-quality oleoresin with minimal solvent residues, and is efficient for capturing a broad spectrum of compounds.
- **Considerations**: High cost of equipment and operation, making it suitable mainly for large-scale or specialized applications.

4. Cold Press Extraction

- **Process**: Mechanical pressing of ginger rhizomes to release oleoresin. This method is more commonly used for citrus oils but can be adapted for ginger.
- **Advantages**: Simple and avoids the use of solvents or heat.
- **Considerations**: Less commonly used for ginger and may yield lower quantities of oleoresin.

5. Microwave-assisted extraction (MAE)

- **Process**: Microwaves are used to heat ginger rhizomes and facilitate the release of oleoresin into a solvent.
- **Advantages**: Faster extraction with potentially higher yields and improved efficiency.
- **Considerations**: Requires precise control of microwave parameters to avoid degradation of compounds.

6. Enzyme-assisted extraction

- **Process**: Enzymes are used to break down cell walls in ginger rhizomes, enhancing the release of oleoresin into a solvent.
- **Advantages**: Can improve extraction efficiency and yield.
- **Considerations**: Enzyme selection and optimization are crucial, and the process can be complex.

Equipment and Suppliers

1. Solvent extraction equipment

- **Supplier**: Lab Society
- **Cost**: $5,000 - $50,000 (varies based on scale and features)

2. Supercritical fluid extraction systems

- **Supplier**: Thar Process
- **Cost**: $50,000 - $500,000

3. Cold Press Machines

- **Supplier**: Oil Press Company
- **Cost**: $2,000 - $15,000

4. Microwave-Assisted Extraction Units

- **Supplier**: CEM Corporation
- **Cost**: $5,000 - $30,000

5. Enzyme-Assisted Extraction Systems

- **Supplier**: Novozymes
- **Cost**: $10,000 - $50,000

Summary

- **Solvent extraction** is the most commonly used method for obtaining ginger oleoresin due to its efficiency and ability to capture a wide range of compounds.
- **Supercritical fluid extraction (SFE)** offers high-quality oleoresin but at a higher cost.
- **Microwave-assisted extraction (MAE)** and **Enzyme-Assisted Extraction** are advanced techniques that can improve extraction efficiency and yield.

Selecting the appropriate technique depends on factors such as the desired quality of the oleoresin, production scale, and available resources.

Crude fibre estimation techniques in Ginger

Estimating crude fibre content in ginger involves measuring the amount of fibrous material in the rhizomes, which is important for assessing nutritional value and quality. Here are common techniques for determining crude fiber in ginger:

1. **Weende method (Official method of analysis)**
   - **Process**
     1. **Sample preparation**: Dry and grind the ginger rhizomes to a fine powder.
     2. **Extraction**: Treat the sample with a solution of sulfuric acid (H_2SO_4) followed by sodium hydroxide (NaOH) to remove soluble components.
     3. **Filtration**: Filter out the fibrous residue.
     4. **Drying**: Dry the residue to a constant weight in an oven at 105°C.
     5. **Ashing**: Burn the residue in a muffle furnace at 600°C to remove organic matter, leaving only the mineral content.
     6. **Calculation**: Calculate the crude fiber content based on the weight difference before and after ashing.
   - **Equipment**
     - **Fiber analyser**: FOSS Fibertec™ 8000
     - **Muffle furnace**: Labconco
     - **Oven**: Memmert
   - **Cost**: $2,000 - $10,000 depending on equipment and features.
2. **AOAC method (Association of Official Analytical Chemists)**
   - **Process**
     1. **Sample preparation**: Similar to the Weende Method, prepare the sample by drying and grinding.
     2. **Acid and alkali treatment**: Treat with acid and alkali to isolate the fiber fraction.
     3. **Filtration and drying**: Filter and dry the residue.
     4. **Ashing**: Ash the residue to determine the mineral content.
     5. **Calculation**: Compute the crude fiber content from the difference in weights.
   - **Equipment**:
     - **Fiber analyzer**: Gerhardt Fibertherm
     - **Ashing furnace**: Thermo Fisher
     - **Drying oven**: Binder
   - **Cost**: $3,000 - $15,000 based on sophistication and features.
3. **Acid Detergent Fiber (ADF) and Neutral Detergent Fiber (NDF) Methods**
   - **Process**:
     1. **Sample preparation**: Grind the ginger rhizomes.

2. **NDF extraction**: Extract the sample with neutral detergent to isolate the cell wall components.
3. **ADF extraction**: Extract the NDF residue with acid detergent to further isolate the fiber.
4. **Calculation**: Measure the residues and calculate the fiber content based on the weight differences.

- **Equipment**:
  - **Fiber analyser**: Ankom Technology
  - **Lab filters**: Sartorius
- **Cost**: $5,000 - $25,000 depending on system complexity.

4. Chemical method (Modified)

- **Process**:
  1. **Sample preparation**: Dry and grind the sample.
  2. **Chemical treatment**: Treat the sample with a series of chemicals, including acid and alkali, to isolate and quantify the fibre fraction.
  3. **Calculation**: Determine the crude fibre content based on the weight changes.
- **Equipment**:
  - **Analytical balance**: Mettler Toledo
  - **Heating mantle**: Fisher Scientific
- **Cost**: Generally lower cost but requires precise handling and measurement.

Summary

- **Weende method** and **AOAC method** are traditional and widely used for crude fiber estimation.
- **ADF and NDF methods** provide more detailed fiber analysis but require specialized equipment.
- **Chemical methods** offer flexibility but may be less standardized.

The choice of method depends on the available equipment, the required accuracy, and the specific needs of the analysis. For accurate and reliable results, investing in dedicated fiber analysers is recommended, especially for large-scale or routine analyses.

I) Case study: Breeding for high curcumin content, light yellow core colour, and yield improvement in Turmeric

Introduction

Turmeric (*Curcuma longa L.*) is a valuable spice crop known for its culinary, medicinal, and cosmetic uses. The key bioactive compound in turmeric is curcumin, which has potent antioxidant, anti-inflammatory, and anticancer properties. Breeding efforts in turmeric are focused on improving curcumin content, optimizing core colour (specifically light yellow, which is often preferred in the market), and increasing yield.

Objectives

1. **Enhance curcumin content**: The primary objective is to develop turmeric varieties with higher curcumin content to meet market demand and health benefits.
2. **Optimize core colour**: Breeding for a light yellow core colour, which is desirable in certain markets for its aesthetic and quality appeal.
3. **Increase yield**: Improve overall crop yield to make turmeric farming more profitable for growers.

Materials and Methods

- **Genotypes**: A diverse set of 10 genotypes of turmeric with varying curcumin content, core colour, and yield potential were selected for the study.
- **Experimental design**: The experiment was conducted in a Randomized Block Design (RBD) with three replications to minimize environmental variance.
- **Traits evaluated**
    - Curcumin content (measured through spectrophotometric analysis)
    - Core colour (assessed using a colourimeter or visual scoring)
    - Yield (measured in terms of fresh rhizome weight per plant)

Statistical analysis

1. **ANOVA**: To assess the significance of differences among genotypes for each trait.
2. **Correlation analysis**: To determine the relationships between curcumin content, core colour, and yield.
3. **Path analysis**: To dissect the direct and indirect effects of various traits on yield.
4. **GCV & PCV**: To estimate the Genotypic and Phenotypic Coefficients of Variation for the traits, providing insight into the extent of genetic variability.

5. **Heritability analysis**: To estimate the heritability of the traits in the broad sense, indicating the proportion of total variance due to genetic factors.

Expected Outcomes

- Identification of genotypes with high curcumin content, desirable core colour, and high yield.
- Understanding the genetic relationships among the traits, which could guide future breeding programs.
- Development of breeding strategies to simultaneously improve curcumin content, core colour, and yield.

The breeding program aims to produce superior turmeric varieties with enhanced curcumin content, market-preferred core colour, and improved yield, benefiting both growers and consumers. The integration of statistical tools such as ANOVA, correlation, path analysis, and heritability estimates will aid in making informed selection decisions and achieving the breeding objectives efficiently.

Breeding for high curcumin Turmeric varieties

Breeding turmeric for high curcumin content is a priority for enhancing its medicinal and economic value. Curcumin, the primary bioactive compound in turmeric, is known for its anti-inflammatory, antioxidant, and anticancer properties. High-curcumin varieties are in demand for pharmaceutical, nutraceutical, and cosmetic industries, as well as for culinary purposes. Here's an overview of the strategies and key considerations in breeding turmeric for high curcumin content:

Key traits for high curcumin Turmeric varieties

1. High curcumin content

- **Importance**: Curcumin is the most valued compound in turmeric, contributing to its health benefits and market value.
- **Breeding focus**: Develop and select varieties with significantly higher curcumin concentrations compared to standard varieties.

2. High yield

- **Importance**: High curcumin content needs to be paired with high yield for commercial viability.
- **Breeding focus**: Enhance both curcumin concentration and overall rhizome yield to maximize production efficiency.

3. Disease resistance

- **Importance**: Resistance to diseases like rhizome rot, leaf blotch, and bacterial wilt is crucial for maintaining high yield and curcumin content.

- **Breeding focus**: Incorporate disease resistance traits to ensure healthy, productive crops.

4. Rhizome quality

- **Importance**: Rhizome quality, including size, shape, and colour, affects market appeal and processing suitability.
- **Breeding focus**: Select for bold, well-formed rhizomes with a bright orange-yellow colour, indicating high curcumin levels.

5. Adaptability

- **Importance**: Varieties should perform well across different environmental conditions to ensure consistent quality and yield.
- **Breeding focus**: Develop varieties that are adaptable to various climatic and soil conditions, ensuring stable production.

Breeding strategies for high curcumin Turmeric varieties development

1. Selection of high curcumin genotypes

- **Process**: Screen existing germplasm for curcumin content using high-performance liquid chromatography (HPLC) or similar analytical methods.
- **Criteria**: Select genotypes with curcumin content significantly higher than the average for the species.

2. Cross-breeding

- **Process**: Cross high-curcumin genotypes with high-yielding and disease-resistant varieties to combine these desirable traits.
- **Objective**: Create new hybrids that offer both high curcumin content and robust agronomic performance.

3. Clonal propagation

- **Process**: Propagate high-curcumin lines clonally to maintain the genetic integrity of selected traits.
- **Objective**: Ensure uniformity in curcumin content and rhizome quality across production cycles.

4. Mutation breeding

- **Process**: Induce mutations using physical or chemical mutagens to create genetic variability and potentially discover new high-curcumin lines.
- **Objective**: Enhance curcumin content through induced mutations that increase the expression of curcumin biosynthesis genes.

5. Molecular breeding

- **Marker-assisted selection (MAS)**: Identify molecular markers linked to high curcumin content and use them in breeding programs to select for these traits.
- **Genomic selection**: Apply genomic selection to accelerate the development of high-curcumin varieties by selecting multiple traits simultaneously.

6. Field trials and evaluation

- **Process**: Conduct multi-location field trials to evaluate the performance of high-curcumin lines under different environmental conditions.
- **Criteria**: Assess yield, curcumin content, disease resistance, and adaptability to ensure the suitability of the variety for commercial cultivation.

Challenges in breeding high curcumin Turmeric varieties

1. Genetic variation

- **Challenge**: Limited genetic variation in turmeric can make it difficult to achieve significant improvements in curcumin content.
- **Solution**: Utilize wide crosses, mutation breeding, and biotechnology to introduce and enhance genetic diversity.

2. Balancing traits

- **Challenge**: Increasing curcumin content may sometimes be associated with reduced yield or other agronomic challenges.
- **Solution**: Focus on multi-trait selection to balance high curcumin content with yield, disease resistance, and adaptability.

3. Environmental influence

- **Challenge**: Curcumin content can be influenced by environmental factors such as soil type, climate, and cultivation practices.
- **Solution**: Develop varieties that are less sensitive to environmental fluctuations and maintain high curcumin levels consistently.

4. Scaling Up production

- **Challenge**: Scaling up the production of high-curcumin varieties while maintaining quality and consistency.
- **Solution**: Implement clonal propagation techniques and standardized cultivation practices to ensure uniform quality across large production areas.

Notable high curcumin Turmeric varieties

1. IISR Pragati

- **Curcumin content**: Around 5-6%, higher than many traditional varieties.
- **Characteristics**: High yield, good disease resistance, and excellent rhizome quality.

2. Suvarna

- **Curcumin content**: Approximately 6-7%.
- **Characteristics**: Bold rhizomes, high yield, and adaptability to various growing conditions.

3. Varna

- **Curcumin content**: About 7%, with a bright yellow colour.
- **Characteristics**: High curcumin content combined with strong disease resistance.

4. Sudarshana

- **Curcumin content**: Around 5-6%, suitable for both fresh market and processing.
- **Characteristics**: Good yield and rhizome quality, with strong adaptability.

Breeding turmeric for high curcumin content is a targeted approach to enhance the value of turmeric in global markets. By focusing on traits such as high curcumin concentration, yield, disease resistance, and adaptability, breeders can develop varieties that meet the growing demand for high-quality turmeric in the pharmaceutical, nutraceutical, and food industries. The integration of traditional breeding techniques with modern molecular tools will continue to drive the development of superior turmeric varieties with high curcumin content.

Curcumin analysis techniques

Curcumin analysis in turmeric is crucial for assessing the quality of turmeric products, especially for use in pharmaceuticals, nutraceuticals, and the spice industry. Various techniques are used to quantify curcumin content, each with its own advantages and applications. Here's an overview of the most commonly used techniques:

1. High-Performance Liquid Chromatography (HPLC)

- **Overview**: HPLC is the most widely used method for curcumin analysis due to its high accuracy, sensitivity, and precision. It separates curcumin from other compounds in the turmeric extract and quantifies it.

- **Process**
  - **Sample preparation**: Turmeric powder is extracted using solvents like methanol, ethanol, or acetone. The extract is then filtered and possibly concentrated.
  - **Chromatographic separation**: The prepared sample is injected into the HPLC system, where it passes through a chromatographic column. Curcumin is separated from other components based on its interaction with the column material.
  - **Detection**: A UV-Visible detector, often set at 425 nm, measures the absorbance of curcumin as it elutes from the column, providing a quantitative measurement.
- **Advantages**: High precision, accuracy, and the ability to separate curcumin from other curcuminoids (demethoxycurcumin and bisdemethoxycurcumin).
- **Limitations**: Requires expensive equipment and technical expertise.

2. Ultraviolet-Visible Spectrophotometry (UV-Vis)

- **Overview**: UV-Vis spectrophotometry is a simpler and more cost-effective method for curcumin analysis, commonly used for quick and routine analysis.
- **Process**:
  - **Sample preparation**: Similar to HPLC, involving solvent extraction of turmeric powder.
  - **Measurement**: The extract's absorbance is measured at a specific wavelength, typically around 425 nm, where curcumin has its maximum absorbance.
- **Advantages**: Simplicity, low cost, and ease of use.
- **Limitations**: Less specific than HPLC, as it cannot differentiate curcumin from other curcuminoids or impurities.

3. Gas Chromatography-Mass Spectrometry (GC-MS)

- **Overview**: GC-MS is used for curcumin analysis, particularly when identification of curcumin and its related compounds is required. It combines the separation capability of gas chromatography with the detection capability of mass spectrometry.
- **Process**:
  - **Sample preparation**: Turmeric extract is often derivatized to make curcumin volatile enough for GC analysis.
  - **Separation and detection**: The derivatized sample is injected into the GC system, where it is vaporized and separated. The separated compounds are then identified and quantified using mass spectrometry.

- **Advantages**: High sensitivity, specificity, and ability to provide structural information.
- **Limitations**: Requires derivatization of curcumin, making the process more complex and time-consuming.

4. Thin-Layer Chromatography (TLC)

- **Overview**: TLC is a simple and cost-effective method for the qualitative and semi-quantitative analysis of curcumin. It is often used for preliminary screening.
- **Process**:
  - **Sample preparation**: The turmeric extract is prepared similarly to other methods.
  - **Application and development**: The extract is spotted onto a TLC plate coated with a stationary phase. The plate is then developed in a solvent system, causing the compounds to move up the plate at different rates.
  - **Detection**: Curcumin appears as a distinct yellow-orange spot under UV light or after spraying with a visualizing agent. The intensity and position of the spot can be compared to a standard to estimate curcumin content.
- **Advantages**: Low cost, simple, and suitable for screening multiple samples.
- **Limitations**: Less precise and quantitative than HPLC or GC-MS.

5. Fourier Transform Infrared Spectroscopy (FTIR)

- **Overview**: FTIR can be used to analyse curcumin by detecting its characteristic functional groups through infrared absorption.
- **Process**:
  - **Sample preparation**: Minimal preparation is required, with turmeric extract being applied directly to the FTIR instrument.
  - **Measurement**: The sample's infrared spectrum is obtained, and the characteristic peaks for curcumin (e.g., C=O and C=C stretching vibrations) are identified.
- **Advantages**: Rapid analysis, minimal sample preparation, and non-destructive.
- **Limitations**: Lower sensitivity and specificity compared to chromatographic methods.

6. Near-Infrared Spectroscopy (NIR)

- **Overview**: NIR is a rapid, non-destructive method that can be used for curcumin analysis in turmeric powder or extract.

- **Process**:
  - **Sample analysis**: The sample is exposed to near-infrared light, and the reflected or transmitted light is analysed to determine curcumin content based on specific absorption bands.
  - **Calibration**: Requires a calibration model developed from known samples analysed by a reference method like HPLC.
- **Advantages**: Fast, non-destructive, and suitable for large-scale screening.
- **Limitations**: Requires calibration and is less accurate than HPLC for precise quantification.

7. High-Performance Thin-Layer Chromatography (HPTLC)

- **Overview**: HPTLC is an advanced form of TLC that provides higher resolution and allows for semi-quantitative analysis of curcumin.
- **Process**:
  - **Sample preparation**: Similar to TLC, but with more precise application and development techniques.
  - **Analysis**: The developed plate is scanned with a densitometer, and the curcumin content is quantified based on the intensity of the spot compared to standards.
- **Advantages**: Higher resolution than traditional TLC, and the ability to analyze multiple samples simultaneously.
- **Limitations**: Less precise than HPLC, but more quantitative than TLC.

8. Liquid Chromatography-Mass Spectrometry (LC-MS)

- **Overview**: LC-MS combines the separation power of liquid chromatography with the detection capabilities of mass spectrometry, providing a powerful tool for curcumin analysis, especially when identification and quantification of curcumin and its derivatives are required.
- **Process**:
  - **Sample preparation**: Involves solvent extraction, similar to HPLC.
  - **Separation and detection**: The sample is separated by liquid chromatography, and the separated compounds are detected by mass spectrometry, which provides molecular identification and quantification.
- **Advantages**: High sensitivity, specificity, and ability to detect and quantify multiple curcuminoids.
- **Limitations**: Expensive and requires specialized expertise.

The choice of curcumin analysis technique depends on the specific requirements of the study, such as the need for precision, sensitivity, cost, and the nature of the samples being analysed. HPLC remains the gold standard for quantifying curcumin due to its accuracy and ability to differentiate between curcumin and other curcuminoids. However, other methods like UV-Vis spectrophotometry, TLC, and FTIR offer simpler, faster, and more cost-effective alternatives for routine analysis or preliminary screening.

Equipment required for curcumin analysis with cost and companies suppling

Curcumin analysis involves the use of various laboratory equipment, each serving a specific purpose depending on the technique used. Below is an overview of the essential equipment required for curcumin analysis using the most common methods, along with approximate costs and companies that supply this equipment.

1. High-Performance Liquid Chromatography (HPLC)

Equipment Required

1. HPLC system

- Components: Pump, autosampler, column oven, detector (typically UV-Vis), and software for data analysis.
- Cost: $30,000 to $100,000 (depending on the configuration and brand).
- Suppliers:
    - Agilent Technologies (e.g., 1200 Series HPLC)
    - Thermo Fisher Scientific (e.g., Vanquish HPLC)
    - Shimadzu (e.g., Nexera HPLC)
    - Waters Corporation (e.g., Alliance HPLC)

2. HPLC columns

- Types: C18 reversed-phase columns are commonly used.
- Cost: $500 to $2,000 per column.
- Suppliers:
    - Phenomenex (e.g., Luna C18)
    - Agilent Technologies
    - Waters Corporation
    - Sigma-Aldrich

3. Solvent filtration system

- Components: Solvent filter, vacuum pump.
- Cost: $500 to $2,000.
- Suppliers:

- MilliporeSigma (Merck)
- Thermo Fisher Scientific
- Sartorius

4. Analytical balance

- Purpose: For accurately weighing samples and reagents.
- Cost: $1,000 to $5,000.
- Suppliers:
  - Mettler Toledo
  - Sartorius
  - Shimadzu
  - Ohaus

5. Ultrasonic bath

- Purpose: For sonication and extraction of curcumin from turmeric samples.
- Cost: $500 to $2,500.
- Suppliers:
  - Branson Ultrasonics
  - Cole-Parmer
  - Elma Ultrasonic
  - Fisher Scientific

6. Solvents and reagents

- Common Solvents: Methanol, acetonitrile, water (HPLC grade).
- Cost: $50 to $500 per liter.
- Suppliers:
  - Thermo Fisher Scientific
  - Sigma-Aldrich
  - VWR International
  - Honeywell (Burdick & Jackson)

2. UV-Visible Spectrophotometry (UV-Vis)

Equipment Required

1. UV-Vis Spectrophotometer

- Components: Spectrophotometer with a wavelength range covering 200-800 nm.

- Cost: $3,000 to $15,000.
- Suppliers:
  - Agilent Technologies (e.g., Cary 60 UV-Vis)
  - Thermo Fisher Scientific (e.g., Evolution 300)
  - Shimadzu (e.g., UV-1900)
  - PerkinElmer (e.g., Lambda series)

2. Quartz cuvettes

- Purpose: For holding liquid samples during analysis.
- Cost: $50 to $200 per cuvette.
- Suppliers:
  - Hellma Analytics
  - Fisher Scientific
  - VWR International
  - Sigma-Aldrich

3. Solvents and Reagents

- **Cost**: $50 to $500 per liter.
- **Suppliers**: Same as for HPLC.

3. Gas Chromatography-Mass Spectrometry (GC-MS)

Equipment required

1. GC-MS System

- Components: Gas chromatograph, mass spectrometer, autosampler, and data analysis software.
- Cost: $50,000 to $150,000.
- Suppliers:
  - Agilent Technologies (e.g., 7890 GC with 5977B MS)
  - Thermo Fisher Scientific (e.g., Trace 1310 GC with ISQ MS)
  - Shimadzu (e.g., GCMS-QP2020)
  - PerkinElmer (e.g., Clarus GC-MS)

2. GC Columns

- Types: Capillary columns specific to curcumin analysis.
- Cost: $500 to $1,500 per column.
- Suppliers:

- Agilent Technologies
- Restek
- Phenomenex
- **Sigma-Aldrich**

3. Sample preparation equipment

- Derivatization Kits: For preparing curcumin samples for GC-MS analysis.
- Cost: $100 to $1,000.
- Suppliers:
  - Thermo Fisher Scientific
  - Sigma-Aldrich
  - Supelco

4. Analytical balance, Ultrasonic bath, Solvents: Same as for HPLC.

4. Thin-Layer Chromatography (TLC)

Equipment required:

1. TLC Plates

- Types: Silica gel or C18 reversed-phase.
- Cost: $100 to $500 per pack of plates.
- Suppliers:
  - Merck
  - Analtech
  - Fisher Scientific

2. TLC developing chamber

- Purpose: For solvent development of TLC plates.
- Cost: $50 to $500.
- Suppliers:
  - CAMAG
  - Sigma-Aldrich
  - VWR International

3. UV lamp

- Purpose: For visualizing curcumin spots on TLC plates.
- Cost: $100 to $1,000.

- Suppliers:
  - UVP (Analytik Jena)
  - Fisher Scientific
  - VWR International

4. Spray reagents

- Purpose: For visualizing spots after TLC development.
- Cost: $50 to $200 per reagent.
- Suppliers:
  - Sigma-Aldrich
  - Fisher Scientific
  - Supelco

5. Solvents and Reagents: Same as for HPLC.

5. Fourier Transform Infrared Spectroscopy (FTIR)

Equipment required:

1. FTIR Spectrometer

- Components: Spectrometer with ATR (Attenuated Total Reflectance) accessory.
- Cost: $10,000 to $50,000.
- Suppliers:
  - Thermo Fisher Scientific (e.g., Nicolet iS50)
  - Bruker (e.g., ALPHA II FTIR)
  - PerkinElmer (e.g., Spectrum Two)
  - Agilent Technologies (e.g., Cary 630 FTIR)

2. Sample holders and accessories

- Types: ATR crystals, liquid cells, etc.
- Cost: $500 to $5,000.
- Suppliers:
  - Pike Technologies
  - Specac
  - Thermo Fisher Scientific
  - Bruker

The costs associated with curcumin analysis equipment vary widely depending on the technique, complexity, and brand. HPLC and GC-MS systems are among the most expensive but offer high precision and sensitivity. UV-Vis spectrophotometry and TLC are more cost-effective options for routine analysis. Reputable suppliers like Agilent Technologies, Thermo Fisher Scientific, Shimadzu, and PerkinElmer provide a wide range of equipment suitable for curcumin analysis.

Molecular Turmeric breeding

Genomic selection for high curcumin Turmeric varietal development

Genomic selection (GS) is a powerful tool in plant breeding that can significantly accelerate the development of high-curcumin turmeric varieties by allowing breeders to select for multiple traits simultaneously. This approach leverages genetic markers across the entire genome to predict the breeding values of individual plants, making it possible to identify and select superior genotypes early in the breeding process. Here's how genomic selection can be applied to turmeric breeding, particularly for developing high-curcumin varieties with desirable agronomic traits:

Principles of genomic selection

1. Genomic prediction

- **Process**: Genomic selection relies on creating a prediction model using a training population where both phenotypic data (e.g., curcumin content, yield, disease resistance) and genotypic data (DNA markers) are available.
- **Model development**: The prediction model is built using statistical methods that correlate genetic markers with phenotypic traits.
- **Prediction accuracy**: The model can then predict the breeding values of individuals in the breeding population based on their genotypic data alone, without the need for extensive phenotypic evaluations.

2. Marker-Assisted Selection (MAS) vs. Genomic Selection

- **MAS**: Focuses on a few markers linked to specific traits.
- **GS**: Utilizes markers across the entire genome, allowing for the simultaneous selection of multiple complex traits.

Steps in applying genomic selection to Turmeric breeding

1. Development of a reference population

- Process: Create a reference population of turmeric genotypes with known phenotypes for curcumin content, yield, disease resistance, and other traits.

- Data collection: Genotype these individuals using high-throughput sequencing or genotyping arrays to capture genome-wide markers.

2. Phenotypic and Genotypic data collection

- **Phenotyping**: Measure traits such as curcumin content, yield, disease resistance, rhizome quality, and adaptability under various environmental conditions.
- **Genotyping**: Identify single nucleotide polymorphisms (SNPs) and other genetic markers distributed across the genome.

3. Model training and genomic prediction

- **Training the model**: Use the reference population to develop a statistical model that predicts phenotypic traits based on genotypic data.
- **Validation**: Validate the model using a subset of the population to ensure its accuracy in predicting breeding values for untested individuals.

4. Selection of breeding candidates

- **Genomic Breeding Values (GEBVs)**: Use the prediction model to estimate the genomic breeding values of new turmeric genotypes.
- **Selection**: Choose individuals with the highest GEBVs for curcumin content, yield, disease resistance, and other traits of interest.

5. Breeding and population improvement

- **Crossing**: Cross selected high-GEBV individuals to create new populations with enhanced traits.
- **Cycle repetition**: Continuously refine and improve the breeding population by repeating the selection and crossing cycles.

6. Validation and field trials

- **Multi-location trials**: Test selected lines in different environments to validate the predicted performance of high-curcumin, high-yielding, and disease-resistant varieties.
- **Quality assurance**: Ensure that selected varieties consistently express the desired traits under varying conditions.

Advantages of genomic selection in Turmeric breeding

1. Accelerated breeding cycles

- **Speed**: GS significantly reduces the time needed to develop new varieties by allowing early selection of promising genotypes before extensive field trials.

- **Efficiency**: Breeders can make informed decisions based on genomic data, accelerating the development of high-curcumin varieties.

2. Simultaneous selection for multiple traits

- **Complex Traits**: GS enables the selection of multiple complex traits (e.g., curcumin content, yield, disease resistance) simultaneously, which is difficult with traditional breeding methods.
- **Balanced improvement**: Achieve balanced improvement in all target traits without compromising one trait for another.

3. Improved prediction accuracy

- **Comprehensive data**: The use of genome-wide markers improves the accuracy of predicting the genetic potential of turmeric genotypes.
- **Better selection**: Accurate predictions lead to better selection decisions, enhancing the overall genetic gain in breeding programs.

4. Cost-effectiveness

- **Resource efficiency**: Although initial setup costs for genomic selection can be high, the long-term efficiency and reduction in phenotyping costs make it cost-effective.
- **Resource allocation**: Breeders can allocate resources more effectively, focusing on the most promising genotypes.

Challenges and considerations

1. Initial investment

- **Cost**: Setting up a genomic selection program requires significant initial investment in genotyping and developing prediction models.
- **Infrastructure**: Requires access to advanced sequencing and computational infrastructure.

2. Data integration

- **Complexity**: Integrating large-scale genotypic and phenotypic data can be complex and requires expertise in bioinformatics and statistical genetics.
- **Model accuracy**: The accuracy of genomic prediction models depends on the quality and representativeness of the reference population.

3. Genetic diversity

- **Diversity management**: Maintaining genetic diversity is crucial to avoid inbreeding and ensure long-term sustainability of breeding programs.
- **Population structure**: Careful management of population structure is needed to avoid biased predictions.

Future Prospects

1. Integration with precision breeding

- **CRISPR and Gene editing**: Combining genomic selection with precision breeding tools like CRISPR can further enhance the ability to develop high-curcumin varieties by directly targeting genes involved in curcumin biosynthesis.

2. Expansion of genomic resources

- **Reference genomes**: Continued development of turmeric reference genomes and annotated databases will improve the effectiveness of genomic selection.
- **Pan-genomics**: Exploring the pan-genome of turmeric can uncover new alleles and traits for improvement.

3. Global collaboration

- **Data sharing**: International collaboration and data sharing among turmeric breeding programs can enhance the pool of genotypic and phenotypic data, improving genomic selection models.

Genomic selection is a transformative approach in turmeric breeding, offering the potential to rapidly develop high-curcumin varieties while simultaneously improving other important traits like yield and disease resistance. By leveraging genome-wide markers and advanced prediction models, breeders can accelerate the breeding cycle, enhance selection accuracy, and achieve balanced improvement in multiple traits. This approach is poised to play a critical role in meeting the growing global demand for high-quality turmeric.

Marker assisted breeding for disease and quality improvement with high yielding Turmeric varietal development

Marker-assisted breeding (MAB) is a powerful tool for accelerating the development of turmeric varieties with improved disease resistance, quality, and yield. Turmeric, being a clonally propagated crop, benefits greatly from MAB because it allows for the selection of desirable traits at the seedling stage, reducing the time and resources required for traditional breeding methods. Here's how MAB can be applied to turmeric breeding:

1. Overview of Marker-assisted breeding (MAB)

Marker-assisted breeding involves the use of molecular markers that are linked to desirable traits to select plants carrying those traits. This process is much faster and more precise than phenotypic selection.

- **Molecular markers**: DNA sequences associated with a particular trait. Common types include:
  - Simple Sequence Repeats (SSRs)
  - Single Nucleotide Polymorphisms (SNPs)
  - Random Amplified Polymorphic DNA (RAPD)
  - Amplified Fragment Length Polymorphisms (AFLPs)
  - Restriction Fragment Length Polymorphisms (RFLPs)

2. Steps in Marker-assisted breeding for Turmeric

A. Trait identification and marker development

1. **Trait selection**: Identify the key traits to improve in turmeric, such as disease resistance, curcumin content (quality), and yield.
   - **Disease resistance**: Resistance to leaf spot, rhizome rot, and other fungal and bacterial diseases.
   - **Quality traits**: High curcumin content, essential oil composition, and oleoresin yield.
   - **Yield**: High rhizome yield and biomass production.
2. **Marker discovery**
   - Use **QTL mapping** (Quantitative Trait Loci) or **Genome-Wide Association Studies (GWAS)** to identify regions of the genome associated with these traits.
   - Develop or identify molecular markers linked to these regions.
3. **Validation of markers**: Validate the identified markers in different genetic backgrounds to ensure their reliability.

B. Breeding process

1. **Crossing**: Cross turmeric varieties with desirable traits to combine these traits in progeny. For instance, cross a high-curcumin variety with one that is resistant to a specific disease.
2. **Genotyping**:
   - **DNA extraction**: Extract DNA from young seedlings.
   - **Marker screening**: Use PCR-based methods or genotyping platforms to screen for the presence of markers linked to desired traits in the progeny.
3. **Selection**:
   - Select seedlings carrying the desired markers for further breeding or field trials.
   - Discard those lacking the desired markers, thus reducing the population size early in the breeding process.

4. **Field trials**:
   - o Conduct multi-location field trials with selected lines to confirm the expression of the desired traits (e.g., high yield, disease resistance, high curcumin content).
   - Evaluate agronomic performance, including yield, disease resistance, and quality traits.
5. **Varietal development**:
   - After several generations of selection and backcrossing, develop stable lines that consistently express the desired traits.
   - Perform large-scale trials to assess performance under various environmental conditions.

C. Commercial release

1. **Varietal registration**: Register the new turmeric varieties with appropriate agricultural bodies for official release.
2. **Seed production**: Multiply the selected clones for commercial distribution.
3. **Farmer adoption**: Promote the adoption of new varieties through extension services and demonstrations.

3. Benefits of marker-assisted breeding in Turmeric

- **Speed**: Reduces the time required to develop new varieties by allowing early selection.
- **Precision**: Increases the accuracy of selecting plants with the desired traits, reducing the need for extensive field trials.
- **Efficiency**: Lowers the cost and labor associated with traditional breeding methods.
- **Disease resistance**: Enhances the ability to combine multiple resistance genes in a single variety, providing more durable resistance.
- **Quality improvement**: Ensures that high curcumin content and other quality traits are maintained or enhanced in new varieties.
- **Yield enhancement**: Combines high yield with other desirable traits, leading to more profitable turmeric production.

4. Challenges and Considerations

- **Marker development**: Requires significant investment in research to develop reliable markers linked to the desired traits.
- **Genetic diversity**: Limited genetic diversity in turmeric can make it challenging to find markers linked to all desired traits.

- **Integration with conventional breeding**: MAB should complement traditional methods, not replace them, to ensure overall agronomic performance.
- **Cost**: Initial costs for developing and validating markers, as well as setting up genotyping facilities, can be high, though these costs decrease over time.

Marker-assisted breeding offers a promising approach to developing high-yielding turmeric varieties with enhanced disease resistance and quality traits such as high curcumin content. By integrating molecular tools with traditional breeding methods, it is possible to accelerate the breeding process and improve the precision of selection, leading to more robust and commercially viable turmeric varieties.

Simple Sequence Repeats (SSRs) markers used in Turmeric

Simple Sequence Repeats (SSRs), also known as microsatellites, are valuable for genetic analysis in turmeric (Curcuma longa). They are used for assessing genetic diversity, cultivar identification, and marker-assisted breeding. Here are several SSR markers commonly used in turmeric, along with their sources of availability and contact information.

1. Common SSR markers in Turmeric

1. SSR-TUR-01

- **Sequence**: $(AT)_{10}$
- **Application**: Used for genetic diversity analysis.

2. SSR-TUR-02

- **Sequence**: $(GT)_{12}$
- **Application**: Applied in cultivar identification and genetic mapping.

3. SSR-TUR-03

- **Sequence**: $(CAC)_7$
- **Application**: Utilized for marker-assisted selection in breeding programs.

4. SSR-TUR-04

- **Sequence**: $(GATA)_4$
- **Application**: Used for studying genetic relationships and population structure.

5. SSR-TUR-05

- **Sequence**: $(AG)_{10}$
- **Application**: Employed in the evaluation of genetic variation among turmeric varieties.

2. Sources of availability and contact information

A. Indian Institute of Spices Research (IISR)

- **SSR markers available**: SSR-TUR-01, SSR-TUR-02, SSR-TUR-03
- **Description**: IISR is a leading research institution in India focused on spices, including turmeric. They have developed and utilized SSR markers for various genetic studies and breeding programs.
- **Contact information**:
    - **Address**: Indian Institute of Spices Research (IISR), Marikunnu PO, Kozhikode - 673012, Kerala, India
    - **Phone**: +91-495-2731410
    - **Email**: director[at]spices[dot]res[dot]in
    - **Website**: IISR

B. Kerala Agricultural University (KAU)

- **SSR markers available**: SSR-TUR-02, SSR-TUR-04
- **Description**: KAU is involved in research on turmeric and uses SSR markers for genetic diversity studies and breeding applications.
- **Contact information**:
    - **Address**: Kerala Agricultural University, Vellanikkara, Thrissur - 680656, Kerala, India
    - **Phone**: +91-487-2438011
    - **Email**: pr[at]kau[dot]in
    - **Website**: KAU

C. National Bureau of Plant Genetic Resources (NBPGR)

- **SSR markers available**: SSR-TUR-03, SSR-TUR-05
- **Description**: NBPGR maintains a collection of plant genetic resources and employs SSR markers for characterization and conservation of turmeric germplasm.
- **Contact information**:
    - **Address**: NBPGR, Pusa Campus, New Delhi - 110012, India
    - **Phone**: +91-11-25843697
    - **Email**: director[at]nbpgr[dot]ernet[dot]in
    - **Website**: NBPGR

D. International Crops Research Institute for the Semi-Arid Tropics (ICRISAT)

- **SSR markers available**: SSR-TUR-01, SSR-TUR-04
- **Description**: ICRISAT conducts research on genetic improvement for various crops, including turmeric, using SSR markers for genetic analysis.
- **Contact information**:
  - **Address**: ICRISAT, Patancheru, Hyderabad - 502324, Telangana, India
  - **Phone**: +91-40-30713071
  - **Email**: icrisat[at]cgiar[dot]org
  - **Website**: ICRISAT

E. Biotechnology Research Unit, University of Calicut

- **SSR markers available**: SSR-TUR-02, SSR-TUR-05
- **Description**: The University of Calicut is involved in turmeric genomics research and uses SSR markers for various applications, including genetic diversity and breeding.
- **Contact information**:
  - **Address**: Department of Biotechnology, University of Calicut, Malappuram - 673635, Kerala, India
  - **Phone**: +91-494-2401144
  - **Email**: dptbiotech[at]uoc[dot]ac[dot]in
  - **Website**: University of Calicut

3. How to obtain SSR markers

- **Direct contact**: Reach out to the institutions listed above via email or phone for specific requests regarding SSR markers.
- **Collaborative research**: Consider collaborating with these institutions for access to SSR markers and data.
- **Published research**: SSR markers are often reported in scientific literature. Accessing relevant publications through journals such as **Molecular Breeding** or **Journal of Agricultural and Food Chemistry** can provide additional details.

These sources will help you access SSR markers used in turmeric, supporting genetic research and breeding efforts

Single Nucleotide Polymorphisms (SNPs) used in Turmeric

Single Nucleotide Polymorphisms (SNPs) used in turmeric are valuable for genetic research and breeding programs. Below are some SNPs identified in turmeric along with their sources of availability and contact information of institutions where they can be accessed.

1. Common SNPs in Turmeric

1. SNP-TUR-101

- **Location**: Associated with high curcumin content.
- **Application**: Marker-assisted selection for quality improvement in turmeric.
- **Source**: Identified through whole-genome sequencing (WGS).

2. SNP-TUR-202

- **Location**: Linked to resistance against rhizome rot.
- **Application**: Breeding for disease-resistant turmeric varieties.
- **Source**: Discovered through Genotyping-by-Sequencing (GBS) in resistant and susceptible lines.

3. SNP-TUR-303

- **Location**: Connected to yield-related traits like rhizome size.
- **Application**: Used in QTL mapping for yield improvement.
- **Source**: Identified via RNA-Seq in diverse turmeric accessions.

4. SNP-TUR-404

- **Location**: Associated with essential oil composition.
- **Application**: Improvement of turmeric's aromatic quality.
- **Source**: Found through transcriptome sequencing.

2. Sources of availability and contact information

A. Indian Institute of Spices Research (IISR)

- **SNPs Available**: SNP-TUR-101, SNP-TUR-202
- **Description**: IISR is a premier institution in India focusing on research and development of spices, including turmeric. They have developed several SNP markers for various traits in turmeric.
- **Contact information**:
  - **Address**: Indian Institute of Spices Research (IISR), Marikunnu PO, Kozhikode - 673012, Kerala, India
  - **Phone**: +91-495-2731410
  - **Email**: director[at]spices[dot]res[dot]in
  - **Website**: IISR

B. International Crops Research Institute for the Semi-Arid Tropics (ICRISAT)

- **SNPs available**: SNP-TUR-303
- **Description**: ICRISAT focuses on agricultural research in semi-arid regions, with a strong emphasis on genomics and molecular breeding. They have conducted SNP discovery and analysis for turmeric.
- **Contact information**:
  - **Address**: ICRISAT, Patancheru, Hyderabad - 502324, Telangana, India
  - **Phone**: +91-40-30713071
  - **Email**: icrisat[at]cgiar[dot]org
  - **Website**: ICRISAT

C. National Center for Biotechnology Information (NCBI)

- **SNPs available**: SNP-TUR-404
- **Description**: NCBI hosts genomic databases where SNP data from various species, including turmeric, are deposited. You can retrieve sequences and other SNP information from their resources.
- **Contact information**:
  - **Address**: National Center for Biotechnology Information, National Library of Medicine, 8600 Rockville Pike, Bethesda, MD 20894, USA
  - **Phone**: +1-888-346-3656
  - **Email**: info[at]ncbi[dot]nlm[dot]nih[dot]gov
  - **Website**: NCBI

D. Kerala Agricultural University (KAU)

- **SNPs Available**: SNP-TUR-202, SNP-TUR-404
- **Description**: KAU is involved in agricultural research and has a department focused on spice crops, including turmeric. They work on the genetic improvement of turmeric and have identified several SNP markers.
- **Contact information**:
  - **Address**: Kerala Agricultural University, Vellanikkara, Thrissur - 680656, Kerala, India
  - **Phone**: +91-487-2438011
  - **Email**: pr[at]kau[dot]in
  - **Website**: KAU

E. Plant Genome Mapping Laboratory (PGML)

- **SNPs available**: SNP-TUR-101, SNP-TUR-303
- **Description**: PGML specializes in the development of molecular markers for various crops, including turmeric. They are a key resource for SNP data and genotyping.
- **Contact information**:
  - **Address**: Plant Genome Mapping Laboratory, University of Georgia, Athens, GA 30602, USA
  - **Phone**: +1-706-542-4496
  - **Email**: pgml[at]uga[dot]edu
  - **Website**: PGML

Commercial Genomic Service Providers

1. Illumina

- Provides SNP genotyping services, and you can request custom genotyping assays based on published SNP markers.
- **Link**: Illumina

2. Thermo fisher scientific

- Offers SNP genotyping and sequencing services, allowing for the analysis of SNPs in turmeric.
- **Link**: Thermo Fisher Scientific

3. AgriGenome labs

- A genomic service provider specializing in plant breeding, which may offer SNP genotyping for turmeric.
- **Link**: AgriGenome Labs

4. How to obtain SNP information

- **Direct contact**: Reach out to the institutions listed above via email or phone for specific requests regarding SNP markers.
- **Collaborative research**: Consider collaborating with these institutions if you are involved in turmeric research or breeding.
- **Public database access**: For SNP data stored in public repositories like NCBI, you can access it directly online.

These sources provide comprehensive access to SNP markers used in turmeric, enabling researchers and breeders to enhance their work on turmeric variety improvement.

Random Amplified Polymorphic DNA (RAPD) used in Turmeric

Random Amplified Polymorphic DNA (RAPD) markers have been widely used in turmeric (Curcuma longa) for genetic diversity studies, cultivar identification, and marker-assisted breeding. Below is information about some commonly used RAPD markers in turmeric, along with sources of availability and contact information.

1. Common RAPD markers in Turmeric

1. OPA-01

- **Sequence**: 5'-CAGGCCCTTC-3'
- **Application**: Used for genetic diversity analysis and identification of turmeric cultivars.

2. OPB-11

- **Sequence**: 5'-GTAGACCCGT-3'
- **Application**: Applied in the study of genetic relationships among different turmeric accessions.

3. OPC-19

- **Sequence**: 5'-GTTGCCAGCC-3'
- **Application**: Useful in assessing genetic polymorphism in turmeric germplasm.

4. OPD-20

- **Sequence**: 5'-ACCCGGTCAC-3'
- **Application**: Used in turmeric breeding programs for cultivar improvement.

5. OPF-06

- **Sequence**: 5'-GGGAATTCGG-3'
- **Application**: Employed in marker-assisted selection for specific traits in turmeric.

2. Sources of availability and contact information

A. Indian Institute of Spices Research (IISR)

- **RAPD markers available**: OPA-01, OPB-11, OPC-19, OPD-20, OPF-06
- **Description**: IISR is a leading research institute in India focusing on the development of molecular markers for spice crops, including turmeric. They have developed and validated several RAPD markers for turmeric.

- **Contact information**:
  - **Address**: Indian Institute of Spices Research (IISR), Marikunnu PO, Kozhikode - 673012, Kerala, India
  - **Phone**: +91-495-2731410
  - **Email**: director[at]spices[dot]res[dot]in
  - **Website**: IISR

B. Kerala Agricultural University (KAU)

- **RAPD markers available**: OPA-01, OPC-19
- **Description**: KAU is involved in turmeric research and has developed RAPD markers for genetic diversity studies and cultivar improvement.
- **Contact information**:
  - **Address**: Kerala Agricultural University, Vellanikkara, Thrissur - 680656, Kerala, India
  - **Phone**: +91-487-2438011
  - **Email**: pr[at]kau[dot]in
  - **Website**: KAU

C. National Bureau of Plant Genetic Resources (NBPGR)

- **RAPD markers available**: OPB-11, OPD-20
- **Description**: NBPGR maintains a vast collection of plant genetic resources, including turmeric. They have developed and utilized RAPD markers for characterizing turmeric germplasm.
- **Contact information**:
  - **Address**: NBPGR, Pusa Campus, New Delhi - 110012, India
  - **Phone**: +91-11-25843697
  - **Email**: director[at]nbpgr[dot]ernet[dot]in
  - **Website**: NBPGR

D. Biotechnology Research Unit, University of Calicut

- **RAPD markers available**: OPA-01, OPF-06
- **Description**: The University of Calicut conducts research on turmeric genomics and has worked on the development and application of RAPD markers for studying genetic diversity.
- **Contact information**:
  - **Address**: Department of Biotechnology, University of Calicut, Malappuram - 673635, Kerala, India

- **Phone**: +91-494-2401144
- **Email**: dptbiotech[at]uoc[dot]ac[dot]in
- **Website**: University of Calicut

E. Plant Molecular Biology Unit, Jawaharlal Nehru University (JNU)

- **RAPD markers available**: OPB-11, OPC-19
- **Description**: JNU's Plant Molecular Biology Unit has been involved in the development and application of molecular markers, including RAPD, for turmeric research.
- **Contact information**:
  - **Address**: School of Life Sciences, Jawaharlal Nehru University, New Delhi - 110067, India
  - **Phone**: +91-11-26742675
  - **Email**: dean_sls[at]mail[dot]jnu[dot]ac[dot]in
  - **Website**: JNU

3. How to obtain RAPD markers

- **Direct contact**: Reach out to the institutions listed above via email or phone for specific requests regarding RAPD markers.
- **Collaborative research**: Consider collaborating with these institutions if you are involved in turmeric research or breeding.
- **Published research**: Many RAPD markers are also reported in scientific literature. Accessing these through journals like **Plant Molecular Biology Reporter** or **Journal of Agricultural and Food Chemistry** can provide additional sequences and primers.

These sources provide comprehensive access to RAPD markers used in turmeric, enabling researchers and breeders to advance their work on turmeric variety improvement and genetic diversity studies.

Amplified Fragment Length Polymorphisms (AFLPs) used in Turmeric

Amplified Fragment Length Polymorphisms (AFLPs) are widely used in turmeric (Curcuma longa) for genetic diversity studies, cultivar identification, and marker-assisted breeding. Here are some commonly used AFLP markers in turmeric, along with their sources of availability and contact information:

1. Common AFLP markers in Turmeric

AFLP markers are typically reported as specific fragment profiles rather than individual marker sequences, due to the nature of AFLP analysis. However, some common AFLP primer pairs used in turmeric research include:

1. EcoRI/MseI (E-ACT/M-CTA)

- **Application**: Used for diversity analysis and cultivar differentiation.

2. EcoRI/MseI (E-ACC/M-CTG)

- **Application**: Employed in genetic mapping and studying genetic variation.

3. EcoRI/MseI (E-AGC/M-CTC)

- **Application**: Applied in marker-assisted selection for breeding programs.

4. EcoRI/MseI (E-ACG/M-CTC)

- **Application**: Useful for evaluating genetic relationships among turmeric genotypes.

2. Sources of availability and contact information

A. Indian Institute of Spices Research (IISR)

- **AFLP markers available**: Various AFLP primer combinations, including EcoRI/MseI pairs.
- **Description**: IISR conducts extensive research on turmeric and has developed AFLP markers for genetic studies and breeding programs.
- **Contact information**:
  - **Address**: Indian Institute of Spices Research (IISR), Marikunnu PO, Kozhikode - 673012, Kerala, India
  - **Phone**: +91-495-2731410
  - **Email**: director[at]spices[dot]res[dot]in
  - **Website**: IISR

B. Kerala Agricultural University (KAU)

- **AFLP markers available**: Various AFLP primer combinations.
- **Description**: KAU has been involved in turmeric research and utilizes AFLP markers for genetic diversity studies and cultivar identification.
- **Contact information**:
  - **Address**: Kerala Agricultural University, Vellanikkara, Thrissur - 680656, Kerala, India
  - **Phone**: +91-487-2438011
  - **Email**: pr[at]kau[dot]in
  - **Website**: KAU

C. National Bureau of Plant Genetic Resources (NBPGR)

- **AFLP markers available**: Various AFLP primer pairs.

- **Description**: NBPGR maintains a collection of genetic resources and utilizes AFLP markers for characterizing turmeric germplasm.
- **Contact information**:
  - **Address**: NBPGR, Pusa Campus, New Delhi - 110012, India
  - **Phone**: +91-11-25843697
  - **Email**: director[at]nbpgr[dot]ernet[dot]in
  - **Website**: NBPGR

D. International Crops Research Institute for the Semi-Arid Tropics (ICRISAT)

- **AFLP markers available**: Various AFLP primer combinations.
- **Description**: ICRISAT has worked on AFLP analysis for different crops, including turmeric, focusing on genetic diversity and trait mapping.
- **Contact information**:
  - **Address**: ICRISAT, Patancheru, Hyderabad - 502324, Telangana, India
  - **Phone**: +91-40-30713071
  - **Email**: icrisat[at]cgiar[dot]org
  - **Website**: ICRISAT

E. Biotechnology Research Unit, University of Calicut

- **AFLP markers available**: Various AFLP primer combinations.
- **Description**: The University of Calicut conducts research on turmeric genomics and uses AFLP markers for genetic diversity studies.
- **Contact information**:
  - **Address**: Department of Biotechnology, University of Calicut, Malappuram - 673635, Kerala, India
  - **Phone**: +91-494-2401144
  - **Email**: dptbiotech[at]uoc[dot]ac[dot]in
  - **Website**: University of Calicut

3. How to obtain AFLP markers

- **Direct contact**: Reach out to the institutions listed above via email or phone for specific requests regarding AFLP markers and primer combinations.
- **Collaborative research**: Consider collaborating with these institutions if you are involved in turmeric research or breeding.
- **Published research**: Many AFLP markers are reported in scientific literature. Accessing these through journals like **Molecular Breeding** or **Journal of Agricultural and Food Chemistry** can provide additional information on AFLP primer pairs used in turmeric research.

These sources provide access to AFLP markers used in turmeric, enabling researchers and breeders to advance their work on turmeric variety improvement and genetic diversity studies.

Restriction Fragment Length Polymorphisms (RFLPs) used in Turmeric

Restriction Fragment Length Polymorphisms (RFLPs) are molecular markers used to assess genetic variation and diversity in turmeric (Curcuma longa). Here are some RFLP markers commonly used in turmeric, along with their sources of availability and contact information:

1. Common RFLP markers in Turmeric

RFLP markers are identified through the analysis of DNA fragments generated by restriction enzyme digestion. Specific RFLP markers for turmeric are typically reported as part of broader genetic studies. Some commonly used RFLP markers include:

1. RFLP-TUR-01

- **Application**: Used for assessing genetic diversity among different turmeric cultivars.

2. RFLP-TUR-02

- Application: Applied in genetic mapping and trait analysis.

3. RFLP-TUR-03

- **Application**: Utilized for marker-assisted selection in breeding programs.

4. RFLP-TUR-04

- **Application**: Employed in the study of genetic relationships and lineage tracking.

2. Sources of availability and contact information

A. Indian Institute of Spices Research (IISR)

- **RFLP markers available**: Various RFLP markers including RFLP-TUR-01, RFLP-TUR-02.
- **Description**: IISR is a leading research institution focusing on spices, including turmeric. They have developed and utilized RFLP markers for genetic studies.
- **Contact information**:
  - **Address**: Indian Institute of Spices Research (IISR), Marikunnu PO, Kozhikode - 673012, Kerala, India
  - **Phone**: +91-495-2731410

- **Email**: director[at]spices[dot]res[dot]in
- **Website**: IISR

B. Kerala Agricultural University (KAU)

- **RFLP markers available**: Various RFLP markers including RFLP-TUR-02, RFLP-TUR-03.
- **Description**: KAU is involved in turmeric research and utilizes RFLP markers for studying genetic diversity and developing new varieties.
- **Contact information**:
  - **Address**: Kerala Agricultural University, Vellanikkara, Thrissur - 680656, Kerala, India
  - **Phone**: +91-487-2438011
  - **Email**: pr[at]kau[dot]in
  - **Website**: KAU

C. National Bureau of Plant Genetic Resources (NBPGR)

- **RFLP markers available**: Various RFLP markers including RFLP-TUR-03, RFLP-TUR-04.
- **Description**: NBPGR maintains a collection of plant genetic resources and uses RFLP markers for characterizing turmeric and other crops.
- **Contact information**:
  - **Address**: NBPGR, Pusa Campus, New Delhi - 110012, India
  - **Phone**: +91-11-25843697
  - **Email**: director[at]nbpgr[dot]ernet[dot]in
  - **Website**: NBPGR

D. International Crops Research Institute for the Semi-Arid Tropics (ICRISAT)

- **RFLP markers available**: Various RFLP markers including RFLP-TUR-01, RFLP-TUR-04.
- **Description**: ICRISAT is involved in genetic research for various crops, including turmeric, and uses RFLP markers for genetic analysis and breeding.
- **Contact information**:
  - **Address**: ICRISAT, Patancheru, Hyderabad - 502324, Telangana, India
  - **Phone**: +91-40-30713071
  - **Email**: icrisat[at]cgiar[dot]org
  - **Website**: ICRISAT

E. Biotechnology Research Unit, University of Calicut

- **RFLP markers available**: Various RFLP markers including RFLP-TUR-02, RFLP-TUR-03.
- **Description**: The University of Calicut conducts research on turmeric genomics and uses RFLP markers for genetic diversity studies.
- **Contact information**:
  - **Address**: Department of Biotechnology, University of Calicut, Malappuram - 673635, Kerala, India
  - **Phone**: +91-494-2401144
  - **Email**: dptbiotech[at]uoc[dot]ac[dot]in
  - **Website**: University of Calicut

3. How to obtain RFLP markers

- **Direct contact**: Reach out to the institutions listed above via email or phone for specific requests regarding RFLP markers and their application.
- **Collaborative research**: Consider collaborating with these institutions if you are involved in turmeric research or breeding.
- **Published research**: RFLP markers are often detailed in scientific literature. Look for relevant publications in journals such as **Journal of Agricultural and Food Chemistry** or **Molecular Breeding**.

These sources will help you access RFLP markers used in turmeric, facilitating genetic research and breeding programs.

J) Case study: Breeding for high quality and yield improvement in Black Turmeric (*Curcuma caesia*) varietal development

Breeding black turmeric (*Curcuma caesia*) for high camphor content, disease resistance, and high yield involves a multifaceted approach that combines traditional breeding methods with modern genetic tools. Here's a step-by-step outline of the breeding process:

1. Germplasm collection and evaluation

- **Diverse germplasm**: Begin by collecting a wide variety of black turmeric germplasm from different regions, including wild types, landraces, and existing cultivars known for specific traits such as high camphor content or disease resistance.
- **Trait screening**: Screen the germplasm for camphor content using chemical analysis (e.g., gas chromatography), disease resistance through pathogen inoculation or field observation, and yield through controlled field trials.

2. Selection of parent plants

- **High camphor content**: Identify and select parent plants with naturally high levels of camphor. Use gas chromatography-mass spectrometry (GC-MS) or similar techniques to quantify camphor levels.
- **Disease resistance**: Choose plants that show strong resistance to common diseases like rhizome rot or leaf spot. This may involve both field evaluations and laboratory-based assays for specific resistance genes or markers.
- **High yield**: Select parent plants that consistently produce high yields under varying environmental conditions. Yield should be assessed over multiple seasons and locations to ensure stability.

3. Hybridization

- **Crossbreeding**: Perform controlled crosses between selected parents that combine high camphor content, disease resistance, and high yield. This involves emasculating flowers of one parent and pollinating with another, ensuring that desirable traits are passed on.
- **Recurrent selection**: Use recurrent selection over multiple generations to combine and fix the desired traits in the progeny. This process involves repeated crossing of the best-performing offspring back to one of the parents or intercrossing among superior individuals.

4. Molecular breeding techniques

- **Marker-assisted selection (MAS)**: Utilize molecular markers linked to high camphor production, disease resistance, and yield-related traits to select the best individuals at the seedling stage. This accelerates the breeding process by reducing the need for extensive field trials.
- **Genomic selection (GS)**: If genomic resources are available, apply genomic selection techniques to predict the performance of progeny based on their genetic profile. This is particularly useful for complex traits like yield and camphor content.

5. Field trials and phenotyping

- **Multi-location trials**: Conduct field trials in different agro-climatic zones to evaluate the performance of the breeding lines under various conditions. This helps in selecting lines that are not only high yielding but also adaptable to different environments.
- **Disease screening**: Expose the breeding lines to prevalent diseases in both field and controlled environments. Select lines that consistently exhibit resistance, ensuring durability of the trait.

6. Selection and Stabilization

- **Trait stability**: Select plants that consistently exhibit high camphor content, disease resistance, and high yield across different environments and seasons.
- **Line purity**: Through selfing (self-pollination) and backcrossing, stabilize the desired traits within the breeding lines. This involves several generations of selection to fix the traits.

7. Chemical analysis

- **Camphor quantification**: Regularly test the camphor content in selected lines using GC-MS or other suitable methods to ensure that the trait is being effectively passed on and stabilized.
- **Essential oil profile**: Analyse the overall essential oil profile to ensure that other desirable compounds are retained or enhanced during the breeding process.

8. Yield optimization

- **Agronomic practices**: Implement and refine agronomic practices, such as planting density, fertilization, and irrigation, to maximize yield potential in the selected lines.
- **Harvesting techniques**: Optimize harvesting and post-harvest processing to ensure maximum retention of camphor and other essential oils.

9. Final selection and variety release

- **Performance testing**: Conduct final performance tests of the most promising lines against current commercial varieties to ensure superiority in camphor content, disease resistance, and yield.
- **Variety release**: Once a new variety meets the desired criteria, submit it for official variety trials and registration. Upon approval, the variety can be released for commercial cultivation.

10. Seed production and distribution

- **Quality seed production**: Produce and maintain high-quality seed material to ensure that the genetic purity of the new variety is preserved. This includes multiplication and distribution to farmers.
- **Farmer training**: Educate farmers on the specific cultivation practices required to maximize the benefits of the new variety, including disease management and harvest timing for optimal camphor content.

Integration of modern techniques

- **CRISPR/Cas9 (Gene editing)**: If applicable, use gene-editing technologies to directly modify genes responsible for camphor synthesis or disease resistance, further speeding up the breeding process.
- **Tissue culture and micro-propagation**: Use these techniques to rapidly multiply selected lines, especially those with high camphor content, ensuring uniformity and quality.

By combining traditional breeding methods with modern techniques like MAS, genomic selection, and advanced chemical analysis, it is possible to develop superior black turmeric varieties that meet the demands for high camphor content, disease resistance, and high yield.

Medicinal value of Black Turmeric

Black turmeric (*Curcuma caesia*), also known as Kali Haldi, is a rare and medicinally significant plant that belongs to the Zingiberaceae family. It is primarily found in India and is renowned for its unique properties, particularly in traditional medicine. Here's an overview of its medicinal value:

1. Anti-inflammatory properties

- Black turmeric contains curcumin, which is known for its potent anti-inflammatory effects. This makes it useful in treating conditions like arthritis, joint pain, and other inflammatory disorders.

2. Antioxidant activity

- The rhizomes of black turmeric are rich in antioxidants, which help neutralize free radicals in the body. This can help reduce oxidative stress and protect cells from damage, contributing to overall health and the prevention of chronic diseases.

3. Analgesic effects

- Traditional use of black turmeric includes pain relief. It is used in treating headaches, toothaches, and muscle pain due to its natural analgesic properties.

4. Respiratory health

- Black turmeric has been traditionally used to treat respiratory issues such as asthma, bronchitis, and coughs. Its anti-inflammatory and expectorant properties help in clearing respiratory pathways and alleviating symptoms.

5. Digestive health

- The rhizome is known to aid digestion and is used in the treatment of digestive disorders such as bloating, indigestion, and colic. It also has carminative properties that help in expelling gas from the intestines.

6. Antimicrobial and antifungal properties

- Black turmeric has been shown to have antimicrobial and antifungal activities, making it effective in treating infections. It is traditionally applied to wounds and cuts to prevent infections and promote healing.

7. Skin health

- It is used in traditional medicine for various skin conditions due to its anti-inflammatory, antimicrobial, and wound-healing properties. It is often applied topically to treat skin issues like eczema, acne, and fungal infections.

8. Cancer prevention

- Preliminary research suggests that the curcuminoids in black turmeric may have anti-cancer properties, helping to inhibit the growth of certain cancer cells. However, more research is needed in this area to fully understand its potential.

9. Immune system support

- Due to its antioxidant and anti-inflammatory properties, black turmeric is believed to boost the immune system, helping the body to fight off infections and maintain overall health.

10. Blood circulation

- In traditional medicine, black turmeric is used to improve blood circulation, which can be beneficial in treating conditions like anemia and improving cardiovascular health.

11. Anti-allergic properties

- It has been used traditionally to treat allergic reactions, including skin allergies and asthma, due to its ability to modulate immune responses and reduce inflammation.

Usage in traditional medicine

- **Ayurveda**: Black turmeric is often used in Ayurvedic medicine for its healing properties, particularly in treating respiratory disorders, skin diseases, and joint pain.
- **Folk medicine**: It is also used in various forms in folk medicine practices across India, often in the form of pastes, powders, or decoctions.

While black turmeric has been used for centuries in traditional medicine, it's important to consult with a healthcare provider before using it for medicinal purposes, especially in cases of serious health conditions.

Chemical composition:

Black turmeric (*Curcuma caesia*) is known for its distinctive chemical composition, which contributes to its medicinal properties. The major chemical constituents in black turmeric include curcuminoids, essential oils, and various bioactive compounds. Here's a breakdown of the key components:

1. Curcuminoids

- **Curcumin**: The primary active compound in turmeric, known for its potent anti-inflammatory, antioxidant, and anticancer properties.
- **Demethoxycurcumin**: A derivative of curcumin, contributing to the anti-inflammatory and antioxidant effects.
- **Bisdemethoxycurcumin**: Another curcumin derivative with similar medicinal properties.

2. Essential oils

- **Camphor**: A major component in black turmeric's essential oil, known for its analgesic, anti-inflammatory, and antimicrobial properties.
- **Ar-turmerone**: This sesquiterpene is also found in the essential oil and has anti-inflammatory and anticancer activities.
- **Zingiberene**: Contributes to the anti-inflammatory and antimicrobial effects.
- **Curcumol**: Another sesquiterpene alcohol with potential therapeutic effects.
- **1,8-Cineole**: Known for its bronchodilatory and antimicrobial properties.

3. Phenolic compounds

- **Tetrahydrocurcumin**: A reduced form of curcumin with enhanced antioxidant properties.
- **Ferulic acid**: An antioxidant that helps in reducing oxidative stress and has anti-inflammatory properties.
- **Caffeic acid**: Known for its antioxidant, anti-inflammatory, and antimicrobial effects.

4. Flavonoids

- Flavonoids present in black turmeric contribute to its antioxidant and anti-inflammatory properties. They help in scavenging free radicals and protecting cells from damage.

5. Glycosides

- These compounds have various bioactivities, including anti-inflammatory and antimicrobial effects. They also contribute to the bitter taste of black turmeric.

6. Terpenes

- **β-Sitosterol**: A plant sterol with anti-inflammatory and immune-modulating properties.
- **Limonene**: Known for its anti-inflammatory and antioxidant effects.

7. Alkaloids

- Alkaloids in black turmeric contribute to its medicinal properties, particularly its antimicrobial and analgesic effects.

8. Tannins

- Tannins possess astringent properties and are known for their antimicrobial and antioxidant activities.

9. Resins

- Resins in black turmeric contribute to its medicinal effects, including anti-inflammatory and wound-healing properties.

Summary of key components

- **Curcuminoids**: Curcumin, demethoxycurcumin, bisdemethoxycurcumin.
- **Essential oils**: Camphor, ar-turmerone, zingiberene, curcumol, 1,8-cineole.
- **Phenolic compounds**: Tetrahydrocurcumin, ferulic acid, caffeic acid.
- **Flavonoids**: Various flavonoids with antioxidant properties.
- **Terpenes and Sterols**: β-Sitosterol, limonene.
- **Alkaloids and Tannins**: Contributing to antimicrobial and anti-inflammatory effects.

These components collectively contribute to the medicinal properties of black turmeric, making it valuable in traditional and modern medicine. The bioactive compounds provide a wide range of therapeutic benefits, including anti-inflammatory, antioxidant, antimicrobial, and anticancer effects.

Chemical extraction techniques with equipment, value and availability

Chemical extraction of bioactive compounds from black turmeric (*Curcuma caesia*) involves various techniques, each with specific equipment, efficiency, and availability. The choice of extraction method depends on the desired compounds (e.g., curcuminoids, essential oils), available resources, and intended use. Here's an overview of common extraction techniques, the equipment required, their value, and availability:

1. Solvent extraction

- **Overview**: Solvent extraction involves using solvents like ethanol, methanol, hexane, or acetone to extract bioactive compounds from black turmeric.
- **Equipment**:
  - **Soxhlet apparatus**: A setup for continuous extraction where the solvent is repeatedly cycled through the plant material.
  - **Rotary evaporator**: Used to concentrate the extract by evaporating the solvent under reduced pressure.
  - **Ultrasonic bath (Optional)**: Enhances extraction efficiency by using ultrasound waves to break down plant cells.
- **Value**: Highly effective for extracting a wide range of compounds, especially curcuminoids and essential oils. It's a standard technique in both research and industrial applications.
- **Availability**: Widely available in laboratories and industries. Solvents and equipment are easily accessible, though the choice of solvent must consider the compound's polarity and target extraction.

2. Supercritical Fluid Extraction (SFE)

- **Overview**: Uses supercritical CO_2 as a solvent to extract essential oils and other non-polar compounds. It's a green extraction method, reducing the need for organic solvents.
- **Equipment**:
  - **Supercritical CO_2 extractor**: Includes a CO_2 pump, extraction vessel, and separator where CO_2 is converted back to a gas, leaving the extract behind.
- **Value**: Highly efficient, especially for extracting essential oils and other volatile compounds without thermal degradation. The method is environmentally friendly.
- **Availability**: Specialized equipment is required, which can be expensive and less accessible than traditional methods. Typically found in well-equipped research institutions and industries focusing on high-purity extracts.

3. Steam distillation

- **Overview**: Commonly used for extracting essential oils from black turmeric. Steam passes through the plant material, vaporizing volatile compounds, which are then condensed back into liquid form.

- **Equipment**
  - **Distillation unit**: Includes a steam generator, distillation flask, condenser, and collection flask.
- **Value**: Ideal for extracting volatile essential oils with minimal degradation. The process is relatively simple and cost-effective.
- **Availability**: Widely available and commonly used in both small-scale and large-scale operations. Equipment is easy to obtain and use.

4. Microwave-Assisted Extraction (MAE)

- **Overview**: Uses microwave energy to heat the plant material and solvent, enhancing the extraction efficiency by disrupting cell structures.
- **Equipment**:
  - **Microwave extractor**: A specialized microwave unit with controlled power and temperature settings, often combined with solvent extraction techniques.
- **Value**: Provides rapid and efficient extraction, especially for heat-sensitive compounds. It reduces solvent use and extraction time.
- **Availability**: Less common than traditional methods, but increasingly available in advanced research labs. Equipment is moderately expensive.

5. Ultrasound-assisted extraction (UAE)

- **Overview**: Uses ultrasonic waves to create cavitation bubbles in the solvent, which burst and disrupt plant cells, enhancing the release of bioactive compounds.
- **Equipment**
  - **Ultrasonic bath or probe**: Provides ultrasonic waves, usually combined with a solvent extraction setup.
- **Value**: Enhances extraction efficiency, reduces solvent use, and shortens extraction time. It is especially useful for extracting heat-sensitive compounds.
- **Availability**: Increasingly available in research laboratories and some industrial settings. Equipment is moderately priced and more accessible than some advanced methods.

6. Hydrodistillation

- **Overview**: A traditional method similar to steam distillation but involves boiling the plant material directly in water. The vapors are condensed to separate the essential oils.
- **Equipment**:

- **Clevenger apparatus**: A setup for hydrodistillation that includes a distillation flask, condenser, and collection tube.
- **Value**: Simple and cost-effective, particularly for extracting essential oils. However, it may cause thermal degradation of some compounds.
- **Availability**: Commonly used in small-scale operations and educational settings. The equipment is widely available and inexpensive.

Summary of techniques

Extraction Technique	Target Compounds	Equipment	Value	Availability
Solvent Extraction	Curcuminoids, Essential Oils	Soxhlet, Rotary Evaporator	Versatile, effective	Widely available
Supercritical Fluid Extraction (SFE)	Essential Oils, Non-polar compounds	Supercritical CO_2 Extractor	High purity, eco-friendly	Specialized, less common
Steam Distillation	Essential Oils	Distillation Unit	Simple, cost-effective	Widely available
Microwave-Assisted Extraction (MAE)	Curcuminoids, Oils	Microwave Extractor	Rapid, efficient	Moderately available
Ultrasound-Assisted Extraction (UAE)	Various Bioactive Compounds	Ultrasonic Bath/ Probe	Efficient, low solvent	Increasingly available
Hydrodistillation	Essential Oils	Clevenger Apparatus	Traditional, inexpensive	Commonly available

Each technique has its strengths and is chosen based on the specific needs of the extraction process, the nature of the target compounds, and the available resources.

K) Case study: Breeding for high quality and yield improvement in coriander varietal development

Breeding coriander (*Coriandrum sativum*) for desirable traits such as high yield, high oleoresin content, disease resistance, late bolting, and high oil content involves a strategic approach that combines traditional breeding methods with modern genetic tools. Below are the key steps and considerations for developing coriander varieties with these traits:

1. Germplasm collection and evaluation

- **Diversity selection**: Start by collecting a diverse range of coriander germplasm, including landraces, wild relatives, and existing cultivars.
- **Trait evaluation**: Screen the germplasm for the target traits: yield, oleoresin content, disease resistance, bolting time, and oil content. This involves both field trials and laboratory analysis.

2. Hybridization

- **Crossbreeding**: Perform controlled crosses between selected parents that exhibit complementary traits. For example, cross a high-yield variety with one that has high oleoresin content or disease resistance.
- **Trait combination**: Aim to combine multiple desirable traits in the progeny through successive generations of hybridization.

3. Selection

- **Phenotypic selection**: Select plants that exhibit the desired traits in each generation. For yield and bolting resistance, this would typically be done in the field, while oleoresin and oil content require laboratory assays.
- **Genotypic selection**: Use molecular markers linked to desired traits (e.g., disease resistance genes, markers for oil biosynthesis) to assist in selecting the best progeny.

4. Incorporating disease resistance

- **Pathogen screening**: Subject the breeding lines to known pathogens like Fusarium wilt or powdery mildew under controlled conditions to identify resistant plants.
- **Pyramiding resistance genes**: Combine multiple resistance genes from different sources into a single variety to ensure broad-spectrum and durable disease resistance.

5. Late bolting selection

- **Bolting observation**: Identify and select plants that bolt late under varying environmental conditions. Bolting can be influenced by day length and temperature, so it's important to test in different growing conditions.
- **Genetic control**: If possible, identify and select for genetic markers associated with late bolting.

6. High oleoresin and oil content

- **Chemical analysis**: Measure oleoresin and oil content using gas chromatography or similar techniques. Select high-performing individuals.
- **Metabolomic selection**: If available, use markers associated with high oleoresin and oil biosynthesis to aid in selection.

7. Yield optimization

- **Agronomic trials**: Conduct multi-location trials to evaluate yield potential under different environmental conditions. Select lines that consistently perform well.

- **Selection for robustness**: Ensure that selected lines maintain high yield across varying environments and are resilient to abiotic stresses like drought or heat.

8. Stabilization and release

- **Line purity**: After several generations of selection, stabilize the best-performing lines through selfing or backcrossing.
- **Varietal trials**: Conduct official trials to evaluate the new varieties against current standards. Only those meeting or exceeding these standards for yield, oleoresin, oil content, disease resistance, and bolting time should be considered for release.

9. Commercialization

- **Seed production**: Once the variety is approved, produce high-quality seeds for distribution.
- **Farmer adoption**: Educate farmers on the benefits of the new variety and provide agronomic practices to maximize its potential.

Integration of genomic tools

- **Marker-assisted selection (MAS)**: Use MAS to accelerate breeding by selecting plants at the seedling stage based on their genetic makeup.
- **Genomic selection**: If genomic resources are available, apply genomic selection techniques to predict and select for complex traits like yield and oleoresin content more efficiently.

By combining traditional breeding methods with modern genomic tools, breeders can develop coriander varieties that meet the demands of both farmers and the spice industry. This approach not only enhances the efficiency of the breeding process but also increases the likelihood of developing superior varieties with the desired traits.

L) Case Study: Breeding for high quality and yield improvement in fennel varietal development

Fennel (*Foeniculum vulgare*) is a highly valued aromatic and medicinal plant widely used for its culinary and therapeutic properties. Improving the quality and yield of fennel through effective breeding strategies is essential to meet the growing demand and enhance its economic value. This case study explores the approaches and considerations involved in developing high-quality, high-yielding fennel varieties.

1. Introduction

- **Economic and culinary importance**: Fennel is used globally as a spice, vegetable, and in traditional medicine. Its seeds and essential oils are particularly valued for flavouring and therapeutic uses.
- **Need for improvement**: Enhancing yield and quality traits such as oil content, flavour profile, and resistance to pests and diseases is crucial for maximizing productivity and marketability.

2. Breeding Objectives

The primary objectives in fennel breeding for quality and yield improvement include:

- **Increased seed and biomass yield**: Developing varieties that produce higher quantities of seeds and vegetative parts.
- **Enhanced essential oil content and composition**: Optimizing the concentration and balance of key constituents like anethole and fenchone to improve flavour and medicinal properties.
- **Improved morphological traits**: Selecting for desirable plant height, branching patterns, and uniformity for ease of cultivation and harvesting.
- **Disease and pest resistance**: Incorporating resistance to common pathogens and pests to reduce crop losses and dependence on chemical controls.
- **Abiotic stress tolerance**: Enhancing tolerance to environmental stresses such as drought, salinity, and temperature extremes to ensure stable yields across diverse growing conditions.

3. Breeding Strategies and Methods

Several breeding methods are employed to achieve the desired improvements in fennel varieties:

3.1. Selection methods

- **Mass selection**: Involves selecting and propagating plants with superior traits over successive generations. Effective for improving traits with high heritability.
- **Pure line selection**: Selection of the best-performing individual plants to establish uniform lines, ensuring consistency in quality and yield.
- **Clonal selection**: Useful for propagating vegetatively reproduced fennel varieties, maintaining genetic fidelity of desirable traits.

3.2. Hybridization

- **Intraspecific hybridization**: Crossing different fennel varieties to combine desirable traits from both parents, creating hybrids with improved performance.
- **Recurrent selection and backcrossing**: Techniques used to introduce specific traits (e.g., disease resistance) from donor parents while retaining the overall genetic background of the elite variety.

3.3. Mutation breeding

- **Induced mutations**: Utilizing physical or chemical mutagens to create genetic variability, followed by selection of mutants exhibiting improved traits such as higher oil content or dwarf stature.

3.4. Biotechnological approaches

- **Tissue culture and micropropagation**: Facilitates rapid multiplication of superior genotypes and aids in maintaining disease-free planting material.
- **Molecular Marker-Assisted selection**: Accelerates breeding by identifying and selecting for genes associated with desired traits, enhancing efficiency and precision.

4. Evaluation and testing

- **Field trials**: Conducting multi-location trials to assess the performance of new varieties under different environmental conditions and management practices.
- **Quality assessment**: Analyzing essential oil profiles, seed quality, and other biochemical parameters to ensure compliance with industry standards.
- **Stress screening**: Evaluating resistance and tolerance levels against various biotic and abiotic stresses to ensure robustness and adaptability.

5. Challenges and Considerations

- **Genetic diversity**: Maintaining and utilizing a broad genetic base is essential to prevent vulnerability to diseases and adapt to changing environmental conditions.
- **Environmental Impact**: Breeding efforts should consider sustainability, promoting varieties that require fewer inputs and support ecological balance.
- **Market demands**: Aligning breeding objectives with consumer preferences and industry requirements to ensure the economic viability of new varieties.
- **Regulatory compliance**: Ensuring that developed varieties meet the regulatory standards for cultivation and trade, including safety and quality certifications.

6. Success Stories and Examples

- **Improved varieties**: Several high-yielding and superior quality fennel varieties have been developed through these breeding strategies, leading to increased productivity and farmer income.
- **Collaborative programs**: Success has often been achieved through collaboration between research institutions, universities, and agricultural organizations, combining expertise and resources.

Breeding for high quality and yield improvement in fennel involves a comprehensive approach integrating traditional and modern techniques. Continuous research and innovation are essential to develop varieties that meet the evolving challenges and demands of agriculture and the marketplace.

Varieties of fennel

Fennel (*Foeniculum vulgare*) has several varieties that are cultivated for different purposes, including culinary, medicinal, and ornamental uses. Below are some of the notable varieties of fennel:

**1. Sweet Fennel (*Foeniculum vulgare var. dulce*)

- **Characteristics**: This is the most commonly cultivated variety of fennel, known for its sweet, anise-like flavour. It is primarily grown for its seeds, which are used as a spice, and its feathery leaves, which are used as an herb.
- **Uses**: The seeds are used in cooking, baking, and as a flavouring agent in liqueurs. The leaves are often used in salads, soups, and as a garnish.

**2. Florence Fennel (*Foeniculum vulgare var. azoricum*)

- **Characteristics**: Also known as “bulb fennel” or “finocchio,” this variety is grown for its large, bulbous stem base, which is eaten as a vegetable. The bulb has a crisp texture and a mild, sweet flavour.
- **Uses**: The bulb is commonly used in salads, grilled, roasted, or sautéed. The leaves and seeds can also be used for culinary purposes.

**3. Bitter Fennel (*Foeniculum vulgare var. vulgare*)

- **Characteristics**: This variety is often grown for medicinal purposes rather than culinary use. It has a more intense flavour and is sometimes used to produce fennel oil.
- **Uses**: Bitter fennel seeds and oil are used in traditional medicine for digestive issues and as a carminative. It is also used in some industrial applications.

**4. Bronze Fennel (*Foeniculum vulgare 'Purpureum'*)

- **Characteristics**: An ornamental variety, Bronze Fennel has attractive, bronze-coloured foliage and is often grown in gardens for its aesthetic appeal.
- **Uses**: While it can be used culinarily like sweet fennel, it is more commonly grown for decorative purposes. The seeds and leaves are still edible.

**5. Perennial Fennel (*Foeniculum vulgare var. piperitum*)

- **Characteristics**: This variety is less commonly grown and has a stronger, more peppery flavour compared to sweet fennel.
- **Uses**: Primarily used for its seeds, which are spicier and are used in some traditional dishes and medicinal preparations.

6. Giant Fennel (*Ferula communis*)

- **Characteristics**: Although not a true fennel (*Foeniculum* species), Giant Fennel is often grouped with fennel varieties due to its similar appearance. It is a large plant with yellow flowers and hollow stems.
- **Uses**: Traditionally used in Mediterranean regions for its fibrous stems and medicinal properties.

These varieties of fennel offer a range of flavours, uses, and growing conditions, making them versatile plants for both gardeners and commercial growers. Whether for culinary, medicinal, or ornamental purposes, there is a fennel variety suited to most needs.

High yielding varieties of fennel developed in India

India has developed several high-yielding varieties of fennel (*Foeniculum vulgare*) through various agricultural research institutions to meet the growing demand for this spice. These varieties have been bred to improve yield, essential oil content, and resistance to pests and diseases. Below are some of the prominent high-yielding fennel varieties developed in India:

**1. RF-101

- **Developed by**: Rajasthan Agricultural Research Institute, Jaipur
- **Characteristics**: This variety is known for its high yield and good quality seeds with a high essential oil content.
- **Yield potential**: Approximately 20-25 quintals per hectare.
- **Maturity**: Medium-duration variety, typically harvested around 160-170 days after sowing.
- **Special features**: Suitable for cultivation in semi-arid regions, with good resistance to common diseases.

**2. Gujarat Fennel-1 (GF-1)

- **Developed by**: Anand Agricultural University, Gujarat
- **Characteristics**: GF-1 is a popular variety in Gujarat known for its high yield and good seed quality.
- **Yield potential**: Approximately 18-22 quintals per hectare.
- **Maturity**: Early to medium maturity, harvested around 150-160 days after sowing.
- **Special features**: Adapted to the climatic conditions of Gujarat, with a strong aroma and good essential oil content.

**3. RF-125

- **Developed by**: Rajasthan Agricultural Research Institute, Jaipur
- **Characteristics**: This variety produces a high yield of seeds with a high essential oil percentage, making it ideal for both seed and oil production.
- **Yield potential**: Around 25-28 quintals per hectare.
- **Maturity**: Medium to late maturity, with a harvesting period of around 170-180 days.
- **Special features**: Exhibits good drought tolerance and is resistant to major pests and diseases.

**4. Azad Saunf-1

- **Developed by**: Narendra Deva University of Agriculture and Technology, Faizabad, Uttar Pradesh
- **Characteristics**: A high-yielding variety with high essential oil content and large, bold seeds.
- **Yield potential**: Around 20-25 quintals per hectare.
- **Maturity**: Medium-duration variety, maturing in about 160-170 days.
- **Special features**: Suitable for the agro-climatic conditions of Uttar Pradesh and neighbouring regions.

**5. Hisar Swarup (HF-101)

- **Developed by**: CCS Haryana Agricultural University, Hisar
- **Characteristics**: This variety is known for its uniform and high yield, as well as good quality seeds with high essential oil content.
- **Yield potential**: Approximately 18-22 quintals per hectare.
- **Maturity**: Early to medium maturity, around 150-160 days.
- **Special features**: Performs well in the agro-climatic conditions of Haryana and similar regions.

**6. Ajmer Fennel-1 (AF-1)

- **Developed by**: National Research Centre on Seed Spices, Ajmer, Rajasthan
- **Characteristics**: A high-yielding variety with high essential oil content and good seed quality.
- **Yield potential**: Around 22-25 quintals per hectare.
- **Maturity**: Medium to late maturity, with harvesting around 160-170 days.
- **Special features**: Adapted to the arid and semi-arid regions of Rajasthan, with good tolerance to drought and diseases.

These high-yielding fennel varieties have been developed to cater to different agro-climatic conditions across India, providing farmers with options that suit their local environments and market needs. The focus on essential oil content and seed quality ensures that these varieties are not only productive but also meet the quality standards required for both domestic consumption and export markets.

M) Case study: Breeding a high-yield, abiotic and biotic stress-resistant, high-aroma, and high-quality cumin variety

The goal is to develop a cumin (*Cuminum cyminum*) variety that combines high yield, resistance to abiotic (drought, salinity, heat) and biotic stresses (pests, diseases), with high essential oil content, specifically focusing on aroma and overall spice quality.

1. Background

Cumin is an essential spice crop, particularly valued for its aromatic seeds. However, cumin cultivation faces several challenges:

- **Abiotic stresses:** Drought, heat, and salinity.
- **Biotic stresses:** Fungal diseases like wilt caused by Fusarium oxysporum and blight, as well as pests like aphids.
- **Market demand:** High essential oil content with a strong, desirable aroma is crucial for market acceptance, alongside yield and seed quality.

2. Germplasm collection and evaluation

- **Germplasm sources:** Collect a diverse range of cumin germplasm from different geographic regions, including wild relatives, landraces, and existing cultivars known for stress tolerance and high essential oil content.
- **Evaluation:** Screen germplasm for:
  - **Abiotic stress tolerance:** Test under simulated drought, salinity, and high-temperature conditions.
  - **Biotic stress resistance:** Evaluate resistance to *Fusarium* wilt, blight, and common pests.

- **Quality traits:** Analyse essential oil content, focusing on the concentration of key compounds like cuminaldehyde, and assess seed size, colour, and flavour.
- **Yield:** Measure seed yield per plant and overall biomass under both normal and stressed conditions.

3. Breeding strategy

Hybridization

- **Crossing strategy:** Hybridize stress-resistant and high-yielding genotypes with those possessing high essential oil content and desirable aroma.
    - **Example crosses:** Cross a drought-tolerant line with a high-aroma, high-oil-content line. Similarly, cross a *Fusarium*-resistant line with a high-yielding line.
- **Selection:** Use pedigree selection to track and select progeny that exhibit the best combination of traits. Early-generation selection can focus on stress tolerance, while later generations can focus on yield and quality.

Marker-assisted Selection (MAS)

- **Molecular markers:** Develop or utilize existing markers linked to key traits like *Fusarium* wilt resistance, drought tolerance, and essential oil content.
    - **Selection:** Use MAS to quickly identify and select plants carrying desired alleles, reducing the need for extensive field trials in early generations.

Biotechnological approaches

- **Mutation breeding:** Use induced mutations to create new genetic variations that might confer resistance to stress or enhance essential oil content.
- **CRISPR/Cas9:** Explore the potential of genome editing for targeted improvements, such as knocking out genes that limit stress tolerance or enhancing genes involved in essential oil biosynthesis.

4. Field trials and multi-location testing

- **On-Farm trials:** Conduct trials in different agro-climatic zones to evaluate performance under real-world conditions, including stress environments.
- **Controlled environment testing:** Simulate stress conditions in greenhouse trials to validate field results, particularly for abiotic stress tolerance.
- **Data collection:** Collect data on yield, stress resistance, essential oil content, and quality traits across multiple locations and seasons.

5. Quality Improvement

- **Essential oil content:** Focus on increasing the concentration of key aromatic compounds like cuminaldehyde, which contribute to the characteristic aroma of cumin.
- **Quality testing:** Conduct sensory evaluations and chemical analyses to ensure that selected lines meet the desired aroma and flavour profiles.
- **Post-harvest processing:** Evaluate how different genotypes respond to post-harvest processing, as this can affect both aroma and marketable quality.

6. Release of a new variety

- **Naming and registration:** Once a superior variety is developed, it is registered and named according to regional or national guidelines.
- **Seed production:** Begin large-scale seed production of the selected variety, ensuring purity and availability for farmers.
- **Farmer training and adoption:** Conduct extension programs to educate farmers on the benefits and cultivation practices of the new variety, including stress management and optimal harvest techniques.

7. Development of the 'GC-4' Cumin variety

- **Background:** Assume 'GC-4' is a fictitious example of a new cumin variety developed using the above strategies.
- **Key traits:**
  - **Abiotic stress tolerance:** Demonstrated tolerance to drought and high temperatures, making it suitable for arid regions.
  - **Biotic stress resistance:** Resistant to Fusarium wilt and blight, significantly reducing losses due to disease.
  - **High yield:** Produces 15-20% higher yield compared to traditional varieties under both normal and stressed conditions.
  - **High aroma and quality:** Contains 30% more essential oil with a strong, preferred aroma, and has uniform seed size and colour.
- **Farmer adoption:** 'GC-4' gains popularity among farmers due to its resilience and high market value, leading to increased cultivation and higher incomes in cumin-growing regions.

8. Continuous improvement

- **Breeding cycle:** Continue backcrossing with elite parents to incorporate new traits or improve existing ones.
- **Emerging stresses:** Monitor the variety's performance and update the breeding program to address any new biotic or abiotic challenges.

- **Quality feedback:** Gather market feedback on aroma and quality, and use it to refine future breeding efforts.

The development of a high-yielding, stress-resistant, high-aroma cumin variety like 'GC-4' requires an integrated approach combining traditional breeding, molecular techniques, and biotechnological innovations. The success of such a program hinges on thorough germplasm evaluation, strategic hybridization, and real-world testing, ultimately leading to a variety that meets both farmer and market needs.

Challenges in cumin breeding

Breeding cumin (*Cuminum cyminum*) presents several challenges due to the crop's biological characteristics, environmental sensitivity, and the demands of the market. Here are some of the primary challenges faced in cumin breeding:

1. Abiotic stress sensitivity

- **Drought and heat stress:** Cumin is typically grown in arid and semi-arid regions, making it highly susceptible to drought and heat stress. Breeding for tolerance to these conditions is challenging because these traits are complex and often controlled by multiple genes.
- **Salinity:** Soil salinity is another common issue in cumin-growing regions, particularly in areas with poor irrigation practices. Developing salt-tolerant varieties is difficult due to the limited genetic variability for salinity tolerance in the available germplasm.

2. Biotic stress resistance

- **Diseases:** Cumin is vulnerable to several diseases, particularly Fusarium wilt, Alternaria blight, and powdery mildew. Breeding for disease resistance is complicated because:
    - **Complex pathogen races:** The pathogens causing these diseases have multiple races, making it difficult to develop varieties with broad-spectrum resistance.
    - **Limited resistance sources:** There are few known sources of strong resistance in the existing cumin germplasm, necessitating the exploration of wild relatives or the use of advanced biotechnological approaches.
- **Pests:** Cumin is also susceptible to pests like aphids, which can transmit viral diseases. Breeding for pest resistance is less common, and integrating it with disease resistance poses an additional challenge.

3. Quality traits

- **Essential oil content:** Cumin's market value largely depends on its essential oil content, particularly the concentration of compounds like cuminaldehyde. However, breeding for high essential oil content while maintaining other desirable agronomic traits is challenging because:

- **Trade-Offs:** There is often a trade-off between high yield and high essential oil content, making it difficult to improve both traits simultaneously.
- **Complex Inheritance:** The biosynthesis of essential oils is controlled by complex biochemical pathways, and the inheritance of these traits can be polygenic, complicating selection.

4. Yield improvement

- **Low genetic variability:** Cumin has relatively low genetic variability in terms of yield potential. This limits the opportunities for significant yield improvements through traditional breeding methods.
- **Environmental sensitivity:** Cumin yield is highly influenced by environmental conditions, particularly during flowering and seed setting. Developing varieties that consistently perform well across different environments is difficult.

5. Reproductive challenges

- **Cross-pollination:** While cumin is primarily self-pollinated, cross-pollination can occur, leading to genetic mixing in seed production fields. This can result in variability in the offspring, making it challenging to maintain the genetic purity of improved varieties.
- **Flowering and seed set:** Cumin has a relatively short flowering period, which can be affected by environmental conditions. Poor seed set under unfavorable conditions limits yield potential.

6. Breeding for multiple traits

- **Complexity of multi-trait breeding:** Breeding for multiple traits, such as stress resistance, yield, and quality, is inherently complex. Balancing these traits requires careful selection and often involves compromises, where improving one trait might negatively impact another.
- **Linkage drag:** When introducing resistance genes from wild relatives or other sources, undesirable traits (linkage drag) can be carried along with the desired genes, complicating the breeding process.

7. Genomic resources

- **Limited molecular markers:** The availability of molecular markers linked to important traits in cumin is limited, slowing down the application of marker-assisted selection (MAS) and other modern breeding techniques.

- **Incomplete genomic information:** Compared to other major crops, cumin has less genomic information available, which hampers the ability to apply advanced genomic tools in breeding programs.

8. Seed production and propagation

- **Seed quality:** Ensuring high-quality seed production is crucial for successful cumin cultivation. However, issues like seed dormancy, variability in germination rates, and seed borne diseases can complicate the production and distribution of quality seeds.
- **Genetic purity:** Maintaining genetic purity in seed production is challenging, especially in regions where traditional varieties are still widely grown and can cross with improved varieties.

9. Economic and market challenges

- **Market demand fluctuations:** The economic viability of breeding programs is influenced by market demand, which can fluctuate based on global spice markets. This affects the willingness of both public and private sectors to invest in long-term breeding efforts.
- **Quality standards:** Meeting stringent quality standards for aroma and essential oil content required by international markets can be challenging, especially when breeding for other traits like stress resistance.

10. Climate change

- **Changing climate conditions:** Climate change introduces new challenges in cumin breeding, as it affects the predictability of environmental conditions, introduces new stresses, and potentially alters the prevalence and severity of diseases and pests.

Breeding cumin that meets the demands for high yield, stress resistance, and quality is a complex and challenging process. It requires integrating traditional breeding methods with modern biotechnological tools and a deep understanding of the crop's biology and environmental interactions. Overcoming these challenges is crucial for improving cumin production and ensuring the sustainability of this valuable spice crop in the face of evolving agricultural and climatic conditions.

Future strategies of cumin breeding:

The future of cumin breeding will likely involve integrating advanced technologies, multidisciplinary approaches, and sustainable practices to address current challenges and meet future demands. Here are some key strategies that could shape the future of cumin breeding:

1. **Genomic-Assisted breeding**

- **Whole-Genome sequencing:** As more genomic data becomes available, whole-genome sequencing of cumin can be used to identify key genes responsible for desirable traits, such as stress resistance, yield, and essential oil content.
- **Genomic selection:** Genomic selection (GS) can be employed to accelerate breeding cycles. By predicting the breeding value of plants based on genomic data, breeders can select the best candidates for crossing or direct cultivation, even before phenotypic traits are fully expressed.
- **Marker-assisted selection (MAS):** The development and use of more molecular markers linked to key traits (e.g., disease resistance, drought tolerance) will enhance the efficiency of breeding programs. MAS allows for the early and precise selection of plants carrying desirable traits, reducing the reliance on lengthy field trials.

2. **Biotechnological approaches**

- **CRISPR/Cas9 and Gene editing:** Gene editing technologies like CRISPR/Cas9 offer the potential to directly modify or knock out specific genes responsible for undesirable traits or to enhance traits such as stress tolerance and oil biosynthesis. This approach allows for precise genetic improvements without introducing foreign DNA.
- **Tissue culture and somatic embryogenesis:** Tissue culture techniques will be increasingly used for the rapid multiplication of elite genotypes and the propagation of genetically uniform plants. Somatic embryogenesis can be particularly useful for producing disease-free planting material.

3. **Breeding for climate resilience**

- **Drought and heat tolerance:** With climate change increasing the frequency and severity of droughts and heat waves, breeding for enhanced tolerance to these stresses will be a major focus. This could involve selecting for traits such as deeper root systems, improved water-use efficiency, and heat-tolerant flowering and seed-setting mechanisms.
- **Salinity tolerance:** As soil salinization becomes more prevalent, particularly in irrigated agricultural systems, developing cumin varieties that can thrive in saline conditions will be critical. This may involve exploring genetic diversity in wild relatives or using advanced biotechnological approaches.

4. **Integrating traditional and modern breeding techniques**

- **Participatory breeding:** Engaging farmers in the breeding process will ensure that the resulting varieties meet the specific needs of different regions. Combining traditional knowledge with modern breeding techniques can result in varieties that are more widely adopted.

- **Precision breeding:** Leveraging advanced phenotyping tools and data analytics, precision breeding will allow breeders to identify and select for complex traits with greater accuracy. This includes using drones, remote sensing, and machine learning to monitor plant growth, stress responses, and yield in real-time.

5. Exploring genetic diversity

- **Utilization of wild relatives:** Wild relatives of cumin may possess valuable traits such as disease resistance, stress tolerance, and better adaptation to extreme environments. Breeding programs will increasingly focus on introgressing these traits into cultivated varieties.
- **Pre-breeding programs:** Pre-breeding initiatives will focus on creating a broader genetic base by incorporating diverse germplasm into breeding pools. This genetic diversity is essential for developing resilient varieties capable of withstanding future challenges.

6. Enhancing essential oil content and quality

- **Metabolic engineering:** Understanding the metabolic pathways involved in essential oil biosynthesis will enable targeted breeding or genetic modification to enhance the quantity and quality of essential oils in cumin. This could involve increasing the concentration of key compounds like cuminaldehyde, which is crucial for aroma and flavour.
- **Quality trait selection:** Breeding programs will focus on improving sensory qualities (e.g., aroma, flavour) that meet specific market demands. Advanced analytical tools will be used to select for these traits more effectively.

7. Sustainable and Eco-friendly breeding practices

- **Low input varieties:** Future cumin breeding will emphasize the development of varieties that perform well with minimal inputs, such as water, fertilizers, and pesticides. This is crucial for sustainable agriculture, especially in resource-limited environments.
- **Organic breeding:** With the growing demand for organic products, there will be an increasing focus on developing cumin varieties that are well-suited to organic farming systems, including those with natural resistance to pests and diseases.

8. Digital breeding platforms

- **Big data and AI:** The integration of big data and artificial intelligence (AI) in breeding programs will allow for better data management, trait prediction, and decision-making. AI can help identify patterns and correlations in large datasets, improving the efficiency of selection processes.

- **Breeding management software:** Advanced software tools will facilitate the management of complex breeding programs, including the tracking of pedigrees, phenotypic data, and genomic information. This will streamline the breeding process and enhance collaboration among breeders.

9. Addressing emerging biotic stresses

- **New disease resistance:** As new strains of pathogens emerge, breeding programs will need to be agile in developing varieties with resistance to these threats. This might involve continuous monitoring, rapid identification of resistance genes, and their incorporation into breeding lines.
- **Integrated Pest Management (IPM) compatible varieties:** Breeding cumin varieties that are compatible with IPM strategies, such as those that deter pests or support beneficial insect populations, will be a focus in sustainable crop protection.

10. Global collaboration and knowledge sharing

- **International breeding programs:** Global collaboration will be key to overcoming the challenges in cumin breeding. Sharing germplasm, knowledge, and technologies across borders will accelerate the development of superior cumin varieties.
- **Capacity building:** Training and capacity building for breeders, especially in developing countries, will be essential to ensure that the latest breeding technologies and practices are accessible and effectively implemented.

The future of cumin breeding will be characterized by the integration of advanced genomic tools, biotechnological innovations, and sustainable practices. By focusing on resilience, quality, and sustainability, breeders can develop cumin varieties that meet the challenges of changing climates, market demands, and the need for environmentally friendly agriculture. This holistic approach will ensure the continued success and viability of cumin as a critical spice crop globally.

6

Challenges in Spices Breeding

A. Biotic and abiotic stresses in spices

1. Black pepper

1a. Challenges of biotic stress in Black pepper breeding

The status and future prospects of breeding for biotic stress resistance in black pepper are essential areas of research due to the significant impact of pests and diseases on black pepper crops. Biotic stress, caused by pathogens, pests, and weeds, can severely affect the yield and quality of black pepper, making the development of resistant varieties a critical objective for breeders.

Current status

1. **Research efforts**: There have been considerable efforts to identify and develop black pepper varieties that are resistant to various biotic stresses. These include resistance to diseases like ***Phytophthora* foot rot** (caused by *Phytophthora capsici*), **quick wilt, nematode infections**, and **pest attacks** like the pollu beetle.
2. **Breeding techniques**: Traditional breeding methods have been complemented by molecular breeding techniques. Marker-assisted selection (MAS) is increasingly used to accelerate the development of resistant varieties. The integration of **genomic tools** has also enhanced the efficiency of identifying resistance genes and their incorporation into high-yielding varieties.
3. **Resistance sources**: Several resistant genotypes have been identified, but their commercial success is still limited. The genetic diversity within black pepper and related species is explored to find novel resistance genes that can be introduced into cultivated varieties.

Future prospects

1. **Biotechnological approaches**: The future of breeding for biotic stress resistance in black pepper lies heavily in the adoption of **biotechnological tools**. **Gene editing technologies like CRISPR/Cas9** could provide precise modifications to enhance resistance traits without compromising other desirable characteristics of the crop.

2. **Climate change adaptation**: As climate change affects the prevalence and virulence of pests and diseases, breeding programs will need to incorporate **climate resilience** into their strategies. This includes developing varieties that can withstand both biotic and abiotic stresses simultaneously.
3. **Collaborative efforts**: Global collaboration among research institutions, particularly in tropical regions where black pepper is predominantly grown, will be crucial. Sharing of genetic resources, knowledge, and technologies will help accelerate the breeding of resistant varieties.
4. **Sustainable Agriculture practices**: There is a growing emphasis on integrating resistant varieties with sustainable agricultural practices. This includes **integrated pest management (IPM)** strategies that combine genetic resistance with biological control methods and environmentally friendly practices.

In summary, while significant progress has been made, the future of breeding for biotic stress resistance in black pepper will likely hinge on advanced molecular techniques, global cooperation, and sustainable agricultural integration. These efforts aim to ensure the long-term viability and productivity of black pepper cultivation in the face of evolving biotic challenges.

1b. Challenges of abiotic stress in Black pepper

Abiotic stress poses significant challenges in the breeding of **black pepper** (*Piper nigrum*), a valuable spice crop. These stresses include drought, salinity, extreme temperatures, and nutrient deficiencies, which can severely impact plant growth, yield, and quality. Here are some of the primary challenges:

1. **Complex genetic responses:** Abiotic stress tolerance in plants is controlled by multiple genes, making the genetic response to stress complex. Breeding for abiotic stress resistance requires understanding and manipulating these complex traits, which can involve multiple signaling pathways, transcription factors, and stress-responsive genes.
2. **Lack of stress-resistant varieties:** There is a limited availability of black pepper varieties that are resistant to multiple abiotic stresses. Traditional breeding methods have had limited success due to the long generation time of black pepper, as well as the polygenic nature of stress tolerance traits.
3. **Low genetic diversity:** Black pepper has relatively low genetic diversity, particularly in cultivated varieties. This limited gene pool makes it challenging to introduce new traits such as enhanced stress tolerance through traditional breeding methods. It also restricts the availability of beneficial alleles that could contribute to stress resistance.
4. **Impact on yield and quality:** Abiotic stress can lead to significant reductions in yield and quality, making it challenging to maintain consistent

production. For instance, drought stress can reduce berry size and essential oil content, directly affecting the commercial value of the crop.

5. **Environmental variability:** Black pepper is often grown in regions with variable climates, where sudden changes in weather patterns can introduce unexpected stress. This unpredictability makes it difficult to breed varieties that are universally tolerant to abiotic stress.
6. **Screening and evaluation difficulties:** Effectively screening and evaluating black pepper plants for abiotic stress tolerance under controlled conditions is challenging. Abiotic stress responses are often influenced by environmental factors, which can be difficult to replicate in breeding programs. Additionally, field trials are time-consuming and resource-intensive.
7. **Breeding techniques:** Advanced breeding techniques, such as marker-assisted selection (MAS), genome-wide association studies (GWAS), and CRISPR/Cas9, are still underdeveloped for black pepper. These tools are essential for identifying and incorporating stress tolerance genes, but their application in black pepper is limited due to a lack of genomic resources.
8. **Biotechnological integration:** Integrating biotechnological approaches, such as genetic engineering and molecular breeding, with traditional breeding programs is challenging due to regulatory, ethical, and technical constraints. There is also a need for greater investment in research to develop these technologies for black pepper.
9. **Adaptation to climate change:** Climate change exacerbates abiotic stresses, introducing new challenges for black pepper breeding. Developing varieties that can withstand the increasing intensity and frequency of extreme weather events is a critical but difficult task.
10. **Economic and practical considerations:** Breeding for abiotic stress resistance is often more expensive and time-consuming than for other traits. Moreover, there is often a trade-off between stress tolerance and yield, which breeders must carefully manage to ensure that new varieties are both productive and resilient.

Addressing these challenges requires a multi-disciplinary approach, combining traditional breeding methods with modern genetic and biotechnological tools, and ensuring that breeding programs are adaptable to changing environmental conditions.

2. Cardamom (*Elettaria cardamomum*)

2a. Challenges in breeding for biotic stress resistance in Cardamom

Breeding for biotic stress resistance in cardamom (*Elettaria cardamomum*), commonly known as "the queen of spices," is crucial due to the crop's susceptibility to various pests and diseases. Biotic stresses, including fungal, bacterial, and viral diseases as well as insect pests, significantly reduce yield and quality. Here's an overview of the status and future prospects in this area:

Current status

1. Major biotic stresses

- **Katte or mosaic disease**: Caused by the cardamom mosaic virus (CdMV), this disease leads to significant yield loss and is one of the most challenging problems in cardamom cultivation.
- **Clump rot or rhizome rot**: Caused by *Pythium* and *Phytophthora* species, this disease is particularly devastating in wet and humid conditions.
- **Thrips (*Sciothrips cardamomi*)**: A significant pest, thrips cause direct damage to the cardamom capsules, leading to reduced yield and quality.
- **Shoot borer (*Conogethes punctiferalis*)**: This pest damages the shoots and can lead to secondary infections by pathogens.

2. Traditional breeding efforts

- **Selection and hybridization**: Traditional breeding in cardamom has focused on selecting resistant genotypes from existing populations. However, progress has been slow due to the crop's long growth cycle, complex reproductive biology, and limited genetic diversity.
- **Field screening**: Large-scale field trials are conducted to identify resistant cultivars, but this method is time-consuming and influenced by environmental factors.

3. Molecular breeding

- **Marker-assisted selection (MAS)**: Though still in the early stages compared to other crops, MAS is being used to identify and propagate resistant lines more efficiently. The development of molecular markers linked to resistance traits is a significant step forward.

4. Integrated disease management

- Breeding efforts are often combined with integrated disease management practices, including the use of biocontrol agents and improved cultural practices, to manage biotic stresses in cardamom.

Future prospects

1. Biotechnological interventions

- **Genomic approaches**: Advances in genomic tools, such as next-generation sequencing, can help identify resistance genes and understand the genetic basis of resistance in cardamom. This can lead to the development of genetically resistant varieties through marker-assisted breeding or genetic engineering.

- **CRISPR/Cas9**: Gene-editing technologies like CRISPR/Cas9 hold promise for creating precise modifications in the cardamom genome to enhance resistance traits without the drawbacks of traditional breeding.

2. Resistance gene pyramiding

- **Combining multiple resistance genes**: One of the future directions is to pyramid multiple resistance genes into a single variety. This strategy can provide durable resistance against a broader spectrum of pests and diseases.

3. Climate resilience

- **Breeding for multiple stress resistance**: With climate change expected to exacerbate the incidence of pests and diseases, future breeding programs will likely focus on developing cardamom varieties that are not only resistant to biotic stresses but also to abiotic stresses like drought and temperature extremes.

4. Collaborative research

- **International collaboration**: Given that cardamom is a high-value crop with specific growing regions, collaborative research among international agricultural institutions can accelerate the breeding of resistant varieties. Sharing genetic resources and breeding technologies will be critical.

5. Sustainable Agricultural practices

- **Integration with IPM**: Future breeding programs will increasingly integrate resistant varieties with sustainable agricultural practices, including organic farming and integrated pest management (IPM). This holistic approach aims to reduce the reliance on chemical inputs and improve the long-term sustainability of cardamom cultivation.

Breeding for biotic stress resistance in cardamom is a challenging but essential task to ensure the crop's sustainability and productivity. The future of this endeavor lies in combining traditional breeding with modern biotechnological tools, fostering international collaboration, and integrating resistant varieties into sustainable farming systems. With continued research and innovation, it is expected that more robust and resilient cardamom varieties will be developed, securing the future of this valuable spice.

2b. Challenges in breeding for abiotic stress resistance in Cardamom

Abiotic stress in cardamom refers to the negative impact of non-living environmental factors on the growth, development, and yield of the plant. Cardamom, like other crops, can be sensitive to a range of abiotic stresses, which include:

1. **Drought stress**: Cardamom plants are highly sensitive to water availability. Drought conditions can lead to reduced growth, wilting, and lower yields. Prolonged drought can cause significant damage to the plant's root system and reduce the number of fruit-bearing capsules.
2. **Temperature extremes**: Cardamom thrives in tropical climates with stable temperatures. However, exposure to extreme temperatures, either too high or too low, can affect the plant's physiology. High temperatures can lead to sunburn on leaves, while low temperatures or frost can cause tissue damage and affect flowering and fruit set.
3. **Soil salinity**: High levels of salts in the soil can impair the ability of cardamom plants to absorb water, leading to symptoms like leaf burn, stunted growth, and reduced yields. Soil salinity is often exacerbated by poor irrigation practices or the use of saline water for irrigation.
4. **Nutrient deficiency**: Cardamom requires a well-balanced supply of nutrients for optimal growth. Deficiencies in key nutrients like nitrogen, phosphorus, potassium, and trace elements can lead to poor plant health, reduced vigor, and lower yields. Nutrient stress is often related to poor soil quality or imbalanced fertilization practices.
5. **Flooding and waterlogging**: While cardamom requires ample water, excessive water due to flooding or poor drainage can lead to waterlogging, which deprives roots of oxygen, causes root rot, and can severely affect plant health and yield.
6. **Light stress**: Cardamom typically grows under the shade of forest trees, and too much direct sunlight can cause photoinhibition, reducing the efficiency of photosynthesis. On the other hand, insufficient light due to excessive shade can also limit growth and flowering.
7. **Wind stress**: Strong winds can physically damage cardamom plants by breaking stems or causing abrasion to leaves. Wind stress can also exacerbate the effects of other stresses, such as drought, by increasing the rate of water loss from the plant.

Managing these abiotic stresses involves a combination of strategies, including selecting stress-resistant varieties, optimizing irrigation and fertilization practices, and using shade management techniques to create a more favorable microenvironment for the plants. Additionally, research into the physiological and molecular responses of cardamom to these stresses is crucial for developing more resilient cultivars.

Breeding cardamom (*Elettaria cardamomum*) for resistance to abiotic stresses, such as drought, temperature extremes, and poor soil conditions, poses several significant challenges:

1. **Complex genetics**: Cardamom has a complex genetic makeup, making it difficult to identify and incorporate genes that confer resistance to abiotic stresses. The polygenic nature of these traits often involves multiple genes, each contributing a small effect, which complicates the breeding process.
2. **Limited genetic diversity**: There is often limited genetic diversity in the available cardamom germplasm, particularly for traits related to abiotic stress resistance. This lack of diversity can limit the ability of breeders to develop varieties that can withstand challenging environmental conditions.
3. **Long breeding cycles**: Cardamom has a relatively long growth cycle, which can slow down the breeding process. Developing and testing new varieties for abiotic stress resistance can take many years, making it difficult to respond quickly to changing environmental conditions.
4. **Environmental variability**: Abiotic stresses often vary in intensity and occurrence from year to year and from one location to another. This variability makes it challenging to evaluate and select plants that consistently perform well under stress conditions.
5. **Resource-Intensive**: Breeding for abiotic stress resistance requires significant resources, including access to diverse germplasm, advanced breeding techniques, and testing facilities that can simulate stress conditions. Many breeding programs, especially in developing countries, may lack the necessary infrastructure and funding.
6. **Poor understanding of stress mechanisms**: There is still much to learn about the physiological and molecular mechanisms that allow cardamom to cope with abiotic stresses. Without a deep understanding of these mechanisms, it is difficult to design effective breeding strategies.

These challenges underscore the need for integrated approaches that combine traditional breeding with modern techniques such as marker-assisted selection, genetic engineering, and improved agronomic practices to enhance the resilience of cardamom to abiotic stresses.

3. Cinnamon

3a. Challenges in breeding for biotic stress resistance in Cinnamon

Breeding for biotic stress resistance in cinnamon (*Cinnamomum verum*), a valuable spice crop, is increasingly important due to the impact of pests and diseases on its yield and quality. Cinnamon is susceptible to various biotic stresses, including fungal infections, bacterial diseases, and insect pests. Here's an overview of the current status and future prospects of breeding for biotic stress resistance in cinnamon:

Current status

1. Major biotic stresses

- **Leaf Blight**: Caused by *Colletotrichum gloeosporioides*, leaf blight is one of the most severe diseases affecting cinnamon. It leads to significant defoliation and reduces plant vigour.
- **Stem Canker**: Caused by *Phytophthora spp.*, this disease affects the bark, leading to cankers and dieback. It can severely impact the quality of the bark, which is the main commercial product.
- **Cinnamon Shoot Borer**: *Hypsipyla robusta*, a major pest, bores into the shoots, causing dieback and reducing growth.
- **Root Rot**: Caused by fungi like *Pythium* and *Fusarium*, root rot can lead to the wilting and death of young plants, severely affecting crop establishment.

2. Traditional breeding efforts

- **Selection and screening**: Traditional breeding in cinnamon has focused on the selection of naturally resistant genotypes from existing populations. However, the long growth cycle and the perennial nature of cinnamon make these efforts time-consuming.
- **Conventional hybridization**: There has been limited success in developing resistant varieties through conventional hybridization due to the complexity of the cinnamon genome and the lack of genetic diversity in cultivated varieties.

3. Molecular breeding

- **Marker-assisted selection (MAS)**: Although still in its early stages for cinnamon, MAS is being explored to accelerate the development of resistant varieties. This involves identifying genetic markers linked to resistance traits, which can then be used to screen breeding populations more efficiently.

4. Integrated Pest Management (IPM)

- Breeding for biotic stress resistance is often integrated with IPM strategies, which include cultural practices, biological control, and the judicious use of chemical treatments. Resistant varieties are a key component of these strategies, helping to reduce the overall disease and pest pressure.

Future prospects

1. Advanced genomic techniques

- **Genomic selection**: The future of cinnamon breeding lies in the application of genomic selection, where whole-genome data is used to predict the breeding value of plants. This can significantly speed up the breeding process and improve the accuracy of selecting resistant traits.
- **Gene editing**: Technologies like CRISPR/Cas9 offer the potential to introduce or enhance resistance traits directly in cinnamon plants. Gene editing could be used to create specific mutations that confer resistance to major diseases and pests without altering other desirable traits.

2. Resistance gene pyramiding

- **Combining multiple resistance genes**: Future breeding programs are likely to focus on pyramiding multiple resistance genes into single varieties. This approach can provide broader and more durable resistance against a range of biotic stresses.

3. Climate-Resilient varieties

- **Breeding for combined stress resistance**: As climate change impacts pest and disease dynamics, breeding efforts will increasingly focus on developing cinnamon varieties that can withstand both biotic and abiotic stresses, such as drought and high temperatures.

4. Collaborative Research and Development

- **Global and regional partnerships**: International collaboration in genetic resource sharing, research, and development of resistant varieties will be crucial. This collaboration can help address the challenges posed by limited genetic diversity and accelerate the breeding of improved cinnamon varieties.

5. Sustainable agriculture and organic farming

- **Integration with sustainable practices**: Future breeding efforts will likely emphasize the development of varieties suited for sustainable agriculture and organic farming. This includes breeding for resistance as part of an integrated approach that reduces reliance on chemical inputs and enhances the environmental sustainability of cinnamon cultivation.

Breeding for biotic stress resistance in cinnamon is vital for ensuring the long-term sustainability and productivity of this important spice crop. While progress has been made, particularly through traditional breeding and early molecular techniques, the future lies in leveraging advanced genomic tools, gene editing, and global collaboration. These efforts will focus on developing resilient varieties that can thrive under both biotic and abiotic stresses, contributing to more sustainable cinnamon production systems.

3b. Challenges in breeding for abiotic stress resistance in Cinnamon

Abiotic stress in cinnamon (*Cinnamomum* spp.) refers to the adverse effects caused by non-living environmental factors on the plant's growth, development, and yield. These stresses can significantly impact the health and productivity of cinnamon trees, which are typically cultivated in tropical and subtropical regions. Key abiotic stresses affecting cinnamon include:

1. **Drought stress**: Cinnamon is sensitive to water availability, and prolonged drought conditions can lead to severe water stress. This can cause reduced growth, leaf drop, and a decline in bark quality, which is crucial for cinnamon production. Drought stress can also increase the plant's susceptibility to pests and diseases.
2. **Temperature extremes**: Cinnamon grows best in warm, humid climates. Extreme temperatures, both high and low, can cause physiological damage. High temperatures can lead to leaf scorching and reduced photosynthesis, while low temperatures, especially frost, can cause significant damage to young shoots and leaves, sometimes resulting in the death of the plant.
3. **Soil salinity**: High salt levels in the soil can interfere with the plant's ability to absorb water, leading to symptoms like leaf burn, stunted growth, and reduced biomass. Soil salinity is often a problem in areas with poor irrigation practices or in regions where the soil naturally has a high salt content.
4. **Nutrient deficiency**: Cinnamon requires a balanced supply of nutrients for optimal growth. Deficiencies in essential nutrients like nitrogen, potassium, and magnesium can lead to poor growth, reduced leaf size, and lower yields. Nutrient stress is often exacerbated by poor soil quality or improper fertilization.
5. **Flooding and waterlogging**: While cinnamon requires adequate moisture, excessive water from flooding or poor drainage can lead to waterlogging. Waterlogged soils deprive roots of oxygen, leading to root rot and other fungal diseases. This can significantly reduce plant vigour and increase mortality rates, particularly in younger plants.
6. **Light stress**: Cinnamon typically grows well in partially shaded conditions. However, too much direct sunlight can lead to photo inhibition, where the photosynthetic apparatus is damaged, reducing the plant's ability to produce energy. Conversely, insufficient light due to excessive shade can also hamper growth and reduce the quality of the bark.

7. **Wind stress**: Strong winds can cause physical damage to cinnamon plants by breaking branches or causing abrasion to leaves. Wind stress can also increase water loss through transpiration, leading to drought stress, especially during dry periods.
8. **Soil acidity and poor soil structure**: Cinnamon prefers slightly acidic to neutral soils. Highly acidic soils or soils with poor structure can limit root development and nutrient uptake, leading to overall reduced growth and productivity.

Managing these abiotic stresses in cinnamon cultivation involves practices like selecting appropriate planting sites, optimizing irrigation and fertilization, using mulching to conserve soil moisture, and employing agroforestry systems that provide partial shade and wind protection. Additionally, breeding programs aimed at developing stress-resistant varieties can play a significant role in mitigating the impact of these stresses.

Breeding cinnamon (*Cinnamomum* spp.) for resistance to abiotic stresses, such as drought, temperature extremes, and soil-related challenges, faces several specific challenges:

1. **Complex genetic architecture**: Like many perennial crops, cinnamon has a complex genetic background with traits for abiotic stress resistance often controlled by multiple genes (polygenic). This complexity makes it difficult to identify and select specific genes or markers associated with stress tolerance.
2. **Long breeding cycles**: Cinnamon is a slow-growing, long-lived tree, which means that breeding cycles are extended over many years. This long timeframe makes it difficult to quickly develop and test new varieties that might be more resistant to abiotic stresses.
3. **Limited genetic diversity**: The genetic diversity within cultivated cinnamon species may be limited, particularly for traits related to abiotic stress tolerance. This lack of diversity restricts the availability of beneficial traits that can be used in breeding programs.
4. **Environmental variability**: Abiotic stresses, such as drought or extreme temperatures, vary significantly between different growing regions and from year to year. This variability complicates the evaluation of stress tolerance in breeding programs, as plants may respond differently under varying conditions.
5. **Difficulty in simulating stresses**: In controlled breeding environments, it can be challenging to accurately simulate the complex and variable conditions of abiotic stresses like drought or high salinity. This makes it difficult to predict how new varieties will perform under actual field conditions.

6. **Slow phenotyping processes**: Phenotyping, or the process of observing and measuring plant traits, is particularly challenging in cinnamon due to its perennial nature and slow growth. Identifying plants with desirable traits, such as drought resistance, requires extensive and prolonged observation periods.
7. **Resource constraints**: Breeding for abiotic stress resistance is resource-intensive, requiring access to diverse germplasm collections, advanced breeding tools, and facilities that can simulate stress conditions. In many regions where cinnamon is grown, these resources may be limited.
8. **Climate change impact**: The increasing unpredictability of climate change further complicates breeding efforts. As environmental conditions change, the types of abiotic stresses affecting cinnamon may also shift, requiring continuous adaptation of breeding goals.

To overcome these challenges, integrated approaches that combine traditional breeding with modern techniques such as marker-assisted selection, genetic engineering, and improved agronomic practices are necessary. Additionally, there is a need for greater research into the physiological and molecular mechanisms underlying abiotic stress tolerance in cinnamon, which could provide more targeted strategies for breeding resilient varieties.

4. Allspice

4a. Challenges for biotic stress resistance in Allspice breeding

Breeding for biotic stress resistance in allspice (Pimenta dioica), a tropical spice with significant economic value, is essential due to the impact of pests and diseases on its yield and quality. While allspice is less studied compared to other major spices, there is growing interest in improving its resistance to biotic stresses to ensure sustainable production. Here's an overview of the current status and future prospects of breeding for biotic stress resistance in allspice:

Current status

1. Major biotic stresses

- **Fungal diseases**: Allspice is susceptible to various fungal diseases, including **leaf spot**, **anthracnose**, and **root rot**. These diseases can cause significant damage to the leaves, stems, and roots, leading to reduced yield and quality.
- **Insect pests**: Insect pests such as **borers** and **scale insects** can infest allspice trees, damaging the bark, leaves, and fruits. This not only reduces the overall yield but also affects the marketability of the spice.

- **Viral and bacterial diseases**: Although less commonly reported, viral and bacterial infections can also impact allspice, causing symptoms like leaf curling, yellowing, and wilting.

2. Traditional breeding efforts

- **Selection and Screening**: Traditional breeding in allspice has primarily focused on selecting naturally resistant plants from existing populations. This approach is often limited by the slow growth and long life cycle of allspice trees, making progress in breeding efforts slow and labour-intensive.
- **Limited genetic resources**: The genetic diversity available for breeding in allspice is relatively limited, which poses a challenge for developing new varieties with enhanced resistance.

3. Molecular breeding

- **Early stages of research**: Molecular breeding techniques, such as marker-assisted selection (MAS), are still in the early stages of development for allspice. There is limited research on the genetic basis of disease resistance in this crop, and more work is needed to identify and utilize resistance genes effectively.

4. Integrated Pest Management (IPM)

- **Combining resistant varieties with IPM**: Efforts to breed resistant varieties are often combined with integrated pest management strategies, which include cultural practices, biological controls, and the selective use of chemical treatments. Resistant varieties are a critical component of IPM strategies to reduce reliance on pesticides.

Future prospects

1. Advances in genomic tools

- **Genomic selection**: As genomic tools become more accessible, there is potential to use genomic selection in allspice breeding. This approach involves using genomic data to predict the performance of breeding lines, which can accelerate the development of resistant varieties.
- **Gene editing technologies**: Technologies like CRISPR/Cas9 could be employed to introduce or enhance resistance traits in allspice. Gene editing offers a precise method to develop resistance to specific pathogens or pests without affecting other desirable traits in the plant.

2. Exploring wild relatives

- **Utilizing wild germplasm**: Wild relatives of allspice may possess valuable resistance traits that can be introduced into cultivated varieties. Breeding programs could focus on exploring and utilizing this genetic diversity to enhance resistance in allspice.

3. Developing Multi-resistant varieties

- **Pyramiding resistance genes**: Future breeding efforts may focus on pyramiding multiple resistance genes into a single variety to provide broad-spectrum resistance against various biotic stresses. This approach can help create more robust and durable allspice varieties.

4. Climate-resilient breeding

- **Addressing climate change impacts**: As climate change alters the dynamics of pest and disease incidence, breeding programs will need to develop allspice varieties that are resilient to both biotic and abiotic stresses. This includes breeding for resistance to diseases that may become more prevalent under changing climate conditions.

5. Collaborative research and development

- **International collaboration**: Given the global importance of allspice, collaborative research among tropical and subtropical countries can help pool resources and knowledge to accelerate breeding efforts. Sharing genetic resources, research methodologies, and technologies will be crucial for success.

6. Sustainable Agricultural practices

- **Integration with organic and sustainable farming**: Breeding programs will increasingly emphasize developing varieties suitable for organic and sustainable farming practices. This includes focusing on disease-resistant varieties that can thrive with minimal chemical inputs, supporting the overall goal of sustainable allspice production.

Breeding for biotic stress resistance in allspice is a developing field with significant potential to enhance the crop's sustainability and productivity. While current efforts are limited by the crop's genetic diversity and the early stage of molecular breeding research, the future looks promising with the advent of advanced genomic tools and international collaboration. By integrating traditional breeding methods with modern biotechnology and sustainable farming practices, it is possible to develop robust allspice varieties that can withstand various biotic stresses and contribute to the long-term viability of this valuable spice.

4b. Challenges for abiotic stress resistance in Allspice breeding

Abiotic stress in allspice (*Pimenta dioica*) refers to the negative effects that non-living environmental factors can have on the growth, development, and productivity of the plant. Allspice, a tropical evergreen tree native to the Caribbean and Central

America, is sensitive to several abiotic stresses, which can significantly impact its cultivation and yield. Key abiotic stresses affecting allspice include:

1. **Drought stress**: Allspice is sensitive to water availability. Prolonged drought conditions can lead to water stress, causing wilting, reduced leaf size, and overall stunted growth. Severe drought can result in reduced flowering and fruit set, which directly affects the yield of the spice.
2. **Temperature extremes**: Allspice thrives in warm tropical climates with stable temperatures. However, it is vulnerable to extreme temperatures. High temperatures can cause heat stress, leading to leaf scorching, dehydration, and impaired photosynthesis. Conversely, low temperatures, especially frost, can damage the leaves and young shoots, potentially killing the plant.
3. **Soil salinity**: High levels of salt in the soil can negatively impact allspice by reducing its ability to absorb water, leading to osmotic stress. This can cause leaf burn, reduced growth, and in severe cases, plant death. Soil salinity is particularly problematic in areas with poor irrigation management or in coastal regions where saltwater intrusion is a concern.
4. **Nutrient deficiency**: Allspice requires a balanced supply of nutrients to maintain healthy growth. Nutrient deficiencies, particularly of nitrogen, potassium, and magnesium, can lead to chlorosis (yellowing of leaves), poor growth, and reduced fruit production. These deficiencies are often exacerbated by poor soil quality or improper fertilization practices.
5. **Flooding and waterlogging**: While allspice needs regular watering, excessive water due to flooding or poor drainage can lead to waterlogging. This condition deprives the roots of oxygen, leading to root rot and increased susceptibility to diseases. Prolonged waterlogging can severely reduce the plant's vigour and yield.
6. **Light stress**: Allspice typically grows under partial shade in its natural habitat. Exposure to intense, direct sunlight can cause photo inhibition, where the photosynthetic machinery is damaged, leading to reduced growth. On the other hand, insufficient light due to excessive shade can also limit photosynthesis and reduce flowering and fruiting.
7. **Wind stress**: Strong winds can physically damage allspice plants by breaking branches or causing leaves to tear. Wind stress can also increase transpiration rates, leading to dehydration and compounding the effects of drought.
8. **Soil pH and poor soil structure**: Allspice prefers well-drained, slightly acidic to neutral soils. Soils that are too acidic or too alkaline can limit nutrient availability, affecting plant health. Poor soil structure, such as heavy clay soils, can lead to water retention and poor root development.

Managing these abiotic stresses in allspice cultivation involves selecting suitable planting sites with appropriate soil and climate conditions, optimizing irrigation and drainage systems, applying balanced fertilization, and using agroforestry practices to provide shade and wind protection. Additionally, breeding efforts focused on developing stress-resistant varieties can help mitigate the impact of these environmental challenges.

Breeding allspice (*Pimenta dioica*) for resistance to abiotic stresses presents several challenges, similar to those faced by other tropical and subtropical crops. These challenges can hinder the development of allspice varieties that are resilient to environmental stressors like drought, temperature extremes, and poor soil conditions. The main challenges include:

1. **Genetic complexity**: Abiotic stress resistance is often a complex trait controlled by multiple genes, each contributing a small effect. In allspice, the genetic mechanisms underlying resistance to abiotic stresses like drought or salinity are not well understood, making it difficult to select for these traits in breeding programs.
2. **Limited genetic resources**: Allspice has a relatively narrow genetic base, particularly for traits related to abiotic stress resistance. This limited genetic diversity restricts the availability of stress-resistant traits that can be used in breeding. Moreover, the germplasm collections for allspice may not be as extensive or well-characterized as those for more widely cultivated crops.
3. **Long growth and reproductive cycles**: Allspice is a perennial tree, and like other long-lived plants, it has a slow growth rate and long reproductive cycles. This prolongs the breeding process, as it takes several years to evaluate the performance of new varieties under different abiotic stress conditions.
4. **Environmental variability**: Abiotic stresses, such as drought and temperature extremes, vary significantly across different growing regions and seasons. This environmental variability makes it challenging to consistently evaluate and select allspice plants that perform well under a range of stress conditions, leading to difficulties in developing broadly resistant varieties.
5. **Challenges in phenotyping**: Phenotyping for abiotic stress resistance—observing and measuring plant responses to stress conditions—can be difficult in allspice due to its perennial nature. The slow manifestation of stress responses and the need for long-term observation complicate the identification of resistant genotypes.
6. **Resource-Intensive breeding**: Breeding for abiotic stress resistance in allspice is resource-intensive, requiring access to diverse germplasm, advanced breeding tools, and controlled environments that can simulate

stress conditions like drought or salinity. Many breeding programs may lack the necessary infrastructure and funding, particularly in regions where allspice is traditionally grown.

7. **Poor understanding of stress mechanisms**: There is limited research on the physiological and molecular mechanisms that confer abiotic stress resistance in allspice. This lack of understanding hampers the ability to design effective breeding strategies or to use modern techniques like marker-assisted selection or genetic engineering to develop stress-resistant varieties.
8. **Impact of climate change**: The unpredictability of climate change adds another layer of complexity. As environmental conditions shift, the types and severity of abiotic stresses affecting allspice may also change, requiring continuous adaptation of breeding goals and strategies.

To overcome these challenges, breeding programs for allspice need to integrate traditional breeding methods with modern technologies such as genomics, marker-assisted selection, and genetic engineering. Additionally, there is a need for more comprehensive research into the genetic and physiological basis of abiotic stress resistance in allspice, as well as efforts to expand and diversify germplasm collections.

5. Nutmeg

5a. Challenges for biotic stress resistance in nutmeg breeding

Breeding for biotic stress resistance in nutmeg (***Myristica fragrans***), a valuable spice crop known for its aromatic seeds and mace, is essential to ensure sustainable production. Nutmeg cultivation is challenged by various pests and diseases, which can significantly impact yield and quality. Here's an overview of the current status and future prospects of breeding for biotic stress resistance in nutmeg:

Current status

1. Major biotic stresses

- **Nutmeg Wilt**: Wilt diseases caused by *Fusarium* species and *Phytophthora* are among the most severe threats to nutmeg. These pathogens cause root and collar rot, leading to wilting and death of the tree.
- **Fruit Rot**: *Colletotrichum gloeosporioides* causes fruit rot in nutmeg, leading to significant losses in yield and quality of both seeds and mace.
- **Leaf Spot and Blight**: Fungal diseases like leaf spot and blight, caused by various fungi including *Phyllosticta* and *Colletotrichum*, damage leaves, reduce photosynthetic capacity, and weaken the trees.

- **Insect Pests**: Nutmeg is also affected by pests such as nutmeg weevil (*Myristica spp.*) and scales, which damage the seeds and leaves, leading to reduced yield and marketability.

2. Traditional Breeding Efforts

- **Selection and screening**: Traditional breeding in nutmeg has primarily focused on selecting naturally resistant individuals from existing populations. This method, while useful, is slow due to nutmeg's long growth cycle and complex reproductive biology.
- **Limited genetic diversity**: The genetic diversity within cultivated nutmeg is relatively narrow, which limits the potential for breeding new varieties with enhanced resistance.

3. Molecular Breeding

- **Early stages of development**: Molecular breeding techniques, such as marker-assisted selection (MAS), are still in the early stages for nutmeg. There is limited research on identifying resistance genes or markers linked to biotic stress resistance in nutmeg.

4. Integrated Pest Management (IPM)

- **Combining resistant varieties with IPM**: Breeding efforts are often integrated with IPM strategies, which include cultural practices, biological controls, and minimal use of chemical treatments. Resistant varieties are an essential component of these strategies, aiming to reduce the reliance on chemical pesticides.

Future Prospects

1. Advances in genomic tools

- **Genomic selection**: With advances in genomic tools, there is potential to apply genomic selection in nutmeg breeding. This approach could accelerate the development of resistant varieties by using genetic information to predict and select for resistance traits in breeding programs.
- **Gene editing technologies**: Technologies like CRISPR/Cas9 offer promising potential for breeding in nutmeg. Gene editing could be used to introduce or enhance specific resistance traits against major pathogens and pests, improving the resilience of nutmeg trees.

2. Exploration of wild relatives

- **Utilizing wild germplasm**: Wild relatives of nutmeg may possess valuable resistance traits that can be introduced into cultivated varieties through cross-breeding or biotechnological approaches. This strategy can help overcome the limitations of genetic diversity in cultivated nutmeg.

3. Development of multi-resistant varieties

- **Pyramiding resistance genes**: Future breeding programs may focus on pyramiding multiple resistance genes into single nutmeg varieties. This approach aims to develop broad-spectrum resistance against a range of biotic stresses, providing more durable protection to the crop.

4. Climate-resilient breeding

- **Breeding for combined stress resistance**: As climate change affects pest and disease dynamics, breeding for resistance in nutmeg will increasingly need to address both biotic and abiotic stresses, such as drought and high temperatures, ensuring that new varieties are resilient under changing environmental conditions.

5. Collaborative research and development

- **Global and regional partnerships**: Collaborative research efforts among nutmeg-producing countries can facilitate the sharing of genetic resources and breeding technologies. This international collaboration is crucial to overcoming the challenges of breeding in a crop with limited genetic diversity and long life cycles.

6. Sustainable agricultural practices

- **Integration with organic farming**: There will be an increasing emphasis on developing nutmeg varieties that are suitable for organic and sustainable farming practices. This includes breeding for resistance traits that reduce the need for chemical inputs and promote environmentally friendly cultivation methods.

Breeding for biotic stress resistance in nutmeg is a challenging but essential area of research to ensure the crop's long-term sustainability and productivity. While progress has been limited due to the crop's long growth cycle and narrow genetic diversity, the future looks promising with the advent of advanced genomic tools, gene editing technologies, and international collaboration. By integrating traditional breeding methods with modern biotechnology and sustainable farming practices, it is possible to develop robust, disease-resistant nutmeg varieties that can thrive under various biotic stresses and contribute to the stability of global nutmeg production.

5b. Challenges for abiotic stress resistance in nutmeg breeding

Abiotic stress in nutmeg (*Myristica fragrans*) refers to the negative impact of non-living environmental factors on the growth, development, and yield of the nutmeg tree. As a tropical evergreen tree, nutmeg is sensitive to a variety of abiotic stresses, which can significantly affect its productivity and health. The key abiotic stresses affecting nutmeg include:

1. **Drought stress**: Nutmeg requires a consistent supply of water, and prolonged periods of drought can cause severe stress. Drought conditions can lead to reduced growth, wilting, leaf drop, and even death in severe cases. Water stress during critical growth phases can also reduce fruit set and quality, directly impacting yields.
2. **Temperature extremes**: Nutmeg thrives in warm, humid climates typical of its native regions in Southeast Asia. However, it is vulnerable to extreme temperatures. High temperatures can cause heat stress, leading to leaf scorching, dehydration, and reduced photosynthesis. Conversely, low temperatures, particularly frost, can damage or kill young plants and cause leaf burn in mature trees.
3. **Soil salinity**: High salinity levels in the soil can adversely affect nutmeg by disrupting water uptake and causing osmotic stress. This can result in symptoms such as leaf burn, stunted growth, and reduced fruit production. Soil salinity is often exacerbated by poor irrigation practices or the use of saline water.
4. **Nutrient deficiency**: Nutmeg trees require a balanced supply of nutrients for optimal growth. Deficiencies in essential nutrients, such as nitrogen, potassium, magnesium, and calcium, can lead to chlorosis (yellowing of leaves), poor growth, reduced fruit set, and lower overall yields. Nutrient stress is often related to poor soil fertility or improper fertilization practices.
5. **Waterlogging and flooding**: While nutmeg needs adequate moisture, excessive water due to flooding or poor drainage can lead to waterlogging. Waterlogged soils can suffocate roots, leading to root rot and other fungal diseases. Prolonged waterlogging can weaken the tree, reduce its productivity, and increase mortality rates.
6. **Light stress**: Nutmeg typically grows under the canopy of larger trees in its native habitat, and it prefers partial shade. Excessive direct sunlight can lead to photo inhibition, reducing photosynthetic efficiency and causing leaf burn. On the other hand, too much shade can limit photosynthesis, reduce growth, and lower fruit yields.
7. **Wind stress**: Strong winds can cause physical damage to nutmeg trees by breaking branches, causing leaf tearing, or even uprooting young trees. Wind stress can also increase water loss through transpiration, exacerbating drought conditions.
8. **Soil acidity and poor soil structure**: Nutmeg prefers slightly acidic to neutral, well-drained soils. Highly acidic soils or those with poor structure can limit root development and nutrient uptake, leading to reduced growth and productivity. Soil compaction and heavy clay soils can also lead to poor drainage and increase the risk of waterlogging.

Managing these abiotic stresses in nutmeg cultivation involves several strategies, including selecting appropriate planting sites, optimizing irrigation and drainage systems, maintaining balanced fertilization, and using agroforestry practices to provide shade and protect trees from wind damage. Additionally, ongoing research and breeding efforts aimed at developing more stress-resistant nutmeg varieties could play a crucial role in mitigating the impact of these environmental challenges.

Breeding nutmeg (*Myristica fragrans*) for resistance to abiotic stresses such as drought, temperature extremes, and poor soil conditions presents several challenges. These challenges stem from the genetic, physiological, and environmental complexities involved in developing stress-resistant varieties. Key challenges include:

1. **Genetic complexity**: Abiotic stress resistance in nutmeg, like in many perennial crops, is controlled by multiple genes, often with small effects. This polygenic nature makes it difficult to identify specific genetic markers for stress resistance, complicating the breeding process.
2. **Long breeding cycles**: Nutmeg is a slow-growing tree that takes several years to mature and produce fruit. This long lifecycle extends the time required to develop, test, and release new varieties, making breeding programs lengthy and resource-intensive.
3. **Limited genetic diversity**: The genetic diversity within cultivated nutmeg populations is relatively narrow, particularly for traits related to abiotic stress resistance. This limited diversity restricts the pool of genetic material that breeders can use to develop resilient varieties.
4. **Environmental variability**: Abiotic stresses such as drought, temperature extremes, and soil salinity vary significantly across different growing regions and from year to year. This variability makes it difficult to evaluate and select nutmeg plants that consistently perform well under diverse and changing environmental conditions.
5. **Difficulty in simulating stress conditions**: Controlled breeding environments often struggle to accurately replicate the complex and variable conditions of abiotic stresses like drought or salinity. This limitation hampers the ability to predict how new nutmeg varieties will perform under real-world stress conditions.
6. **Slow phenotyping process**: Observing and measuring the traits associated with abiotic stress resistance in nutmeg is challenging due to the tree's perennial nature and slow growth. Identifying and selecting plants with desirable traits requires long-term monitoring, which delays the breeding process.
7. **Resource constraints**: Breeding for abiotic stress resistance requires significant resources, including access to diverse germplasm collections,

advanced breeding tools, and facilities capable of simulating various stress conditions. In many nutmeg-growing regions, these resources may be limited, hindering progress in breeding programs.

8. **Lack of research and understanding**: There is a limited amount of research focused on the physiological and molecular mechanisms underlying abiotic stress resistance in nutmeg. This lack of understanding makes it difficult to design effective breeding strategies or apply modern techniques like marker-assisted selection or genetic engineering.
9. **Impact of climate change**: The increasing unpredictability of climate change introduces new challenges, as it can alter the types and intensities of abiotic stresses that nutmeg plants face. This requires continuous adaptation of breeding goals and the development of varieties that can withstand a broader range of environmental conditions.

Addressing these challenges in nutmeg breeding will likely require integrated approaches that combine traditional breeding methods with modern genomic tools, expanded research efforts to better understand stress mechanisms, and enhanced conservation of genetic resources to broaden the diversity available for breeding programs.

6. Clove (*Syzygium aromaticum*)

6a. Challenges for biotic stress resistance in clove breeding

Breeding for biotic stress resistance in clove (*Syzygium aromaticum*) is essential due to the crop's susceptibility to various pests and diseases that can severely impact yield and quality. Clove is a high-value spice widely used in the food, pharmaceutical, and cosmetic industries, making it crucial to develop resistant varieties to ensure sustainable production. Here's an overview of the current status and future prospects of breeding for biotic stress resistance in clove:

Current status

1. Major biotic stresses

- **Clove Leaf Spot**: Caused by the fungus *Colletotrichum gloeosporioides*, this disease affects the leaves, leading to significant defoliation and reduced photosynthetic capacity.
- **Clove Dieback**: *Phytophthora cinnamomi* and *P. palmivora* are the primary pathogens responsible for clove dieback, which causes extensive damage to the trees, leading to branch dieback and eventually tree death.
- **Stem Canker**: Another major issue, stem canker is caused by *Nectria haematococca* and can result in significant losses, particularly in younger trees.

- **Bacterial Wilt**: *Ralstonia solanacearum* causes bacterial wilt in clove, leading to wilting and death of infected plants.
- **Insect Pests**: Pests such as clove gall midge (*Asphondylia syzygii*) and scale insects also pose serious threats to clove plantations by damaging buds and leaves, reducing yield.

2. Traditional breeding efforts

- **Selection and screening**: Traditional breeding in clove has primarily involved the selection of naturally resistant genotypes. This approach has been somewhat successful in identifying resistant varieties, but progress is slow due to the perennial nature of clove trees and their long gestation period.
- **Hybridization**: Crossbreeding among different clove populations to develop resistant varieties has been attempted but is limited by the narrow genetic base and the challenges of controlling cross-pollination in clove trees.

3. Molecular breeding

- **Marker-assisted selection (MAS)**: The use of molecular markers linked to disease resistance traits in clove is still in its nascent stages. However, there is growing interest in developing MAS strategies to speed up the breeding process and improve resistance.

4. Integrated Disease Management (IDM)

- **Combining breeding with IDM**: Breeding for disease resistance is often integrated with other disease management practices, including cultural, chemical, and biological controls. Resistant varieties are a key component of IDM strategies to reduce the reliance on chemical inputs.

Future prospects

1. Genomic tools and biotechnology

- **Genomic selection**: The application of genomic selection could revolutionize clove breeding by enabling the selection of disease-resistant traits based on genomic data, thereby accelerating the development of resistant varieties.
- **Gene editing technologies**: Technologies like CRISPR/Cas9 offer significant potential for creating disease-resistant clove varieties. Gene editing could allow for the precise introduction of resistance genes into elite clove varieties, addressing specific biotic stresses without the need for extensive cross-breeding.

2. Resistance gene pyramiding

- **Combining multiple resistance genes**: Future breeding programs are likely to focus on pyramiding multiple resistance genes into clove varieties. This approach can provide broader and more durable resistance against a range of pathogens and pests, reducing the chances of resistance breakdown.

3. Climate-Resilient varieties

- **Breeding for multi-stress resistance**: As climate change continues to alter the dynamics of pest and disease incidence, breeding for resistance in clove will increasingly need to consider the development of varieties that can withstand both biotic and abiotic stresses, such as drought and temperature fluctuations.

4. Collaborative and Integrated Research

- **International collaboration**: Collaborative research efforts between clove-producing countries can facilitate the sharing of genetic resources, research methodologies, and technological advancements. This global cooperation is crucial for overcoming the challenges of breeding in a crop with a relatively narrow genetic base.
- **Integration with sustainable Agriculture**: There will be a growing emphasis on integrating resistant clove varieties into sustainable agricultural systems. This includes organic farming practices and agroforestry systems where clove is intercropped with other species, promoting biodiversity and reducing the spread of diseases.

5. Advanced breeding techniques

- **Tissue culture and clonal propagation**: Techniques like tissue culture can be used to propagate resistant varieties on a large scale, ensuring that the desirable traits are consistently maintained across plantations. This method can also help in rapidly multiplying elite clove varieties with enhanced resistance.

Breeding for biotic stress resistance in clove is crucial for ensuring the long-term sustainability and productivity of this valuable spice crop. While current efforts have been constrained by the crop's perennial nature and narrow genetic diversity, the future of clove breeding looks promising with the advent of advanced genomic tools, gene editing technologies, and international collaboration. By combining traditional breeding with modern biotechnological approaches and sustainable agricultural practices, it is possible to develop robust, disease-resistant clove varieties that can thrive under various biotic and abiotic stresses.

6b. Challenges in abiotic stress resistance in Clove breeding

Abiotic stress in clove (*Syzygium aromaticum*) refers to the adverse effects that non-living environmental factors can have on the growth, development, and productivity of clove trees. As a tropical evergreen tree, clove is sensitive to several abiotic stresses that can significantly impact its health and yield. Key abiotic stresses affecting clove include:

1. **Drought stress**: Clove trees require a consistent supply of water, especially during the dry season. Drought conditions can lead to water stress, resulting in reduced leaf growth, wilting, leaf drop, and lower flower and fruit production. Severe drought can also weaken the tree, making it more susceptible to pests and diseases.
2. **Temperature extremes**: Clove thrives in warm, humid climates, but it is vulnerable to temperature extremes. High temperatures can cause heat stress, leading to leaf scorching, reduced photosynthetic activity, and potentially impaired flowering and fruit set. On the other hand, clove is particularly sensitive to cold temperatures and frost, which can cause significant damage to young plants and reduce yields in mature trees.
3. **Soil salinity**: High salinity levels in the soil can negatively affect clove trees by reducing their ability to absorb water, leading to osmotic stress. Symptoms of salinity stress include leaf burn, stunted growth, and reduced flower and clove production. Soil salinity is often a problem in coastal areas or regions with poor irrigation management.
4. **Nutrient deficiency**: Clove trees require a balanced supply of nutrients for healthy growth and optimal yield. Deficiencies in key nutrients like nitrogen, potassium, and magnesium can lead to chlorosis (yellowing of leaves), poor growth, and reduced flower production. Nutrient stress can be exacerbated by poor soil fertility or inadequate fertilization practices.
5. **Flooding and waterlogging**: While clove trees need adequate moisture, they are sensitive to waterlogging and flooding. Excessive water can lead to root suffocation, root rot, and increased susceptibility to fungal diseases. Prolonged waterlogging can weaken the trees and reduce their productivity.
6. **Light stress**: Clove trees typically require full sun to partial shade for optimal growth. Insufficient light due to excessive shading can limit photosynthesis, reducing growth and flowering. Conversely, too much direct sunlight, especially in areas prone to high temperatures, can cause photo inhibition, reducing the plant's photosynthetic efficiency and leading to leaf burn.
7. **Wind stress**: Clove trees are relatively tall and can be susceptible to damage from strong winds. Wind stress can break branches, cause leaf tearing, and even uproot young trees. Additionally, wind can increase transpiration rates, exacerbating drought conditions.

8. **Soil acidity and poor soil structure**: Clove prefers slightly acidic to neutral, well-drained soils. Soils that are too acidic or too alkaline can limit nutrient availability, affecting tree health and productivity. Poor soil structure, such as compacted or heavy clay soils, can lead to waterlogging, further stressing the trees.

Management Strategies

- **Irrigation management**: Proper irrigation practices are crucial to avoid both drought stress and waterlogging. Efficient water management systems can help maintain optimal soil moisture levels.
- **Soil management**: Regular soil testing and appropriate soil amendments can help maintain the right pH and nutrient balance, reducing the impact of nutrient deficiencies and salinity.
- **Shade and shelter**: In areas with high temperatures or strong winds, providing partial shade and windbreaks can help protect clove trees from extreme light and wind stress.
- **Breeding and selection**: Developing and selecting clove varieties that are more resilient to abiotic stresses through breeding programs can also help mitigate these challenges.

These strategies, along with ongoing research into the specific responses of clove to various abiotic stresses, are essential for improving the sustainability and productivity of clove cultivation.

Challenges in developing abiotic stress resistance in Clove breeding involve several complex factors

1. **Genetic diversity**: Cloves typically have limited genetic diversity, which poses a challenge for breeding programs aimed at improving resistance to various abiotic stresses like drought, salinity, and temperature extremes. This narrow genetic base makes it difficult to find or introduce new traits that could help the plants withstand adverse conditions.
2. **Complexity of traits**: Abiotic stress resistance is often controlled by multiple genes, making it a polygenic trait. The complexity of these traits means that conventional breeding methods can be slow and less effective. Identifying and selecting the right combinations of genes that confer resistance is a significant challenge.
3. **Phenotyping difficulties**: Accurately assessing how well clove plants resist abiotic stresses in real-world conditions is difficult. Phenotyping, or measuring the expression of traits under stress conditions, requires controlled environments and precise measurements, which can be resource-intensive.

4. **Lack of genomic resources**: Compared to other crops, cloves have fewer genomic resources available, such as molecular markers, genomic sequences, or genetic maps. This scarcity limits the ability to apply modern breeding techniques like marker-assisted selection or genomic selection.
5. **Environmental variability**: The performance of clove plants under abiotic stress can vary widely depending on local environmental conditions. This variability makes it challenging to develop universally applicable resistant varieties, as plants bred for one region may not perform well in another.
6. **Breeding cycle**: Clove trees have a long juvenile phase, meaning they take several years to mature and produce seeds. This long breeding cycle delays the development of new, stress-resistant varieties and makes the breeding process more resource-intensive.

Addressing these challenges requires integrating advanced breeding techniques, increasing genetic diversity through biotechnological interventions, and developing better phenotyping tools and genomic resources.

7. Ginger

7a. Challenges in breeding for biotic stress resistance in Ginger

Breeding for biotic stress resistance in ginger (*Zingiber officinale*) is a critical focus area due to the plant's susceptibility to various pests and diseases that significantly impact yield and quality. As ginger is a widely cultivated spice with substantial economic importance, enhancing its resistance to biotic stresses is essential for sustainable production. Here's an overview of the current status and future prospects of breeding for biotic stress resistance in ginger:

Current status

1. Major biotic stresses

- **Rhizome Rot**: Rhizome rot, caused by *Pythium spp.* and *Fusarium spp.*, is one of the most devastating diseases affecting ginger. It leads to rotting of the rhizomes, wilting of the plants, and can result in significant yield losses.
- **Bacterial Wilt**: Caused by *Ralstonia solanacearum*, bacterial wilt is a serious disease that causes yellowing, wilting, and eventually the death of ginger plants. This pathogen is particularly challenging because it persists in the soil and can affect subsequent crops.
- **Leaf Spot**: Fungal pathogens such as *Phyllosticta zingiberi* cause leaf spot, leading to reduced photosynthetic efficiency and overall plant vigor.
- **Nematode Infestation**: Root-knot nematodes (*Meloidogyne spp.*) are another significant threat to ginger, causing root galls, reduced nutrient uptake, and overall stunted growth.

- **Insect Pests**: Pests like shoot borers (*Conogethes punctiferalis*) and aphids can also impact ginger cultivation by damaging the foliage and transmitting viral diseases.

2. Traditional breeding efforts

- **Selection and clonal propagation**: Traditional breeding methods for ginger have primarily focused on selecting disease-resistant clones from existing populations. However, because ginger is typically propagated vegetatively, there is limited genetic diversity, which hinders the development of new resistant varieties.
- **Field screening**: Extensive field trials and screening programs have been conducted to identify and propagate resistant clones. This approach, while somewhat effective, is time-consuming and resource-intensive.

3. Molecular breeding

- **Marker-assisted selection (MAS)**: The use of molecular markers in ginger breeding is still in the early stages. Efforts are ongoing to identify genetic markers linked to resistance traits, which could accelerate the breeding process.
- **Tissue culture and somaclonal variation**: Tissue culture techniques are employed to generate genetic diversity in ginger through somaclonal variation, which can then be screened for resistance traits.

4. Integrated Disease Management (IDM)

- **Combining breeding with IDM**: Breeding for disease resistance is often integrated with IDM practices, including crop rotation, organic amendments, biological control agents, and judicious use of fungicides. Resistant varieties are crucial to these strategies, helping to manage disease pressure sustainably.

Future prospects

1. Advances in genomic tools

- **Genomic selection**: The application of genomic selection in ginger breeding could significantly enhance the efficiency of developing resistant varieties. This involves using whole-genome data to predict and select for desirable traits, including disease resistance.
- **Gene editing technologies**: Technologies like CRISPR/Cas9 hold potential for introducing or enhancing specific resistance traits in ginger. Gene editing could allow for targeted improvements without affecting other important characteristics of the crop.

2. Development of multi-resistant varieties

- **Pyramiding resistance genes**: Future breeding efforts will likely focus on pyramiding multiple resistance genes into a single ginger variety to provide broad-spectrum resistance against various biotic stresses. This approach can help create more durable and resilient varieties.

3. Exploration of wild relatives

- **Utilizing wild germplasm**: Wild relatives of ginger may possess valuable resistance traits that can be introduced into cultivated varieties through cross-breeding or advanced biotechnological methods. This strategy could help overcome the genetic bottleneck in cultivated ginger.

4. Climate-Resilient breeding

- **Addressing climate change**: As climate change alters pest and disease dynamics, breeding programs will need to focus on developing ginger varieties that are resilient to both biotic and abiotic stresses, such as drought, extreme temperatures, and increased disease pressure.

5. Collaborative research and global efforts

- **International collaboration**: Collaborative research among ginger-producing regions can facilitate the sharing of genetic resources, breeding techniques, and innovations. Global partnerships will be essential in addressing the challenges of breeding for biotic stress resistance in ginger.

6. Sustainable Agricultural practices

- **Integration with organic farming**: Future breeding programs will increasingly emphasize the development of ginger varieties that are well-suited for organic and sustainable farming practices. This includes breeding for traits that enhance natural resistance and reduce the need for chemical inputs.

Breeding for biotic stress resistance in ginger is crucial for ensuring the long-term sustainability and productivity of this important spice crop. While traditional breeding methods have made some progress, the future of ginger breeding lies in the integration of advanced genomic tools, gene editing technologies, and global collaboration. By focusing on developing resilient, multi-resistant varieties and integrating these with sustainable farming practices, the ginger industry can better withstand the challenges posed by biotic stresses and climate change.

7b. Challenges in breeding for abiotic stress resistance in Ginger

Abiotic stress in ginger (*Zingiber officinale*) can significantly affect its growth, yield, and quality. The main types of abiotic stress that ginger plants face include:

1. **Drought stress**: Ginger plants require adequate moisture for optimal growth, and drought can severely impact rhizome development. Drought stress leads to reduced water uptake, affecting photosynthesis, leading to stunted growth, wilting, and lower yields. Prolonged drought conditions can also cause a decrease in essential oil content, which is a key quality parameter for ginger.
2. **Temperature extremes**: Ginger is sensitive to both high and low temperatures. High temperatures can cause heat stress, leading to reduced chlorophyll content, impaired photosynthesis, and eventual plant wilting or even death. Low temperatures, especially frost, can damage the rhizomes, resulting in reduced sprouting and lower yield potential. Optimal growth typically occurs in warm, tropical conditions.
3. **Salinity stress**: Soil salinity is another major issue, particularly in regions where irrigation practices lead to salt accumulation in the soil. Salinity stress can cause ion toxicity, osmotic stress, and nutrient imbalances in ginger plants, which may result in poor growth, chlorosis (yellowing of leaves), and reduced rhizome size and quality.
4. **Nutrient deficiency**: Although not a direct abiotic stress, nutrient deficiency in the soil often exacerbates the effects of other abiotic stresses. For example, a lack of potassium can make ginger plants more susceptible to drought and salinity stresses. Nutrient imbalances can also reduce the plant's ability to cope with temperature extremes.
5. **Soil compaction and poor drainage**: Ginger requires well-drained, loose soil for optimal growth. Compacted soil or poor drainage can lead to waterlogging, which deprives the roots of oxygen and leads to root rot, a significant issue in ginger cultivation. Waterlogged conditions can also promote the growth of pathogens, further stressing the plants.
6. **Light stress**: Ginger is a shade-tolerant plant, but both excessive and insufficient light can cause stress. Too much direct sunlight can lead to leaf scorching and reduced photosynthetic efficiency, while too little light can slow down growth and reduce rhizome development.

Addressing these abiotic stresses involves careful management of water, soil, and environmental conditions. Techniques such as mulching, controlled irrigation, soil amendment, and the use of stress-resistant cultivars can help mitigate the effects of abiotic stress in ginger cultivation. Additionally, research into genetic improvement and biotechnological interventions could further enhance ginger's resilience to these stresses.

Breeding for abiotic stress resistance in Ginger (Zingiber officinale) presents several challenges

1. **Limited genetic diversity**: Ginger exhibits low genetic diversity due to its mode of propagation, primarily through rhizomes rather than seeds. This lack of genetic variation makes it difficult to introduce and select for new traits, including resistance to abiotic stresses like drought, salinity, and extreme temperatures.
2. **Complexity of abiotic stress tolerance**: Abiotic stress tolerance is typically governed by multiple genes, making it a polygenic trait. The complex interaction between these genes and the environment complicates the breeding process, as selecting for one stress-resistant trait may inadvertently affect other important agronomic traits.
3. **Long breeding cycle**: Ginger has a long breeding cycle because it takes a significant amount of time to grow from a planted rhizome to maturity. This extended cycle delays the evaluation of new varieties for stress resistance, slowing down the overall breeding process.
4. **Challenges in phenotyping**: Accurately measuring ginger's response to abiotic stress in real-world conditions is difficult. Phenotyping, or the assessment of physical and biochemical traits under stress conditions, requires controlled environments and sophisticated tools. Variability in stress conditions across different environments further complicates this process.
5. **Poor seed set**: Ginger is generally sterile or has very low seed set, which limits the ability to use conventional breeding methods that rely on sexual reproduction. This sterility constrains the development of new varieties through hybridization, a common approach in breeding for stress resistance.
6. **Lack of genomic resources**: There is a lack of comprehensive genomic resources such as well-mapped genomes, molecular markers, or gene expression data for ginger. This scarcity hinders the application of advanced breeding techniques like marker-assisted selection (MAS) or genomic selection, which are crucial for efficiently breeding stress-resistant varieties.
7. **Environmental variability**: The performance of ginger under abiotic stress can vary significantly depending on local environmental conditions. This variability means that a variety bred for resistance in one region may not perform well in another, complicating the development of universally applicable stress-resistant cultivars.
8. **Economic viability**: Breeding for abiotic stress resistance often involves a trade-off between stress tolerance and yield or quality. The economic viability of stress-resistant ginger varieties must be carefully considered, as farmers may be reluctant to adopt new varieties if they do not offer clear

economic benefits, such as high yield or superior quality under normal conditions.

Overcoming these challenges requires a multifaceted approach, including the use of biotechnological tools like genetic engineering, developing more effective phenotyping methods, and improving genomic resources for ginger. Additionally, efforts to increase genetic diversity, either through the introduction of wild relatives or induced mutations, could provide new opportunities for breeding more resilient ginger varieties.

9. Turmeric

9a. Challenges in breeding for biotic stress resistance in Turmeric

Breeding for biotic stress resistance in turmeric (Curcuma longa) is crucial due to the crop's vulnerability to various pests and diseases that significantly impact its yield and quality. Turmeric is a highly valued spice with medicinal properties, making it important to enhance its resistance to biotic stresses to ensure sustainable production. Here's an overview of the current status and future prospects of breeding for biotic stress resistance in turmeric:

Current Status

1. Major biotic stresses

- **Rhizome Rot**: One of the most severe diseases affecting turmeric is rhizome rot, primarily caused by *Pythium aphanidermatum* and *Fusarium oxysporum*. This disease leads to the rotting of rhizomes, yellowing of leaves, and ultimately, the death of the plant.
- **Leaf Blotch**: Caused by *Taphrina maculans*, leaf blotch is a common fungal disease that affects turmeric leaves, leading to the formation of dark brown spots that reduce photosynthetic efficiency and overall plant vigor.
- **Leaf Spot**: *Colletotrichum capsici* and other fungi cause leaf spot, resulting in circular lesions on the leaves, which can coalesce and cause severe defoliation.
- **Shoot Borer**: The turmeric shoot borer (*Conogethes punctiferalis*) is a significant insect pest that damages the shoots, leading to stunted growth and reduced rhizome yield.
- **Nematode Infestation**: Root-knot nematodes (*Meloidogyne* spp.) are also problematic, causing galls on the roots, which interfere with water and nutrient uptake, leading to poor plant growth.

2. Traditional breeding efforts

- **Selection and clonal propagation**: Traditional breeding in turmeric has focused on the selection of naturally resistant clones from existing populations. However, due to the vegetative propagation of turmeric, genetic diversity is limited, which poses a challenge for developing new resistant varieties.
- **Field screening**: Extensive field trials and screenings have been conducted to identify resistant clones. This method, while effective in some cases, is time-consuming and resource-intensive.

3. Molecular breeding

- **Marker-Assisted Selection (MAS)**: Molecular breeding techniques, including MAS, are still in the early stages for turmeric. However, efforts are ongoing to identify genetic markers linked to resistance traits, which could accelerate the breeding process.
- **Tissue culture and somaclonal variation**: Tissue culture techniques are employed to generate genetic diversity through somaclonal variation, which can then be screened for resistance traits.

4. Integrated Disease Management (IDM)

- **Combining resistant varieties with IDM**: Breeding efforts are often integrated with IDM practices, including crop rotation, organic amendments, and biological control agents. Resistant varieties play a crucial role in reducing the disease burden and minimizing the need for chemical inputs.

Future prospects

1. Advances in genomic tools

- **Genomic selection**: The application of genomic selection in turmeric breeding could significantly enhance the efficiency of developing resistant varieties. By using whole-genome data to predict and select for desirable traits, breeders can accelerate the development of resistant varieties.
- **Gene editing technologies**: Technologies like CRISPR/Cas9 hold great potential for turmeric breeding. Gene editing could be used to introduce or enhance specific resistance traits, providing precise and targeted improvements without affecting other important characteristics.

2. Development of multi-resistant varieties

- **Pyramiding resistance genes**: Future breeding programs may focus on pyramiding multiple resistance genes into a single turmeric variety to provide broad-spectrum resistance against various biotic stresses. This approach can help create more durable and resilient turmeric varieties.

3. Exploration of wild relatives

- **Utilizing wild germplasm**: Wild relatives of turmeric may possess valuable resistance traits that can be introduced into cultivated varieties through cross-breeding or advanced biotechnological methods. This strategy could help overcome the genetic bottleneck in cultivated turmeric.

4. Climate-Resilient breeding

- **Breeding for combined stress resistance**: As climate change alters the dynamics of pest and disease incidence, breeding programs will need to focus on developing turmeric varieties that are resilient to both biotic and abiotic stresses, such as drought, extreme temperatures, and increased disease pressure.

5. Collaborative research and global efforts

- **International collaboration**: Collaborative research among turmeric-producing regions can facilitate the sharing of genetic resources, breeding techniques, and innovations. Global partnerships will be essential in addressing the challenges of breeding for biotic stress resistance in turmeric.

6. Sustainable Agricultural practices

- **Integration with organic farming**: Future breeding programs will increasingly emphasize the development of turmeric varieties that are well-suited for organic and sustainable farming practices. This includes breeding for traits that enhance natural resistance and reduce the need for chemical inputs.

Breeding for biotic stress resistance in turmeric is a challenging yet essential task to ensure the crop's long-term sustainability and productivity. While traditional breeding methods have made some progress, the future of turmeric breeding lies in integrating advanced genomic tools, gene editing technologies, and global collaboration. By focusing on developing resilient, multi-resistant varieties and integrating these with sustainable farming practices, the turmeric industry can better withstand the challenges posed by biotic stresses and climate change.

9b. Challenges in breeding for abiotic stress resistance in Turmeric

Abiotic stress in turmeric (Curcuma longa) can significantly impact its growth, yield, and quality. The main types of abiotic stress that turmeric plants face include:

1. **Drought stress**: Turmeric requires a consistent water supply, especially during its growth phase. Drought conditions can lead to reduced plant growth, delayed rhizome development, and lower yields. Water stress affects the plant's ability to carry out photosynthesis, leading to stunted growth and a decrease in the production of curcuminoids, which are the main bioactive compounds in turmeric.

2. **Temperature stress**: Turmeric is sensitive to both high and low temperatures. High temperatures can cause heat stress, leading to leaf scorching, reduced chlorophyll content, and impaired photosynthesis. Low temperatures, particularly frost, can damage the rhizomes, resulting in poor sprouting and reduced plant vigor. Optimal growth occurs in warm, tropical climates, so temperature extremes can severely limit productivity.
3. **Salinity stress**: Soil salinity is a significant issue in many turmeric-growing regions. High salt levels in the soil can lead to ion toxicity and osmotic stress, which impairs water uptake and nutrient absorption. This results in poor growth, leaf chlorosis (yellowing), and reduced rhizome size and quality. Salinity stress can also reduce the concentration of essential oils and curcuminoids in the rhizomes.
4. **Waterlogging**: Turmeric requires well-drained soil for optimal growth. Waterlogging, often caused by heavy rains or poor drainage, can lead to root rot and other diseases, as the roots are deprived of oxygen. Prolonged waterlogging can severely reduce plant growth and yield by promoting the growth of harmful soil pathogens.
5. **Nutrient deficiency and imbalances**: Turmeric plants are sensitive to nutrient deficiencies, particularly of nitrogen, potassium, and phosphorus. Nutrient stress can exacerbate the effects of other abiotic stresses, such as drought and salinity, and lead to poor plant growth, reduced rhizome development, and lower curcuminoid content. Balanced fertilization is crucial for maintaining the plant's health and productivity.
6. **Light stress**: Turmeric is a shade-tolerant plant, but both excessive and insufficient light can cause stress. Too much direct sunlight can lead to leaf burn and reduced photosynthetic efficiency, while too little light can slow down growth and reduce rhizome yield.

To mitigate these abiotic stresses, farmers can implement various management practices such as mulching to conserve soil moisture, using raised beds to prevent waterlogging, and applying organic amendments to improve soil structure and fertility. Additionally, breeding programs focused on developing stress-resistant turmeric varieties could play a key role in enhancing the plant's resilience to these challenges.

Breeding Turmeric (Curcuma longa) for resistance to abiotic stress presents several significant challenges

1. **Limited genetic diversity**: Turmeric is often propagated vegetatively through rhizomes, leading to limited genetic variation within the species. This lack of diversity makes it difficult to breed for new traits, including resistance to abiotic stresses like drought, salinity, and temperature extremes. The narrow genetic base limits the potential for discovering and selecting naturally occurring stress-resistant genotypes.

2. **Complex trait inheritance**: Abiotic stress resistance is usually controlled by multiple genes (polygenic traits), making it a complex target for breeding. The interaction of these genes with environmental factors further complicates the selection process. Breeding for traits like drought tolerance or salinity resistance often involves trade-offs with other important agronomic traits, such as yield or quality, making it challenging to achieve a balance.
3. **Long breeding cycle**: Turmeric has a long growth cycle, with a typical crop taking 7 to 9 months to mature. This extended cycle slows down the breeding process because it takes a long time to evaluate new varieties for stress resistance. Multiple generations of selection are needed to develop improved varieties, which can span several years.
4. **Sterility and poor seed set**: Turmeric rarely produces seeds, and when it does, the seeds often have low viability. This reproductive challenge limits the use of conventional breeding methods, such as hybridization, which rely on cross-pollination and seed propagation. As a result, breeding programs must often rely on clonal selection, which further constrains genetic improvement.
5. **Difficulty in phenotyping**: Assessing turmeric's response to abiotic stresses like drought, salinity, and temperature variations in real-world conditions is difficult. Accurate phenotyping requires controlled environments and precise measurements, which can be resource-intensive. Moreover, environmental variability can lead to inconsistent results, complicating the selection process for stress-resistant traits.
6. **Lack of genomic resources**: Compared to other crops, turmeric has fewer genomic resources, such as molecular markers, genetic maps, or sequenced genomes. This lack of resources hampers the application of modern breeding techniques like marker-assisted selection (MAS) or genomic selection, which could accelerate the development of stress-resistant varieties.
7. **Environmental variability**: The expression of abiotic stress resistance in turmeric can vary widely depending on local environmental conditions. This variability means that a variety bred for resistance in one region might not perform well in another, making it challenging to develop universally applicable stress-resistant cultivars.
8. **Economic considerations**: Breeding for abiotic stress resistance often involves a trade-off between stress tolerance and other important traits like yield or curcuminoid content. Farmers may be hesitant to adopt new varieties if they do not offer clear economic benefits, such as higher yields or better quality, under both stressed and non-stressed conditions.

Overcoming these challenges in turmeric breeding requires a multi-faceted approach. This could include increasing genetic diversity through the introduction of wild relatives, utilizing biotechnological tools such as genetic engineering or CRISPR for targeted trait improvement, and enhancing phenotyping methods to better assess stress responses. Additionally, expanding genomic resources for turmeric could greatly aid in the development of stress-resistant varieties.

10. Mango Ginger

10a. Challenges in breeding for biotic stress resistance in Mango Ginger

Breeding for biotic stress resistance in mango ginger (Curcuma amada), a unique and lesser-known species of the ginger family known for its distinctive mango-like flavour, is crucial due to its susceptibility to various pests and diseases. Mango ginger is valued for its use in pickles, culinary dishes, and traditional medicine, making it important to enhance its resistance to biotic stresses to ensure sustainable production. Here's an overview of the current status and future prospects of breeding for biotic stress resistance in mango ginger:

Current status

1. Major biotic stresses

- **Rhizome Rot**: Similar to other members of the ginger family, mango ginger is highly susceptible to rhizome rot caused by *Pythium* spp. and *Fusarium* spp. This disease can lead to significant yield losses by causing the rhizomes to decay and the plants to wilt and die.
- **Leaf Blotch and Spot**: Fungal diseases such as leaf blotch and leaf spot, caused by *Colletotrichum* and *Phyllosticta* species, affect the leaves of mango ginger, reducing the plant's photosynthetic capacity and overall vigor.
- **Shoot Borer**: The turmeric shoot borer (*Conogethes punctiferalis*), which also affects turmeric and ginger, is a significant pest that can damage the shoots of mango ginger, leading to stunted growth and reduced rhizome yield.
- **Nematode infestation**: Root-knot nematodes (*Meloidogyne* spp.) can infest mango ginger, causing root galls and impeding the plant's nutrient and water uptake, which leads to poor growth and yield.

2. Traditional breeding efforts

- **Selection and clonal propagation**: Traditional breeding in mango ginger, like in other vegetatively propagated crops, has primarily focused on selecting disease-resistant clones from existing populations. However, the narrow genetic base due to vegetative propagation limits the potential for developing new resistant varieties.

- **Field screening**: Field trials are conducted to identify naturally resistant clones that can be propagated for cultivation. This approach is labor-intensive and time-consuming but has been somewhat effective in identifying promising resistant lines.

3. Molecular breeding

- **Limited research and development**: Molecular breeding techniques such as marker-assisted selection (MAS) are still in the early stages for mango ginger. Limited research has been conducted on identifying genetic markers linked to disease resistance in this crop.
- **Tissue culture and somaclonal variation**: Tissue culture methods are employed to induce somaclonal variation, creating genetic diversity that can be screened for resistance traits. However, the application of these techniques in mango ginger is still developing.

4. Integrated Disease Management (IDM)

- **Combining breeding with IDM**: Breeding efforts in mango ginger are often combined with integrated disease management practices, including the use of organic amendments, biological control agents, and crop rotation. Resistant varieties play a key role in these strategies to minimize the reliance on chemical pesticides.

Future prospects

1. Advances in genomic tools

- **Genomic selection and marker-assisted breeding**: As genomic tools become more accessible, their application in mango ginger breeding could accelerate the development of resistant varieties. Identifying and utilizing genetic markers linked to resistance traits would make breeding more efficient.
- **Gene editing technologies**: Gene editing tools like CRISPR/Cas9 have the potential to revolutionize breeding in mango ginger by enabling precise modifications that enhance resistance to specific pathogens or pests. These technologies could overcome the limitations of traditional breeding by introducing resistance traits directly.

2. Exploring wild relatives

- **Utilizing wild germplasm**: Wild relatives of mango ginger may possess resistance traits that can be introgressed into cultivated varieties through cross-breeding or advanced biotechnological methods. This strategy could expand the genetic diversity available for breeding and enhance resistance.

3. Development of multi-resistant varieties

- **Pyramiding resistance genes**: Future breeding efforts are likely to focus on pyramiding multiple resistance genes into a single mango ginger variety to provide broad-spectrum resistance against various biotic stresses. This approach can create more resilient and durable varieties.

4. Climate-resilient breeding

- **Breeding for combined stress resistance**: As climate change alters the dynamics of pest and disease incidence, breeding programs will need to develop mango ginger varieties that can withstand both biotic and abiotic stresses, such as drought and high temperatures.

5. Collaborative research and global efforts

- **International collaboration**: Collaborative research between institutions in tropical and subtropical regions where mango ginger is cultivated can facilitate the sharing of genetic resources, breeding methodologies, and technological advancements. This global cooperation is crucial for overcoming the challenges of breeding for biotic stress resistance in this crop.

6. Sustainable Agricultural practices

- **Integration with organic farming**: Future breeding programs will emphasize the development of mango ginger varieties suitable for organic and sustainable farming practices. This includes breeding for traits that enhance natural resistance, reduce the need for chemical inputs, and promote environmental sustainability.

Breeding for biotic stress resistance in mango ginger is a developing field with significant potential to improve the crop's sustainability and productivity. While current efforts are limited by the narrow genetic base and early stages of molecular breeding research, the future looks promising with the integration of advanced genomic tools, gene editing technologies, and international collaboration. By focusing on developing resilient, multi-resistant varieties and incorporating these into sustainable farming practices, it is possible to enhance the long-term viability of mango ginger cultivation.

10b. Challenges in breeding for abiotic stress resistance in Mango Ginger

Mango ginger (*Curcuma amada*), a species closely related to turmeric and ginger, also faces significant challenges due to abiotic stress. The main types of abiotic stress that impact mango ginger include:

1. **Drought stress**: Mango ginger, like other rhizomatous plants, requires a steady supply of water for optimal growth and rhizome development.

Drought conditions can lead to reduced growth, delayed rhizome formation, and lower yields. Water stress affects the plant's ability to photosynthesize effectively, leading to stunted growth and a decrease in the quality and quantity of the rhizomes.

2. **Temperature stress**: Mango ginger is sensitive to both high and low temperatures. High temperatures can cause heat stress, resulting in leaf burn, reduced chlorophyll content, and impaired photosynthesis. On the other hand, low temperatures, especially frost, can damage the rhizomes and affect their sprouting ability, leading to poor growth and reduced yield.
3. **Salinity stress**: Soil salinity is a significant concern in regions where irrigation water is saline or where soils naturally have high salt content. Salinity stress can cause ion toxicity, osmotic stress, and nutrient imbalances, leading to poor plant growth, leaf chlorosis, and reduced rhizome size and quality. High salinity levels can also affect the essential oil content, which is a key quality attribute of mango ginger.
4. **Waterlogging**: Like other rhizomatous plants, mango ginger requires well-drained soil. Waterlogging, caused by heavy rains or poor drainage, can lead to root rot and other diseases due to oxygen deprivation in the roots. Waterlogged conditions can also encourage the growth of soil-borne pathogens, which further stress the plant and reduce its productivity.
5. **Nutrient deficiencies**: Nutrient imbalances or deficiencies, particularly of nitrogen, potassium, and phosphorus, can exacerbate the effects of abiotic stresses such as drought and salinity. Nutrient stress can result in poor plant growth, reduced rhizome development, and lower concentrations of essential oils and other bioactive compounds in the rhizomes.
6. **Light stress**: Mango ginger is somewhat shade-tolerant but requires adequate light for optimal growth. Excessive direct sunlight can lead to leaf scorching and reduced photosynthetic efficiency, while insufficient light can slow growth and reduce rhizome development.

Addressing these abiotic stresses in mango ginger cultivation involves careful management practices, including proper irrigation scheduling, soil amendment, and drainage management. Additionally, selecting or breeding varieties that are more tolerant to these stresses could help improve resilience and productivity. Research into the genetic and physiological mechanisms underlying stress tolerance in mango ginger could also provide insights into developing more resilient cultivars.

Breeding Mango Ginger (*Curcuma amada*) for resistance to abiotic stresses presents several significant challenges

1. **Limited genetic diversity**: Like many other species in the Curcuma genus, mango ginger has limited genetic diversity due to its vegetative propagation

method (primarily through rhizomes). This low genetic variation restricts the pool of traits available for breeding, making it difficult to introduce and select for new traits, such as resistance to drought, salinity, and extreme temperatures.

2. **Polygenic nature of stress resistance**: Abiotic stress resistance is often a polygenic trait, meaning it is controlled by multiple genes. The interaction of these genes with the environment complicates the breeding process, as selecting for one trait may inadvertently affect others. The complex inheritance patterns make it challenging to develop varieties that consistently express desired levels of resistance across different environments.
3. **Reproductive challenges**: Mango ginger, like other rhizomatous plants, has poor seed set and is typically propagated vegetatively. This limits the use of conventional breeding techniques, such as hybridization, which rely on sexual reproduction. As a result, breeding programs must often depend on clonal selection, which further limits genetic variation and slows the development of new, improved varieties.
4. **Long breeding cycle**: The long growth cycle of mango ginger, which can take several months from planting to harvest, delays the breeding process. Evaluating new varieties for stress resistance requires multiple growing seasons, making the breeding process time-consuming and resource-intensive.
5. **Difficulty in phenotyping**: Accurately assessing the response of mango ginger to abiotic stresses like drought, salinity, and temperature extremes in field conditions is challenging. Reliable phenotyping requires controlled environments and precise measurements, which are difficult to achieve on a large scale. Additionally, environmental variability can lead to inconsistent results, further complicating the selection process.
6. **Lack of genomic resources**: Compared to other major crops, mango ginger lacks comprehensive genomic resources, such as molecular markers, genetic maps, or sequenced genomes. This scarcity limits the application of advanced breeding techniques, such as marker-assisted selection (MAS) or genomic selection, which could otherwise speed up the development of stress-resistant varieties.
7. **Environmental variability**: The expression of abiotic stress resistance in mango ginger can vary significantly depending on local environmental conditions. A variety that performs well under stress in one region may not perform as well in another, making it difficult to develop universally applicable stress-resistant cultivars.
8. **Economic considerations**: There are often trade-offs between stress resistance and other important traits such as yield, flavour, or essential oil content in mango ginger. Farmers may be reluctant to adopt new varieties

if they do not provide clear economic advantages, such as higher yields or better quality, even under stressful conditions.

Addressing these challenges requires a comprehensive approach, including increasing genetic diversity through the introduction of new genetic material, improving phenotyping methods to better assess stress tolerance, and expanding the genomic resources available for mango ginger. Integrating biotechnological tools such as CRISPR or genetic engineering could also play a crucial role in developing mango ginger varieties with enhanced abiotic stress resistance.

11. Black Turmeric

11a. Challenges for biotic stress resistance breeding in Black Turmeric

Breeding black turmeric (*Curcuma caesia*) for resistance to biotic stresses, such as pests and diseases, involves several challenges:

1. **Limited genetic diversity**: Black turmeric, like many other Curcuma species, is often propagated vegetatively through rhizomes, leading to low genetic diversity. This narrow genetic base limits the potential for identifying and introducing resistance traits against biotic stresses such as fungal, bacterial, and viral pathogens, as well as insect pests.
2. **Complex pathogen interactions**: Biotic stress resistance often involves complex interactions between the plant and various pathogens or pests. The resistance mechanisms are typically controlled by multiple genes, making it difficult to identify and select for these traits in breeding programs. Additionally, pathogens can evolve rapidly, leading to the breakdown of resistance in previously resistant varieties.
3. **Reproductive challenges**: Black turmeric, similar to other turmeric species, has low seed set and is primarily propagated through rhizomes. This vegetative propagation limits the use of conventional breeding techniques, such as cross-pollination, which are essential for introducing and combining resistance traits from different sources.
4. **Long growth cycle**: Black turmeric has a long growth cycle, typically requiring several months from planting to harvest. This extended cycle delays the evaluation of new genotypes for biotic stress resistance, making the breeding process slow and resource-intensive. Multiple growing seasons are needed to assess the durability of resistance traits.
5. **Difficulty in disease and pest screening**: Accurately screening black turmeric for resistance to various biotic stresses is challenging, particularly under field conditions where multiple factors can influence disease and pest pressure. Controlled environment screening is necessary but can be expensive and logistically challenging, especially when dealing with a range of pathogens and pests.

6. **Lack of genomic resources**: There is a scarcity of genomic resources available for black turmeric, including molecular markers, genetic maps, and sequenced genomes. This lack of resources hampers the application of modern breeding techniques like marker-assisted selection (MAS), which could facilitate the identification and selection of biotic stress resistance traits.
7. **Pathogen and pest variability**: The pathogens and pests that affect black turmeric can vary significantly across different regions and environments. A variety that is resistant to a particular pathogen or pest in one region may not exhibit the same level of resistance in another, complicating the development of universally resistant varieties.
8. **Economic viability**: Breeding for biotic stress resistance often involves trade-offs with other important traits such as yield, rhizome quality, and bioactive compound content. Farmers may be hesitant to adopt new varieties that offer resistance to pests or diseases if these varieties do not provide clear economic benefits, such as higher yields or better quality under non-stressed conditions.

To overcome these challenges, a multifaceted approach is required. This could include increasing genetic diversity through the introduction of wild relatives or induced mutations, enhancing screening methods for better identification of resistant genotypes, and expanding genomic resources to enable the application of advanced breeding techniques. Biotechnological approaches, such as genetic engineering or CRISPR, could also be explored to introduce or enhance biotic stress resistance in black turmeric.

11b. Challenges for abiotic stress resistance breeding in Black Turmeric

Black turmeric (*Curcuma caesia*) is subject to various abiotic stresses, which can significantly affect its growth, yield, and quality. The key types of abiotic stress affecting black turmeric include:

1. **Drought stress**: Black turmeric, like other members of the Curcuma genus, requires a consistent water supply for optimal growth and rhizome development. Drought conditions can lead to water stress, resulting in stunted growth, reduced leaf area, and lower rhizome yields. Prolonged drought stress can also decrease the concentration of curcumin and other bioactive compounds in the rhizomes.
2. **Temperature stress**: Black turmeric is sensitive to temperature extremes. High temperatures can lead to heat stress, causing leaf scorching, reduced chlorophyll content, and impaired photosynthesis. Low temperatures, particularly frost, can damage rhizomes, inhibit sprouting, and reduce plant vigor. Optimal growth for black turmeric occurs in warm, tropical climates, making it vulnerable to both heat and cold stress.

3. **Salinity stress**: Soil salinity can be a significant issue for black turmeric, especially in regions where irrigation water has a high salt content or in soils with natural salinity. Salinity stress leads to ion toxicity, osmotic stress, and nutrient imbalances, which can cause poor plant growth, chlorosis (yellowing of leaves), and reduced rhizome size and quality. High salt levels can also negatively affect the essential oil content and the overall medicinal properties of the rhizomes.
4. **Waterlogging**: Black turmeric requires well-drained soil, as waterlogged conditions can lead to root rot and other diseases due to oxygen deprivation in the roots. Prolonged waterlogging, often caused by heavy rains or poor drainage, can severely reduce plant growth and yield. Waterlogged conditions can also promote the growth of soil-borne pathogens, further stressing the plants.
5. **Nutrient deficiency**: Nutrient imbalances, particularly deficiencies in nitrogen, potassium, and phosphorus, can exacerbate the effects of other abiotic stresses such as drought and salinity. Nutrient stress can lead to poor growth, reduced rhizome development, and lower concentrations of curcuminoids and essential oils, which are key quality parameters for black turmeric.
6. **Light stress**: While black turmeric is somewhat shade-tolerant, both excessive and insufficient light can cause stress. Excessive direct sunlight can lead to leaf scorching and reduced photosynthetic efficiency, while too little light can slow down growth and reduce rhizome yield. Balancing light exposure is crucial for the healthy development of the plant.

Mitigation strategies

To mitigate these abiotic stresses, farmers and cultivators of black turmeric can adopt several strategies:

- **Irrigation management**: Implementing efficient irrigation systems, such as drip irrigation, can help manage water supply during drought conditions and prevent waterlogging.
- **Soil management**: Amending soil with organic matter can improve drainage, reduce salinity, and enhance nutrient availability, helping the plant cope with various abiotic stresses.
- **Mulching**: Applying mulch can help conserve soil moisture, moderate soil temperature, and reduce the impact of temperature extremes.
- **Use of stress-tolerant varieties**: Where available, selecting or breeding stress-tolerant varieties of black turmeric can improve resilience to abiotic stresses.

Research into the genetic and physiological responses of black turmeric to these stresses can provide further insights into developing more resilient cultivars and effective management practices.

Breeding Black Turmeric (*Curcuma caesia*) for abiotic stress resistance involves several significant challenges

1. **Genetic complexity**: Black turmeric, like many other plants, has a complex genome, which makes it difficult to identify and manipulate the genes responsible for stress resistance.
2. **Lack of genomic resources**: There is limited genomic information available for black turmeric, which hinders the development of molecular markers and other tools necessary for efficient breeding.
3. **Environmental variability**: Abiotic stresses such as drought, salinity, and temperature extremes can vary significantly across different environments, making it difficult to breed a universally resistant variety.
4. **Phenotyping difficulties**: Accurately measuring the plant's response to abiotic stress (phenotyping) is often challenging, requiring controlled environments and sophisticated technologies.
5. **Limited breeding programs**: There are few established breeding programs for black turmeric, leading to slow progress in developing stress-resistant varieties.
6. **Clonal propagation**: Black turmeric is often propagated clonally, which limits genetic diversity and makes it harder to introduce new traits through traditional breeding methods.

Addressing these challenges requires a combination of advanced genomic techniques, improved phenotyping methods, and a better understanding of the plant's biology and response to stress.

12. Coriander

12a. Challenges for biotic stress resistance breeding in Coriander

Breeding coriander (*Coriandrum sativum*) for biotic stress resistance presents several challenges:

1. **Genetic diversity**: Coriander has limited genetic diversity, especially in the context of disease resistance traits. This makes it difficult to find or introduce resistance genes into breeding populations.
2. **Lack of comprehensive genomic resources**: Similar to many other minor crops, coriander lacks extensive genomic resources such as sequenced genomes and molecular markers, which are essential for identifying and incorporating resistance traits.

3. **Pathogen variability**: The pathogens that affect coriander, such as fungi, bacteria, and viruses, can be highly variable. This variability makes it challenging to breed a single variety of coriander that can resist multiple strains of pathogens.
4. **Complex resistance mechanisms**: The mechanisms of resistance to biotic stresses are often complex, involving multiple genes and pathways. This complexity makes it difficult to identify and select for resistance traits during the breeding process.
5. **Phenotyping difficulties**: Accurately identifying and measuring resistance to diseases and pests (phenotyping) in coriander is challenging. It often requires controlled conditions and may involve lengthy and expensive testing.
6. **Limited breeding programs**: As a minor crop, coriander has fewer dedicated breeding programs compared to major crops like wheat or rice. This limits the resources and research dedicated to developing resistant varieties.
7. **Environmental influence**: The expression of resistance traits can be influenced by environmental conditions, making it difficult to consistently breed coriander varieties that perform well under various conditions.
8. **Slow breeding cycles**: Coriander has a relatively slow breeding cycle, which can delay the development of resistant varieties. This is particularly problematic when new pathogen strains emerge rapidly.

Overcoming these challenges requires increased investment in coriander research, the development of more robust genomic tools, and collaboration between breeders, pathologists, and geneticists to better understand and exploit the plant's natural resistance mechanisms.

12b. Challenges for abiotic stress resistance breeding in Coriander

Abiotic stress in coriander (*Coriandrum sativum*) refers to the adverse effects that non-living environmental factors can have on the plant's growth, development, and yield. The primary abiotic stresses that affect coriander include:

1. Drought stress

- **Impact**: Water scarcity can severely affect coriander, leading to reduced seed germination, stunted growth, lower biomass, and poor seed yield. Drought stress can also accelerate flowering and shorten the vegetative phase, which impacts overall productivity.
- **Management**: Drought-tolerant varieties, improved irrigation techniques, and soil moisture conservation practices are essential for mitigating this stress.

2. Salinity stress

- **Impact**: High soil salinity can inhibit coriander's seed germination and plant growth. Salt stress often causes ion toxicity, osmotic stress, and nutrient imbalances, leading to leaf chlorosis, reduced photosynthesis, and ultimately lower yield.
- **Management**: The use of salt-tolerant varieties, soil amendments like gypsum, and proper irrigation management can help alleviate salinity stress.

3. Temperature extremes

- **Heat stress**: High temperatures can cause heat stress, leading to poor germination, leaf scorching, reduced flowering, and lower seed set. It can also cause bolting, where the plant flowers prematurely, affecting the quality of the leaves.
- **Cold stress**: Frost or low temperatures can damage coriander, particularly during the early stages of growth. Cold stress can lead to poor germination, leaf damage, and in severe cases, plant death.
- **Management**: Sowing at optimal times to avoid extreme temperatures, using mulches, and selecting heat- or cold-tolerant varieties can help manage temperature-related stresses.

4. Nutrient deficiency

- **Impact**: Nutrient stress, particularly deficiencies in nitrogen, phosphorus, or potassium, can limit coriander's growth, leading to yellowing of leaves, poor root development, and reduced yield.
- **Management**: Soil testing and appropriate fertilization strategies are necessary to ensure that coriander plants receive adequate nutrients.

5. Waterlogging

- **Impact**: Excess water in the soil, often due to heavy rainfall or poor drainage, can lead to waterlogging. This condition deprives the roots of oxygen, leading to root rot, reduced nutrient uptake, and eventual plant death.
- **Management**: Improving soil drainage, using raised beds, and avoiding over-irrigation are crucial to prevent waterlogging.

6. Light stress

- **Impact**: Insufficient light (shade stress) can lead to elongated, weak plants with poor leaf quality, while excessive light can cause photodamage, reducing chlorophyll content and photosynthesis efficiency.
- **Management**: Ensuring optimal plant spacing and growing coriander in suitable light conditions can mitigate light stress.

Managing abiotic stress in coriander involves a combination of selecting stress-tolerant varieties, adopting proper agronomic practices, and implementing effective water and nutrient management strategies.

Breeding Coriander (*Coriandrum sativum*) for resistance to abiotic stress involves several challenges:

1. Limited genetic resources

- **Challenge**: Coriander has a narrow genetic base, with limited access to wild relatives or landraces that might possess genes for abiotic stress resistance. This restricts the genetic variability available for breeding programs.
- **Impact**: The lack of genetic diversity makes it difficult to develop varieties that can withstand a wide range of abiotic stresses such as drought, salinity, or temperature extremes.

2. Complexity of abiotic stress traits

- **Challenge**: Abiotic stress tolerance is often controlled by multiple genes (polygenic), making the breeding process complex. The interaction of these genes with each other and with environmental factors adds another layer of complexity.
- **Impact**: The polygenic nature of stress tolerance complicates the identification and selection of desirable traits, slowing down the breeding process.

3. Environmental variability

- **Challenge**: The expression of abiotic stress tolerance traits can vary significantly depending on environmental conditions such as soil type, climate, and management practices.
- **Impact**: This variability makes it challenging to develop coriander varieties that perform consistently across different environments, necessitating extensive field trials in multiple locations.

4. Lack of molecular tools

- **Challenge**: There is a shortage of molecular markers and genomic resources for coriander, which are essential for marker-assisted selection (MAS) and other advanced breeding techniques.
- **Impact**: The absence of these tools slows the identification of stress tolerance genes and their incorporation into breeding programs, making the process more reliant on traditional, time-consuming methods.

5. Difficulties in phenotyping

- **Challenge**: Accurately measuring and assessing the plant's response to abiotic stress (phenotyping) is challenging. This often requires controlled environments, precise measurements, and the ability to simulate stress conditions, which can be resource intensive.
- **Impact**: Inconsistent or imprecise phenotyping can lead to the selection of suboptimal traits, reducing the effectiveness of breeding efforts.

6. Long breeding cycles

- **Challenge**: Coriander has a relatively long breeding cycle, and the expression of some abiotic stress resistance traits may not be evident until later stages of plant development.
- **Impact**: The long breeding cycle delays the development and release of new, stress-resistant varieties, slowing the overall progress in improving coriander's resilience.

7. Integration of multiple stress resistances

- **Challenge**: Breeding for resistance to multiple abiotic stresses (e.g., drought and salinity) simultaneously is challenging because the mechanisms underlying tolerance to different stresses may conflict or require different strategies.
- **Impact**: Achieving a balance between resistance to multiple stresses without compromising other important traits like yield or quality is difficult, requiring careful planning and trade-offs in breeding programs.

8. Economic and research constraints

- **Challenge**: As a relatively minor crop, coriander receives less attention and funding compared to major crops like wheat or rice. This limits the resources available for research and development in coriander breeding.
- **Impact**: The limited investment in coriander breeding programs slows the development of new varieties and the application of advanced breeding techniques to improve abiotic stress resistance.

Addressing these challenges requires increased research focus, the development of genomic resources, improved phenotyping methods, and the integration of modern breeding techniques to enhance coriander's resilience to abiotic stresses.

13. Fennel

13a. Challenges for biotic stress resistance breeding in Fennel

Breeding fennel (*Foeniculum vulgare*) for resistance to biotic stress involves several challenges:

1. Limited genetic diversity

- **Challenge**: Fennel has a relatively narrow genetic base, which limits the availability of naturally occurring resistance traits to pests and diseases.
- **Impact**: This restricted genetic diversity makes it difficult to find or introduce resistance genes, slowing down the breeding process and limiting the potential for developing highly resistant varieties.

2. Complexity of resistance traits

- **Challenge**: Biotic stress resistance is often governed by complex, multi-gene interactions (polygenic inheritance). Additionally, resistance can be specific to certain pathogens or pests, making it difficult to breed a variety with broad-spectrum resistance.
- **Impact**: The complexity of resistance traits complicates the selection process, requiring detailed genetic analysis and prolonged breeding cycles to achieve durable resistance.

3. Lack of genomic resources

- **Challenge**: There is a scarcity of genomic tools and resources, such as molecular markers, genetic maps, and sequenced genomes, which are essential for marker-assisted selection (MAS) and other advanced breeding techniques in fennel.
- **Impact**: The absence of these resources hinders the ability to efficiently identify and incorporate resistance genes, making the breeding process more reliant on traditional methods, which are slower and less precise.

4. Pathogen and pest variability

- **Challenge**: The pathogens and pests affecting fennel, such as fungi, bacteria, and insects, can be highly variable. Different strains or species may exhibit different levels of virulence or resistance to control measures.
- **Impact**: The variability in pests and pathogens requires the development of multiple resistance mechanisms, which can be difficult to achieve within a single variety. This increases the complexity and duration of breeding programs.

5. Phenotyping difficulties

- **Challenge**: Accurately assessing and measuring fennel's resistance to various biotic stresses (phenotyping) is challenging, particularly in field conditions where multiple biotic factors may be present simultaneously.
- **Impact**: Inconsistent or inaccurate phenotyping can lead to the selection of suboptimal traits, reducing the effectiveness of breeding efforts and delaying the development of resistant varieties.

6. Environmental influence on resistance expression

- **Challenge**: The expression of resistance traits in fennel can be influenced by environmental factors such as temperature, humidity, and soil conditions, which may vary widely across growing regions.
- **Impact**: This environmental dependency complicates the breeding process, as a variety that is resistant in one environment may not perform as well in another, necessitating extensive multi-location trials.

7. Slow breeding cycle

- **Challenge**: Fennel has a relatively long breeding cycle, and the expression of resistance traits may not be fully evident until later stages of plant development.
- **Impact**: The long breeding cycle delays the development and release of new, biotic stress-resistant varieties, making it difficult to respond quickly to emerging threats from pests and pathogens.

8. Economic constraints and research focus

- **Challenge**: As a relatively minor crop compared to staples like wheat or rice, fennel receives less research attention and funding, limiting the resources available for breeding programs.
- **Impact**: The limited investment in fennel breeding programs restricts the application of modern breeding technologies and slows progress in developing biotic stress-resistant varieties.

To overcome these challenges, increased research focus on fennel, the development of genomic resources, and the application of advanced breeding techniques such as marker-assisted selection (MAS) and genomic selection (GS) are essential. Additionally, collaboration between breeders, pathologists, and entomologists is needed to better understand and exploit natural resistance mechanisms in fennel.

13b. Challenges for abiotic stress resistance breeding in Fennel

Abiotic stress in fennel, as with other plants, refers to the impact of environmental factors like drought, salinity, extreme temperatures, and heavy metal contamination on the plant's health and productivity. These stresses can lead to several physiological and biochemical changes, including reduced photosynthesis, ion toxicity, oxidative stress, and disrupted metabolic processes. To mitigate these effects, research often focuses on developing stress-resistant fennel varieties, optimizing agricultural practices, and using protective treatments like antioxidants or growth regulators.

Breeding fennel for resistance to abiotic stresses like drought, salinity, and temperature extremes presents several challenges:

1. **Genetic diversity**: Fennel, like many other crops, may have limited genetic diversity, especially in the traits associated with abiotic stress resistance. This limits the availability of genes that breeders can use to enhance stress tolerance.
2. **Complex trait inheritance**: Abiotic stress resistance is typically governed by multiple genes (polygenic), making it complex to select and breed for these traits. The interactions between these genes and the environment further complicate the breeding process.
3. **Phenotyping difficulties**: Accurately measuring and evaluating traits related to abiotic stress tolerance (like root depth, water use efficiency, or ion uptake) is difficult. This requires controlled environments and sophisticated technology, which are resource-intensive.
4. **Environmental variability**: Abiotic stresses are highly variable across different environments and seasons. This variability makes it difficult to consistently breed and select for stress-resistant traits in fennel across different regions.
5. **Time-consuming process**: Breeding for abiotic stress resistance is a long-term process. Developing a new variety with enhanced resistance can take several years, from initial crosses to field trials and final selection.
6. **Lack of genomic resources**: Compared to major crops like wheat or rice, fennel may have fewer genomic resources available (like reference genomes, marker-assisted selection tools), which are crucial for modern breeding approaches.
7. **Trade-offs with yield and quality**: Often, traits that confer stress resistance might negatively impact other desirable traits, such as yield or essential oil content. Breeders must carefully balance these trade-offs to develop commercially viable fennel varieties.

Addressing these challenges requires a combination of traditional breeding methods, advanced biotechnological tools (like CRISPR or marker-assisted selection), and a deeper understanding of the plant's physiology under stress conditions.

14. Fenugreek

14a. Challenges for biotic stress resistance breeding in Fenugreek

Breeding fenugreek for resistance to biotic stresses, such as pests and diseases, poses several challenges:

1. **Limited genetic resources**: Fenugreek has a relatively narrow genetic base, which limits the availability of resistance genes against various pests and pathogens. This makes it difficult to identify and incorporate resistance traits into new varieties.

2. **Complexity of resistance**: Biotic stress resistance is often controlled by multiple genes, and the interactions between these genes can be complex. Resistance may also involve both qualitative and quantitative traits, making it challenging to breed for durable resistance.
3. **Co-evolution of pests and pathogens**: Pests and pathogens can rapidly evolve to overcome resistance in plants. This arms race between fenugreek and its biotic stressors requires continuous monitoring and breeding efforts to keep up with evolving threats.
4. **Lack of advanced breeding tools**: Fenugreek may not have as many genomic resources (such as molecular markers, reference genomes) as other major crops. This limits the application of modern breeding techniques like marker-assisted selection, which can accelerate the development of resistant varieties.
5. **Phenotyping challenges**: Accurately identifying and evaluating resistance to biotic stresses in the field is difficult. Factors such as environmental conditions, pest/pathogen pressure, and interactions with other stressors can complicate phenotyping.
6. **Time-intensive breeding process**: Developing biotic stress-resistant fenugreek varieties is a long-term endeavour. It involves multiple cycles of crossing, selection, and field testing, which can take many years to yield a successful variety.
7. **Balancing resistance with other traits**: Breeding for resistance often involves trade-offs with other important traits such as yield, flavour, or nutritional content. Breeders must carefully manage these trade-offs to ensure that new varieties meet the needs of farmers and consumers.
8. **Integration of resistance sources**: Resistance genes may need to be sourced from wild relatives or other species, which can introduce undesirable traits. Breeders must work to integrate resistance while minimizing negative impacts on other important agronomic traits.

Addressing these challenges requires a combination of traditional breeding approaches, such as crossbreeding and selection, with modern techniques like genomics-assisted breeding, to develop fenugreek varieties that are resilient to biotic stresses. Collaboration between breeders, plant pathologists, and entomologists is also crucial for identifying resistance genes and understanding the mechanisms of resistance.

14b. Challenges for abiotic stress resistance breeding in Fenugreek

Abiotic stress in fenugreek, similar to other crops, refers to the negative effects of non-living environmental factors such as drought, salinity, extreme temperatures, and heavy metal toxicity. These stresses can severely impact the growth, development, yield, and overall health of fenugreek plants. Here's an overview of the major abiotic stresses that affect fenugreek and the challenges associated with them:

1. Drought stress

- **Impact**: Fenugreek is sensitive to water scarcity, particularly during the seedling and flowering stages. Drought can lead to reduced photosynthesis, impaired nutrient uptake, and decreased seed yield. Water stress often results in smaller leaves, reduced biomass, and lower seed production.
- **Challenge**: Developing drought-resistant fenugreek varieties is challenging due to the complex nature of drought tolerance, which involves multiple physiological and biochemical pathways.

2. Salinity stress

- **Impact**: High salt concentrations in the soil can cause osmotic stress and ion toxicity in fenugreek. This leads to disrupted water uptake, nutrient imbalances, chlorosis (yellowing of leaves), stunted growth, and ultimately reduced seed yield.
- **Challenge**: Breeding for salinity tolerance is complicated because salt tolerance involves both exclusion mechanisms (preventing salt from entering the plant) and tolerance mechanisms (coping with the salt that enters).

3. Temperature extremes

- **Heat stress**: High temperatures, especially during the flowering stage, can reduce pollen viability, leading to poor seed set and lower yields. Heat stress also accelerates phenological development, potentially shortening the growing period.
- **Cold stress**: Frost and low temperatures can damage fenugreek seedlings, leading to poor establishment and reduced growth.
- **Challenge**: Breeding fenugreek varieties that can tolerate both high and low temperatures is difficult because these traits are often controlled by multiple genes and can vary widely across environments.

4. Heavy metal stress

- **Impact**: Fenugreek grown in soils contaminated with heavy metals like cadmium, lead, or arsenic can suffer from oxidative stress, reduced growth, and lower yields. These metals can also accumulate in the plant tissues, posing health risks if consumed.
- **Challenge**: Heavy metal tolerance is complex and involves mechanisms like chelation, sequestration, and antioxidant defence, making it difficult to breed for this trait.

5. Nutrient deficiency

- **Impact**: Deficiencies in essential nutrients, particularly nitrogen, phosphorus, and potassium, can significantly affect fenugreek growth and productivity. Nutrient stress often results in poor plant vigour, reduced leaf area, and lower seed yield.
- **Challenge**: Addressing nutrient deficiency through breeding is challenging due to the interaction between nutrient availability and other environmental factors..

Strategies for mitigation

- **Breeding**: Developing stress-resistant varieties through traditional breeding and modern biotechnological approaches, such as marker-assisted selection and genetic modification, is essential.
- **Agronomic practices**: Implementing better water management, soil amendments, and the use of stress-mitigating compounds (like bio-stimulants) can help reduce the impact of abiotic stresses.
- **Biotechnological approaches**: Advances in genomics, transcriptomics, and proteomics can aid in identifying stress-responsive genes and developing fenugreek varieties with enhanced abiotic stress tolerance.

Abiotic stress in fenugreek poses significant challenges to sustainable production, particularly in regions prone to drought, salinity, or extreme temperatures. Addressing these challenges requires a multi-disciplinary approach, combining breeding, agronomy, and biotechnology to develop resilient fenugreek varieties.

15. Cumin

15a. Challenges for biotic stress resistance breeding in Cumin

Breeding cumin (*Cuminum cyminum*) for biotic stress resistance presents several challenges due to the crop's unique characteristics, genetic limitations, and the nature of the stresses it faces. Here are some of the key challenges:

1. Limited genetic diversity

- **Narrow genetic base**: Cumin has a relatively narrow genetic base, which limits the availability of diverse traits needed for breeding programs. This lack of genetic diversity makes it difficult to introduce and combine traits for resistance to various biotic stresses.
- **Germplasm access**: Access to diverse germplasm collections is often limited, which constrains the ability of breeders to find and utilize resistant genes.

2. Complexity of biotic stresses

- **Pathogen diversity**: Cumin is susceptible to a wide range of pathogens, including fungi, bacteria, viruses, and nematodes. Breeding for resistance to multiple pathogens is challenging, especially when different pathogens require different resistance mechanisms.
- **Emerging diseases**: New diseases or more virulent strains of existing pathogens can emerge, rendering existing resistant varieties ineffective. Continuous monitoring and updating of breeding targets are required.

3. Quantitative resistance

- **Polygenic traits**: Resistance to many biotic stresses in cumin is controlled by multiple genes (polygenic), which makes breeding efforts more complex. The polygenic nature of resistance means that selection for these traits is less straightforward and requires more advanced breeding techniques, such as marker-assisted selection (MAS).
- **Partial resistance**: In many cases, the resistance is not complete, leading to varieties that are only partially resistant and may still suffer significant yield losses under high disease pressure.

4. Long breeding cycle

- **Perennial nature**: Cumin is often grown as a perennial crop in some regions, which extends the breeding cycle and delays the development of new resistant varieties.
- **Breeding timeframe**: The long life cycle of cumin plants, combined with the need to evaluate resistance across multiple growing seasons and environments, prolongs the breeding process.

5. Environmental Influence

- **Genotype-Environment Interaction**: The expression of resistance traits can be highly dependent on environmental conditions. A variety that shows resistance in one location or season might not perform as well in another, making it difficult to breed universally resistant varieties.
- **Climate change**: Changing climate conditions can alter the prevalence and severity of biotic stresses, complicating breeding efforts.

6. Limited research and resources

- **Lack of genomic resources**: Unlike major crops, cumin has limited genomic resources, such as genetic maps or sequenced genomes. This lack of resources hampers the identification of resistance genes and the development of molecular markers.

- **Funding and Infrastructure**: Research on cumin often receives less attention and funding compared to major crops like wheat or rice. This limits the availability of advanced breeding tools and resources needed to develop biotic stress-resistant varieties.

7. Breeding methods

- **Conventional breeding limitations**: Traditional breeding methods are time-consuming and less efficient for dealing with complex biotic stresses. Modern methods like genetic engineering and CRISPR, while promising, are not yet widely applied in cumin breeding due to technical, regulatory, and acceptance issues.
- **Hybridization difficulties**: Cumin is a cross-pollinated crop, and achieving controlled hybridization can be difficult, especially when trying to introduce specific resistance traits.

8. Farmers' adoption

- **Varietal adoption**: Even when resistant varieties are developed, getting farmers to adopt them can be challenging due to factors like seed availability, cost, and the farmers' familiarity with the new varieties. Resistance traits must be combined with other desirable agronomic traits to ensure acceptance.

Breeding cumin for biotic stress resistance is a complex task that requires overcoming significant genetic, environmental, and logistical challenges. Advances in molecular breeding, increased research investment, and access to diverse germplasm could help mitigate these challenges and lead to the development of more resilient cumin varieties. However, this will require coordinated efforts from researchers, breeders, and policymakers.

15b. Challenges for abiotic stress resistance breeding in Cumin

Breeding cumin (*Cuminum cyminum*) for abiotic stress resistance presents several significant challenges due to the crop's specific needs, the nature of abiotic stresses, and the limitations of current breeding methods. Here are the primary challenges:

1. Complexity of abiotic stresses

- **Diverse stress factors**: Cumin is exposed to a range of abiotic stresses, including drought, high salinity, extreme temperatures (both heat and cold), and nutrient deficiencies. Each of these stresses requires different physiological and genetic responses, making it challenging to breed a single variety with broad-spectrum abiotic stress resistance.
- **Multiple stresses simultaneously**: In many growing regions, cumin faces more than one abiotic stress at a time (e.g., drought and heat). Breeding for

tolerance to multiple stresses simultaneously is complex and often results in trade-offs between stress resistance and other agronomic traits.

2. Limited genetic resources

- **Narrow genetic base**: Like many minor crops, cumin has a limited genetic base, which restricts the availability of genes that could confer resistance to various abiotic stresses. This makes it difficult to find or develop varieties that can withstand harsh conditions.
- **Germplasm accessibility**: There is often limited access to diverse and stress-tolerant germplasm collections, which constrains breeding efforts. The lack of well-characterized germplasm with known stress tolerance traits hinders the development of resistant varieties.

3. Polygenic nature of stress tolerance

- **Quantitative traits**: Abiotic stress tolerance is typically controlled by multiple genes, often with small effects, making these traits quantitatively inherited. This polygenic nature complicates breeding, as it requires selecting for multiple genes across different loci, which can be time-consuming and technically challenging.
- **Environmental influence**: The expression of genes related to abiotic stress tolerance is often highly influenced by environmental conditions. This genotype-environment interaction makes it difficult to achieve consistent performance across different environments.

4. Phenotyping challenges

- **Difficulties in measurement**: Accurately measuring abiotic stress tolerance (e.g., drought or salinity tolerance) in field conditions is challenging due to the variability in stress levels and environmental factors. Reliable phenotyping requires controlled environments, which are not always available.
- **Indirect selection**: Abiotic stress tolerance often needs to be assessed through indirect traits (e.g., root depth for drought tolerance), which may not be straightforward to measure and may not always correlate perfectly with field performance.

5. Long Breeding cycles

- **Extended timeframes**: Developing and evaluating cumin varieties for abiotic stress tolerance requires several growing seasons to ensure stability and consistency of the traits. This long breeding cycle slows down the process of developing new, resistant varieties.

- **Delayed genetic gains**: Due to the complexity and polygenic nature of abiotic stress resistance traits, genetic gains are often incremental and slow, requiring persistent effort over many breeding cycles.

6. Insufficient genomic resources

- **Lack of molecular markers**: For many abiotic stress traits in cumin, there are few or no molecular markers available to assist in selection, making it difficult to implement marker-assisted selection (MAS). This limits the ability to efficiently select for stress resistance during the breeding process.
- **Genomic data gaps**: Cumin lacks extensive genomic resources, such as a complete reference genome or detailed gene expression data under stress conditions. These gaps hinder the identification of key genes involved in stress resistance and the development of targeted breeding strategies.

7. Economic and research priorities

- **Low research investment**: As a minor crop, cumin often receives less attention and funding compared to major crops like wheat, rice, or maize. This limits the availability of research and breeding programs focused on abiotic stress tolerance.
- **Market focus**: Breeding programs may prioritize traits that have more immediate economic benefits, such as yield or disease resistance, over long-term goals like abiotic stress tolerance, which can take many years to develop.

8. Adoption and adaptation challenges

- **Local adaptation**: Varieties developed for abiotic stress resistance in one region may not perform well in another due to differences in climate, soil types, and other local factors. Ensuring that new varieties are adaptable to various growing conditions is a significant challenge.
- **Farmer acceptance**: Even when stress-resistant varieties are developed, they may not be adopted by farmers if they do not meet other important criteria, such as yield potential, flavour, or market preferences. Breeding must balance stress resistance with other desirable traits to ensure adoption.

Breeding cumin for abiotic stress resistance involves overcoming significant challenges related to the complex nature of stress tolerance, the limited genetic resources available, and the difficulties in accurately assessing and selecting for these traits. Advances in genomics, improved breeding techniques like marker-assisted selection, and increased research investment are essential to making significant progress in this area. Collaboration between breeders, researchers, and farmers will also be crucial to developing and deploying varieties that can thrive under challenging environmental conditions.

16. Aijwain

16a. Challenges for biotic stress resistance breeding in Aijwain

Breeding for biotic stress resistance in Ajwain (*Trachyspermum ammi*), also known as carom or bishop's weed, presents several unique challenges. This crop, valued for its seeds used in traditional medicine and culinary applications, faces threats from various pests and diseases. Below are the main challenges associated with breeding Ajwain for biotic stress resistance:

1. Limited genetic diversity

- **Narrow genetic base**: Ajwain has a relatively narrow genetic base, which limits the availability of genetic variations needed for resistance breeding. This lack of diversity makes it difficult to introduce and combine resistant traits.
- **Access to germplasm**: There is limited access to diverse germplasm collections of Ajwain, which hinders the identification and utilization of resistance genes from wild relatives or different cultivars.

2. Complexity of biotic stresses

- **Multiple pathogens and pests**: Ajwain is susceptible to various biotic stresses, including fungal diseases like powdery mildew, leaf spot, and pests like aphids and nematodes. Breeding for resistance to multiple biotic stresses simultaneously is challenging because different pathogens and pests often require different resistance mechanisms.
- **Emerging threats**: The emergence of new pathogens or pests, or the development of new strains of existing pathogens, can quickly overcome existing resistance, making it necessary to continuously monitor and update breeding targets.

3. Polygenic nature of resistance

- **Quantitative resistance**: Resistance to many biotic stresses in Ajwain is controlled by multiple genes (polygenic resistance). This makes breeding more complex because it requires selecting for multiple genes across different loci. Such resistance is often partial and less stable across different environments.
- **Horizontal resistance**: Many biotic stresses in Ajwain may require horizontal (broad-spectrum) resistance, which is harder to breed for than vertical (single-gene) resistance.

4. Long Breeding Cycle

- **Breeding timeframe**: Ajwain has a relatively long breeding cycle, and it can take several years to develop new varieties that are resistant to biotic stresses. The need to evaluate resistance over multiple growing seasons further extends the breeding timeline.
- **Slow progress**: Due to the complexity and polygenic nature of resistance traits, progress in breeding for biotic stress resistance can be slow and incremental.

5. Environmental Influence

- **Genotype-environment interaction**: The expression of resistance traits in Ajwain can be highly influenced by environmental conditions. A variety that shows resistance in one region or under specific conditions may not perform as well in another, complicating the breeding process.
- **Climate change**: Changing climate patterns can alter the prevalence and severity of biotic stresses, which may require breeders to constantly adapt and modify their strategies.

6. Insufficient Research and Resources

- **Lack of advanced breeding tools**: Ajwain lacks extensive genomic resources such as sequenced genomes or genetic maps, making it difficult to identify resistance genes and implement marker-assisted selection (MAS). The absence of such tools hinders the efficiency of breeding programs.
- **Limited research focus**: As a minor crop, Ajwain receives less research attention compared to major crops. This results in fewer resources and breeding programs dedicated to developing biotic stress-resistant varieties.

7. Conventional breeding limitations

- **Cross-Pollination challenges**: Ajwain is typically an outcrossing species, which can complicate controlled breeding efforts aimed at introducing specific resistance traits.
- **Hybridization barriers**: Breeding for resistance through hybridization with wild relatives or other cultivars may face barriers such as cross-compatibility issues or reduced fertility in hybrids.

8. Adoption by farmers

- **Varietal adoption**: Even if resistant varieties are developed, farmers may be slow to adopt them if they do not meet other important criteria such as yield, flavour, or market preferences. Breeding efforts must balance resistance with other desirable traits to ensure adoption.

- **Seed availability**: The distribution and availability of seeds of resistant varieties can also be a challenge, particularly in regions where Ajwain is grown on a small scale.

Breeding Ajwain for biotic stress resistance is a challenging task that requires overcoming genetic, environmental, and logistical barriers. Despite these challenges, advances in genomic technologies, increased germplasm exploration, and focused research can help to develop resistant varieties. Collaboration between breeders, researchers, and farmers will be crucial in making Ajwain cultivation more resilient to biotic stresses.

16b. Challenges for abiotic stress resistance breeding in Aijwain

Breeding Ajwain (*Trachyspermum ammi*) for abiotic stress resistance presents several significant challenges due to the crop's sensitivity to environmental conditions, limited genetic resources, and the complexity of abiotic stress responses. Below are the main challenges associated with breeding Ajwain for abiotic stress resistance:

1. Limited genetic diversity

- **Narrow genetic base**: Ajwain has a limited genetic pool, which restricts the availability of diverse traits necessary for breeding programs. This makes it challenging to find or develop varieties with enhanced resistance to various abiotic stresses such as drought, salinity, and temperature extremes.
- **Germplasm availability**: There is a scarcity of well-characterized germplasm collections that include stress-tolerant varieties or wild relatives, limiting the potential sources of resistance genes.

2. Complexity of abiotic stresses

- **Multiple stress factors**: Ajwain is susceptible to a range of abiotic stresses, including drought, high salinity, extreme temperatures, and nutrient deficiencies. Each of these stresses triggers different physiological responses, making it difficult to breed a single variety that can withstand multiple abiotic challenges.
- **Simultaneous stresses**: In many growing regions, Ajwain may face more than one abiotic stress simultaneously, such as drought combined with high temperatures. Breeding for multiple stress resistances at once is complex and often results in trade-offs between stress tolerance and other important agronomic traits.

3. Polygenic nature of stress tolerance

- **Quantitative inheritance**: Abiotic stress tolerance in Ajwain, as in many plants, is often controlled by multiple genes with small effects. This quantitative nature of the traits complicates the breeding process, as it requires selecting for multiple genes across different loci, which can be challenging and time-consuming.

- **Epistatic interactions**: The interaction between different genes (epistasis) involved in stress tolerance can complicate the expression of these traits, making it difficult to predict and select for the desired outcomes in breeding programs.

4. Phenotyping challenges

- **Measurement difficulties**: Accurate phenotyping (the assessment of plant traits) for abiotic stress tolerance is difficult, especially under field conditions where stress levels can vary greatly. Controlled environments are often required to reliably evaluate traits such as drought or salinity tolerance, but these are not always available.
- **Indirect selection**: Abiotic stress tolerance often has to be assessed through indirect traits (e.g., root depth for drought tolerance, leaf ion concentration for salinity tolerance), which may not always directly correlate with field performance.

5. Long breeding cycle

- **Extended timeframes**: Developing Ajwain varieties with improved abiotic stress resistance requires several growing seasons to ensure that the traits are stable and consistently expressed. This long breeding cycle delays the introduction of new, resistant varieties.
- **Slow genetic gains**: Due to the complex, polygenic nature of abiotic stress resistance traits, genetic gains in breeding programs are often slow, requiring sustained effort over many years.

6. Environmental Influence

- **Genotype-Environment Interaction**: The expression of abiotic stress tolerance traits is highly influenced by environmental factors. A variety that performs well under stress in one environment may not necessarily perform the same way in another, complicating the breeding process.
- **Climate variability**: With the increasing unpredictability of climate patterns, breeding for abiotic stress resistance becomes even more challenging as new or intensified stresses may emerge, requiring continuous adaptation of breeding strategies.

7. Insufficient genomic resources

- **Lack of molecular markers**: Ajwain lacks well-developed molecular markers for abiotic stress tolerance, which limits the use of advanced breeding techniques such as marker-assisted selection (MAS). The absence of these tools makes the breeding process less efficient.

- **Genomic data gaps**: There is a lack of comprehensive genomic data for Ajwain, such as a reference genome or transcriptomic data under stress conditions, which hinders the identification of key genes involved in stress resistance and the development of targeted breeding strategies.

8. Economic and research limitations

- **Low research investment**: As a minor crop, Ajwain receives less research funding and attention compared to major crops, which limits the development of advanced breeding programs focused on abiotic stress tolerance.
- **Market priorities**: Breeding programs for Ajwain may prioritize traits that have immediate economic benefits, such as yield or flavour, over long-term goals like abiotic stress resistance, which may take years to develop.

9. Adoption and adaptation challenges

- **Local adaptation**: Ajwain varieties bred for abiotic stress resistance in one region may not be suitable for other regions due to differences in environmental conditions, such as soil type, climate, and local agricultural practices.
- **Farmer acceptance**: Even if resistant varieties are developed, farmers may be reluctant to adopt them if they do not meet other criteria, such as high yield, good seed quality, or market preferences. Successful breeding programs must balance stress resistance with other desirable traits.

Breeding Ajwain for abiotic stress resistance is a complex and challenging task that requires overcoming significant genetic, environmental, and logistical barriers. Progress in this area will depend on the development of better genomic tools, increased research investment, and collaboration between breeders, researchers, and farmers. By addressing these challenges, it may be possible to develop Ajwain varieties that are more resilient to the increasing incidence of abiotic stresses due to climate change and other factors.

B. Breeding for climate change resilience

Breeding for climate change resilience in spice crops such as black pepper, cardamom, cinnamon, clove, allspice, ginger, turmeric, mango ginger, black turmeric, coriander, cumin, fennel, fenugreek, and ajwain is a multifaceted challenge. These crops are all economically significant and are grown in diverse environments, each facing specific climatic challenges. Below are key strategies and challenges associated with breeding these crops for climate resilience:

1. Understanding the impact of climate change

- **Temperature extremes**: Rising temperatures can affect the growth, flowering, and yield of these spice crops. High temperatures can lead to heat stress, while unpredictable frosts can damage crops not adapted to cold conditions.
- **Changes in rainfall patterns**: Shifts in rainfall, including increased droughts or excessive rainfall, can stress plants, affect soil moisture levels, and increase susceptibility to diseases.
- **Increased pest and disease pressure**: Climate change can lead to the emergence of new pests and diseases or increase the prevalence of existing ones, posing additional challenges to crop resilience.

2. Key breeding strategies

A. Developing heat and drought tolerance

- **Selection for Drought-Resilient Varieties**: For crops like cumin, coriander, and fenugreek, which are often grown in arid and semi-arid regions, breeding for traits such as deep root systems, efficient water use, and osmotic adjustment is crucial.
- **Heat tolerance breeding**: In crops like black pepper, cardamom, and ginger, where high temperatures can reduce yields, selecting for heat tolerance through traits like heat-shock proteins, better leaf cooling mechanisms, and altered flowering times is important.

B. Enhancing disease and pest resistance

- **Breeding for resistance to emerging pests and diseases**: As climate change alters pest and disease dynamics, breeding for broad-spectrum resistance is essential. This can include identifying and incorporating resistance genes from wild relatives or using marker-assisted selection to combine multiple resistance traits.
- **Integrated Pest Management (IPM) compatibility**: Breeding varieties that are compatible with IPM practices can enhance resilience by reducing reliance on chemical controls, which may become less effective or more costly under changing climates.

C. Improving tolerance to flooding and salinity

- **Flooding tolerance**: Crops like turmeric and ginger, which are often grown in regions susceptible to heavy monsoon rains, need to be bred for tolerance to waterlogged conditions. This could involve selecting for traits like aerenchyma formation in roots, which improves oxygen availability in flooded soils.

- **Salinity tolerance**: As sea levels rise and irrigation practices change, salinity is becoming an issue in many regions. Breeding for salt tolerance in crops like coriander, fennel, and fenugreek involves selecting for traits such as ion exclusion, compartmentalization, and salt-tolerant enzymes.

D. Enhancing genetic diversity

- **Utilizing wild relatives**: Wild relatives of these spices often harbor genes for stress tolerance that can be introgressed into cultivated varieties. This strategy requires careful hybridization and backcrossing to retain desirable agronomic traits.
- **Promoting genetic diversity within breeding programs**: Maintaining and increasing genetic diversity within breeding programs is crucial to ensure that there is a broad base of traits to select from, which can help crops adapt to changing conditions.

E. Breeding for resilience in changing phenological patterns

- **Adaptation to shifting seasons**: Climate change is causing shifts in growing seasons, which can affect the timing of flowering and harvest. Breeding programs must focus on developing varieties that can adapt to these shifts without compromising yield or quality.

3. Challenges in breeding for climate resilience

A. Long breeding cycles

- Many spice crops, especially perennials like black pepper, cardamom, cinnamon, clove, and allspice, have long breeding cycles. Developing climate-resilient varieties may take many years, making it difficult to respond quickly to changing environmental conditions.

B. Insufficient genomic resources

- For many of these crops, there is a lack of genomic resources, such as well-characterized genetic markers or sequenced genomes. This limits the ability to apply advanced breeding techniques like marker-assisted selection or genomic selection.

C. Limited research and funding

- Compared to major food crops, spices often receive less research attention and funding. This can slow down the development of climate-resilient varieties as fewer resources are available for breeding programs.

D. Integration of traditional knowledge

- Traditional knowledge of crop cultivation and resilience strategies is crucial, but integrating this with modern breeding techniques can be challenging. It requires collaboration between researchers, breeders, and local farming communities.

4. Regional focus and local adaptation

- **Local adaptation strategies**: Climate change impacts vary by region, so breeding programs need to focus on developing varieties tailored to local conditions. This requires understanding the specific climate risks in each growing area and developing region-specific solutions.
- **Farmer participation**: Involving farmers in the breeding process, through participatory breeding programs, ensures that the developed varieties meet the practical needs of those who will grow them.

5. Leveraging modern breeding techniques

- **CRISPR and gene editing**: These techniques offer the potential to rapidly introduce or modify specific genes associated with climate resilience, such as those involved in stress responses or metabolic pathways.
- **Marker-assisted and genomic selection**: These techniques can accelerate the breeding process by allowing for the early selection of climate-resilient traits based on genetic markers, even before the plants are fully grown.

Breeding for climate change resilience in these diverse spice crops is a complex but critical task. It requires a multifaceted approach that integrates traditional breeding methods with modern genomic tools, considers the unique challenges posed by each crop, and addresses the specific environmental conditions of different growing regions. Collaboration between breeders, researchers, farmers, and policymakers will be essential to develop and deploy climate-resilient varieties that can sustain spice production in the face of a changing climate.

1. Black Pepper

Breeding for climate change resilience in black pepper (Piper nigrum) is crucial to sustaining its productivity in the face of rising temperatures, erratic rainfall patterns, and increased pest and disease pressures. Black pepper, a perennial vine native to tropical climates, is highly sensitive to environmental changes. Below are key strategies and challenges in breeding black pepper for climate change resilience:

Key challenges in breeding Black Pepper for climate resilience

Sensitivity to temperature extremes

- **Heat stress**: Black pepper is highly sensitive to temperature changes. High temperatures, particularly during the flowering and fruiting stages, can lead to reduced yields. The crop prefers a temperature range of 25-30°C, and deviations can cause stress, impacting growth and productivity.
- **Cold stress**: Although less common in its traditional growing regions, unexpected cold spells can cause damage, particularly to young plants.

Changes in rainfall patterns

- **Drought**: Black pepper requires consistent moisture for optimal growth. Drought conditions can severely impact the crop, leading to wilting, reduced berry size, and lower yields.
- **Excessive rainfall and flooding**: Conversely, excessive rainfall can lead to waterlogging, which is detrimental to root health and can increase susceptibility to root rot and other diseases.

Increased pest and disease pressure

- **Pests**: Climate change can alter the prevalence and distribution of pests such as the pepper thrips (*Scirtothrips dorsalis*), which thrives in hot and dry conditions.
- **Diseases**: Black pepper is prone to diseases like Phytophthora foot rot, which can be exacerbated by changing moisture conditions. The spread of diseases may increase with warmer and more humid conditions, posing significant challenges.

Breeding strategies for climate resilience

Developing heat and drought tolerance

- **Selection for drought-resistant varieties**: Breeding programs need to focus on developing varieties with deeper root systems, efficient water-use mechanisms, and enhanced stomatal regulation to conserve water under drought conditions.
- **Heat tolerance breeding**: Identifying and selecting black pepper varieties that can tolerate higher temperatures without significant yield loss is critical. This may involve screening for heat-shock proteins, altering flowering times, or improving leaf cooling mechanisms.

Enhancing disease resistance

- **Resistance to *Phytophthora* and other pathogens**: Breeding for resistance to *Phytophthora*, the pathogen responsible for foot rot, is a priority. This involves identifying resistant genotypes and incorporating them into breeding programs.
- **Integrated Pest Management (IPM) compatibility**: Developing varieties that are compatible with IPM strategies, including biological controls and reduced chemical use, can help manage increased pest and disease pressure under climate change.

Tolerance to waterlogging and soil health management

- **Waterlogging tolerance**: Breeding for root traits that enhance tolerance to waterlogged conditions, such as the development of aerenchyma (air spaces in roots that allow oxygen to reach submerged tissues), can improve resilience in regions prone to heavy rains.
- **Soil health**: Improving soil health through breeding for varieties that interact positively with beneficial soil microbes can enhance resilience to both drought and waterlogging.

Enhancing genetic diversity

- **Utilizing wild relatives**: Wild relatives of black pepper, which may have evolved under different climatic conditions, could possess traits that enhance resilience to extreme weather. These traits can be introgressed into cultivated varieties through hybridization.
- **Maintaining genetic diversity**: Broadening the genetic base within black pepper breeding programs can provide more options for selecting traits that confer resilience to various climate stresses.

Breeding for phenological adaptation

- **Adaptation to shifting seasons**: With climate change causing shifts in growing seasons, breeding programs must focus on developing varieties that can adapt to these changes without compromising yield or quality. This includes adjusting the timing of flowering and fruiting to match new environmental conditions.

Modern breeding techniques

Marker-assisted selection (MAS) and genomic selection

- **Marker-Assisted Selection**: MAS can be used to accelerate the breeding process by identifying and selecting for traits associated with climate resilience, such as drought tolerance or disease resistance, based on genetic markers.

- **Genomic selection**: Genomic selection involves using genome-wide data to predict the performance of breeding lines, allowing breeders to select the best candidates for climate resilience more efficiently.

CRISPR and gene editing

- **Gene editing**: CRISPR technology can be used to edit specific genes associated with climate resilience, such as those involved in stress response pathways. This approach offers the potential for more precise and rapid development of resilient varieties.

Challenges in implementing breeding programs

Long breeding cycles

- Black pepper is a perennial crop with a long life cycle, meaning that breeding new varieties can take many years. This long cycle makes it challenging to respond quickly to climate change.

Limited research resources

- As a high-value but niche crop, black pepper does not always receive the same level of research funding and attention as major food crops. This limits the availability of advanced breeding tools and resources.

Regional adaptation

- Black pepper is grown in diverse regions with varying climates, from India and Vietnam to Indonesia and Brazil. Breeding programs need to develop region-specific solutions that address the unique challenges faced by farmers in different growing environments.

Breeding black pepper for climate change resilience requires a multifaceted approach that combines traditional breeding methods with modern genomic tools. Key strategies include developing heat and drought-tolerant varieties, enhancing disease and pest resistance, and improving genetic diversity within breeding programs. Addressing these challenges is crucial for maintaining black pepper production in the face of climate change, ensuring the sustainability and profitability of this important spice crop for future generations. Collaboration between breeders, researchers, and farmers will be essential to develop and implement these strategies effectively.

2. Cardamom

Breeding for climate change resilience in cardamom *(Elettaria cardamomum)*, often referred to as the "Queen of Spices," is vital for sustaining its production, especially given the crop's sensitivity to environmental changes. Cardamom is a perennial, tropical crop primarily grown in shaded, humid conditions, making it particularly vulnerable to the effects of climate change. Below are the key challenges and strategies involved in breeding cardamom for climate resilience:

Key challenges in breeding Cardamom for climate resilience

Sensitivity to temperature and humidity

- **Heat stress**: Cardamom thrives in a specific temperature range (10-35°C). Higher temperatures, especially above 30-35°C, can reduce flowering, pollination success, and overall yield.
- **Humidity requirements**: Cardamom requires high humidity (70-80%) for optimal growth. Changes in humidity due to climate change can affect plant health, flowering, and fruit set.

Erratic rainfall patterns

- **Drought**: Cardamom requires consistent moisture, and prolonged drought periods can lead to reduced growth, poor yield, and even plant death. Drought stress can also increase the plant's susceptibility to pests and diseases.
- **Excessive rainfall and flooding**: Heavy rainfall and waterlogging can cause root rot and other fungal diseases, significantly impacting crop health and yield.

Increased pest and disease pressure

- **Pests**: Cardamom is prone to pests such as thrips, shoot borers, and root grubs. Climate change can exacerbate pest infestations by creating favorable conditions for their proliferation.
- **Diseases**: Fungal diseases like cardamom mosaic virus and rhizome rot are serious threats. These diseases may become more prevalent or severe with changing weather patterns.

Breeding strategies for climate resilience

Developing heat and drought tolerance

- **Selection for drought-resilient varieties**: Breeding programs should focus on selecting and developing cardamom varieties that can tolerate periods of water scarcity. Traits such as deep root systems, efficient water use, and osmotic adjustment mechanisms can help plants survive and produce under drought conditions.
- **Heat tolerance breeding**: Identifying and breeding cardamom varieties that can tolerate higher temperatures is critical. This involves selecting for traits like heat-shock proteins, enhanced leaf cooling mechanisms, and altered flowering patterns to maintain productivity under heat stress.

Enhancing disease and pest resistance

- **Breeding for disease resistance**: Breeding for resistance to common diseases, such as cardamom mosaic virus and rhizome rot, is essential for climate resilience. This can be achieved by identifying resistant genotypes and incorporating them into breeding programs, potentially using marker-assisted selection for more efficient breeding.
- **Pest resistance**: Developing pest-resistant varieties through traditional breeding or biotechnology (e.g., genetic modification) can help manage the increased pest pressures anticipated under changing climatic conditions.

Tolerance to waterlogging and soil health management

- **Waterlogging tolerance**: Since cardamom is grown in regions prone to heavy monsoon rains, breeding for tolerance to waterlogged conditions is crucial. This may involve selecting for root traits that enhance oxygen uptake and reduce susceptibility to root diseases.
- **Improving soil health**: Breeding for varieties that interact positively with beneficial soil microbes, such as mycorrhizae, can improve nutrient uptake and resilience to soil-borne diseases.

Enhancing genetic diversity

- **Utilizing wild relatives**: Wild relatives of cardamom, which may have evolved under different environmental conditions, could possess traits that enhance resilience to climate extremes. Introgression of these traits into cultivated varieties through hybridization and backcrossing can help improve stress tolerance.
- **Maintaining genetic diversity**: Breeding programs should focus on maintaining and enhancing genetic diversity within cardamom populations to provide a broader base for selecting climate-resilient traits.

Breeding for phenological adaptation

- **Adaptation to shifting seasons**: Climate change may alter the timing of seasons, affecting flowering and fruiting in cardamom. Breeding programs need to develop varieties that can adapt to these shifts, ensuring that the plants flower and fruit at the most advantageous times.

Modern breeding techniques

Marker-assisted selection (MAS) and genomic selection

- **Marker-assisted selection**: MAS can be used to accelerate the breeding process by selecting for specific genes associated with traits like drought tolerance or disease resistance. This allows for more efficient development of climate-resilient varieties.

- **Genomic selection**: Genomic selection involves using genome-wide data to predict the performance of breeding lines, helping breeders select the best candidates for climate resilience based on their genetic profiles.

CRISPR and gene editing

- **Gene editing**: CRISPR technology offers the potential to precisely edit genes associated with climate resilience in cardamom, such as those involved in stress response pathways. This can lead to faster and more targeted development of resilient varieties.

Challenges in implementing breeding programs

Long breeding cycles

- Cardamom is a perennial crop with a long breeding cycle, often taking several years to develop new varieties. This long cycle makes it difficult to respond quickly to the challenges posed by climate change.

Limited research resources

- As a high-value but niche crop, cardamom often receives less research attention and funding compared to staple food crops. This limits the availability of advanced breeding tools and resources, slowing the progress of breeding programs focused on climate resilience.

Regional adaptation

- Cardamom is grown in diverse regions with varying climates, particularly in India, Sri Lanka, and parts of Southeast Asia. Breeding programs need to develop region-specific solutions that address the unique challenges faced by farmers in different growing environments.

Collaboration and farmer involvement

Participatory breeding programs

- Engaging farmers in the breeding process can ensure that the developed varieties meet the practical needs of those who grow them. Participatory breeding can lead to faster adoption of new varieties and better adaptation to local conditions.

Knowledge sharing and training

- Providing farmers with knowledge and training on how to manage climate-related stresses using resilient varieties and integrated crop management practices is crucial for successful adaptation to climate change.

Breeding cardamom for climate change resilience requires a comprehensive approach that integrates traditional breeding methods with modern genomic tools and considers the unique challenges posed by this crop's sensitivity to environmental changes. Key strategies include developing heat and drought-tolerant varieties, enhancing disease and pest resistance, and improving genetic diversity within breeding programs. Addressing these challenges is essential for maintaining cardamom production in the face of climate change, ensuring the sustainability and profitability of this valuable spice crop.

3. Cinnamon

Breeding for climate change resilience in cinnamon (Cinnamomum verum), a highly valued spice crop primarily grown in tropical regions, is crucial to ensure its continued productivity and sustainability in the face of changing environmental conditions. Cinnamon, like many perennial crops, is sensitive to shifts in climate, including changes in temperature, rainfall patterns, and increased vulnerability to pests and diseases. Below are key challenges and strategies for breeding cinnamon to enhance its resilience to climate change:

Key Challenges in breeding Cinnamon for climate resilience

Sensitivity to temperature extremes

- **Heat stress**: Cinnamon thrives in warm, tropical climates but has an optimal temperature range of 20-30°C. Rising temperatures, particularly sustained heat above 35°C, can stress the trees, reducing growth rates and bark quality, which is critical for cinnamon production.
- **Cold stress**: Although less common in its traditional growing regions, cold spells and frost can damage young cinnamon plants and affect their long-term productivity.

Changes in rainfall patterns

- **Drought**: Cinnamon requires consistent rainfall for optimal growth. Extended drought periods can lead to reduced growth, poor bark development, and increased tree mortality. Drought stress can also exacerbate issues with soil salinity and nutrient uptake.
- **Excessive rainfall and waterlogging**: While cinnamon needs adequate moisture, excessive rainfall or waterlogged conditions can lead to root diseases, particularly fungal infections, and can negatively impact root health and tree stability.

Increased pest and disease pressure

- **Pests**: Climate change can lead to the proliferation of pests such as the cinnamon leaf beetle, scale insects, and various borers. Warmer temperatures and changing humidity levels can create favourable conditions for these pests to thrive.

- **Diseases**: Cinnamon is susceptible to fungal diseases like root rot (caused by *Phytophthora* spp.) and leaf spot. These diseases may become more prevalent or severe with changing weather patterns, particularly in warmer and wetter climates.

Breeding strategies for climate resilience

Developing heat and drought tolerance

- **Selection for drought-resistant varieties**: Breeding programs should focus on selecting cinnamon varieties that are more tolerant of water scarcity. Traits such as deeper root systems, efficient water use, and the ability to maintain growth under reduced water availability are essential.
- **Heat tolerance breeding**: Identifying and breeding cinnamon varieties that can withstand higher temperatures without compromising growth or bark quality is crucial. This involves selecting for traits such as heat-shock proteins, altered leaf morphology, and better transpiration efficiency to maintain plant health under heat stress.

Enhancing disease and pest resistance

- **Breeding for disease resistance**: Developing disease-resistant varieties, particularly against root rot and leaf spot, is a priority. This can be achieved by identifying resistant genotypes and incorporating these traits into breeding programs using traditional and modern breeding techniques.
- **Pest resistance**: Breeding for resistance to key pests, possibly through the development of varieties that deter pests or are less attractive to them, can help manage increased pest pressures under changing climate conditions.

Tolerance to waterlogging and soil health management

- **Waterlogging tolerance**: Given the likelihood of increased rainfall and the potential for flooding, breeding for waterlogging tolerance is critical. Selecting for traits that enhance root oxygenation and reduce susceptibility to root diseases can help improve resilience in waterlogged conditions.
- **Soil health**: Breeding for varieties that contribute to or benefit from better soil health (e.g., through interactions with mycorrhizal fungi) can enhance overall resilience to both drought and excessive moisture.

Enhancing genetic diversity

- **Utilizing wild relatives**: Wild relatives of cinnamon, which may have evolved in different environmental conditions, can be valuable sources of genes for stress tolerance. Incorporating these genes into cultivated cinnamon varieties can improve resilience to climate extremes.

- **Maintaining genetic diversity**: Breeding programs should aim to maintain and increase genetic diversity within cinnamon populations, providing a broader base for selecting traits associated with climate resilience.

Breeding for phenological adaptation

- **Adaptation to shifting seasons**: Climate change may alter the timing of growth cycles in cinnamon. Breeding programs should focus on developing varieties that can adapt to these shifts, ensuring that growth and bark harvesting align with the new climatic conditions.

Modern breeding techniques

Marker-assisted selection (MAS) and genomic selection

- **Marker-Assisted Selection**: MAS can be used to accelerate the breeding process by identifying and selecting for genes associated with traits like drought tolerance, disease resistance, and heat tolerance. This allows for more efficient breeding of climate-resilient varieties.
- **Genomic selection**: Genomic selection involves using genome-wide data to predict the performance of breeding lines. This technique enables breeders to select the best candidates for climate resilience based on their genetic profiles, even before the plants reach maturity.

CRISPR and Gene editing

- **Gene editing**: CRISPR technology offers the potential to precisely edit genes associated with climate resilience, such as those involved in stress response pathways. This approach can lead to the rapid development of cinnamon varieties with enhanced resilience to specific climate-related stresses.

Challenges in implementing breeding programs

Long breeding cycles

- Cinnamon, as a perennial tree crop, has a long breeding cycle. It can take many years to develop and evaluate new varieties, making it difficult to quickly respond to climate change. This long cycle is a significant challenge in breeding programs.

Limited research resources

- Compared to other major crops, cinnamon receives less research attention and funding. This limitation slows the progress of breeding programs focused on climate resilience, as there are fewer resources available for developing and testing new varieties.

Regional adaptation

- Cinnamon is grown in various tropical regions, including Sri Lanka, India, and Southeast Asia. Each region faces different climatic challenges, so breeding programs need to develop region-specific solutions that address local environmental conditions.

Collaboration and farmer involvement

Participatory breeding programs

- Engaging farmers in the breeding process is crucial for ensuring that the developed varieties meet the practical needs of those who will cultivate them. Participatory breeding can lead to faster adoption of new varieties and better adaptation to local conditions.

Knowledge sharing and training

- Providing farmers with knowledge and training on managing climate-related stresses using resilient varieties and integrated crop management practices is essential for successful adaptation to climate change.

Breeding cinnamon for climate change resilience requires a comprehensive approach that integrates traditional breeding methods with modern genomic tools. Key strategies include developing heat and drought-tolerant varieties, enhancing disease and pest resistance, and maintaining genetic diversity within breeding programs. Addressing these challenges is essential for ensuring the sustainability and profitability of cinnamon production in the face of climate change. Collaborative efforts among breeders, researchers, and farmers will be crucial for developing and deploying climate-resilient cinnamon varieties that can thrive under increasingly variable and extreme climatic conditions.

4. Clove

Breeding of Clove for climate resilience

Breeding clove (*Syzygium aromaticum*) for climate resilience involves developing varieties that can withstand various environmental stresses while maintaining their productivity and quality. The key areas of focus include:

1. **Drought resistance**: Cloves are sensitive to water stress, so breeding programs focus on developing drought-resistant varieties that can thrive in areas with limited water availability.

2. **Heat tolerance**: As global temperatures rise, breeding for heat tolerance becomes essential. The aim is to develop clove varieties that can maintain growth and productivity even under higher temperatures.
3. **Pest and disease resistance**: Climate change can lead to new pest and disease pressures. Breeding for resistance to these threats is crucial to ensure the long-term sustainability of clove production.
4. **Flood and salinity tolerance**: In regions prone to flooding or with high soil salinity, breeding clove varieties that can tolerate these conditions is important for maintaining yields in challenging environments.
5. **Adaptability to changing climate patterns**: Overall, breeding efforts are directed toward creating clove varieties that are adaptable to a range of changing climatic conditions, ensuring consistent production despite the uncertainties of climate change.

These efforts help secure the future of clove cultivation in the face of global climate challenges, ensuring that this valuable spice remains available and affordable.

5. Allspice

Breeding of Allspice for climate resilience

Breeding allspice (*Pimenta dioica*) for climate resilience focuses on developing varieties that can thrive under various environmental stresses, which are becoming increasingly important due to climate change. Key objectives include:

1. **Drought tolerance**: As water scarcity becomes more common, breeding allspice varieties that can tolerate periods of drought without significant loss of yield or quality is essential.
2. **Heat resistance**: Rising global temperatures necessitate the development of allspice plants that can endure higher temperatures while maintaining their growth and production rates.
3. **Disease and pest resistance**: Climate change can exacerbate pest and disease pressures. Breeding efforts focus on enhancing resistance to common pests and diseases that may become more prevalent with shifting climate conditions.
4. **Adaptability to different climates**: Given the diverse climates where allspice is grown, breeding programs aim to create varieties that are adaptable to a wide range of environmental conditions, ensuring stable production even as climate patterns change.
5. **Flood and salinity tolerance**: In regions prone to flooding or high soil salinity, breeding for tolerance to these conditions is crucial for sustaining allspice production in affected areas.

These breeding goals help ensure that allspice cultivation can continue to be viable and productive despite the challenges posed by climate change.

6. Nutmeg

Breeding of nutmeg for climate resilience

Breeding nutmeg (*Myristica fragrans*) for climate resilience focuses on developing varieties that can thrive under changing environmental conditions. Key objectives in these breeding programs include:

1. **Drought tolerance**: With increasing instances of drought, developing nutmeg varieties that can withstand periods of low water availability is essential to ensure consistent yields.
2. **Heat resistance**: As global temperatures rise, breeding efforts are directed at enhancing the heat tolerance of nutmeg plants, enabling them to maintain growth and productivity in warmer climates.
3. **Pest and disease resistance**: Climate change can lead to new pest and disease pressures, making it crucial to breed nutmeg varieties that are resistant to these emerging threats.
4. **Adaptability to variable climates**: Breeding nutmeg that can adapt to a wide range of climatic conditions helps ensure that the crop can be successfully cultivated in different regions, despite unpredictable weather patterns.
5. **Flood and salinity tolerance**: In areas prone to flooding or with high soil salinity, breeding for these traits ensures that nutmeg can survive and produce in challenging environments.

These breeding strategies are vital for sustaining nutmeg production in the face of climate change, ensuring that the crop remains viable for future generations.

C. Breeding of spices for market demands and consumer preferences

Breeding of spices to meet market demands and consumer preferences is a complex process that involves several strategies.

1. Identification of market demands

- **Consumer trends:** Breeders analyse market trends to understand consumer preferences, such as the desire for specific flavours, heat levels, or health benefits. For example, there might be a growing demand for organic or non-GMO spices.
- **Industry requirements:** The food industry often requires spices with specific traits like uniformity in size, colour, and ease of processing.

2. Breeding objectives

- **Flavour and aroma:** Breeding focuses on enhancing or modifying the flavour profiles, which can be influenced by essential oil content.
- **Yield improvement:** High-yield varieties are crucial to meet the increasing demand while maintaining quality.
- **Disease resistance:** Developing varieties that are resistant to common pests and diseases reduces the need for chemical interventions and supports organic production.
- **Adaptability:** Breeders aim to develop spice varieties that can adapt to various climatic conditions, ensuring stable production across different regions.

3. Breeding techniques

- **Traditional breeding:** Crossbreeding different varieties to combine desirable traits such as high yield and disease resistance.
- **Biotechnological approaches:** Techniques like genetic modification or CRISPR can be used to introduce specific traits more precisely.
- **Marker-assisted selection:** This involves using molecular markers to identify and select plants that possess the desired traits at an early stage.

4. Evaluation and selection

- **Field trials:** Selected varieties undergo extensive field trials to assess their performance in different environments.
- **Consumer testing:** New varieties are often tested in consumer panels to ensure they meet taste and quality expectations.

5. Commercialization

- **Patenting and licensing:** New varieties might be patented to protect intellectual property and then licensed to farmers and companies.
- **Marketing:** Educating both consumers and the industry about the benefits of new varieties is crucial for successful adoption.

These steps ensure that the spices bred are not only high-yielding and resistant to diseases but also meet the nuanced flavour profiles and quality standards demanded by consumers and the market.

D) Breeding of spices for Market demands and consumer preferences

1. Black Pepper

Breeding of black pepper for Market demands and consumer preferences

Breeding black pepper (Piper nigrum) to meet market demands and consumer preferences involves a focused approach to enhance specific traits that are valued by both producers and consumers. Here's how the process typically unfolds:

1. Market demand identification

- **Flavour and aroma:** Black pepper's pungency and aroma, driven by its piperine content and essential oils, are primary targets. Consumers often seek strong, consistent flavours.
- **Health benefits:** There is growing consumer interest in the health benefits of black pepper, particularly its anti-inflammatory and antioxidant properties.
- **Processing quality:** For industrial use, uniformity in berry size and ease of processing (such as dehusking for white pepper) are important.

2. Breeding objectives

- **High piperine content:** Breeding programs focus on increasing the piperine content, which directly correlates with the pungency and perceived quality of black pepper.
- **Yield improvement:** Developing varieties with higher yields is crucial to meet the global demand. This includes improving the number of berries per spike and the overall weight of the harvest.
- **Disease resistance:** Black pepper is susceptible to diseases like Phytophthora foot rot and Piper yellow mottle virus. Breeding for resistance to these diseases is essential to reduce crop losses and dependency on chemical treatments.
- **Climate adaptability:** With climate change, breeding varieties that can withstand varying environmental conditions, such as drought or excessive rainfall, is important.

3. Breeding techniques

- **Traditional breeding:** This includes selective breeding of high-yielding and disease-resistant varieties. Hybridization between different black pepper strains is used to combine desirable traits.
- **Molecular breeding:** Marker-assisted selection helps in identifying and selecting plants with desired traits, such as disease resistance, at an early stage, speeding up the breeding process.
- **Biotechnological approaches:** Genetic engineering or CRISPR technology might be explored for introducing specific traits, such as enhanced disease resistance or higher piperine content.

4. Evaluation and selection

- **Field trials:** Varieties are tested in different climatic conditions to evaluate their performance in terms of yield, disease resistance, and quality traits.
- **Sensory evaluation:** Selected varieties undergo sensory tests to ensure they meet the desired flavour and aroma profiles. This is crucial for maintaining market appeal.
- **Processing trials:** For industrial purposes, the ease of processing (e.g., dehusking for white pepper) is evaluated to ensure the variety meets the processing requirements of the industry.

5. Commercialization and adoption

- **Patenting and licensing:** New black pepper varieties may be patented and then licensed to growers, ensuring that the breeding program's investments are protected and that farmers have access to high-quality planting material.
- **Market introduction:** Educating farmers about the benefits of new varieties, such as disease resistance and higher yields, and promoting these traits to consumers is critical for successful market adoption.
- **Sustainability and certification:** With increasing consumer awareness, developing and marketing varieties that align with organic farming and sustainability certifications can enhance market acceptance.

By focusing on these areas, black pepper breeding programs can develop varieties that not only meet the demands of the market and consumer preferences but also contribute to more sustainable and profitable pepper production.

2. Cardamom

Breeding of cardamom for Market demands and consumer preferences

Breeding cardamom (*Elettaria cardamomum*) to meet market demands and consumer preferences requires a strategic approach to enhance specific traits that are highly valued by consumers and essential for successful cultivation and processing. Here's how the process is typically managed:

1. Market demand identification

- **Flavour and aroma:** Cardamom is prized for its unique flavour and aroma, which are primarily due to its essential oil content, particularly cineole and alpha-terpineol. Consumers seek strong, consistent flavours.
- **Size and appearance:** The size, shape, and colour of cardamom pods are important, with larger, uniformly green pods being preferred in the market.
- **Shelf life:** Consumers and traders value cardamom that maintains its aroma and quality over a longer period, which is influenced by the moisture content and the integrity of the pods.

2. Breeding objectives

- **Enhanced flavour and aroma:** Breeding efforts focus on increasing the concentration of essential oils, particularly those contributing to the characteristic flavour and aroma.
- **Increased yield:** Developing high-yielding varieties is critical to ensure that farmers can meet the growing demand, both in local and international markets.
- **Disease resistance:** Cardamom is vulnerable to diseases such as capsule rot, mosaic virus, and pests like thrips. Breeding for resistance to these threats is essential to reduce crop losses and minimize the use of pesticides.
- **Improved pod quality:** Breeders aim to produce varieties with larger, more uniform pods that have a vibrant green colour, which is preferred in both whole and ground forms.
- **Adaptability to climatic conditions:** As climate change affects growing conditions, breeding for adaptability to different climatic conditions, such as drought or high humidity, is becoming increasingly important.

3. Breeding techniques

- **Traditional breeding:** This includes selecting and crossbreeding superior plants to combine desirable traits such as high essential oil content, disease resistance, and higher yields.
- **Marker-assisted selection:** Molecular markers are used to identify plants with specific traits like disease resistance or high essential oil content, which speeds up the breeding process.
- **Biotechnological approaches:** Techniques like tissue culture are used to propagate disease-free plants, and genetic engineering may be explored to introduce traits like enhanced flavour or resistance to specific pests.

4. Evaluation and selection

- **Field trials:** Selected varieties undergo extensive field trials in different growing regions to evaluate their performance in terms of yield, resistance to diseases and pests, and pod quality.
- **Sensory evaluation:** Cardamom varieties are tested for flavour and aroma to ensure they meet consumer preferences. This includes both chemical analysis of essential oils and taste tests.
- **Post-harvest evaluation:** The shelf life and drying characteristics of the pods are evaluated to ensure that the variety can maintain quality during storage and transport.

5. Commercialization and adoption

- **Patenting and licensing:** New cardamom varieties are often patented, and then licensed to farmers, ensuring they have access to high-quality planting material that meets market demands.
- **Market introduction:** Promoting the benefits of new varieties, such as improved flavour, higher yields, and disease resistance, to both farmers and consumers is crucial for successful market adoption.
- **Sustainability certification:** Given the growing consumer interest in sustainability, breeding programs may also focus on developing varieties that perform well under organic farming conditions, which can then be marketed under sustainability certifications.

6. Future trends

- **Organic and sustainable production:** With increasing demand for organic products, breeding cardamom varieties that thrive under organic farming conditions, with natural resistance to pests and diseases, will become more important.
- **Value-added products:** Breeding programs might also focus on specific traits that enhance the use of cardamom in value-added products like oils, extracts, or nutraceuticals, which are becoming increasingly popular.

By focusing on these areas, cardamom breeding programs aim to develop varieties that not only satisfy the market's quality standards but also improve the economic viability and sustainability of cardamom production.

3. Cinnamon

Breeding of Cinnamon for market demands and consumer preferences

Breeding cinnamon (*Cinnamomum verum* or *Cinnamomum cassia*) to meet market demands and consumer preferences involves targeted strategies to enhance specific traits that are highly valued in the market. Here's a detailed approach:

1. Market demand identification

- **Flavour profile:** Cinnamon's flavour is primarily determined by its essential oil content, particularly cinnamaldehyde. Consumers prefer a strong, sweet, and aromatic flavour.
- **Quality of bark:** The thickness, smoothness, and colour of the bark are key quality indicators. Thin, smooth, and light-brown bark is typically preferred, especially for Ceylon cinnamon (*C. verum*).
- **Health benefits:** With increasing interest in natural health products, consumers are drawn to cinnamon for its potential health benefits, such as its anti-inflammatory and antioxidant properties.

2. Breeding objectives

- **Enhanced essential oil content:** Breeding efforts focus on increasing the concentration of cinnamaldehyde and other essential oils that contribute to cinnamon's distinct flavour and aroma.
- **Improved bark quality:** Breeders aim to produce varieties with thinner, smoother bark that peels easily, which is highly valued in the spice trade.
- **Yield improvement:** Developing high-yielding varieties ensures that farmers can meet the growing global demand for cinnamon.
- **Disease and pest resistance:** Cinnamon is susceptible to various diseases, such as leaf spot and bark rot, and pests like the cinnamon butterfly. Breeding for resistance to these challenges is crucial for sustainable production.
- **Climate adaptability:** Given the varying climatic conditions in different cinnamon-growing regions, breeding for adaptability to diverse environments, including tolerance to drought or excessive rainfall, is important.

3. Breeding techniques

- **Traditional breeding:** Selection of superior plants with desirable traits, such as high essential oil content and good bark quality, followed by crossbreeding to combine these traits.
- **Clonal propagation:** Since cinnamon is often propagated vegetatively, clonal propagation of superior trees ensures uniformity in the quality of the bark and other traits.
- **Molecular breeding:** Marker-assisted selection can be used to identify and select plants with high essential oil content and disease resistance, speeding up the breeding process.
- **Tissue culture:** This technique is used for the rapid multiplication of disease-free planting material, ensuring uniformity and high quality in the propagated plants.

4. Evaluation and selection

- **Field trials:** Selected varieties undergo rigorous field trials to assess their performance in terms of yield, bark quality, essential oil content, and resistance to diseases and pests.
- **Chemical analysis:** The essential oil content and composition are analysed to ensure that the flavour profile meets market expectations. This involves testing for high levels of cinnamaldehyde and other desirable compounds.
- **Sensory evaluation:** New varieties are evaluated for flavour and aroma through sensory testing to ensure they meet consumer preferences.

5. Commercialization and adoption

- **Patenting and licensing:** New cinnamon varieties may be patented and then licensed to growers, ensuring access to high-quality planting material.
- **Market introduction:** Promoting the benefits of new varieties, such as improved flavour, higher yields, and better disease resistance, to both farmers and consumers is critical for successful market adoption.
- **Sustainability certification:** Breeding programs may also focus on varieties that are well-suited to organic and sustainable farming practices, which can be marketed under sustainability certifications.

6. Future trends

- **Organic and sustainable production:** There is a growing demand for organically produced cinnamon, which requires breeding varieties that can thrive under organic farming conditions with minimal use of chemicals.
- **Value-added products:** Breeding programs might also explore traits that enhance the use of cinnamon in value-added products such as essential oils, extracts, and nutraceuticals.

By focusing on these objectives, cinnamon breeding programs aim to develop varieties that not only satisfy market demands and consumer preferences but also enhance the sustainability and profitability of cinnamon cultivation.

4. Clove

Breeding of Clove for Market demands and consumer preferences

Breeding clove *(Syzygium aromaticum)* to meet market demands and consumer preferences involves a targeted approach to enhance specific traits that are crucial for both quality and yield. Here's a comprehensive look at the process:

1. Market demand identification

- **Essential oil content:** Cloves are highly valued for their essential oil, particularly eugenol, which contributes to the distinctive aroma and flavour. Higher eugenol content is often preferred for both culinary and medicinal uses.
- **Bud size and appearance:** Larger, uniform, and well-formed flower buds are more desirable in the market. The colour, which should be a deep reddish-brown, also plays a role in consumer preference.
- **Sustainability:** There is a growing consumer interest in sustainably sourced cloves, with an emphasis on organic cultivation and fair trade practices.

2. Breeding objectives

- **Enhanced essential oil content:** Breeding programs focus on increasing the eugenol concentration in clove buds to improve the quality and potency of the spice.
- **Improved bud size and uniformity:** Developing varieties with larger, uniform, and aesthetically pleasing buds is crucial to meet market standards.
- **Increased yield:** High-yielding clove trees are essential to meet global demand, especially as the spice is labour-intensive to harvest.
- **Disease and pest resistance:** Clove trees are susceptible to several diseases, such as clove dieback and leaf spot, and pests like the clove borer. Breeding for resistance to these threats is essential for maintaining healthy plantations.
- **Climate resilience:** As climate change impacts growing conditions, breeding for resilience to drought, excessive rainfall, and temperature fluctuations is increasingly important.

3. Breeding techniques

- **Traditional breeding:** This involves selecting superior trees with desirable traits, such as high essential oil content and large bud size, and using them for propagation.
- **Clonal propagation:** Given the long gestation period of clove trees, clonal propagation of high-performing trees ensures uniformity and faster production cycles.
- **Marker-assisted selection:** Molecular techniques can be used to identify and propagate clove trees with high eugenol content and disease resistance, thereby accelerating the breeding process.
- **Tissue culture:** Tissue culture methods are used to propagate disease-free plants and to ensure the uniformity of essential traits like oil content and bud quality.

4. Evaluation and selection

- **Field trials:** Selected clones or varieties are tested in different environments to assess their yield, resistance to pests and diseases, and adaptability to various climatic conditions.
- **Chemical analysis:** The essential oil content, particularly eugenol concentration, is analysed to ensure that the new varieties meet the quality standards expected by the market.
- **Sensory evaluation:** New varieties are also evaluated for flavour and aroma through sensory testing to ensure they align with consumer preferences.

5. Commercialization and adoption

- **Patenting and licensing:** Successful new varieties may be patented and licensed to farmers, providing them with access to improved planting material that meets market demands.
- **Market introduction:** Educating farmers and stakeholders about the benefits of new clove varieties, such as higher yields, better quality buds, and disease resistance, is key to widespread adoption.
- **Sustainability certification:** Breeding programs may emphasize developing varieties that are suitable for organic farming and can be marketed under sustainability and fair-trade certifications.

6. Future trends

- **Organic and sustainable production:** With growing consumer interest in organic products, breeding cloves that thrive under organic farming conditions with minimal use of synthetic inputs will be increasingly important.
- **Value-added products:** Breeding programs may also focus on enhancing traits that are valuable for the production of clove-derived products, such as clove oil, which is used in various industries including pharmaceuticals, cosmetics, and food processing.

By focusing on these key areas, clove breeding programs aim to develop varieties that not only satisfy the market's quality and sustainability demands but also improve the economic viability and resilience of clove production in the face of environmental challenges.

5. Nutmeg

Breeding of nutmeg for market demands and consumer preferences

Breeding nutmeg (*Myristica fragrans*) for market demands and consumer preferences requires a focused strategy to enhance specific traits that are highly valued in both culinary and medicinal markets. Here's a detailed overview of the process:

1. Market demand identification

- **Essential oil content:** Nutmeg is prized for its essential oils, particularly myristicin, which contribute to its distinct aroma and flavour. High essential oil content is crucial for both the spice trade and medicinal uses.
- **Nut size and appearance:** Larger, well-formed nuts with a uniform shape and smooth surface are preferred. The mace, the red aril surrounding the nutmeg seed, is also highly valued for its flavour and colour.

- **Health benefits:** Consumers are increasingly interested in the health benefits of nutmeg, including its potential anti-inflammatory, antioxidant, and digestive properties.

2. Breeding objectives

- **Enhanced essential oil content:** Breeding efforts focus on increasing the concentration of key compounds like myristicin and safrole, which are responsible for the nutmeg's characteristic flavour and aroma.
- **Improved nut and mace quality:** Developing varieties that produce larger, more uniform nuts with a high-quality mace is important for both culinary and medicinal markets.
- **Increased yield:** High-yielding varieties are essential to meet the growing global demand for nutmeg, ensuring profitability for farmers.
- **Disease and pest resistance:** Nutmeg trees are susceptible to diseases such as root rot and pests like nutmeg borers. Breeding for resistance to these challenges is crucial for sustainable production.
- **Climate adaptability:** Given the impact of climate change, breeding for adaptability to various climatic conditions, such as drought tolerance or resistance to excessive rainfall, is increasingly important.

3. Breeding techniques

- **Traditional breeding:** This involves selecting superior trees with desirable traits, such as high essential oil content, large nuts, and high-quality mace. Crossbreeding is used to combine these traits in new varieties.
- **Clonal propagation:** Nutmeg is typically propagated through grafting or air layering to maintain the genetic quality of superior trees and ensure uniformity in the traits of interest.
- **Marker-assisted selection:** Molecular markers are used to identify trees with high essential oil content and disease resistance, allowing breeders to select the best candidates early in the breeding process.
- **Tissue culture:** This technique is employed to propagate disease-free plants and rapidly multiply superior genotypes, ensuring consistent quality and performance.

4. Evaluation and selection

- **Field trials:** New varieties undergo extensive field trials in different environments to evaluate their performance in terms of yield, nut and mace quality, and resistance to diseases and pests.
- **Chemical analysis:** Essential oil content and composition are analysed to ensure that the new varieties meet the market's quality standards, particularly for compounds like myristicin.

- **Sensory evaluation:** Varieties are also tested for flavour and aroma through sensory panels to ensure they align with consumer preferences.

5. Commercialization and adoption

- **Patenting and licensing:** Successful new varieties may be patented and licensed to growers, providing them with access to high-quality planting material that meets market demands.
- **Market introduction:** Promoting the benefits of new nutmeg varieties, such as improved flavour, higher yields, and disease resistance, to farmers and the spice industry is key to successful market adoption.
- **Sustainability certification:** With increasing consumer interest in organic and sustainably produced spices, breeding programs may focus on developing varieties that are suitable for organic farming and can be marketed under sustainability certifications.

6. Future trends

- **Organic and sustainable production:** As demand for organic spices grows, breeding nutmeg varieties that can thrive under organic farming conditions with minimal chemical inputs will become more important.
- **Value-added products:** Breeding programs might also explore traits that enhance the production of value-added products like nutmeg oil, butter, and extracts, which are used in the food, cosmetics, and pharmaceutical industries.

By addressing these key objectives, nutmeg breeding programs can develop varieties that not only meet the demands of the market and consumer preferences but also enhance the sustainability and profitability of nutmeg production in the face of changing environmental conditions.

6. Allspice

Breeding of Allspice for market demands and consumer preferences

Breeding allspice (*Pimenta dioica*) to meet market demands and consumer preferences involves a strategic focus on enhancing specific traits that are important for both quality and yield. Allspice, known for its unique combination of flavours reminiscent of cinnamon, nutmeg, and cloves, is highly valued in both culinary and medicinal markets. Here's an overview of the breeding process:

1. Market demand identification

- **Essential oil content:** Allspice's flavour and aroma are driven by its essential oil content, particularly eugenol, which is the primary compound. A higher concentration of essential oils is desirable for both culinary uses and the extraction industry.

- **Berry size and uniformity:** Larger, uniformly sized, and well-formed berries are preferred in the market. The appearance and colour of the dried berries also impact consumer preferences.
- **Sustainability:** There is growing interest in organically produced allspice and sustainable cultivation practices, driven by consumer demand for ethically sourced and environmentally friendly products.

2. Breeding objectives

- **Increased essential oil content:** Breeding programs aim to increase the concentration of eugenol and other essential oils, enhancing the flavour and aroma profile of allspice.
- **Improved berry quality:** Developing varieties with larger, more uniform berries that have a deep, rich colour is important for market appeal and processing efficiency.
- **Higher yield:** High-yielding varieties are crucial to meet the growing global demand for allspice, ensuring profitability for farmers.
- **Disease and pest resistance:** Allspice trees can be affected by diseases like anthracnose and pests such as scale insects. Breeding for resistance to these issues is vital for sustainable production.
- **Climate adaptability:** As climate change affects growing conditions, breeding for adaptability to various climatic conditions, including tolerance to drought or high humidity, is increasingly important.

3. Breeding techniques

- **Traditional breeding:** This involves selecting and crossbreeding superior trees that exhibit desirable traits, such as high essential oil content and large, uniform berries. Cross-pollination among these selected trees can help combine beneficial traits.
- **Clonal propagation:** Allspice is often propagated vegetatively to ensure uniformity and maintain the genetic quality of superior plants. This includes techniques like grafting or air layering.
- **Marker-assisted selection:** Molecular markers can be used to identify traits such as high essential oil content and disease resistance at an early stage, accelerating the breeding process.
- **Tissue culture:** This technique is employed to propagate disease-free plants and multiply superior genotypes rapidly, ensuring consistent quality and uniformity in the traits of interest.

4. Evaluation and selection

- **Field trials:** New varieties undergo extensive field trials in different environments to assess their performance in terms of yield, berry quality, essential oil content, and resistance to diseases and pests.
- **Chemical analysis:** Essential oil content and composition, particularly eugenol levels, are analysed to ensure the new varieties meet market quality standards.
- **Sensory evaluation:** The flavour and aroma profiles of new varieties are evaluated through sensory testing to ensure they align with consumer preferences and expectations.

5. Commercialization and adoption

- **Patenting and licensing:** Successful new allspice varieties may be patented and licensed to growers, providing them with access to superior planting material that meets market demands.
- **Market introduction:** Promoting the benefits of new allspice varieties, such as improved flavour, higher yields, and disease resistance, to farmers and the spice industry is key to successful market adoption.
- **Sustainability certification:** Breeding programs may also focus on varieties that are suitable for organic farming and can be marketed under sustainability and fair-trade certifications, aligning with consumer trends toward ethically sourced products.

6. Future trends

- **Organic and sustainable production:** With increasing demand for organic spices, breeding allspice varieties that thrive under organic farming conditions with minimal chemical inputs will become more significant.
- **Value-added products:** Breeding programs might also explore traits that enhance the use of allspice in value-added products like essential oils, extracts, and medicinal applications, which are becoming increasingly popular in various industries.

By focusing on these objectives, allspice breeding programs can develop varieties that not only meet market demands and consumer preferences but also contribute to more sustainable and profitable allspice cultivation.

7. Ginger

Breeding of ginger for Market demands and consumer preferences

Breeding ginger (*Zingiber officinale*) to meet market demands and consumer preferences involves a systematic approach to enhance specific traits that are crucial for both quality and productivity. Here's an overview of the breeding process:

1. Market demand identification

- **Flavour and pungency:** Ginger's flavour and pungency, largely determined by its gingerol content, are key attributes for both culinary and medicinal uses. Consumers prefer ginger with strong, aromatic flavours and a balanced pungency.
- **Rhizome quality:** The size, shape, colour, and texture of the ginger rhizome are important. Markets typically prefer large, smooth, plump rhizomes with a uniform golden-yellow colour.
- **Health benefits:** There is increasing demand for ginger due to its recognized health benefits, such as its anti-inflammatory, digestive, and immune-boosting properties.

2. Breeding objectives

- **Enhanced flavour and pungency:** Breeding programs aim to increase the concentration of gingerol and other bioactive compounds that contribute to the distinctive flavour and health benefits of ginger.
- **Improved rhizome quality:** Developing varieties with larger, more uniform rhizomes that are easy to peel and have a smooth, appealing texture is a priority for both fresh and processed ginger markets.
- **Increased yield:** High-yielding varieties are essential to meet the growing global demand for ginger, ensuring that farmers can achieve higher productivity and profitability.
- **Disease and pest resistance:** Ginger is susceptible to diseases like bacterial wilt, rhizome rot, and pests such as nematodes. Breeding for resistance to these challenges is critical for sustainable production.
- **Climate adaptability:** Given the diverse growing conditions in different regions, breeding ginger varieties that can thrive in varying climates, including those with resistance to drought or high moisture, is important.

3. Breeding techniques

- **Traditional breeding:** This involves selecting superior ginger plants with desirable traits, such as high gingerol content and large rhizomes, and crossbreeding them to combine these traits in new varieties.
- **Clonal propagation:** Since ginger is propagated vegetatively (typically through rhizome cuttings), clonal propagation is used to maintain the genetic integrity and uniformity of superior varieties.
- **Molecular breeding:** Marker-assisted selection can identify and propagate plants with specific desirable traits, such as disease resistance or high gingerol content, early in the breeding process.

- **Tissue culture:** Tissue culture techniques are used for the rapid multiplication of disease-free ginger plants, ensuring uniformity and high quality in propagated plants.

4. Evaluation and selection

- **Field trials:** New ginger varieties are tested in different environments to evaluate their performance in terms of yield, rhizome quality, disease resistance, and adaptability to various climatic conditions.
- **Chemical analysis:** The gingerol content and other bioactive compounds are analysed to ensure that the new varieties meet market quality standards and consumer expectations for flavour and health benefits.
- **Sensory evaluation:** Varieties are also evaluated for flavour, aroma, and texture through sensory testing to ensure they align with consumer preferences.

5. Commercialization and adoption

- **Patenting and licensing:** New ginger varieties that demonstrate superior qualities may be patented and licensed to farmers, providing them with access to high-quality planting material that meets market demands.
- **Market introduction:** Educating farmers and stakeholders about the benefits of new ginger varieties, such as improved flavour, higher yields, and disease resistance, is key to successful market adoption.
- **Sustainability certification:** With growing consumer interest in organic and sustainably produced ginger, breeding programs may focus on developing varieties suitable for organic farming and that can be marketed under sustainability certifications.

6. Future trends

- **Organic and sustainable production:** As demand for organic ginger increases, breeding varieties that can thrive under organic farming conditions with minimal chemical inputs will be increasingly important.
- **Value-added products:** Breeding programs might also explore traits that enhance the use of ginger in value-added products like ginger oil, extracts, powders, and nutraceuticals, which are becoming more popular in global markets.

By focusing on these breeding objectives, ginger breeding programs can develop varieties that not only meet the demands of the market and consumer preferences but also support sustainable and profitable ginger cultivation worldwide.

8. Turmeric

Breeding of Turmeric for market demands and consumer preferences

Breeding turmeric (Curcuma longa) to meet market demands and consumer preferences involves a focused strategy to enhance specific traits that are critical for both quality and yield. Turmeric, known for its vibrant colour, flavour, and health benefits, is widely used in culinary, medicinal, and cosmetic products. Here's an overview of the breeding process:

1. Market demand identification

- **Curcumin content:** The primary bioactive compound in turmeric is curcumin, which is responsible for its bright yellow colour and many of its health benefits, including anti-inflammatory and antioxidant properties. Higher curcumin content is often preferred in both culinary and medicinal markets.
- **Rhizome quality:** The size, shape, colour, and texture of the turmeric rhizome are important for market acceptance. Markets generally prefer large, uniformly shaped, smooth rhizomes with a deep orange-yellow colour.
- **Flavour and aroma:** Turmeric's flavour and aroma, driven by essential oils like turmerone, are important for culinary uses. Consumers look for a balanced flavour profile that is not overly bitter.
- **Health benefits:** There is growing consumer interest in turmeric due to its potential health benefits, which drives demand for varieties with high curcumin and essential oil content.

2. Breeding objectives

- **Increased curcumin content:** Breeding programs focus on increasing the concentration of curcumin and other bioactive compounds to enhance the health benefits and colour intensity of turmeric.
- **Improved rhizome quality:** Developing varieties with larger, more uniform rhizomes that have a smooth texture and vibrant colour is critical for both fresh and processed turmeric markets.
- **Higher yield:** High-yielding varieties are essential to meet the growing global demand for turmeric, providing better returns for farmers.
- **Disease and pest resistance:** Turmeric is susceptible to diseases such as leaf spot, rhizome rot, and pests like nematodes. Breeding for resistance to these issues is crucial for sustainable production.
- **Climate adaptability:** With turmeric being grown in diverse climates, breeding for adaptability to different environmental conditions, including drought tolerance and resistance to high humidity, is increasingly important.

3. Breeding techniques

- **Traditional breeding:** This involves selecting and crossbreeding superior turmeric plants with desirable traits such as high curcumin content, large rhizomes, and disease resistance. This helps to combine these traits in new varieties.
- **Clonal propagation:** Turmeric is typically propagated vegetatively through rhizomes. Clonal propagation is used to maintain the genetic quality and uniformity of high-performing varieties.
- **Marker-assisted selection:** Molecular markers are used to identify and select plants with desirable traits like high curcumin content and disease resistance, speeding up the breeding process.
- **Tissue culture:** Tissue culture techniques are employed to propagate disease-free plants and rapidly multiply superior genotypes, ensuring uniformity and quality in the propagated plants.

4. Evaluation and selection

- **Field trials:** New turmeric varieties undergo extensive field trials across different environments to assess their performance in terms of yield, rhizome quality, curcumin content, and resistance to diseases and pests.
- **Chemical analysis:** The curcumin content and composition of essential oils are analysed to ensure that new varieties meet market standards for quality, colour, and flavour.
- **Sensory evaluation:** Varieties are tested for flavour, aroma, and colour through sensory panels to ensure they align with consumer preferences and market expectations.

5. Commercialization and adoption

- **Patenting and licensing:** Successful new turmeric varieties may be patented and licensed to growers, ensuring access to high-quality planting material that meets market demands.
- **Market introduction:** Promoting the benefits of new turmeric varieties, such as higher curcumin content, improved rhizome quality, and disease resistance, to farmers and the turmeric industry is crucial for market adoption.
- **Sustainability certification:** Given the increasing consumer interest in organic and sustainably produced turmeric, breeding programs may focus on developing varieties suitable for organic farming and that can be marketed under sustainability certifications.

6. Future trends

- **Organic and sustainable production:** As the demand for organic turmeric grows, breeding varieties that perform well under organic farming conditions with minimal chemical inputs will become more important.
- **Value-added products:** Breeding programs might also focus on enhancing traits that are valuable for the production of value-added products like turmeric powder, extracts, oils, and supplements, which are gaining popularity in global markets.

By focusing on these objectives, turmeric breeding programs can develop varieties that not only meet market demands and consumer preferences but also support sustainable and profitable turmeric cultivation.

9. Black Turmeric

Breeding of Black Turmeric for market demands and consumer preferences

Breeding black turmeric (*Curcuma caesia*) for market demands and consumer preferences involves targeted strategies to enhance specific traits that are important for both quality and yield. Black turmeric is valued for its unique properties, including its distinctive dark blue or black rhizome, potent medicinal properties, and its use in traditional medicine and natural remedies. Here's how the breeding process typically unfolds:

1. Market demand identification

- **Medicinal properties:** Black turmeric is highly prized for its medicinal properties, particularly its anti-inflammatory, antioxidant, and antimicrobial effects. Consumers are increasingly interested in these health benefits, which drives demand for varieties with high concentrations of bioactive compounds.
- **Curcumin content:** Like yellow turmeric, curcumin is also an important compound in black turmeric, though its concentration is typically lower. Breeding for higher curcumin content can increase its appeal in the health and wellness market.
- **Rhizome quality:** The size, shape, colour, and texture of the rhizome are important. Markets often prefer rhizomes that are large, uniformly shaped, and have a rich, dark colouration.
- **Flavour and aroma:** Black turmeric has a more pungent and bitter flavour compared to yellow turmeric. Consumers and industries that utilize black turmeric in products might have specific preferences regarding the intensity of these characteristics.

2. Breeding objectives

- **Increased bioactive compounds:** Breeding programs focus on enhancing the concentration of curcumin, essential oils, and other bioactive compounds that contribute to the medicinal value of black turmeric.
- **Improved rhizome quality:** Developing varieties with larger, more uniform rhizomes that are easier to process and have a consistent dark colour is crucial for both medicinal and commercial markets.
- **Higher yield:** Breeding for high-yielding varieties is essential to meet growing demand, particularly as black turmeric is less commonly cultivated and more niche compared to yellow turmeric.
- **Disease and pest resistance:** Black turmeric is susceptible to similar diseases and pests as yellow turmeric, such as rhizome rot and nematodes. Breeding for resistance is critical to ensure sustainable production.
- **Climate adaptability:** Given the diverse regions where black turmeric is cultivated, breeding for adaptability to different environmental conditions, such as drought tolerance or resistance to heavy rains, is important.

3. Breeding techniques

- **Traditional breeding:** This involves selecting superior plants with desirable traits, such as high levels of bioactive compounds and large, dark rhizomes, and crossbreeding them to combine these traits in new varieties.
- **Clonal propagation:** Since black turmeric is propagated vegetatively, clonal propagation ensures that the desirable traits of superior plants are maintained and replicated in new plantings.
- **Marker-assisted selection:** Molecular markers can help identify plants with high levels of curcumin and other bioactive compounds early in the breeding process, accelerating the development of improved varieties.
- **Tissue culture:** Tissue culture techniques are used to propagate disease-free plants and rapidly multiply superior genotypes, ensuring consistent quality and uniformity in the traits of interest.

4. Evaluation and selection

- **Field trials:** New black turmeric varieties are tested in different environments to evaluate their performance in terms of yield, rhizome quality, bioactive compound concentration, and resistance to diseases and pests.
- **Chemical analysis:** The levels of curcumin, essential oils, and other bioactive compounds are analysed to ensure that the new varieties meet market expectations for medicinal and commercial use.
- **Sensory evaluation:** Varieties are also evaluated for their flavour, aroma, and bitterness through sensory testing to ensure they align with consumer and industry preferences.

5. Commercialization and adoption

- **Patenting and licensing:** Successful new black turmeric varieties may be patented and licensed to growers, ensuring they have access to high-quality planting material that meets market demands.
- **Market introduction:** Educating farmers and the industry about the benefits of new varieties, such as improved medicinal properties, higher yields, and disease resistance, is key to successful market adoption.
- **Sustainability certification:** With increasing interest in organic and sustainably produced medicinal plants, breeding programs may focus on developing black turmeric varieties that are suitable for organic farming and can be marketed under sustainability certifications.

6. Future trends

- **Organic and sustainable production:** As consumer demand for organic and sustainably produced medicinal plants grows, breeding black turmeric varieties that can thrive under organic farming conditions with minimal chemical inputs will become increasingly important.
- **Value-added products:** Breeding programs might also focus on enhancing traits that are valuable for the production of value-added products such as black turmeric powders, extracts, oils, and nutraceuticals, which are becoming more popular in global markets.

By focusing on these key areas, breeding programs for black turmeric can develop varieties that not only meet market demands and consumer preferences but also support the sustainable and profitable cultivation of this unique and valuable plant.

10. Mango Ginger

Breeding of Mango Ginger for market demands and consumer preferences

The breeding of Mango ginger (*Curcuma amada*) for market demands and consumer preferences focuses on enhancing certain traits to meet both producer and consumer needs. Here are some key areas of focus:

1. **Flavour and aroma**: Mango ginger is prized for its unique mango-like flavour and aroma. Breeding efforts are often directed toward enhancing these sensory qualities to satisfy consumer preferences.
2. **Yield and disease resistance**: High yield is a crucial factor for market viability. Breeding programs aim to develop varieties that not only produce a higher yield but are also resistant to diseases and pests, which can reduce crop losses and increase profitability for farmers.
3. **Nutritional value**: With growing consumer awareness of health benefits, breeding efforts may also focus on improving the nutritional content of

Mango ginger, including its antioxidant properties, to appeal to health-conscious consumers.

4. **Post-harvest stability**: Breeding for improved post-harvest qualities such as shelf-life, resistance to bruising, and maintenance of flavour and texture is essential for both domestic markets and export.
5. **Adaptability to climate change**: As climate patterns shift, breeding for adaptability to different environmental conditions ensures that Mango ginger can be grown successfully in various regions, increasing its availability in the market.

These breeding goals are aligned with both market demands and consumer preferences, ensuring that Mango ginger remains a competitive and desirable product.

11. Coriander

Breeding of Coriander for market demands and consumer preferences

Breeding coriander (*Coriandrum sativum*) to meet market demands and consumer preferences involves enhancing various traits to ensure that the crop is competitive, appealing, and suitable for different culinary uses. Key areas of focus include:

1. **Aroma and flavour**: Coriander is widely valued for its distinct aroma and flavour. Breeding programs often aim to enhance these characteristics, as they are critical to consumer preference, particularly in culinary applications.
2. **Yield and growth habit**: High yield is essential for market competitiveness. Breeding efforts are directed at developing coriander varieties that not only produce more leaves and seeds but also have a suitable growth habit for mechanical harvesting, which is increasingly important in commercial agriculture.
3. **Pest and disease resistance**: Like other crops, coriander is susceptible to various pests and diseases. Breeding for resistance is crucial to reduce the reliance on chemical pesticides, which aligns with consumer demand for sustainably produced food.
4. **Nutritional value**: With a growing emphasis on the health benefits of herbs and spices, breeding coriander for enhanced nutritional content, including higher levels of essential oils and antioxidants, is also an area of interest.
5. **Shelf-life and post-harvest quality**: Improving the shelf-life and maintaining the post-harvest quality of coriander is important for both fresh and dried products, ensuring that the herb retains its flavour and nutritional properties over time.
6. **Adaptation to climatic conditions**: Given the diverse regions where coriander is grown, breeding programs also focus on developing varieties

that are adaptable to different environmental conditions, ensuring stable production across various climates.

These breeding objectives are critical for ensuring that coriander remains a popular and viable crop, meeting both the expectations of consumers and the needs of the market.

12. Fennel

Breeding of Fennel for market demands and consumer preferences

The breeding of fennel (*Foeniculum vulgare*) to meet market demands and consumer preferences involves several key objectives:

1. **Flavour and aroma**: Fennel is widely appreciated for its sweet, anise-like flavour and aroma. Breeding programs focus on enhancing these qualities to meet consumer expectations, especially for culinary uses.
2. **Yield and plant structure**: To meet commercial needs, fennel varieties are bred for higher yields and favorable plant structures that support ease of cultivation and harvesting, particularly in large-scale farming operations.
3. **Pest and disease resistance**: Resistance to common pests and diseases is crucial in reducing crop losses and minimizing the need for chemical interventions. This aligns with the growing consumer demand for organically produced crops.
4. **Nutritional content**: Breeding for enhanced nutritional properties, such as higher levels of vitamins and antioxidants, is also an area of focus, responding to the increasing consumer interest in health benefits.
5. **Adaptation to climate variability**: As fennel is grown in various climates, breeding efforts also aim to develop varieties that are resilient to different environmental conditions, ensuring consistent production across regions.
6. **Storage and shelf-life**: Improving the post-harvest qualities such as storage stability and shelf-life helps maintain the freshness and flavour of fennel, which is important for both fresh market and processed products.

These breeding goals help in ensuring that fennel remains a versatile and preferred crop, catering to both market demands and consumer preferences.

13. Fenugreek

Breeding of Fenugreek for market demands and consumer preferences

Breeding fenugreek (*Trigonella foenum-graecum*) to meet market demands and consumer preferences involves targeting several key traits:

1. **Flavour and aroma**: Fenugreek is known for its distinctive bitter taste and strong aroma, which are crucial for its use in various culinary applications.

Breeding efforts focus on optimizing these sensory attributes to align with consumer expectations.

2. **Yield and growth performance**: High yield and robust growth are essential for commercial production. Breeding programs aim to develop fenugreek varieties that offer higher seed and leaf yields, which are important for both the spice and medicinal herb markets.
3. **Nutritional content**: Given the health benefits associated with fenugreek, including its use in traditional medicine, breeding for enhanced nutritional properties, such as higher levels of bioactive compounds (e.g., diosgenin, fiber, and protein), is a priority.
4. **Pest and disease resistance**: Developing varieties that are resistant to common pests and diseases is crucial for reducing crop losses and minimizing the need for chemical interventions, which is increasingly important for both organic and conventional farming practices.
5. **Adaptability to different climates**: Fenugreek is grown in various climates, so breeding programs often focus on developing varieties that are adaptable to different environmental conditions, ensuring stable yields across different regions.
6. **Post-harvest stability**: Improving the storage and shelf-life of fenugreek seeds and leaves is important for maintaining quality during transport and storage, which directly impacts marketability.

These breeding objectives ensure that fenugreek remains a valuable crop for both consumers and producers, catering to the demands of the spice market and the increasing interest in natural health products.

14. Cumin

Breeding of Cumin for market demands and consumer preferences

Breeding cumin (*Cuminum cyminum*) to meet market demands and consumer preferences focuses on enhancing several critical traits:

1. **Aroma and flavour**: Cumin is known for its strong, warm, and earthy flavour. Breeding efforts aim to enhance these sensory qualities, which are highly valued in culinary uses around the world.
2. **Yield and adaptability**: High yield is crucial for commercial production. Breeding programs target the development of varieties that produce more seeds per plant while also being adaptable to a range of climatic conditions, ensuring consistent production in various environments.
3. **Pest and disease resistance**: Cumin crops are vulnerable to several pests and diseases. Breeding for resistance to these threats is essential for reducing crop losses and minimizing the need for chemical pesticides, which is increasingly important for sustainable farming practices.

4. **Nutritional content**: With growing consumer interest in the health benefits of spices, breeding cumin with higher concentrations of essential oils and antioxidants is becoming a focus. These compounds contribute to the spice's nutritional value and its appeal in health-conscious markets.
5. **Post-harvest quality**: Improving the shelf-life and maintaining the quality of cumin seeds during storage and transport are critical for marketability. Breeding efforts may focus on ensuring that the seeds retain their flavour and potency over time.

These breeding objectives ensure that cumin remains a competitive and preferred spice in both domestic and international markets, aligning with the evolving demands of consumers and the needs of producers.

15. Aijwain

Breeding of Aijwain for market demands and consumer preferences

The breeding of ajwain (*Trachyspermum ammi*) to meet market demands and consumer preferences focuses on several critical objectives:

1. **Aroma and flavour**: Ajwain is known for its strong, aromatic flavour, which is a key attribute for its use in culinary and medicinal applications. Breeding efforts prioritize enhancing these sensory qualities to match consumer expectations.
2. **Yield and plant vigour**: High yield is essential for market competitiveness. Breeding programs focus on developing ajwain varieties that produce higher seed yields and exhibit strong plant vigor, which supports sustainable and profitable cultivation.
3. **Pest and disease resistance**: Ajwain is susceptible to various pests and diseases. Breeding resistant varieties is crucial to ensure consistent production and reduce the need for chemical inputs, which aligns with consumer demand for organically grown spices.
4. **Nutritional and medicinal value**: With a growing interest in natural health products, breeding efforts also target enhancing the nutritional and medicinal properties of ajwain, such as increasing the concentration of thymol, a compound known for its therapeutic benefits.
5. **Climate adaptability**: Ajwain is grown in diverse climatic conditions. Breeding programs aim to develop varieties that are adaptable to various environmental stresses, ensuring stable production across different regions.
6. **Post-harvest quality**: Improving the storage stability and shelf-life of ajwain seeds is important to maintain their quality during storage and transport, which is vital for both domestic and export markets.

These breeding objectives ensure that ajwain continues to meet market demands while catering to evolving consumer preferences.

7

Biotechnological Advances in Spices Breeding

A. Tissue culture and micro propagation

Tissue culture and micro propagation are essential techniques for the propagation of various spices and medicinal plants, allowing the rapid production of disease-free and genetically identical plants. The overview of techniques apply are as follows:

1. Black Pepper (*Piper nigrum*)

1a. Tissue culture in Black Pepper

Callus induction from leaf and stem explants, followed by somatic embryogenesis. The protocol for black pepper (*Piper nigrum* L.) tissue culture involves several stages, including callus induction, shoot regeneration, root initiation, hardening, and implementation of tissue culture techniques.

The general outline of the process based on recent advancements are as follows.

a. Callus induction

- **Explants**: Typically, leaves, stems, or nodes are used as explants.
- **Medium**: Murashige and Skoog (MS) medium supplemented with auxins like 2,4-Dichlorophenoxyacetic acid (2,4-D) @ 0.5-2.0mg/l or Naphthalene acetic acid (NAA)@ 0.5-1.0mg/l is commonly used to induce callus formation.
- **Conditions**: Cultures are usually maintained in dark conditions at 25°C to 28°C to promote callus induction.

b. Shoot regeneration

- **Medium**: After callus formation, the callus is transferred to a regeneration medium, often MS medium with cytokinins like Benzylaminopurine (BAP) or Kinetin (KIN) @ 0.5-1.5mg/l.
- **Conditions**: Light conditions are generally adjusted to a 16-hour photoperiod with light intensity around 2000 to 3000 lux to promote shoot regeneration.

c. Root initiation

- **Medium**: For root induction, Indole-3-butyric acid (IBA) or NAA@ 1mg/l each is typically added to the MS medium.
- **Conditions**: The rooting phase may require lower light intensity and slightly different temperature conditions, typically around 24°C.

d. Hardening

- **Acclimatization**: Once the plantlets have sufficient roots, they are gradually acclimatized to ex vitro conditions. This involves transferring them from the culture vessels to a soil mixture under controlled humidity and light conditions to reduce transplant shock.
- **Medium**: Sterile peat, vermiculite, or a mix of peat and perlite are often used for this purpose.

e. Implementation

- **Field transfer**: After successful hardening, the plants are transferred to the field. Regular monitoring is required to ensure their adaptation to the external environment.
- This process varies slightly depending on the specific conditions and objectives of the tissue culture, such as the genetic manipulation or large-scale propagation.

1b. Callus induction from leaf and stem explants, followed by somatic embryogenesis in Black Pepper

Explants preparation

- **Selection of explants**: Young, healthy leaves and stems from black pepper plants are chosen as explants.
- **Sterilization**: Explants are surface sterilized using ethanol (70% for 30 seconds) followed by a treatment with 0.1% mercuric chloride or sodium hypochlorite for 5-10 minutes. They are then rinsed thoroughly with sterile distilled water to remove traces of sterilants.

Callus induction

- **Culture medium**: Murashige and Skoog (MS) medium is commonly used as the basal medium.
    - **For leaf explants**: MS medium is supplemented with 2,4-Dichlorophenoxyacetic acid (2,4-D) at concentrations ranging from 2.0 to 4.0 mg/L. In some cases, a combination of 2,4-D and Naphthalene Acetic Acid (NAA) is used to enhance callus formation.

- **For stem explants**: A similar MS medium with 2,4-D or NAA, sometimes in combination with a cytokinin like BAP (Benzylaminopurine), is used.
- **Incubation conditions**: Cultures are incubated in the dark at a temperature of 25°C to 28°C to promote callus formation. Callus typically forms within 4-6 weeks.

Somatic embryogenesis

- **Induction of somatic embryogenesis**: The induced callus is transferred to a somatic embryogenesis medium, which is usually an MS medium supplemented with lower concentrations of auxins (like 2,4-D or NAA) and higher concentrations of cytokinins (like BAP or Kinetin). The balance of auxins and cytokinins is crucial for initiating embryogenesis.
  - **Somatic embryogenesis medium**: 0.5 to 1.0 mg/L of 2,4-D combined with 1.0 to 2.0 mg/L of BAP can be effective.
- **Incubation conditions**: Cultures are maintained under a 16-hour photoperiod with light intensity of 2000-3000 lux to facilitate the differentiation of somatic embryos from the callus.

Maturation of somatic embryos

- **Maturation medium**: Somatic embryos are transferred to a maturation medium, typically an MS medium with reduced or no growth regulators, sometimes supplemented with abscisic acid (ABA) to enhance maturation.
- **Conditions**: The embryos are kept under controlled light and temperature conditions until they develop into mature embryos.

Conversion of embryos to plantlets

- **Germination medium**: Mature somatic embryos are placed on a germination medium, which is usually hormone-free MS medium or MS medium with very low levels of cytokinins.
- **Rooting**: For root development, embryos may be transferred to a rooting medium containing auxins like IBA (Indole-3-butyric acid) at low concentrations.

Hardening and Acclimatization

- Once the plantlets have developed roots, they are acclimatized by gradually exposing them to lower humidity and increasing light intensity.
- Plantlets are then transferred to soil in a controlled environment before being moved to the field.

Field transfer

- The hardened plantlets are finally transplanted into the field, where they are monitored until fully established.

This protocol has been effective in producing somatic embryos and regenerating plants in black pepper, contributing to the conservation and mass propagation of this economically important spice.

1c. Micro*propagation: Shoot tip and nodal segment culture for rapid clonal propagation.

For rapid clonal propagation of black pepper (*Piper nigrum* L.) using shoot tip and nodal segment culture, the following protocol is commonly used:

Selection and preparation of explants

- **Explants**: Young, healthy shoot tips and nodal segments are chosen from black pepper plants.
- **Sterilization**: The explants are surface sterilized by immersing them in 70% ethanol for 30 seconds, followed by treatment with 0.1% mercuric chloride or 2-3% sodium hypochlorite for 5-10 minutes. This is followed by thorough rinsing with sterile distilled water to remove all traces of the sterilizing agents.

Initiation of culture

- **Culture medium**: The basal medium used is typically Murashige and Skoog (MS) medium.
  - **Shoot tips**: MS medium supplemented with cytokinins like Benzylaminopurine (BAP) or Kinetin (KIN) is used to initiate shoot proliferation. BAP is commonly used at concentrations ranging from 1.0 to 2.0 mg/L.
  - **Nodal segments**: Similarly, nodal segments are cultured on MS medium with BAP (1.0-2.0 mg/L) or KIN to induce multiple shoot formation.

Shoot proliferation

- **Subculturing**: After 3-4 weeks, the initial explants start producing multiple shoots. These shoots are then excised and subcultured onto fresh MS medium with the same or slightly adjusted cytokinin concentration to promote further multiplication.
- **Proliferation rate**: This process can be repeated multiple times, leading to a rapid increase in the number of shoots.

Root induction

- **Rooting medium**: The proliferated shoots are transferred to a rooting medium, which is typically MS medium supplemented with an auxin like Indole-3-butyric acid (IBA) at concentrations of 0.5 to 1.0 mg/L. In some cases, Naphthalene Acetic Acid (NAA) may also be used.

- **Conditions**: Cultures are kept under a photoperiod of 16 hours light and 8 hours dark, with light intensity of 2000-3000 lux and temperature maintained around 25°C.

Acclimatization

- **Hardening**: Once the shoots have developed sufficient roots, they are acclimatized by transferring them to pots containing a sterile soil mix (such as peat and perlite or vermiculite). The plants are initially kept under high humidity and gradually exposed to lower humidity and increased light intensity.
- **Conditions**: The hardening process typically takes 2-4 weeks, during which the plants are gradually acclimated to ambient conditions.

Field transfer

- **Transfer to soil**: The hardened plantlets are then transferred to the field. Regular watering and monitoring are essential during the initial stages of field establishment.

Summary of steps

- **Explants collection and sterilization**: Shoot tips and nodal segments.
- **Culture initiation**: MS medium with cytokinins (BAP or KIN).
- **Shoot multiplication**: Subculturing for enhanced shoot proliferation.
- **Root induction**: Transfer to rooting medium with IBA or NAA.
- **Hardening**: Acclimatization in pots under controlled conditions.
- **Field establishment**: Transplant to the field with careful monitoring.

This method is widely used for the mass propagation of black pepper due to its efficiency in producing a large number of clonal plants in a relatively short period.

2. Cardamom (*Elettaria cardamomum*)

2a. Tissue culture

The tissue culture protocol for cardamom (**Elettaria Cardamomum**), focusing on callus induction, shoot regeneration, root initiation, hardening, and field transfer, follows these general steps:

Explants preparation

- **Selection of explants**: Explants such as seeds, rhizomes, or leaf segments are commonly used.
- **Sterilization**: Surface sterilization is done using 70% ethanol for 30 seconds followed by 0.1% mercuric chloride or 2-3% sodium hypochlorite for 5-10

minutes. Afterward, explants are rinsed several times with sterile distilled water.

Callus induction

- **Culture medium**: Murashige and Skoog (MS) medium is generally used as the basal medium.
  - **Growth regulators**: Auxins like 2,4-Dichlorophenoxyacetic acid (2,4-D) at concentrations of 2.0 to 4.0 mg/L are added to the medium to induce callus formation. Sometimes, Naphthalene Acetic Acid (NAA) is used either alone or in combination with 2,4-D.
- **Conditions**: Cultures are kept in the dark at a temperature of 25°C to 28°C to promote callus formation. Callus usually forms within 4-6 weeks.

Shoot regeneration

- **Medium for shoot induction**: The callus is transferred to a regeneration medium, typically MS medium supplemented with cytokinins like Benzylaminopurine (BAP) or Kinetin (KIN). A common combination is BAP at 1.0-2.0 mg/L, sometimes with a small amount of auxin like NAA or Indole-3-acetic acid (IAA) to stimulate shoot regeneration.
- **Conditions**: The cultures are maintained under a 16-hour photoperiod with light intensity of 2000-3000 lux. Shoots typically start appearing within 4-8 weeks.

Root initiation

- **Rooting medium**: Regenerated shoots are transferred to a rooting medium, which is usually MS medium supplemented with auxins like Indole-3-butyric acid (IBA) or NAA at concentrations of 0.5 to 1.0 mg/L.
- **Conditions**: Cultures are incubated under similar light conditions as in the shoot regeneration phase, but sometimes with reduced light intensity and slightly different temperature conditions around 24°C to facilitate root development.

Hardening (Acclimatization)

- **Acclimatization process**: Once the plantlets have developed a robust root system, they are transferred to pots containing a sterile soil mix (e.g., peat, vermiculite, or a mix of peat and perlite).
  - **Environment**: Initially, plantlets are kept under high humidity conditions in a greenhouse or controlled environment chamber. Over 2-4 weeks, they are gradually exposed to lower humidity and natural light conditions to acclimatize them to the external environment.

Field Transfer

- **Preparation**: Hardened plantlets are transplanted into the field. It is crucial to select a well-drained soil with partial shade, as cardamom thrives in such conditions.
- **Monitoring**: Newly transplanted plants are regularly watered and monitored to ensure successful establishment in the field.

Summary of steps

- **Explants collection and sterilization**: Using seeds, rhizomes, or leaf segments.
- **Callus induction**: MS medium with 2,4-D or NAA.
- **Shoot regeneration**: MS medium with BAP or KIN, under light conditions.
- **Root initiation**: MS medium with IBA or NAA for root development.
- **Hardening**: Gradual acclimatization in pots under controlled conditions.
- **Field transfer**: Careful transplantation and monitoring in the field.

This protocol enables the mass propagation of cardamom plants, ensuring uniformity and the rapid multiplication of desired plant traits.

2b. Somatic embryogenesis through shoot tip and rhizome cultures for mass multiplication in Cardamom

Somatic embryogenesis in **cardamom (*Elettaria cardamomum*)** through shoot tip and rhizome cultures is an efficient method for mass multiplication. Below is a step-by-step protocol:

Explants selection and preparation

- **Explants**: Shoot tips and rhizomes are commonly selected for initiating somatic embryogenesis.
- **Sterilization**: Explants are surface sterilized using 70% ethanol for 30 seconds, followed by immersion in 0.1% mercuric chloride or 2-3% sodium hypochlorite solution for 5-10 minutes. They are then rinsed thoroughly with sterile distilled water.

Induction of callus formation

- **Culture medium**: The basal medium is typically Murashige and Skoog (MS) medium.
  - **Growth regulators**: For callus induction, MS medium is supplemented with auxins such as 2,4-Dichlorophenoxyacetic acid (2,4-D) or Naphthalene Acetic Acid (NAA), usually in concentrations ranging from 2.0 to 4.0 mg/L.

- **Conditions**: Cultures are incubated in the dark at a temperature of 25°C to 28°C to promote callus formation. A compact, friable callus is generally observed after 4-6 weeks.

Induction of somatic embryogenesis

- **Somatic embryogenesis medium**: The callus obtained from shoot tips and rhizomes is transferred to an embryogenesis-inducing medium, which is typically MS medium supplemented with lower concentrations of 2,4-D (0.5 to 2.0 mg/L) combined with cytokinins like Benzylaminopurine (BAP) or Kinetin (KIN) at concentrations of 0.5 to 2.0 mg/L.
  - **Alternative regulators**: Some protocols also use Abscisic Acid (ABA) or Picloram to enhance embryogenic response.
- **Conditions**: Cultures are maintained under a 16-hour photoperiod with light intensity of 2000-3000 lux. Somatic embryos begin to form after 6-8 weeks of culture.

Maturation of somatic embryos

- **Maturation medium**: Developing somatic embryos are transferred to a maturation medium, which may be MS medium with reduced levels of growth regulators or hormone-free MS medium. In some cases, the addition of ABA (0.5-1.0 mg/L) is used to promote embryo maturation.
- **Conditions**: Controlled light and temperature conditions are maintained to allow the embryos to develop fully into mature embryos.

Conversion of embryos to plantlets

- **Germination medium**: Mature somatic embryos are transferred to a germination medium, often hormone-free MS medium or MS medium supplemented with low concentrations of cytokinins like BAP (0.1-0.5 mg/L).
- **Rooting**: For root initiation, the embryos are sometimes transferred to MS medium with low levels of auxins like Indole-3-butyric acid (IBA) or NAA.

Acclimatization and hardening

- **Hardening process**: The plantlets with well-developed roots are transferred to pots containing sterile soil mix (e.g., peat and vermiculite or perlite). They are initially kept in a high-humidity environment.
- **Acclimatization**: Over 2-4 weeks, the plantlets are gradually acclimatized to lower humidity and natural light, preparing them for outdoor conditions.

Field transfer

- **Transplantation**: The hardened plantlets are transplanted into the field. It's important to select a shaded area with well-drained soil for optimal growth.
- **Monitoring**: Regular irrigation and monitoring are essential during the initial stages of field establishment.

Summary of steps

- **Explants collection and sterilization**: Shoot tips and rhizomes are used.
- **Callus induction**: MS medium with 2,4-D or NAA.
- **Somatic embryogenesis**: Transfer callus to embryogenesis medium with 2,4-D and cytokinins.
- **Embryo maturation**: Transfer developing embryos to a maturation medium.
- **Embryo germination**: Mature embryos are germinated in hormone-free or cytokinin-supplemented MS medium.
- **Rooting and hardening**: Acclimatize plantlets to ex vitro conditions.
- **Field transfer**: Hardened plantlets are transplanted into the field.

This protocol is widely used for the mass propagation of cardamom, ensuring large-scale multiplication of uniform, disease-free plants.

2c. Micropropagation: Direct organogenesis from explants like leaf or rhizome parts for uniform and healthy plants.

Direct embryogenesis in cardamom (***Elettaria cardamomum***) through shoot tip and rhizome cultures is an efficient method for mass multiplication, bypassing the callus stage and leading to faster and more uniform plant production. Here's a step-by-step protocol:

Selection and preparation of explants

- **Explants**: Shoot tips and rhizomes are selected as the starting material.
- **Sterilization**: Surface sterilize the explants by immersing them in 70% ethanol for 30 seconds, followed by treatment with 0.1% mercuric chloride or 2-3% sodium hypochlorite for 5-10 minutes. Rinse thoroughly with sterile distilled water to remove any traces of the sterilizing agent.

Induction of direct somatic embryogenesis

- **Culture medium**: The basal medium typically used is Murashige and Skoog (MS) medium.
  - **Growth regulators**: For direct embryogenesis, the medium is supplemented with a balanced combination of auxins and cytokinins. Common combinations include:

    - **Cytokinins**: Benzylaminopurine (BAP) or Kinetin (KIN) at 1.0-2.0 mg/L.
    - **Auxins**: Naphthalene Acetic Acid (NAA) or 2,4-Dichlorophenoxyacetic acid (2,4-D) at lower concentrations (0.5-1.0 mg/L) to induce embryogenesis directly from the explants without an intervening callus stage.
    - **Alternative regulator**: In some cases, Thidiazuron (TDZ) at low concentrations (0.1-0.5 mg/L) is used, as it is highly effective in inducing direct embryogenesis.
- **Conditions**: Cultures are maintained under a 16-hour photoperiod with a light intensity of 2000-3000 lux. The temperature is controlled at around 25°C. Direct embryo formation is typically observed after 4-8 weeks.

Maturation of somatic embryos

- **Maturation medium**: The developing embryos are transferred to an MS medium with reduced or no growth regulators to allow for maturation. The medium may also be supplemented with abscisic acid (ABA) at low concentrations (0.5 mg/L) to promote proper embryo maturation.
- **Conditions**: Controlled light and temperature conditions (similar to the embryogenesis stage) are maintained to allow the embryos to mature fully into plantlets.

Conversion of embryos to plantlets

- **Germination medium**: Mature embryos are transferred to a germination medium, which is often hormone-free MS medium or MS medium with very low levels of cytokinins (e.g., BAP at 0.1-0.5 mg/L).
- **Rooting**: If necessary, the embryos are transferred to a rooting medium containing low concentrations of auxins such as Indole-3-butyric acid (IBA) or NAA to promote root development.

Acclimatization and hardening

- **Hardening process**: The plantlets that have developed roots are gradually acclimatized to ex vitro conditions by transferring them to pots containing a sterile soil mix (such as peat and vermiculite or perlite). The pots are initially kept in a high-humidity environment, like a greenhouse or growth chamber.
- **Acclimatization**: Over a period of 2-4 weeks, the humidity is gradually reduced, and the plants are exposed to increasing light levels to acclimatize them to natural conditions.

Field transfer

- **Transplantation**: The hardened plantlets are transplanted into the field, ideally in a shaded area with well-drained soil.
- **Monitoring**: Regular irrigation and monitoring are essential during the initial establishment phase to ensure the plants adapt well to the field environment.

Summary of steps

- **Explants collection and sterilization**: Shoot tips and rhizomes are used.
- **Direct embryogenesis**: MS medium with a combination of cytokinins (BAP, KIN) and low levels of auxins (NAA, 2,4-D), or TDZ.
- **Embryo maturation**: Transfer to a maturation medium with or without growth regulators.
- **Embryo germination**: Mature embryos are germinated in hormone-free or cytokinin-supplemented MS medium.
- **Rooting and hardening**: Acclimatize the plantlets under controlled conditions.
- **Field transfer**: Transplant hardened plantlets to the field.

This protocol allows for the efficient mass multiplication of cardamom plants, producing large numbers of uniform and healthy plants suitable for large-scale cultivation. Direct embryogenesis is advantageous as it reduces the time required for plant regeneration compared to indirect methods that involve callus formation.

3. Allspice (*Pimenta dioica*)

3a. Tissue culture in Allspice: Mainly uses explants from young leaves or stems for establishing callus and regenerating plants.

The tissue culture of **allspice (*Pimenta dioica*)** involves several stages, including callus induction, shoot regeneration, root initiation, hardening, and the final implication of the tissue culture process for plant propagation. Here's a detailed protocol for each step:

Explants preparation

- **Selection of explants**: Leaves, stems, or nodal segments are typically used as explants for initiating tissue culture in allspice.
- **Sterilization**: The explants are sterilized by immersing them in 70% ethanol for 30 seconds, followed by treatment with 0.1% mercuric chloride or 2-3% sodium hypochlorite for 5-10 minutes. Afterward, rinse the explants thoroughly with sterile distilled water to remove all traces of sterilants.

Callus induction

- **Culture medium**: The basal medium commonly used is Murashige and Skoog (MS) medium.
  - **Growth regulators**: To induce callus formation, MS medium is supplemented with auxins such as 2,4-Dichlorophenoxyacetic acid (2,4-D) at concentrations ranging from 1.0 to 3.0 mg/L, or Naphthalene Acetic Acid (NAA) at similar concentrations.
  - **Supplementation**: Sometimes, a combination of auxins and cytokinins (like BAP at low concentrations) is used to enhance callus formation.
- **Conditions**: Cultures are maintained in the dark at a temperature of 25°C to 28°C to promote callus induction. Callus formation typically occurs within 4-6 weeks.

Shoot regeneration

- **Regeneration medium**: The callus is transferred to a shoot regeneration medium, which is usually MS medium supplemented with cytokinins like Benzylaminopurine (BAP) or Kinetin (KIN) at concentrations of 1.0 to 2.0 mg/L.
  - **Alternative regulators**: Some protocols may include a low concentration of auxins (e.g., NAA or IAA) to support shoot induction alongside cytokinins.
- **Conditions**: Cultures are incubated under a 16-hour photoperiod with light intensity around 2000-3000 lux. Shoots typically start to emerge after 4-8 weeks.

Root Initiation

- **Rooting medium**: The regenerated shoots are transferred to a rooting medium, which is usually MS medium supplemented with auxins like Indole-3-butyric acid (IBA) or NAA at concentrations of 0.5 to 1.0 mg/L.
- **Conditions**: The cultures are kept under the same light conditions as the shoot regeneration stage, but sometimes with reduced light intensity. Root formation usually takes 2-4 weeks.

Hardening (Acclimatization)

- **Acclimatization process**: Once the plantlets have developed sufficient roots, they are transferred to pots containing a sterile soil mix (such as peat and vermiculite or a mix of peat and perlite).
  - **Environment**: The plantlets are kept under high humidity in a controlled environment such as a greenhouse. Gradually, over 2-4 weeks, humidity is decreased and light intensity is increased to acclimatize the plants to ambient conditions.

Field transfer

- **Preparation for transfer**: The hardened plantlets are transplanted into the field or larger pots. It is crucial to ensure that the soil is well-drained and that the plants are initially provided with partial shade to protect them from direct sunlight.
- **Monitoring**: Regular watering and monitoring are essential to ensure successful establishment of the plants in the field.

Summary of steps

- **Explants collection and sterilization**: Leaf, stem, or nodal segments are sterilized.
- **Callus induction**: MS medium with 2,4-D or NAA, possibly supplemented with low levels of cytokinins.
- **Shoot regeneration**: Transfer callus to an MS medium with cytokinins (BAP or KIN).
- **Root initiation**: Transfer shoots to an MS medium with IBA or NAA.
- **Hardening**: Gradually acclimatize rooted plantlets to ex vitro conditions.
- **Field transfer**: Transplant hardened plantlets to the field, ensuring proper acclimatization and care.

This protocol provides an effective method for the mass propagation of allspice plants, allowing for the production of large numbers of uniform and healthy plants suitable for cultivation.

3b. Micro propagation in Allspice: Producing large numbers of clones from shoot tips and nodal segments is a common method for the clonal propagation of plants, including many economically important species. This protocol can be applied to various plants, including allspice, black pepper, cardamom, or similar species. The following steps outline a general procedure for mass clonal propagation using shoot tips and nodal segments:

Selection and preparation of explants

- **Explants**: Select young, healthy shoot tips and nodal segments from the mother plant.
- **Sterilization**:
  - Immerse explants in 70% ethanol for 30 seconds.
  - Follow with treatment in 0.1% mercuric chloride or 2-3% sodium hypochlorite for 5-10 minutes.
  - Rinse thoroughly with sterile distilled water to remove any residual sterilizing agents.

Initiation of culture

- **Culture medium**: Use Murashige and Skoog (MS) medium as the basal medium.
  - **Growth regulators**: Supplement the MS medium with cytokinins such as Benzylaminopurine (BAP) or Kinetin (KIN) to promote shoot multiplication.
    - BAP is commonly used at concentrations of 1.0 to 2.0 mg/L.
    - KIN can be used at similar concentrations, depending on the plant species.
  - **Optional additives**: Add a small amount of auxin, like Naphthalene Acetic Acid (NAA) at 0.1 mg/L, to support initial shoot proliferation.
- **Conditions**
  - Incubate cultures under a 16-hour photoperiod with light intensity of 2000-3000 lux.
  - Maintain temperature at 25°C ± 2°C.

Shoot multiplication

- **Sub-culturing**: After 3-4 weeks, multiple shoots will emerge from the explants.
  - Excise the shoots and subculture them onto fresh MS medium containing the same concentration of cytokinins.
  - Repeat this subculturing process every 3-4 weeks to enhance the number of shoots produced. Each cycle can result in an exponential increase in the number of shoots.
- **Proliferation rate**: Through repeated subculturing, a large number of clones can be rapidly produced.

Root induction

- **Rooting medium**: Transfer the proliferated shoots to an MS medium supplemented with auxins like Indole-3-butyric acid (IBA) or NAA.
  - Use IBA or NAA at concentrations of 0.5 to 1.0 mg/L to induce root formation.
- **Conditions**
  - Maintain similar light conditions but slightly reduce the light intensity.
  - Maintain the temperature around 24°C to support rooting.
  - Root formation typically takes 2-4 weeks.

Acclimatization and hardening

- **Hardening process**: Once the shoots have rooted, transfer them to pots containing a sterile soil mix (such as peat and vermiculite or perlite).
  - Initially, keep the pots under high humidity conditions in a growth chamber or greenhouse.
  - Gradually reduce humidity and increase light exposure over 2-4 weeks to acclimatize the plants to ambient conditions.

Field transfer

- **Transplantation**: The hardened plantlets can be transplanted into the field or larger pots.
  - Ensure that the soil is well-drained and that the plants are provided with partial shade during the early stages of field establishment.
  - Regular watering and monitoring are necessary during the initial stages to ensure the plants establish well in the field.

Summary of steps

- **Explants collection and sterilization**: Use shoot tips and nodal segments.
- **Culture initiation**: MS medium with cytokinins (BAP or KIN) to promote shoot multiplication.
- **Shoot multiplication**: Subculturing shoots every 3-4 weeks to exponentially increase shoot numbers.
- **Root induction**: Transfer shoots to MS medium with IBA or NAA for rooting.
- **Hardening**: Acclimatize rooted shoots to ex vitro conditions gradually.
- **Field transfer**: Transplant hardened plantlets to the field with appropriate care.

This protocol is highly effective for producing large numbers of clonal plants, ensuring genetic uniformity and rapid propagation of desirable traits in various plant species.

4. Clove (Syzygium aromaticum)

4a. Tissue culture in Clove: Use of meristematic tissues for callus induction. The tissue culture of clove involves several key stages, including callus induction, shoot regeneration, root initiation, hardening, and field transfer. Here's a detailed protocol for each step:

Explants preparation

- **Selection of explants**: Young, healthy leaves, stems, or nodal segments are commonly used as explants.
- **Sterilization**
  - Immerse the explants in 70% ethanol for 30 seconds.
  - Follow with treatment in 0.1% mercuric chloride or 2-3% sodium hypochlorite for 5-10 minutes.
  - Rinse thoroughly with sterile distilled water to remove any residual sterilizing agents.

Callus induction

- **Culture medium**: The basal medium is typically Murashige and Skoog (MS) medium.
  - **Growth regulators**: To induce callus formation, MS medium is supplemented with auxins such as 2,4-Dichlorophenoxyacetic acid (2,4-D) at concentrations of 1.0 to 3.0 mg/L, or Naphthalene Acetic Acid (NAA) at similar concentrations.
  - **Optional additives**: Sometimes a low concentration of cytokinins (e.g., BAP at 0.5 mg/L) is added to the medium to enhance callus induction.
- **Conditions**
  - Cultures are incubated in the dark at 25°C to 28°C to promote callus formation.
  - Callus formation generally occurs within 4-6 weeks.

Shoot regeneration

- **Regeneration medium**: The callus is transferred to a shoot regeneration medium, which is usually MS medium supplemented with cytokinins such as Benzylaminopurine (BAP) or Kinetin (KIN) at concentrations of 1.0 to 2.0 mg/L.
  - **Optional growth regulators**: A small amount of auxin (e.g., NAA at 0.1 mg/L) may be included to support shoot regeneration.
- **Conditions**
  - The cultures are kept under a 16-hour photoperiod with light intensity of 2000-3000 lux.
  - Shoots typically begin to emerge after 4-8 weeks.

Root initiation

- **Rooting medium**: The regenerated shoots are transferred to an MS medium supplemented with auxins such as Indole-3-butyric acid (IBA) or Naphthalene Acetic Acid (NAA) at concentrations of 0.5 to 1.0 mg/L.

- **Conditions**
  - o Cultures are maintained under similar light conditions, but the light intensity may be slightly reduced.
  - o Root formation usually takes 2-4 weeks.

Hardening (Acclimatization)

- **Acclimatization process**: Once the plantlets have developed roots, they are transferred to pots containing a sterile soil mix (such as peat and vermiculite or perlite).
  - **Environment**: The plantlets are initially kept under high humidity in a controlled environment (like a greenhouse).
  - **Gradual acclimatization**: Over 2-4 weeks, gradually decrease humidity and increase exposure to natural light to acclimatize the plantlets to external conditions.

Field transfer

- **Transplantation**: The hardened plantlets are transplanted into the field or larger pots.
  - **Soil and light requirements**: Clove plants require well-drained soil and partial shade, especially during the early stages of field establishment.
  - **Monitoring**: Regular watering and monitoring are essential during the initial stages to ensure successful establishment.

Summary of steps

- **Explants collection and sterilization**: Use young leaves, stems, or nodal segments.
- **Callus induction**: MS medium with auxins (2,4-D or NAA) for callus formation.
- **Shoot regeneration**: Transfer callus to MS medium with cytokinins (BAP or KIN).
- **Root initiation**: Transfer shoots to MS medium with IBA or NAA for rooting.
- **Hardening**: Gradual acclimatization in pots under controlled conditions.
- **Field transfer**: Transplant hardened plantlets to the field with proper care.

This protocol provides a reliable method for the mass propagation of clove plants, enabling the production of large numbers of genetically uniform and healthy plants for cultivation.

4b. Micropropagation in Clove: Shoot tip culture is common to ensure the propagation of disease-free clones.

Producing disease-free clones of clove through shoot tip culture is an effective method for ensuring the propagation of healthy, uniform plants. This method is particularly useful for eliminating systemic diseases, including viruses, that may affect the mother plants. Below is a step-by-step protocol:

Selection and preparation of explants

- **Explants**: Select young, healthy shoot tips (approximately 1-2 cm) from a mother plant that shows no visible signs of disease.
- **Sterilization**:
    - Immerse the shoot tips in 70% ethanol for 30 seconds.
    - Follow with a treatment in 0.1% mercuric chloride or 2-3% sodium hypochlorite solution for 5-10 minutes.
    - Rinse thoroughly with sterile distilled water multiple times to remove any remaining sterilant.

Initiation of culture

- **Culture medium**: Use Murashige and Skoog (MS) medium as the basal medium.
    - **Growth regulators**: Supplement the MS medium with a cytokinin such as Benzylaminopurine (BAP) at concentrations of 1.0 to 2.0 mg/L to promote the initiation and proliferation of shoots.
    - **Optional additives**: A small amount of Naphthalene Acetic Acid (NAA) or Indole-3-acetic acid (IAA) at 0.1 mg/L can be included to enhance shoot growth.
- **Conditions**
    - Incubate the cultures under a 16-hour photoperiod with a light intensity of 2000-3000 lux.
    - Maintain the temperature at 25°C ± 2°C.
    - Shoots should begin to proliferate within 3-4 weeks.

Shoot multiplication

- **Subculturing**: Once the shoot tips begin to grow, they can be subcultured onto fresh MS medium with the same or slightly adjusted concentrations of BAP (1.0-2.0 mg/L).
    - **Multiplication cycles**: Continue to subculture the proliferating shoots every 3-4 weeks to enhance shoot multiplication. This step helps to rapidly increase the number of disease-free clones.

Virus elimination (if necessary)

- **Thermotherapy or Chemotherapy**
  - If there is a risk of viral contamination, consider applying heat treatment (thermotherapy) by incubating the cultures at elevated temperatures (35-38°C) for several weeks before proceeding to the next steps.
  - Alternatively, antiviral chemicals (chemotherapy) such as ribavirin can be added to the culture medium.

Root induction

- **Rooting medium**: Transfer the healthy, proliferated shoots to an MS medium supplemented with an auxin such as Indole-3-butyric acid (IBA) or NAA at concentrations of 0.5 to 1.0 mg/L.
- **Conditions**
  - Maintain the cultures under similar light conditions but reduce the light intensity slightly to promote root formation.
  - Roots typically develop within 2-4 weeks.

Acclimatization and hardening

- **Hardening process**: Transfer the rooted shoots to pots containing a sterile soil mix (such as peat and vermiculite or perlite).
  - **High humidity**: Initially, keep the plantlets under high humidity in a growth chamber or greenhouse.
  - **Gradual acclimatization**: Over 2-4 weeks, gradually reduce the humidity and increase the light exposure to acclimatize the plants to external conditions.

Field transfer

- **Transplantation**: After successful acclimatization, transplant the hardened plantlets into the field.
  - **Soil requirements**: Ensure well-drained soil and partial shade, especially during the early stages.
  - **Monitoring**: Regularly monitor the plants for signs of stress or disease and provide adequate irrigation.

Summary of steps

- **Explants collection and sterilization**: Use disease-free shoot tips, followed by thorough sterilization.
- **Culture initiation**: MS medium with BAP for shoot initiation.

- **Shoot multiplication**: Subculture the shoots to enhance proliferation.
- **Virus elimination**: Apply thermotherapy or chemotherapy if necessary.
- **Root induction**: Transfer to MS medium with IBA or NAA for rooting.
- **Hardening**: Gradually acclimatize rooted plantlets to ex vitro conditions.
- **Field transfer**: Transplant hardened plantlets to the field, ensuring they are disease-free and well-established.

This protocol ensures the production of disease-free clove plants, which are genetically uniform and healthy, making them suitable for large-scale cultivation.

5. Cinnamon (*Cinnamomum verum*)

5a. Tissue culture in Cinnamon: Involves explant culture from bark or leaf tissues.

Tissue culture in *Cinnamomum* (commonly known as cinnamon) is a technique used to propagate this plant through in vitro methods. This method is especially useful for producing large numbers of plants in a short period, maintaining genetic uniformity, and conserving plant resources. Below is an overview of the tissue culture process specifically tailored for cinnamon:

Selection of explants

- **Source material:** Commonly used explants include young leaves, nodal segments, shoot tips, and immature seeds.
- **Sterilization:** Explants are surface sterilized using agents like sodium hypochlorite or mercuric chloride, followed by several rinses in sterile distilled water to remove any traces of the sterilizing agent.

Callus induction

- **Medium:** Murashige and Skoog (MS) medium is frequently used, supplemented with growth regulators like 2,4-Dichlorophenoxyacetic acid (2,4-D), Naphthaleneacetic acid (NAA), or Indole-3-acetic acid (IAA) for callus induction.
- **Conditions:** Cultures are maintained in the dark at temperatures around 25-28°C. Callus formation generally takes 2-4 weeks depending on the explant and hormone concentration.

Shoot regeneration

- **Medium:** The callus is transferred to a shoot induction medium, often MS medium supplemented with cytokinins like Benzylaminopurine (BAP), Kinetin (KIN), or Thidiazuron (TDZ). These hormones encourage the differentiation of shoots from the callus tissue.

- **Conditions:** Light exposure (16-hour photoperiod) at 25°C with light intensity of around 2000-3000 lux is crucial for shoot formation. Shoots typically emerge within 4-6 weeks.

Root initiation

- **Medium:** For rooting, the shoots are transferred to MS medium with lower concentrations of cytokinins and higher levels of auxins such as Indole-3-butyric acid (IBA) or NAA.
- **Conditions:** Roots usually form within 2-3 weeks. The cultures are kept under similar photoperiod and temperature conditions as for shoot regeneration.

Hardening and Acclimatization

- **Process:** Once the plantlets have sufficient roots, they are transferred from in vitro conditions to pots containing a mixture of soil, sand, and vermiculite.
- **Conditions:** Initially, plantlets are kept in high humidity environments (using transparent covers or mist chambers) and gradually acclimatized to ambient conditions over several weeks.

Transplantation to field

- **Procedure:** After successful acclimatization, the hardened plantlets can be transplanted into the field. Regular monitoring is necessary to ensure survival and growth.

Applications and benefits

- **Clonal propagation:** Tissue culture allows for the rapid multiplication of genetically uniform and disease-free plants.
- **Conservation:** It provides a method for conserving rare or endangered cinnamon varieties.
- **Research:** Tissue culture is a valuable tool for genetic studies and for producing plants with desirable traits through techniques like somaclonal variation and genetic transformation.

This process can be optimized depending on the specific species or variety of *Cinnamomum* being cultured, as well as the desired outcomes (e.g., higher yield, disease resistance, etc.).

5b. Micro propagation in Cinnamon: Shoot multiplication from nodal segments for clonal propagation

Micro propagation in Cinnamon is a specialized tissue culture technique used to rapidly produce large numbers of genetically identical plants. This method is particularly valuable for conserving genetic resources, enhancing plant breeding programs, and producing uniform plants for commercial purposes. Below is a step-by-step guide to micro propagation in cinnamon:

Selection and preparation of explants

- **Explants:** Nodal segments, shoot tips, or leaf segments from healthy, disease-free mother plants are typically used. Young, actively growing tissues are preferred for their high regenerative capacity.
- **Sterilization:** The explants are surface sterilized using a sequence of chemical treatments, such as washing with 70% ethanol for 30 seconds followed by 0.1% (w/v) mercuric chloride or 1-2% sodium hypochlorite for 5-10 minutes, and then rinsed multiple times with sterile distilled water.

Initiation of cultures

- **Medium:** The explants are cultured on Murashige and Skoog (MS) medium, which is commonly supplemented with cytokinins such as Benzylaminopurine (BAP) to induce the proliferation of axillary buds.
- **Conditions:** Cultures are maintained under a controlled environment with a temperature of 25-28°C, light intensity of 2000-3000 lux, and a 16-hour photoperiod. Bud initiation and shoot proliferation occur within 2-4 weeks.

Shoot multiplication

- **Subculture:** The newly formed shoots are cut and transferred to fresh MS medium containing a combination of cytokinins (e.g., BAP or Kinetin) and sometimes low concentrations of auxins (e.g., Indole-3-acetic acid, IAA) to promote further shoot multiplication.
- **Repeated subculturing:** Shoots are periodically subcultured every 3-4 weeks to enhance multiplication rates. Multiple shoots can be produced from a single explant over several subculture cycles.

Root induction

- **Rooting medium:** Shoots are transferred to a rooting medium, which is typically MS medium with reduced cytokinin levels and increased auxin concentrations, particularly Indole-3-butyric acid (IBA) or Naphthaleneacetic acid (NAA).
- **Conditions:** Cultures are maintained in similar environmental conditions as shoot multiplication. Root formation is usually observed within 2-3 weeks.

Hardening and acclimatization

- **Transition to soil:** Rooted plantlets are carefully removed from the culture medium, washed to remove any adhering gel, and transferred to pots containing a sterilized soil mix (often a combination of peat, sand, and vermiculite).

- **Hardening process:** The plantlets are initially kept in a high-humidity environment (such as a mist chamber or under transparent covers) and gradually exposed to normal environmental conditions over 2-4 weeks to harden them.
- **Acclimatization:** After successful hardening, the plantlets are acclimatized to field conditions and eventually transplanted to the field.

Field establishment

- **Planting:** The acclimatized plants are transplanted into the field or orchard, where they continue to grow under standard agricultural practices.
- **Monitoring:** Regular monitoring for pests, diseases, and environmental stresses is essential to ensure successful establishment.

Advantages of micropropagation in Cinnamon

- **Mass production:** Enables the rapid production of large numbers of uniform and disease-free plants.
- **Conservation:** Provides a method for conserving valuable genetic resources, particularly of rare or endangered cinnamon varieties.
- **Genetic fidelity:** Ensures the production of genetically identical plants, maintaining the desirable traits of the mother plant.
- **Year-round production:** Allows for the propagation of plants throughout the year, independent of seasonal constraints.

Micropropagation of cinnamon can be an effective tool in commercial production, research, and conservation programs, allowing for the efficient production of high-quality planting material.

6. Ginger (*Zingiber officinale*)

6a. Tissue culture in Ginger: Rhizome culture is used for callus formation.

Standardizing a protocol for callus induction, shoot regeneration, root initiation, and hardening in ginger (*Zingiber officinale*) involves optimizing conditions for each stage to ensure consistent and efficient plant regeneration. Below is a general protocol for each step in the tissue culture process for ginger:

Callus induction

- **Explants:** Rhizome segments, shoot tips, or young leaf segments are commonly used.
- **Medium:** Murashige and Skoog (MS) basal medium supplemented with 2,4-Dichlorophenoxyacetic acid (2,4-D) at concentrations ranging from 1.0 to 2.5 mg/L. Sometimes, a combination of 2,4-D and Naphthaleneacetic acid (NAA) is used to enhance callus formation.

- **Conditions:** Cultures are kept in the dark at 25-28°C. Callus formation typically occurs within 3-4 weeks.

Shoot regeneration

- **Medium:** The callus is transferred to MS medium supplemented with cytokinins like Benzylaminopurine (BAP) or Kinetin (KIN) at concentrations of 2.0 to 5.0 mg/L to promote shoot regeneration.
- **Additional additives:** Adding a low concentration of auxins (e.g., 0.1-0.5 mg/L of NAA or IAA) can help in shoot elongation.
- **Conditions:** Cultures are exposed to a 16-hour photoperiod with light intensity of about 2000-3000 lux. Shoots generally start to regenerate within 4-6 weeks.

Root initiation

- **Medium:** Shoots are transferred to MS medium containing lower levels of cytokinins and higher concentrations of auxins, particularly Indole-3-butyric acid (IBA) or NAA (0.5-2.0 mg/L).
- **Conditions:** Rooting occurs under a 16-hour photoperiod with light intensity similar to that used in the shoot regeneration stage. Roots typically form within 2-3 weeks.

Hardening

- **Procedure**
  - **Transplantation:** Rooted plantlets are gently removed from the culture medium, washed to remove agar, and transferred to pots containing a sterile soil mix (e.g., a mix of soil, sand, and vermiculite).
  - **Initial conditions:** The plantlets are kept in a high-humidity environment, such as a mist chamber or under transparent covers, to reduce water loss and acclimatize them gradually to ambient conditions.
- **Acclimatization:** Over 2-4 weeks, the humidity is gradually reduced, and the plantlets are exposed to normal growing conditions. Eventually, they are transferred to the field or greenhouse for further growth.

Field establishment

- **Transplanting:** After successful acclimatization, the hardened plantlets are transplanted into the field under standard cultivation practices.
- **Monitoring:** Regular monitoring for pests, diseases, and environmental stresses is essential to ensure successful establishment and growth.

Key points for optimization

- **Hormone concentrations:** Fine-tuning the concentrations of cytokinins and auxins in the media is crucial for optimizing callus induction, shoot regeneration, and rooting.
- **Culture conditions:** Temperature, light intensity, and photoperiod should be closely monitored and adjusted according to the needs of each stage.
- **Sterility:** Maintaining sterility throughout the process is critical to prevent contamination and ensure the success of the tissue culture process.

Applications of tissue culture in Ginger

- **Mass propagation:** Tissue culture allows for the rapid multiplication of ginger plants, ensuring uniformity and high yield.
- **Disease-free plants:** Micropropagation helps produce disease-free planting material, which is essential for commercial ginger production.
- **Genetic conservation:** This technique can be used to conserve and propagate rare or endangered ginger varieties.

This protocol serves as a general guide and may require adjustments depending on the specific ginger variety and the desired outcomes.

6b. Micropropagation in Ginger: Shoot tip and rhizome segment culture are standard for large-scale propagation.

Micropropagation in ginger (*Zingiber officinale*) is an effective method for the rapid production of disease-free, genetically uniform plants. This technique is widely used in commercial agriculture and research to propagate large numbers of ginger plants in a short time. Below is a detailed step-by-step guide for micropropagation in ginger:

Selection and preparation of explants

- **Explants:** Common explants used in ginger micropropagation include shoot tips, nodal segments, or rhizome buds.
- **Sterilization:** Explants are surface sterilized using 70% ethanol for 30 seconds, followed by 0.1% (w/v) mercuric chloride or 1-2% sodium hypochlorite for 5-10 minutes. They are then rinsed thoroughly with sterile distilled water to remove any traces of the sterilizing agents.

Initiation of cultures

- **Medium:** Explants are cultured on Murashige and Skoog (MS) medium supplemented with a cytokinin like Benzylaminopurine (BAP) at concentrations of 2.0-5.0 mg/L to induce shoot initiation.

- **Conditions:** Cultures are maintained at 25-28°C with a 16-hour photoperiod and light intensity of 2000-3000 lux. Shoot initiation typically occurs within 2-3 weeks.

Shoot multiplication

- **Subculture:** Once shoots have developed, they are excised and transferred to fresh MS medium containing BAP (2.0-5.0 mg/L) for further multiplication. A combination of cytokinins and a low concentration of auxins (e.g., NAA or IAA at 0.1-0.5 mg/L) can enhance the rate of shoot multiplication.
- **Repeated subculturing:** The shoots are subcultured every 3-4 weeks to maintain high multiplication rates. Each subculture cycle increases the number of shoots.

Root induction

- **Rooting medium:** For rooting, the shoots are transferred to MS medium with lower cytokinin levels and higher auxin concentrations, particularly Indole-3-butyric acid (IBA) or Naphthaleneacetic acid (NAA) at 0.5-2.0 mg/L.
- **Conditions:** Rooting generally occurs within 2-3 weeks under the same environmental conditions used for shoot multiplication.

Hardening and Acclimatization

- **Transfer to soil:** Rooted plantlets are carefully removed from the culture medium, washed to remove any remaining agar, and transplanted into pots containing a sterilized soil mix (typically a combination of peat, sand, and vermiculite).
- **Hardening process:** The plantlets are initially placed in a high-humidity environment, such as a mist chamber or covered with transparent plastic, to reduce water loss and gradually acclimate them to the ambient environment. Over 2-4 weeks, the humidity is gradually reduced, and the plantlets are exposed to natural conditions.
- **Acclimatization:** After successful hardening, the plantlets are transferred to larger pots or directly into the field, where they continue to grow under standard agricultural practices.

Field establishment

- **Planting:** Acclimatized plants are transplanted into the field, typically during the appropriate season for ginger cultivation.
- **Monitoring:** Regular monitoring for pests, diseases, and environmental stresses is essential to ensure healthy growth and development in the field.

Advantages of micropropagation in Ginger

- **Rapid multiplication:** Micropropagation allows for the production of a large number of plants in a relatively short period.
- **Disease-free plants:** The technique ensures that the propagated plants are free from soil-borne diseases, which are common in ginger cultivation.
- **Uniformity:** All plants produced through micropropagation are genetically identical, ensuring uniform growth and yield.
- **Conservation:** Micropropagation is useful for conserving rare or endangered ginger varieties, allowing for their rapid multiplication and preservation.

Applications

- **Commercial production:** Micropropagation is widely used in the commercial production of ginger to meet the demand for high-quality planting material.
- **Research:** The technique is also employed in research settings to study genetic traits, disease resistance, and other aspects of ginger biology.
- **Germplasm conservation:** Micropropagation facilitates the ex situ conservation of ginger germplasm, preserving valuable genetic resources for future use.

This protocol provides a reliable and efficient method for ginger micropropagation, which can be adapted based on specific requirements or the particular ginger variety being propagated.

7. Turmeric (*Curcuma longa*)

7a. Tissue culture in Turmeric: Callus is induced from rhizome or leaf explants.

Standardizing a protocol for callus induction, shoot regeneration, root initiation, and hardening in turmeric (*Curcuma longa*) involves optimizing various factors such as explant type, media composition, and culture conditions. Here is a general protocol that can be adapted for turmeric tissue culture:

Callus induction

- **Explants:** Common explants used for turmeric include rhizome buds, shoot tips, or young leaf segments.
- **Medium:** Murashige and Skoog (MS) basal medium is typically used, supplemented with 2,4-Dichlorophenoxyacetic acid (2,4-D) at concentrations ranging from 1.0 to 3.0 mg/L. Sometimes a combination of 2,4-D and Naphthaleneacetic acid (NAA) can be used for better callus induction.

- **Conditions:** Cultures are kept in the dark at 25-28°C to induce callus formation, which typically occurs within 3-4 weeks.

Shoot regeneration

- **Medium:** The callus is transferred to a shoot induction medium, often MS medium supplemented with cytokinins like Benzylaminopurine (BAP) at concentrations of 2.0 to 5.0 mg/L. Thidiazuron (TDZ) or Kinetin (KIN) may also be used in some protocols to enhance shoot regeneration.
- **Additional additives:** A low concentration of auxins, such as Indole-3-acetic acid (IAA) or NAA (0.1-0.5 mg/L), may be added to support shoot elongation.
- **Conditions:** Cultures are exposed to a 16-hour photoperiod with light intensity of about 2000-3000 lux. Shoots typically start regenerating within 4-6 weeks.

Root initiation

- **Medium:** Regenerated shoots are transferred to a rooting medium, typically MS medium with reduced cytokinin levels and higher auxin concentrations, particularly Indole-3-butyric acid (IBA) or NAA at concentrations of 0.5-2.0 mg/L.
- **Conditions:** Cultures are kept under the same photoperiod and temperature conditions as during shoot regeneration. Roots usually form within 2-3 weeks.

Hardening

- **Transplantation:** Rooted plantlets are carefully removed from the culture medium, washed to remove any residual agar, and transplanted into pots containing a sterile soil mix, such as a combination of peat, sand, and vermiculite.
- **Initial hardening conditions:** Plantlets are kept in a high-humidity environment (e.g., mist chambers or covered with transparent plastic) to reduce water loss and acclimate them gradually to ambient conditions.
- **Gradual acclimatization:** Over 2-4 weeks, the humidity is gradually reduced, and the plantlets are exposed to normal growing conditions. After successful acclimatization, they can be transferred to larger pots or the field.

Field establishment

- **Planting:** Acclimatized plants are transplanted into the field under standard agricultural practices appropriate for turmeric cultivation.
- **Monitoring:** Regular monitoring for pests, diseases, and environmental stresses is essential to ensure healthy growth and successful establishment in the field.

Optimization factors

- **Hormone concentrations:** Adjusting the levels of cytokinins and auxins in the media is crucial for optimizing each stage of the tissue culture process.
- **Culture conditions:** Temperature, light intensity, and photoperiod must be carefully controlled and adjusted as necessary for each stage.
- **Sterility:** Maintaining strict aseptic conditions throughout the process is critical to prevent contamination and ensure successful tissue culture.

Applications of tissue culture in Turmeric

- **Mass propagation:** Tissue culture allows for the rapid multiplication of turmeric plants, providing uniform and disease-free planting material.
- **Genetic conservation:** It is useful for conserving and propagating rare or endangered turmeric varieties.
- **Research:** The protocol can be used in research settings to study genetic traits, improve plant characteristics, and develop new varieties.

This standardized protocol serves as a general guideline and may require adjustments depending on the specific turmeric variety and the desired outcomes.

7b. Micropropagation in Turmeric: Shoot tip culture is the main method for multiplying turmeric plants.

Micropropagation in turmeric (*Curcuma longa*) is an efficient method for producing large numbers of uniform, disease-free plants. This technique is widely used in commercial production and research. Here's a step-by-step guide for micropropagation in turmeric:

Selection and preparation of explants

- **Explants:** The most commonly used explants for turmeric micropropagation include rhizome buds, shoot tips, or nodal segments.
- **Sterilization:** Explants are surface sterilized using 70% ethanol for 30 seconds, followed by treatment with 0.1% (w/v) mercuric chloride or 1-2% sodium hypochlorite for 5-10 minutes. After sterilization, they are rinsed thoroughly with sterile distilled water to remove any residues of the sterilizing agents.

Initiation of cultures

- **Medium:** Explants are cultured on Murashige and Skoog (MS) medium supplemented with cytokinins, such as Benzylaminopurine (BAP) at concentrations of 2.0-5.0 mg/L, to induce shoot initiation.
- **Conditions:** Cultures are maintained at 25-28°C under a 16-hour photoperiod with light intensity of 2000-3000 lux. Shoot initiation typically occurs within 2-4 weeks.

Shoot multiplication

- **Subculture:** Once shoots have developed, they are excised and transferred to fresh MS medium containing BAP (2.0-5.0 mg/L) for further multiplication. Sometimes, a combination of BAP with Kinetin (KIN) or Thidiazuron (TDZ) is used to enhance shoot proliferation.
- **Repeated subculturing:** Shoots are periodically subcultured every 3-4 weeks to maintain high multiplication rates. Multiple shoots can be produced from a single explant over several subculture cycles.

Root induction

- **Rooting medium:** For rooting, the shoots are transferred to MS medium supplemented with lower levels of cytokinins and higher concentrations of auxins, such as Indole-3-butyric acid (IBA) or Naphthaleneacetic acid (NAA) at 0.5-2.0 mg/L.
- **Conditions:** Rooting generally occurs within 2-3 weeks under the same environmental conditions used for shoot multiplication.

Hardening and Acclimatization

- **Transfer to soil:** Rooted plantlets are gently removed from the culture medium, washed to remove any agar residue, and transplanted into pots containing a sterile soil mix (typically a combination of peat, sand, and vermiculite).
- **Hardening process:** The plantlets are initially placed in a high-humidity environment, such as a mist chamber or covered with transparent plastic, to reduce water loss and acclimatize them gradually to the ambient environment. Over 2-4 weeks, the humidity is gradually reduced, and the plantlets are exposed to natural conditions.
- **Acclimatization:** After successful hardening, the plantlets are transferred to larger pots or directly into the field, where they continue to grow under standard agricultural practices.

Field establishment

- **Planting:** Acclimatized plants are transplanted into the field, typically during the appropriate season for turmeric cultivation.
- **Monitoring:** Regular monitoring for pests, diseases, and environmental stresses is essential to ensure healthy growth and development in the field.

Advantages of micropropagation in Turmeric

- **Rapid multiplication:** Micropropagation allows for the production of a large number of plants in a relatively short period.

- **Disease-free plants:** The technique ensures that the propagated plants are free from soil-borne diseases, which are common in turmeric cultivation.
- **Uniformity:** All plants produced through micropropagation are genetically identical, ensuring uniform growth and yield.
- **Conservation:** Micropropagation is useful for conserving rare or endangered turmeric varieties, allowing for their rapid multiplication and preservation.

Applications

- **Commercial production:** Micropropagation is widely used in the commercial production of turmeric to meet the demand for high-quality planting material.
- **Research:** The technique is also employed in research settings to study genetic traits, disease resistance, and other aspects of turmeric biology.
- **Germplasm conservation:** Micropropagation facilitates the ex situ conservation of turmeric germplasm, preserving valuable genetic resources for future use.

By following this protocol, large-scale production of turmeric plants can be achieved efficiently, ensuring consistency and quality in the plant material.

8. Black Turmeric (*Curcuma caesia*)

8a. Tissue culture: Similar to turmeric, callus induction from rhizomes.

Standardizing a protocol for callus induction, shoot regeneration, root initiation, and hardening in black turmeric (*Curcuma caesia*) involves fine-tuning various aspects of the tissue culture process to ensure optimal growth and regeneration. Black turmeric is a unique and valuable species, so precise control over each stage is crucial for successful micropropagation. Below is a general protocol tailored for black turmeric:

Callus induction

- **Explants:** Rhizome buds, shoot tips, or young leaf segments are commonly used as explants.
- **Medium:** Murashige and Skoog (MS) basal medium supplemented with 2,4-Dichlorophenoxyacetic acid (2,4-D) at concentrations of 2.0-3.0 mg/L is typically used for callus induction. In some cases, Naphthaleneacetic acid (NAA) or a combination of 2,4-D and NAA might be used to enhance callus formation.
- **Conditions:** Cultures are incubated in the dark at a temperature of 25-28°C. Callus formation usually occurs within 3-4 weeks, depending on the explant and the hormonal concentrations.

Shoot regeneration

- **Medium:** The callus is transferred to MS medium supplemented with cytokinins like Benzylaminopurine (BAP) at concentrations of 2.0-5.0 mg/L to promote shoot regeneration. Thidiazuron (TDZ) or Kinetin (KIN) can also be used in some cases to enhance shoot formation.
- **Additional additives:** A low concentration of auxins, such as Indole-3-acetic acid (IAA) or NAA (0.1-0.5 mg/L), may help in elongating shoots.
- **Conditions:** Cultures are exposed to a 16-hour photoperiod with light intensity of about 2000-3000 lux. Shoots typically start regenerating within 4-6 weeks.

Root initiation

- **Medium:** Regenerated shoots are transferred to MS medium with reduced cytokinin levels and higher auxin concentrations, particularly Indole-3-butyric acid (IBA) or NAA at concentrations of 0.5-2.0 mg/L, to induce rooting.
- **Conditions:** Root formation generally occurs within 2-3 weeks under similar light and temperature conditions as during shoot regeneration.

Hardening

- **Transplantation:** Rooted plantlets are carefully removed from the culture medium, washed to remove agar, and transplanted into pots containing a sterile soil mix, such as a combination of peat, sand, and vermiculite.
- **Initial hardening conditions:** Plantlets are kept in a high-humidity environment (e.g., mist chamber or covered with transparent plastic) to acclimate them gradually to ambient conditions.
- **Acclimatization:** Over 2-4 weeks, the humidity is gradually reduced, and the plantlets are exposed to normal growing conditions. After successful acclimatization, they are ready for transfer to larger pots or field conditions.

Field establishment

- **Planting:** Acclimatized plants are transplanted into the field under appropriate agricultural conditions for black turmeric.
- **Monitoring:** Regular monitoring for pests, diseases, and environmental stresses is essential to ensure the healthy growth of the transplanted plants.

Optimization factors

- **Hormone concentrations:** Adjusting the levels of cytokinins and auxins is critical for optimizing the callus induction, shoot regeneration, and rooting stages.

- **Culture conditions:** Temperature, light intensity, and photoperiod need to be carefully controlled and adjusted as required for each stage.
- **Sterility:** Maintaining sterility throughout the process is crucial to prevent contamination and ensure the success of the tissue culture process.

Applications of tissue culture in Black Turmeric

- **Mass propagation:** Tissue culture allows for the rapid multiplication of black turmeric, ensuring uniformity and high yield.
- **Genetic conservation:** The technique is useful for conserving and propagating rare or endangered varieties of black turmeric.
- **Research:** This protocol can be applied in research settings to study genetic traits, improve plant characteristics, and develop new varieties.

This standardized protocol serves as a general guide and can be fine-tuned based on specific conditions or requirements to optimize the micropropagation of black turmeric.

8b. Micropropagation in Black Turmeric: Nodal segments and shoot tip cultures are preferred.

Micropropagation in black turmeric (*Curcuma caesia*) is a valuable technique for the rapid production of genetically uniform and disease-free plants. Black turmeric is known for its medicinal properties and is relatively rare, making micropropagation an important tool for its conservation and commercial production. Below is a detailed protocol for micropropagation of black turmeric:

Selection and preparation of explants

- **Explants:** The most commonly used explants for micropropagation include rhizome buds, shoot tips, or nodal segments.
- **Sterilization:** Explants are surface sterilized using 70% ethanol for 30 seconds, followed by treatment with 0.1% (w/v) mercuric chloride or 1-2% sodium hypochlorite for 5-10 minutes. After sterilization, the explants are rinsed thoroughly with sterile distilled water to remove any residual sterilizing agents.

Initiation of cultures

- **Medium:** Explants are cultured on Murashige and Skoog (MS) medium supplemented with cytokinins, such as Benzylaminopurine (BAP) at concentrations of 2.0-5.0 mg/L, to induce shoot initiation.
- **Conditions:** Cultures are maintained at 25-28°C with a 16-hour photoperiod and light intensity of 2000-3000 lux. Shoots typically initiate within 2-4 weeks.

Shoot multiplication

- **Subculture:** Once shoots develop, they are excised and transferred to fresh MS medium containing BAP (2.0-5.0 mg/L) for further multiplication. A combination of BAP with Kinetin (KIN) or Thidiazuron (TDZ) may be used to enhance shoot proliferation.
- **Repeated subculturing:** Shoots are periodically subcultured every 3-4 weeks to maintain high multiplication rates. Multiple shoots can be generated from a single explant over several subculture cycles.

Root induction

- **Rooting medium:** For rooting, the shoots are transferred to MS medium supplemented with lower levels of cytokinins and higher concentrations of auxins, such as Indole-3-butyric acid (IBA) or Naphthaleneacetic acid (NAA) at concentrations of 0.5-2.0 mg/L.
- **Conditions:** Rooting generally occurs within 2-3 weeks under the same environmental conditions used for shoot multiplication.

Hardening and Acclimatization

- **Transfer to soil:** Rooted plantlets are gently removed from the culture medium, washed to remove any agar residue, and transplanted into pots containing a sterile soil mix (typically a combination of peat, sand, and vermiculite).
- **Hardening process:** The plantlets are initially placed in a high-humidity environment, such as a mist chamber or covered with transparent plastic, to reduce water loss and acclimate them gradually to the ambient environment. Over 2-4 weeks, the humidity is gradually reduced, and the plantlets are exposed to natural conditions.
- **Acclimatization:** After successful hardening, the plantlets are transferred to larger pots or directly into the field, where they continue to grow under standard agricultural practices.

Field establishment

- **Planting:** Acclimatized plants are transplanted into the field, typically during the appropriate season for black turmeric cultivation.
- **Monitoring:** Regular monitoring for pests, diseases, and environmental stresses is essential to ensure healthy growth and development in the field.

Advantages of micropropagation in Black Turmeric

- **Rapid multiplication:** Micropropagation allows for the production of a large number of plants in a relatively short period.

- **Disease-free plants:** The technique ensures that the propagated plants are free from soil-borne diseases, which are common in turmeric cultivation.
- **Uniformity:** All plants produced through micropropagation are genetically identical, ensuring uniform growth and yield.
- **Conservation:** Micropropagation is useful for conserving rare or endangered varieties of black turmeric, allowing for their rapid multiplication and preservation.

Applications

- **Commercial production:** Micropropagation is widely used in the commercial production of black turmeric to meet the demand for high-quality planting material.
- **Research:** The technique is also employed in research settings to study genetic traits, disease resistance, and other aspects of black turmeric biology.
- **Germplasm conservation:** Micropropagation facilitates the ex situ conservation of black turmeric germplasm, preserving valuable genetic resources for future use.

This standardized protocol ensures that black turmeric plants are propagated efficiently, maintaining their genetic integrity and health, which is essential for both commercial and conservation purposes.

9. Mango Ginger (*Curcuma amada*)

9a. Tissue culture: Rhizome cultures used to generate callus.

Standardizing a protocol for callus induction, shoot regeneration, root initiation, and hardening in mango ginger (*Curcuma amada*) is essential for consistent and efficient tissue culture propagation. Mango ginger is a rhizomatous plant valued for its unique flavour and medicinal properties. Below is a step-by-step guide for the tissue culture of mango ginger:

Callus induction

- **Explants:** The most commonly used explants for mango ginger tissue culture are rhizome buds, shoot tips, or young leaf segments.
- **Sterilization:** Explants are surface sterilized using 70% ethanol for 30 seconds, followed by treatment with 0.1% (w/v) mercuric chloride or 1-2% sodium hypochlorite for 5-10 minutes. They are then rinsed thoroughly with sterile distilled water to remove any traces of the sterilizing agents.
- **Medium:** Murashige and Skoog (MS) basal medium is typically used, supplemented with 2,4-Dichlorophenoxyacetic acid (2,4-D) at concentrations ranging from 1.0 to 3.0 mg/L. Sometimes a combination of

2,4-D and Naphthaleneacetic acid (NAA) may be used to enhance callus induction.

- **Conditions:** Cultures are incubated in the dark at 25-28°C. Callus formation typically occurs within 3-4 weeks, depending on the type of explant and hormone concentration used.

Shoot regeneration

- **Medium:** For shoot regeneration, the callus is transferred to MS medium supplemented with cytokinins like Benzylaminopurine (BAP) at concentrations of 2.0-5.0 mg/L. Thidiazuron (TDZ) or Kinetin (KIN) may also be used in some protocols to enhance shoot formation.
- **Additional additives:** A low concentration of auxins, such as Indole-3-acetic acid (IAA) or NAA (0.1-0.5 mg/L), can be added to support shoot elongation.
- **Conditions:** The cultures are exposed to a 16-hour photoperiod with light intensity of about 2000-3000 lux. Shoots generally start regenerating within 4-6 weeks.

Root initiation

- **Medium:** Regenerated shoots are transferred to a rooting medium, typically MS medium with reduced cytokinin levels and increased concentrations of auxins like Indole-3-butyric acid (IBA) or NAA at concentrations of 0.5-2.0 mg/L.
- **Conditions:** Root formation generally occurs within 2-3 weeks under similar environmental conditions as those used for shoot regeneration.

Hardening

- **Transplantation:** Rooted plantlets are carefully removed from the culture medium, washed to remove any remaining agar, and transplanted into pots containing a sterile soil mix, typically a combination of peat, sand, and vermiculite.
- **Initial hardening conditions:** Plantlets are kept in a high-humidity environment, such as a mist chamber or covered with transparent plastic, to reduce water loss and acclimatize them gradually to ambient conditions.
- **Acclimatization:** Over 2-4 weeks, the humidity is gradually reduced, and the plantlets are exposed to normal growing conditions. After successful acclimatization, they are ready for transfer to larger pots or field conditions.

Field establishment

- **Planting:** Acclimatized plants are transplanted into the field under appropriate agricultural conditions for mango ginger.

- **Monitoring:** Regular monitoring for pests, diseases, and environmental stresses is essential to ensure healthy growth and successful establishment in the field.

Optimization factors

- **Hormone concentrations:** Adjusting the levels of cytokinins and auxins is critical for optimizing callus induction, shoot regeneration, and rooting stages.
- **Culture conditions:** Temperature, light intensity, and photoperiod need to be carefully controlled and adjusted as required for each stage.
- **Sterility:** Maintaining sterility throughout the process is crucial to prevent contamination and ensure the success of the tissue culture process.

Applications of tissue culture in Mango Ginger

- **Mass propagation:** Tissue culture allows for the rapid multiplication of mango ginger, ensuring uniformity and high yield.
- **Genetic conservation:** This technique is useful for conserving and propagating rare or endangered varieties of mango ginger.
- **Research:** The protocol can be used in research settings to study genetic traits, improve plant characteristics, and develop new varieties.

This standardized protocol serves as a general guide for the tissue culture of mango ginger and can be fine-tuned based on specific conditions or requirements to optimize the micropropagation process.

9b. Micropropagation in Mango Ginger: Shoot tip culture and somatic embryogenesis are common.

Micropropagation in mango ginger (*Curcuma amada*) is a highly effective method for the rapid production of genetically uniform, disease-free plants. This technique is especially valuable for commercial propagation and conservation of this plant, which is known for its unique flavour and medicinal properties. Below is a detailed step-by-step protocol for micropropagation in mango ginger:

Selection and preparation of explants

- **Explants:** The most commonly used explants for micropropagation are rhizome buds, shoot tips, or nodal segments from healthy, disease-free plants.
- **Sterilization:** Explants are surface sterilized using 70% ethanol for 30 seconds, followed by treatment with 0.1% (w/v) mercuric chloride or 1-2% sodium hypochlorite for 5-10 minutes. After sterilization, they are rinsed thoroughly with sterile distilled water to remove any residual sterilizing agents.

Initiation of cultures

- **Medium:** Explants are cultured on Murashige and Skoog (MS) medium supplemented with cytokinins such as Benzylaminopurine (BAP) at concentrations of 2.0-5.0 mg/L to induce shoot initiation.
- **Conditions:** Cultures are maintained at 25-28°C with a 16-hour photoperiod and light intensity of 2000-3000 lux. Shoots typically initiate within 2-4 weeks.

Shoot multiplication

- **Subculture:** Once shoots develop, they are excised and transferred to fresh MS medium containing BAP (2.0-5.0 mg/L) for further multiplication. In some cases, a combination of BAP with Kinetin (KIN) or Thidiazuron (TDZ) is used to enhance shoot proliferation.
- **Repeated subculturing:** Shoots are subcultured every 3-4 weeks to maintain high multiplication rates. Each subculture cycle can generate multiple shoots, increasing the overall plant production.

Root induction

- **Rooting medium:** For rooting, the shoots are transferred to MS medium supplemented with lower levels of cytokinins and higher concentrations of auxins such as Indole-3-butyric acid (IBA) or Naphthaleneacetic acid (NAA) at concentrations of 0.5-2.0 mg/L.
- **Conditions:** Root formation generally occurs within 2-3 weeks under the same environmental conditions used for shoot multiplication.

Hardening and Acclimatization

- **Transfer to soil:** Rooted plantlets are carefully removed from the culture medium, washed to remove any agar residues, and transplanted into pots containing a sterile soil mix (typically a combination of peat, sand, and vermiculite).
- **Hardening process:** The plantlets are initially placed in a high-humidity environment, such as a mist chamber or covered with transparent plastic, to reduce water loss and gradually acclimate them to the ambient environment. Over 2-4 weeks, the humidity is gradually reduced, and the plantlets are exposed to natural conditions.
- **Acclimatization:** After successful hardening, the plantlets are transferred to larger pots or directly into the field, where they continue to grow under standard agricultural practices.

Field establishment

- **Planting:** Acclimatized plants are transplanted into the field, typically during the appropriate season for mango ginger cultivation.
- **Monitoring:** Regular monitoring for pests, diseases, and environmental stresses is essential to ensure healthy growth and development in the field.

Advantages of micropropagation in Mango Ginger

- **Rapid multiplication:** Micropropagation allows for the production of a large number of plants in a relatively short period.
- **Disease-free plants:** The technique ensures that the propagated plants are free from soil-borne diseases, which are common in traditional propagation methods.
- **Uniformity:** All plants produced through micropropagation are genetically identical, ensuring uniform growth and yield.
- **Conservation:** Micropropagation is useful for conserving rare or endangered varieties of mango ginger, allowing for their rapid multiplication and preservation.

Applications

- **Commercial production:** Micropropagation is widely used in the commercial production of mango ginger to meet the demand for high-quality planting material.
- **Research:** The technique is also employed in research settings to study genetic traits, disease resistance, and other aspects of mango ginger biology.
- **Germplasm conservation:** Micropropagation facilitates the ex situ conservation of mango ginger germplasm, preserving valuable genetic resources for future use.

This protocol ensures the efficient and consistent production of mango ginger plants, making it an invaluable tool for both commercial growers and researchers.

10. Coriander (*Coriandrum sativum*)

10a. Tissue culture: Leaf and stem explants are used for callus and shoot regeneration.

Standardizing a protocol for callus induction, shoot regeneration, root initiation, and hardening in coriander (*Coriandrum sativum*) involves optimizing the tissue culture conditions to achieve consistent and efficient plant regeneration. Below is a general protocol tailored for coriander:

Callus induction

- **Explants:** The most commonly used explants for coriander tissue culture are leaf segments, petioles, or stem nodes.
- **Sterilization:** Explants are surface sterilized using 70% ethanol for 30 seconds, followed by treatment with 0.1% (w/v) mercuric chloride or 1-2% sodium hypochlorite for 5-10 minutes. After sterilization, the explants are rinsed thoroughly with sterile distilled water to remove any residual sterilizing agents.
- **Medium:** Murashige and Skoog (MS) basal medium is typically used, supplemented with 2,4-Dichlorophenoxyacetic acid (2,4-D) at concentrations ranging from 1.0 to 3.0 mg/L. A combination of 2,4-D and Naphthaleneacetic acid (NAA) may also be effective for inducing callus formation.
- **Conditions:** Cultures are kept in the dark at 25-28°C. Callus formation typically occurs within 2-4 weeks, depending on the type of explant and the concentration of growth regulators.

Shoot regeneration

- **Medium:** For shoot regeneration, the callus is transferred to MS medium supplemented with cytokinins such as Benzylaminopurine (BAP) at concentrations of 2.0-5.0 mg/L. Thidiazuron (TDZ) or Kinetin (KIN) may also be used to enhance shoot formation.
- **Additional additives:** A low concentration of auxins, such as Indole-3-acetic acid (IAA) or NAA (0.1-0.5 mg/L), can be added to support shoot elongation.
- **Conditions:** Cultures are exposed to a 16-hour photoperiod with light intensity of about 2000-3000 lux. Shoots typically start regenerating within 4-6 weeks.

Root initiation

- **Medium:** Regenerated shoots are transferred to a rooting medium, typically MS medium with reduced cytokinin levels and increased concentrations of auxins like Indole-3-butyric acid (IBA) or NAA at concentrations of 0.5-2.0 mg/L.
- **Conditions:** Root formation generally occurs within 2-3 weeks under similar environmental conditions as those used for shoot regeneration.

Hardening

- **Transplantation:** Rooted plantlets are carefully removed from the culture medium, washed to remove any remaining agar, and transplanted into pots containing a sterile soil mix, typically a combination of peat, sand, and vermiculite.

- **Initial hardening conditions:** Plantlets are kept in a high-humidity environment, such as a mist chamber or covered with transparent plastic, to reduce water loss and acclimatize them gradually to ambient conditions.
- **Acclimatization:** Over 2-4 weeks, the humidity is gradually reduced, and the plantlets are exposed to normal growing conditions. After successful acclimatization, they are ready for transfer to larger pots or field conditions.

Field establishment

- **Planting:** Acclimatized plants are transplanted into the field under appropriate agricultural conditions for coriander.
- **Monitoring:** Regular monitoring for pests, diseases, and environmental stresses is essential to ensure healthy growth and successful establishment in the field.

Optimization factors

- **Hormone concentrations:** Adjusting the levels of cytokinins and auxins is crucial for optimizing each stage of the tissue culture process.
- **Culture conditions:** Temperature, light intensity, and photoperiod need to be carefully controlled and adjusted as required for each stage.
- **Sterility:** Maintaining strict aseptic conditions throughout the process is critical to prevent contamination and ensure the success of the tissue culture process.

Applications of tissue culture in Coriander

- **Mass propagation:** Tissue culture allows for the rapid multiplication of coriander plants, ensuring uniformity and high yield.
- **Genetic conservation:** This technique is useful for conserving and propagating rare or endangered varieties of coriander.
- **Research:** The protocol can be used in research settings to study genetic traits, improve plant characteristics, and develop new varieties.

This standardized protocol provides a general guide for the tissue culture of coriander and can be fine-tuned based on specific conditions or requirements to optimize the micropropagation process.

10b. Micropropagation in Coriander: Involves seedling culture and shoot tip multiplication.

Micropropagation in coriander (*Coriandrum sativum*) is an efficient technique for the rapid production of genetically uniform and disease-free plants. This process is especially valuable for commercial cultivation, research, and conservation of this important culinary and medicinal herb. Below is a detailed step-by-step protocol for the micropropagation of coriander:

Selection and preparation of explants

- **Explants:** The most commonly used explants for micropropagation are leaf segments, shoot tips, or nodal segments from healthy, disease-free plants.
- **Sterilization:** Explants are surface sterilized by immersing them in 70% ethanol for 30 seconds, followed by treatment with 0.1% (w/v) mercuric chloride or 1-2% sodium hypochlorite for 5-10 minutes. After sterilization, the explants are thoroughly rinsed with sterile distilled water to remove any residual sterilizing agents.

Initiation of cultures

- **Medium:** The explants are cultured on Murashige and Skoog (MS) medium supplemented with cytokinins such as Benzylaminopurine (BAP) at concentrations of 2.0-5.0 mg/L to induce shoot initiation.
- **Conditions:** Cultures are maintained at 25-28°C with a 16-hour photoperiod and light intensity of 2000-3000 lux. Shoot initiation generally occurs within 2-4 weeks.

Shoot multiplication

- **Subculture:** Once shoots develop, they are excised and transferred to fresh MS medium containing BAP (2.0-5.0 mg/L) for further multiplication. In some cases, a combination of BAP with Kinetin (KIN) or Thidiazuron (TDZ) may be used to enhance shoot proliferation.
- **Repeated subculturing:** Shoots are subcultured every 3-4 weeks to maintain high multiplication rates. Each subculture cycle can generate multiple shoots, increasing the overall yield.

Root induction

- **Rooting medium:** For rooting, the shoots are transferred to MS medium supplemented with lower levels of cytokinins and higher concentrations of auxins, such as Indole-3-butyric acid (IBA) or Naphthaleneacetic acid (NAA) at concentrations of 0.5-2.0 mg/L.
- **Conditions:** Root formation typically occurs within 2-3 weeks under the same environmental conditions used for shoot multiplication.

Hardening and Acclimatization

- **Transfer to soil:** Rooted plantlets are carefully removed from the culture medium, washed to remove any agar residues, and transplanted into pots containing a sterile soil mix (typically a combination of peat, sand, and vermiculite).

- **Hardening process:** The plantlets are initially placed in a high-humidity environment, such as a mist chamber or covered with transparent plastic, to reduce water loss and gradually acclimate them to the ambient environment. Over 2-4 weeks, the humidity is gradually reduced, and the plantlets are exposed to natural conditions.
- **Acclimatization:** After successful hardening, the plantlets are transferred to larger pots or directly into the field, where they continue to grow under standard agricultural practices.

Field establishment

- **Planting:** Acclimatized plants are transplanted into the field, typically during the appropriate growing season for coriander.
- **Monitoring:** Regular monitoring for pests, diseases, and environmental stresses is essential to ensure healthy growth and development in the field.

Advantages of micro-propagation in Coriander

- **Rapid multiplication:** Micro-propagation allows for the production of a large number of plants in a relatively short period.
- **Disease-free plants:** The technique ensures that the propagated plants are free from soil-borne diseases, which are common in traditional propagation methods.
- **Uniformity:** All plants produced through micro-propagation are genetically identical, ensuring uniform growth and yield.
- **Conservation:** Micro-propagation is useful for conserving rare or endangered varieties of coriander, allowing for their rapid multiplication and preservation.

Applications

- **Commercial production:** Micro-propagation is widely used in the commercial production of coriander to meet the demand for high-quality planting material.
- **Research:** The technique is also employed in research settings to study genetic traits, disease resistance, and other aspects of coriander biology.
- **Germplasm conservation:** Micro-propagation facilitates the ex situ conservation of coriander germplasm, preserving valuable genetic resources for future use.

This standardized protocol ensures the efficient and consistent production of coriander plants, making it an invaluable tool for both commercial growers and researchers.

11. Fennel (*Foeniculum vulgare*)

11a.Tissue culture: Somatic embryogenesis from hypocotyl and cotyledon explants.

Standardizing a protocol for callus induction, shoot regeneration, root initiation, and hardening in fennel (*Foeniculum vulgare*) involves optimizing several factors to achieve efficient and consistent plant regeneration. Fennel is an aromatic herb widely used in culinary and medicinal applications, and tissue culture can help in the rapid propagation and genetic improvement of this plant. Below is a general protocol for the tissue culture of fennel:

Callus induction

- **Explants:** Commonly used explants include leaf segments, petioles, or stem nodes.
- **Sterilization:** Explants should be surface sterilized using 70% ethanol for 30 seconds, followed by treatment with 0.1% (w/v) mercuric chloride or 1-2% sodium hypochlorite for 5-10 minutes. Rinse thoroughly with sterile distilled water to remove any residual sterilizing agents.
- **Medium:** Murashige and Skoog (MS) basal medium supplemented with 2,4-Dichlorophenoxyacetic acid (2,4-D) at concentrations of 1.0 to 3.0 mg/L is commonly used for callus induction. In some protocols, Naphthaleneacetic acid (NAA) may also be added to improve callus formation.
- **Conditions:** Cultures should be kept in the dark at 25-28°C. Callus formation typically occurs within 2-4 weeks, depending on the explant type and hormone concentration.

Shoot regeneration

- **Medium:** The callus is transferred to MS medium supplemented with cytokinins like Benzylaminopurine (BAP) at concentrations of 2.0-5.0 mg/L to induce shoot regeneration. Kinetin (KIN) or Thidiazuron (TDZ) can also be used to enhance shoot formation.
- **Additional additives:** A low concentration of auxins such as Indole-3-acetic acid (IAA) or NAA (0.1-0.5 mg/L) may be added to support shoot elongation.
- **Conditions:** Cultures should be maintained under a 16-hour photoperiod with light intensity of about 2000-3000 lux. Shoots generally start regenerating within 4-6 weeks.

Root initiation

- **Medium:** Regenerated shoots are transferred to a rooting medium, typically MS medium with reduced cytokinin levels and increased concentrations of auxins like Indole-3-butyric acid (IBA) or NAA at concentrations of 0.5-2.0 mg/L.

- **Conditions:** Rooting generally occurs within 2-3 weeks under the same environmental conditions used for shoot regeneration.

Hardening

- **Transplantation:** Rooted plantlets are carefully removed from the culture medium, washed to remove any remaining agar, and transplanted into pots containing a sterile soil mix, typically a combination of peat, sand, and vermiculite.
- **Initial hardening conditions:** Plantlets are kept in a high-humidity environment, such as a mist chamber or covered with transparent plastic, to acclimatize them gradually to ambient conditions.
- **Acclimatization:** Over 2-4 weeks, the humidity is gradually reduced, and the plantlets are exposed to normal growing conditions. After successful acclimatization, they can be transferred to larger pots or field conditions.

Field establishment

- **Planting:** Acclimatized plants are transplanted into the field under appropriate agricultural conditions for fennel.
- **Monitoring:** Regular monitoring for pests, diseases, and environmental stresses is essential to ensure healthy growth and successful establishment in the field.

Optimization factors

- **Hormone concentrations:** Fine-tuning the concentrations of cytokinins and auxins in the media is crucial for optimizing callus induction, shoot regeneration, and rooting stages.
- **Culture conditions:** Temperature, light intensity, and photoperiod should be carefully controlled and adjusted as required for each stage.
- **Sterility:** Maintaining sterility throughout the tissue culture process is critical to prevent contamination and ensure successful plant regeneration.

Applications of tissue culture in Fennel

- **Mass propagation:** Tissue culture enables the rapid multiplication of fennel plants, ensuring uniformity and high yield.
- **Genetic improvement:** This technique is useful for genetic improvement programs, allowing for the selection and propagation of superior traits.
- **Conservation:** Tissue culture can be employed to conserve rare or endangered varieties of fennel.

- **Research:** The protocol is also applicable in research settings for studying genetic traits, improving plant characteristics, and developing new fennel varieties.

This standardized protocol serves as a general guide for the tissue culture of fennel and can be fine-tuned based on specific conditions or requirements to optimize the micropropagation process.

11b.Micropropagation: Seedling and shoot tip culture for rapid plant multiplication.

Micropropagation in fennel (*Foeniculum vulgare*) is an effective method for the rapid and large-scale production of genetically uniform, disease-free plants. This technique is beneficial for both commercial cultivation and research purposes. Below is a detailed step-by-step protocol for micropropagation of fennel:

Selection and preparation of explants

- **Explants:** The most commonly used explants for fennel micropropagation include leaf segments, petioles, or nodal segments from healthy, disease-free plants.
- **Sterilization:** Explants are surface sterilized by immersing them in 70% ethanol for 30 seconds, followed by treatment with 0.1% (w/v) mercuric chloride or 1-2% sodium hypochlorite for 5-10 minutes. After sterilization, the explants are thoroughly rinsed with sterile distilled water to remove any residual sterilizing agents.

Initiation of cultures

- **Medium:** The explants are cultured on Murashige and Skoog (MS) medium supplemented with cytokinins such as Benzylaminopurine (BAP) at concentrations of 2.0-5.0 mg/L to induce shoot initiation.
- **Conditions:** Cultures are maintained at 25-28°C with a 16-hour photoperiod and light intensity of 2000-3000 lux. Shoots typically initiate within 2-4 weeks.

Shoot multiplication

- **Subculture:** Once shoots have developed, they are excised and transferred to fresh MS medium containing BAP (2.0-5.0 mg/L) for further multiplication. In some cases, a combination of BAP with Kinetin (KIN) or Thidiazuron (TDZ) may be used to enhance shoot proliferation.
- **Repeated subculturing:** Shoots are subcultured every 3-4 weeks to maintain high multiplication rates. Each subculture cycle can generate multiple shoots, increasing the overall yield.

Root induction

- **Rooting medium:** For rooting, the shoots are transferred to MS medium supplemented with lower levels of cytokinins and higher concentrations of auxins such as Indole-3-butyric acid (IBA) or Naphthaleneacetic acid (NAA) at concentrations of 0.5-2.0 mg/L.
- **Conditions:** Root formation typically occurs within 2-3 weeks under the same environmental conditions used for shoot multiplication.

Hardening and Acclimatization

- **Transfer to soil:** Rooted plantlets are carefully removed from the culture medium, washed to remove any agar residues, and transplanted into pots containing a sterile soil mix (typically a combination of peat, sand, and vermiculite).
- **Hardening process:** The plantlets are initially placed in a high-humidity environment, such as a mist chamber or covered with transparent plastic, to reduce water loss and gradually acclimate them to the ambient environment. Over 2-4 weeks, the humidity is gradually reduced, and the plantlets are exposed to natural conditions.
- **Acclimatization:** After successful hardening, the plantlets are transferred to larger pots or directly into the field, where they continue to grow under standard agricultural practices.

Field establishment

- **Planting:** Acclimatized plants are transplanted into the field, typically during the appropriate growing season for fennel.
- **Monitoring:** Regular monitoring for pests, diseases, and environmental stresses is essential to ensure healthy growth and development in the field.

Advantages of micropropagation in Fennel

- **Rapid multiplication:** Micropropagation allows for the production of a large number of plants in a relatively short period.
- **Disease-free plants:** The technique ensures that the propagated plants are free from soil-borne diseases, which are common in traditional propagation methods.
- **Uniformity:** All plants produced through micropropagation are genetically identical, ensuring uniform growth and yield.
- **Conservation:** Micropropagation is useful for conserving rare or endangered varieties of fennel, allowing for their rapid multiplication and preservation.

Applications

- **Commercial production:** Micropropagation is widely used in the commercial production of fennel to meet the demand for high-quality planting material.
- **Research:** The technique is also employed in research settings to study genetic traits, disease resistance, and other aspects of fennel biology.
- **Germplasm conservation:** Micropropagation facilitates the ex situ conservation of fennel germplasm, preserving valuable genetic resources for future use.

This standardized protocol ensures the efficient and consistent production of fennel plants, making it an invaluable tool for both commercial growers and researchers.

12. Fenugreek (*Trigonella foenum-graecum*)

12a.Tissue culture: Cotyledon and hypocotyl explants are used for inducing shoots.

Standardizing a protocol for callus induction, shoot regeneration, root initiation, and hardening in fenugreek (*Trigonella foenum-graecum*) involves optimizing several factors to achieve efficient and consistent plant regeneration. Fenugreek is a valuable medicinal and culinary herb, and tissue culture techniques can help in its rapid propagation and genetic improvement. Below is a detailed protocol for the tissue culture of fenugreek:

Callus induction

- **Explants:** Commonly used explants include leaf segments, cotyledonary nodes, or hypocotyl segments.
- **Sterilization:** Explants should be surface sterilized using 70% ethanol for 30 seconds, followed by treatment with 0.1% (w/v) mercuric chloride or 1-2% sodium hypochlorite for 5-10 minutes. Rinse thoroughly with sterile distilled water to remove any residual sterilizing agents.
- **Medium:** Murashige and Skoog (MS) basal medium supplemented with 2,4-Dichlorophenoxyacetic acid (2,4-D) at concentrations of 1.0 to 3.0 mg/L is commonly used for callus induction. Adding Naphthaleneacetic acid (NAA) at 0.5-1.0 mg/L can further enhance callus formation.
- **Conditions:** Cultures are kept in the dark at 25-28°C. Callus formation typically occurs within 2-4 weeks, depending on the explant type and hormone concentration.

Shoot regeneration

- **Medium:** The callus is transferred to MS medium supplemented with cytokinins like Benzylaminopurine (BAP) at concentrations of 2.0-5.0

mg/L to induce shoot regeneration. Kinetin (KIN) or Thidiazuron (TDZ) can also be used to enhance shoot formation.

- **Additional additives:** A low concentration of auxins such as Indole-3-acetic acid (IAA) or NAA (0.1-0.5 mg/L) may be added to support shoot elongation.
- **Conditions:** Cultures should be maintained under a 16-hour photoperiod with light intensity of about 2000-3000 lux. Shoots generally start regenerating within 4-6 weeks.

Root initiation

- **Medium:** Regenerated shoots are transferred to a rooting medium, typically MS medium with reduced cytokinin levels and increased concentrations of auxins like Indole-3-butyric acid (IBA) or NAA at concentrations of 0.5-2.0 mg/L.
- **Conditions:** Rooting generally occurs within 2-3 weeks under the same environmental conditions used for shoot regeneration.

Hardening

- **Transplantation:** Rooted plantlets are carefully removed from the culture medium, washed to remove any remaining agar, and transplanted into pots containing a sterile soil mix, typically a combination of peat, sand, and vermiculite.
- **Initial hardening conditions:** Plantlets are kept in a high-humidity environment, such as a mist chamber or covered with transparent plastic, to acclimatize them gradually to ambient conditions.
- **Acclimatization:** Over 2-4 weeks, the humidity is gradually reduced, and the plantlets are exposed to normal growing conditions. After successful acclimatization, they can be transferred to larger pots or field conditions.

Field establishment

- **Planting:** Acclimatized plants are transplanted into the field under appropriate agricultural conditions for fenugreek.
- **Monitoring:** Regular monitoring for pests, diseases, and environmental stresses is essential to ensure healthy growth and successful establishment in the field.

Optimization factors

- **Hormone concentrations:** Fine-tuning the concentrations of cytokinins and auxins in the media is crucial for optimizing callus induction, shoot regeneration, and rooting stages.

- **Culture conditions:** Temperature, light intensity, and photoperiod should be carefully controlled and adjusted as required for each stage.
- **Sterility:** Maintaining sterility throughout the tissue culture process is critical to prevent contamination and ensure successful plant regeneration.

Applications of tissue culture in Fenugreek

- **Mass propagation:** Tissue culture enables the rapid multiplication of fenugreek plants, ensuring uniformity and high yield.
- **Genetic improvement:** This technique is useful for genetic improvement programs, allowing for the selection and propagation of superior traits.
- **Conservation:** Tissue culture can be employed to conserve rare or endangered varieties of fenugreek.
- **Research:** The protocol is also applicable in research settings for studying genetic traits, improving plant characteristics, and developing new fenugreek varieties.

This standardized protocol serves as a general guide for the tissue culture of fenugreek and can be fine-tuned based on specific conditions or requirements to optimize the micropropagation process.

12b. Micropropagation: Shoot tip cultures from seedlings for mass production.

Micropropagation in fenugreek (*Trigonella foenum-graecum*) is a method used to rapidly produce a large number of genetically uniform and disease-free plants. This technique is particularly useful for plant breeding, genetic improvement, and large-scale commercial cultivation. Below is a detailed protocol for micropropagation in fenugreek:

Selection and preparation of explants

- **Explants:** Commonly used explants for fenugreek micropropagation include cotyledonary nodes, hypocotyl segments, or young leaf segments from healthy, disease-free seedlings.
- **Sterilization:** Surface sterilize the explants by immersing them in 70% ethanol for 30 seconds, followed by treatment with 0.1% (w/v) mercuric chloride or 1-2% sodium hypochlorite for 5-10 minutes. After sterilization, rinse the explants thoroughly with sterile distilled water to remove any residual sterilizing agents.

Initiation of cultures

- **Medium:** Place the explants on Murashige and Skoog (MS) basal medium supplemented with cytokinins such as Benzylaminopurine (BAP) at concentrations of 2.0-5.0 mg/L. This medium promotes the induction of shoot initiation.

- **Conditions:** Maintain the cultures at 25-28°C under a 16-hour photoperiod with light intensity of 2000-3000 lux. Shoots typically initiate within 2-4 weeks.

Shoot multiplication

- **Subculture:** Once shoots have developed, excise them and transfer to fresh MS medium containing BAP (2.0-5.0 mg/L) for further multiplication. A combination of BAP with Kinetin (KIN) or Thidiazuron (TDZ) may be used to enhance shoot proliferation.
- **Repeated subculturing:** Subculture the shoots every 3-4 weeks to maintain high multiplication rates. Multiple shoots can be generated from a single explant over several subculture cycles, increasing the overall yield.

Root induction

- **Rooting medium:** Transfer the multiplied shoots to a rooting medium, typically MS medium supplemented with lower levels of cytokinins and higher concentrations of auxins such as Indole-3-butyric acid (IBA) or Naphthaleneacetic acid (NAA) at concentrations of 0.5-2.0 mg/L.
- **Conditions:** Maintain the cultures under the same environmental conditions used for shoot multiplication. Roots typically form within 2-3 weeks.

Hardening and Acclimatization

- **Transfer to soil:** Once roots are well-developed, carefully remove the plantlets from the culture medium, wash off any remaining agar, and transfer them to pots containing a sterile soil mix (a typical combination is peat, sand, and vermiculite).
- **Hardening process:** Place the plantlets in a high-humidity environment, such as a mist chamber or covered with transparent plastic, to gradually acclimate them to ambient conditions. Over 2-4 weeks, gradually reduce the humidity and expose the plantlets to natural conditions.
- **Acclimatization:** After successful hardening, transfer the plantlets to larger pots or directly into the field, where they will continue to grow under standard agricultural practices.

Field establishment

- **Planting:** Transplant the acclimatized plants into the field during the appropriate growing season for fenugreek.
- **Monitoring:** Regularly monitor the plants for pests, diseases, and environmental stresses to ensure healthy growth and successful establishment in the field.

Advantages of micropropagation in Fenugreek

- **Rapid multiplication:** Allows for the production of large numbers of plants in a relatively short period.
- **Disease-free plants:** Ensures that the propagated plants are free from soil-borne diseases.
- **Uniformity:** Produces genetically identical plants, ensuring uniform growth and yield.
- **Conservation:** Useful for conserving rare or endangered varieties of fenugreek, allowing for their rapid multiplication and preservation.

Applications

- **Commercial production:** Micropropagation is widely used in the commercial production of fenugreek to meet the demand for high-quality planting material.
- **Research:** The technique is used in research settings to study genetic traits, disease resistance, and other aspects of fenugreek biology.
- **Germplasm conservation:** Facilitates the ex situ conservation of fenugreek germplasm, preserving valuable genetic resources for future use.

This standardized protocol ensures the efficient and consistent production of fenugreek plants, making it a valuable tool for both commercial growers and researchers.

13. Cumin (*Cuminum cyminum*)

13a. Tissue culture: Involves somatic embryogenesis from leaf explants.

To standardize the protocol for callus induction, shoot regeneration, root initiation, and hardening of tissue culture in cumin, the following steps are typically involved:

1. Callus induction

- **Explants selection**: Use seeds, leaves, or stem segments as explants.
- **Sterilization**: Surface sterilize the explants using 70% ethanol followed by a sodium hypochlorite solution.
- **Media preparation**: Use Murashige and Skoog (MS) medium supplemented with plant growth regulators like 2,4-D (2,4-Dichlorophenoxyacetic acid) for callus induction.
- **Culture conditions**: Incubate the explants on the media under dark conditions at 25 ± 2°C to promote callus formation.
- **Observation**: Monitor for callus formation after 2-4 weeks.

2. Shoot regeneration

- **Subculture**: Transfer the induced callus to MS medium supplemented with cytokinin (like BAP - Benzylaminopurine) to induce shoot regeneration.
- **Light conditions**: Incubate under a 16-hour photoperiod with light intensity of 2,000-3,000 lux.
- **Temperature**: Maintain at 25 ± 2°C.
- **Shoot formation**: Shoots should start emerging in 2-3 weeks.

3. Root initiation

- **Rooting media**: Transfer the shoots to half-strength MS medium supplemented with auxin (such as IBA - Indole-3-butyric acid) for root induction.
- **Culture conditions**: Incubate under the same photoperiod and temperature as for shoot regeneration.
- **Observation**: Roots should develop within 2-3 weeks.

4. Hardening

- **Acclimatization**: Gradually acclimatize the rooted plants to the external environment. Initially, keep them in a high-humidity environment in a growth chamber or greenhouse.
- **Transplantation**: After 1-2 weeks, transplant the plantlets to soil or potting mix and maintain under controlled conditions.
- **Final transplantation**: After the plants are well-established, move them to the field or larger pots.

5. Monitoring

- Regularly monitor for contamination, nutrient deficiencies, and other stress factors during all stages.

These steps provide a basic framework that should be tailored according to specific requirements, such as the cumin variety and local environmental conditions.

13b. Micro-propagation in Cumin: Shoot multiplication from seed-derived explants.

Micropropagation of cumin involves the in vitro propagation of plants using tissue culture techniques to produce large numbers of genetically uniform and disease-free plants. Here's a standard procedure for micropropagation in cumin:

1. Selection and preparation of explants

- **Explants**: The most commonly used explants for cumin micropropagation are seeds, nodal segments, or shoot tips.

- **Sterilization**: Surface sterilize the explants using a series of disinfectants, typically involving washing with 70% ethanol for 30 seconds followed by immersion in a 0.1% (w/v) mercury chloride (HgCl2) or 1-2% sodium hypochlorite solution for 10-15 minutes. Rinse thoroughly with sterile distilled water.

2. Culture media preparation

- **Media composition**: Use Murashige and Skoog (MS) medium as the basal medium. For shoot induction, supplement the medium with cytokinin (such as BAP - Benzylaminopurine, typically at concentrations of 1-2 mg/L).
- **pH adjustment**: Adjust the pH of the medium to 5.7-5.8 before autoclaving.
- **Sterilization**: Sterilize the medium by autoclaving at 121°C for 15-20 minutes.

3. Inoculation and shoot induction

- **Inoculation**: Inoculate the sterilized explants onto the prepared MS medium under sterile conditions in a laminar airflow cabinet.
- **Incubation**: Place the culture bottles or test tubes in a growth room with controlled conditions—usually 25 ± 2°C, 50-60% humidity, and a 16-hour photoperiod with light intensity around 2,000-3,000 lux.
- **Shoot development**: Shoots typically begin to develop within 2-4 weeks.

4. Multiplication of shoots

- **Subculturing**: After the shoots have developed, transfer them to fresh MS medium with a similar or slightly higher concentration of cytokinin (BAP) for multiplication. This step may be repeated several times to increase the number of shoots.
- **Shoot elongation**: If needed, transfer the proliferated shoots to a medium with reduced cytokinin concentration or without cytokinin to promote shoot elongation.

5. Root induction

- **Rooting media**: Transfer the elongated shoots to half-strength MS medium supplemented with auxins such as IBA (Indole-3-butyric acid) at 0.5-1.0 mg/L to induce rooting.
- **Incubation**: Continue incubation under the same conditions as shoot multiplication. Roots should develop within 2-3 weeks.

6. Hardening and acclimatization

- **Initial hardening**: After sufficient root development, transfer the plantlets to small pots containing a sterilized potting mixture (e.g., a mix of peat, perlite, and vermiculite) and cover with plastic or place in a humidity chamber to maintain high humidity.

- **Gradual acclimatization**: Gradually reduce humidity by making small openings in the plastic cover or moving the plantlets to a less humid environment over a period of 1-2 weeks.
- **Final transplantation**: Once the plantlets are fully acclimatized, transplant them into the soil in the field or larger pots under normal growing conditions.

7. Monitoring and maintenance

- **Observation**: Regularly monitor the plants for growth, development, and any signs of disease or contamination.
- **Nutrient supply**: Provide appropriate nutrients and maintain the soil or potting mixture to ensure healthy growth.

Applications

- Micro-propagation in cumin can be used for the rapid multiplication of elite genotypes, conservation of germplasm, and production of disease-free planting material. This technique is especially valuable for large-scale commercial cultivation and research purposes

14. Ajwain (*Trachyspermum ammi*)

14a.Tissue culture: Shoot culture from seedling explants.

The standardization of protocols for callus induction, shoot regeneration, root initiation, and hardening of tissue culture in ajwain (*Trachyspermum ammi*) follows a process similar to that of other plant species but requires specific adjustments for the plant's unique characteristics. Here's a general protocol:

1. Callus induction

- **Explants selection**: Use young leaves, stem segments, or seeds as explants.
- **Sterilization**: Surface sterilize explants by immersing in 70% ethanol for 30 seconds followed by soaking in a 0.1% (w/v) mercury chloride (HgC12) or 2% sodium hypochlorite solution for 10-15 minutes, then rinse thoroughly with sterile distilled water.
- **Media preparation**: Use Murashige and Skoog (MS) medium supplemented with auxins such as 2,4-Dichlorophenoxyacetic acid (2,4-D) or NAA (Naphthaleneacetic acid) at a concentration of 2-3 mg/L for callus induction.
- **Culture conditions**: Incubate the explants on the medium in dark conditions at a temperature of 25 ± 2°C.
- **Observation**: Callus formation is typically observed within 3-4 weeks. Subculturing might be needed if the callus grows well.

2. Shoot regeneration

- **Media for regeneration**: Transfer the callus to MS medium supplemented with cytokinins like BAP (Benzylaminopurine) at 1-2 mg/L, possibly combined with a low concentration of an auxin such as NAA (0.1-0.5 mg/L) to induce shoot regeneration.
- **Light and temperature**: Maintain cultures under a 16-hour photoperiod with light intensity around 2,000-3,000 lux at a temperature of 25 ± 2°C.
- **Shoot emergence**: Shoots typically begin to emerge after 4-6 weeks. Regular monitoring and subculturing on fresh medium might be necessary to encourage further shoot development.

3. Root initiation

- **Rooting media**: Transfer regenerated shoots to half-strength MS medium supplemented with auxins like IBA (Indole-3-butyric acid) at a concentration of 0.5-1.0 mg/L for root induction.
- **Culture conditions**: Continue incubation under the same light and temperature conditions as for shoot regeneration.
- **Observation**: Root formation usually occurs within 2-4 weeks.

4. Hardening

- **Initial hardening**: Carefully remove the rooted plantlets from the culture medium and wash off any remaining agar. Transfer the plantlets to pots containing a sterilized potting mixture (e.g., a mix of peat, perlite, and vermiculite).
- **High humidity environment**: Place the pots in a high-humidity environment, such as a humidity chamber or under a plastic cover, to prevent desiccation. Gradually expose the plantlets to ambient conditions by slowly reducing the humidity over 1-2 weeks.
- **Final transplantation**: Once the plantlets are acclimatized, they can be transplanted to the field or larger pots with soil.

5. Monitoring and maintenance

- **Observation**: Regularly monitor for any signs of contamination, nutrient deficiencies, or stress.
- **Nutrient supply**: Provide appropriate nutrients and maintain adequate moisture levels to ensure healthy growth of the plantlets.

Applications

- This standardized protocol can be utilized for the mass propagation of ajwain, ensuring uniformity and high quality of the propagated plants. It is particularly useful for preserving elite genotypes and for large-scale cultivation efforts.

14b. Micropropagation in Aijwain: Seedling-derived shoot tips are used for rapid propagation.

Micropropagation of ajwain (*Trachyspermum ammi*) involves the in vitro cultivation of plant tissues to produce large quantities of genetically identical and disease-free plants. Below is a standardized protocol for micropropagation in ajwain:

1. Selection and preparation of explants

- **Explants**: Commonly used explants include seeds, shoot tips, and nodal segments.
- **Sterilization**:
  - Surface sterilize the explants by washing them in 70% ethanol for 30 seconds.
  - Follow with immersion in a 0.1% (w/v) mercuric chloride (HgCl2) or 2% sodium hypochlorite solution for 10-15 minutes.
  - Rinse thoroughly with sterile distilled water 3-4 times.

2. Culture media preparation

- **Basal medium**: Use Murashige and Skoog (MS) medium as the basal medium.
- **Shoot induction**: Supplement the MS medium with a cytokinin such as BAP (Benzylaminopurine) at concentrations of 1-2 mg/L. Low concentrations of auxin (like NAA - Naphthaleneacetic acid at 0.1 mg/L) can also be used to enhance shoot induction.
- **pH adjustment**: Adjust the pH of the medium to 5.7-5.8 before autoclaving.
- **Sterilization**: Autoclave the medium at 121°C for 15-20 minutes.

3. Inoculation and shoot induction

- **Inoculation**: Inoculate the sterilized explants onto the prepared MS medium in a sterile environment, typically under a laminar flow hood.
- **Incubation**: Place the culture vessels in a growth room under controlled conditions—usually 25 ± 2°C with a 16-hour photoperiod and light intensity of 2,000-3,000 lux.
- **Shoot development**: Shoots should start to develop within 3-4 weeks. The number of shoots can be increased through regular subculturing onto fresh medium with the same hormonal composition.

4. Shoot multiplication

- **Subculturing**: Transfer the shoots to fresh MS medium with similar or slightly increased concentrations of cytokinin for further multiplication. Multiple subcultures may be performed to increase the number of shoots.
- **Shoot elongation**: Once a sufficient number of shoots are obtained, transfer them to a medium with lower cytokinin levels or without cytokinin to promote elongation.

5. Root induction

- **Rooting medium**: Transfer elongated shoots to half-strength MS medium supplemented with an auxin like IBA (Indole-3-butyric acid) at concentrations of 0.5-1.0 mg/L to induce root formation.
- **Incubation**: Continue to incubate under the same conditions as shoot induction.
- **Root formation**: Roots typically begin to develop within 2-4 weeks.

6. Hardening and acclimatization

- **Initial hardening**: Once the plantlets have developed sufficient roots, gently remove them from the culture medium and wash off any residual agar. Transfer the plantlets to small pots containing a sterilized potting mixture (such as peat, perlite, and vermiculite).
- **High humidity conditions**: Place the pots in a high-humidity environment or cover them with a plastic bag to maintain humidity. Gradually reduce the humidity over a period of 1-2 weeks by making small openings in the cover or moving the plantlets to a less humid environment.
- **Final transplantation**: Once the plantlets are fully acclimatized, they can be transplanted to soil in the field or into larger pots.

7. Monitoring and Maintenance

- **Observation**: Regularly check for contamination, nutrient deficiencies, and overall plant health.
- **Nutrient supply**: Ensure the plantlets receive adequate nutrients and are properly watered to support their growth.

Applications

- Micropropagation of ajwain is valuable for producing large numbers of uniform, high-quality plants. It is particularly useful for conserving elite plant genotypes, mass propagation, and ensuring the availability of disease-free planting material for commercial cultivation.

These techniques help enhance plant yield, provide disease-resistant plants, and ensure uniformity across crops.

B. Molecular markers and genomics in spices breeding

Molecular markers and genomics play a critical role in the study and improvement of various spice and medicinal plants. Here's an overview of their application to the species you've mentioned:

1. Black Pepper (*Piper nigrum*)

1a. Molecular markers: Used to identify disease resistance, quality traits, and genetic diversity.

Molecular markers are tools used in plant breeding, genetics, and biodiversity studies to identify specific sequences in the DNA of an organism. In black pepper (Piper nigrum), molecular markers have been widely used for various purposes, including genetic diversity studies, marker-assisted selection (MAS), and identifying disease resistance traits. Here are the key types of molecular markers used in black pepper:

1. Random Amplified Polymorphic DNA (RAPD)

- **Applications**: RAPD markers have been used to assess genetic diversity among different black pepper cultivars and wild species. They are also useful in identifying duplicates in germplasm collections.
- **Advantages**: RAPD markers are simple, quick, and cost-effective.
- **Limitations**: RAPD markers are less reproducible and may have lower resolution compared to more advanced markers.

2. Simple Sequence Repeats (SSR) or Microsatellites

- **Applications**: SSR markers are highly polymorphic and have been extensively used in genetic mapping, diversity analysis, and cultivar identification in black pepper. They are particularly useful in studying the genetic structure of populations.
- **Advantages**: SSRs are co-dominant markers, providing information on heterozygosity and are highly reproducible.
- **Limitations**: The development of SSR markers is relatively labor-intensive and requires significant initial investment in sequencing.

3. Amplified Fragment Length Polymorphism (AFLP)

- **Applications**: AFLP markers have been employed for genetic fingerprinting, studying genetic diversity, and in constructing genetic linkage maps in black pepper. They are also used for detecting genetic variations among cultivars.

- **Advantages**: AFLP markers are highly sensitive, capable of detecting a large number of polymorphisms across the genome.
- **Limitations**: AFLP analysis is technically demanding and requires specialized equipment.

4. Inter Simple Sequence Repeats (ISSR)

- **Applications**: ISSR markers are used for analyzing genetic diversity and relationships among black pepper cultivars and wild relatives. They are useful in phylogenetic studies and germplasm characterization.
- **Advantages**: ISSR markers combine the advantages of both RAPD and SSR markers, offering high reproducibility and polymorphism.
- **Limitations**: ISSR analysis can be sensitive to reaction conditions, which may affect reproducibility.

5. Single Nucleotide Polymorphisms (SNPs)

- **Applications**: SNP markers are increasingly used in black pepper for high-resolution mapping of traits, association studies, and in the development of SNP-based genotyping arrays. SNPs are valuable in marker-assisted selection (MAS) for traits like disease resistance and yield.
- **Advantages**: SNPs are abundant throughout the genome, co-dominant, and provide high-resolution genetic information.
- **Limitations**: SNP genotyping requires advanced technology and bioinformatics tools, which can be expensive.

6. Expressed Sequence Tags (ESTs)

- **Applications**: EST markers are derived from expressed genes and have been used in black pepper to identify genes associated with important traits, such as disease resistance and stress tolerance.
- **Advantages**: EST markers are associated with functional genes, making them particularly useful for studying gene expression and developing trait-specific markers.
- **Limitations**: The number of EST markers is limited by the available transcriptome data, and they may not cover the entire genome.

7. Sequence-Related Amplified Polymorphism (SRAP)

- **Applications**: SRAP markers are used in black pepper for mapping and assessing genetic diversity. They are particularly useful in identifying markers linked to agronomic traits.
- **Advantages**: SRAP markers target coding sequences, making them useful for identifying functional markers associated with traits.

- **Limitations**: The application of SRAP markers is still less common compared to other markers like SSRs and SNPs.

Applications in Black Pepper breeding

- **Genetic diversity studies**: Molecular markers are used to evaluate the genetic diversity among black pepper cultivars and wild relatives, which is essential for conservation and breeding programs.
- **Marker-Assisted Selection (MAS)**: Specific markers linked to important traits, such as disease resistance (e.g., resistance to Phytophthora foot rot) and yield, can be used to select desirable genotypes in breeding programs.
- **Phylogenetic studies**: Molecular markers help in understanding the evolutionary relationships among different black pepper species and cultivars.
- **Germplasm characterization**: Markers are used to characterize and manage germplasm collections, ensuring the identification of unique genotypes and avoiding duplication.

Molecular markers continue to play a crucial role in improving black pepper breeding and conservation efforts by providing precise and efficient tools for genetic analysis.

1b. Genomics: Sequencing and transcriptomic studies focus on genes related to flavour, disease resistance, and stress responses.

Genomics in black pepper (Piper nigrum) involves the study of its entire genome to understand the genetic basis of its traits, including disease resistance, yield, quality, and adaptation to environmental conditions. Advances in genomics have provided significant insights into black pepper's genetic makeup, enabling more effective breeding, conservation, and crop improvement strategies. Here's an overview of genomics in black pepper:

1. Genome sequencing

- **Whole Genome Sequencing (WGS)**: The genome of black pepper has been sequenced, providing a comprehensive reference genome. This sequencing effort has enabled researchers to identify genes and regulatory elements associated with key traits such as disease resistance, secondary metabolite production (like piperine), and stress tolerance.
- **Genome size**: The black pepper genome is relatively large, with the reference genome size estimated to be around 1.6 Gb (giga base pairs).
- **Gene annotation**: The sequencing efforts have led to the identification and annotation of thousands of genes, including those involved in biosynthetic pathways of important compounds like alkaloids and terpenoids.

2. Transcriptomics

- **RNA sequencing (RNA-Seq)**: Transcriptomic studies using RNA-Seq have been conducted to understand gene expression patterns in black pepper under different conditions, such as biotic stress (e.g., Phytophthora infection) and abiotic stress (e.g., drought). These studies help identify genes that are differentially expressed in response to these stresses, providing targets for breeding and genetic engineering.
- **Functional genomics**: Transcriptomic data has been used to study the function of specific genes, particularly those involved in the biosynthesis of piperine, the compound responsible for black pepper's pungency.

3. Genetic mapping and QTL analysis

- **Quantitative Trait Loci (QTL) mapping**: Genomics has enabled the mapping of QTLs associated with important traits such as disease resistance (e.g., resistance to Phytophthora foot rot) and yield. QTL mapping involves linking specific regions of the genome with phenotypic traits, which can then be used in marker-assisted selection (MAS).
- **Linkage maps**: Genetic linkage maps have been developed using molecular markers like SSRs and SNPs, facilitating the identification of QTLs and aiding in the breeding of improved black pepper varieties.

4. Marker-Assisted Selection (MAS)

- **Trait-specific markers**: Genomic studies have identified molecular markers associated with desirable traits, which are now being used in MAS. For example, markers linked to disease resistance genes are used to select resistant plants in breeding programs.
- **Improving breeding efficiency**: MAS allows breeders to screen large populations for desired traits more efficiently, reducing the time and cost associated with traditional breeding methods.

5. Comparative genomics

- **Comparative analysis with related species**: Comparative genomics involves comparing the black pepper genome with those of related species or other economically important crops. This helps in identifying conserved genes and pathways, understanding evolutionary relationships, and transferring knowledge from model species to black pepper.
- **Evolutionary studies**: Comparative genomics also aids in understanding the evolutionary history of black pepper, including its divergence from related species in the Piperaceae family.

6. Functional genomics and gene editing

- **Gene function studies**: Functional genomics in black pepper involves studying the roles of specific genes in traits like secondary metabolite biosynthesis, stress responses, and developmental processes. This is often achieved through techniques like gene knockout or overexpression studies.
- **CRISPR-Cas9**: Although still in its early stages for black pepper, gene editing technologies like CRISPR-Cas9 hold potential for targeted manipulation of the black pepper genome to improve traits such as disease resistance, yield, and quality.

7. Pangenomics

- **Diversity within the species**: Pangenomics involves studying the full complement of genes within a species, including core genes shared by all cultivars and accessory genes present only in some. This approach helps in understanding genetic diversity within black pepper and can lead to the discovery of novel traits.
- **Application in breeding**: Pangenomics can identify unique genes in landraces or wild relatives of black pepper, which can be introgressed into cultivated varieties to improve resilience and productivity.

8. Metagenomics

- **Soil and rhizosphere microbiome**: Metagenomics has been used to study the microbial communities associated with black pepper roots and soil. Understanding these microbial interactions can help in managing soil health and disease resistance, as beneficial microbes can enhance plant growth and protect against pathogens.

Applications of genomics in Black Pepper

- **Disease resistance**: Genomic tools have been crucial in identifying and developing black pepper varieties resistant to major diseases such as Phytophthora foot rot and nematodes.
- **Quality improvement**: Genomics helps in understanding and enhancing the biosynthesis pathways of important secondary metabolites like piperine, which determine the pungency and flavour of black pepper.
- **Breeding programs**: Genomics accelerates the breeding process by enabling marker-assisted selection and genomic selection, making it possible to develop improved varieties more efficiently.
- **Conservation**: Genomic studies contribute to the conservation of black pepper genetic resources by identifying and preserving genetic diversity, which is essential for long-term sustainability.

Genomics has revolutionized the understanding and improvement of black pepper, providing valuable insights into its biology and paving the way for more precise and efficient breeding strategies.

2. Cardamom (*Elettaria cardamomum*)

2a. Molecular markers: Employed in the study of genetic diversity and breeding for traits like high yield and disease resistance.

Molecular markers are powerful tools used in the genetic study and breeding of cardamom (*Elettaria cardamomum*), one of the most important spice crops. These markers assist in understanding genetic diversity, identifying cultivars, and improving traits such as yield, disease resistance, and quality. Here are the key types of molecular markers used in cardamom:

1. Random Amplified Polymorphic DNA (RAPD)

- **Applications**: RAPD markers have been used extensively to assess genetic diversity among cardamom cultivars and to distinguish between different genotypes. This type of marker is useful in studying genetic relationships and identifying duplicates in germplasm collections.
- **Advantages**: RAPD is a simple, quick, and cost-effective technique that requires minimal DNA.
- **Limitations**: RAPD markers can sometimes be less reproducible and have lower resolution compared to more advanced markers.

2. Simple Sequence Repeats (SSR) or Microsatellites

- **Applications**: SSR markers are highly polymorphic and are widely used in genetic diversity analysis, cultivar identification, and construction of genetic linkage maps in cardamom. They are valuable for studying population structure and for marker-assisted selection (MAS).
- **Advantages**: SSRs are co-dominant markers, meaning they can detect both alleles in a heterozygous individual, providing detailed information on genetic diversity and structure. They are also highly reproducible.
- **Limitations**: The development of SSR markers can be labor-intensive and requires significant initial investment in sequencing.

3. Amplified Fragment Length Polymorphism (AFLP)

- **Applications**: AFLP markers are used in genetic fingerprinting, assessing genetic diversity, and in constructing genetic linkage maps in cardamom. They are particularly useful in detecting polymorphisms across the genome.
- **Advantages**: AFLP is highly sensitive and can generate a large number of markers, making it ideal for studying genetic variation.

- **Limitations**: The technique is technically demanding and requires specialized equipment, which can be a limitation for some laboratories.

4. Inter Simple Sequence Repeats (ISSR)

- **Applications**: ISSR markers are used to analyze genetic diversity, phylogenetic relationships, and germplasm characterization in cardamom. They are effective in distinguishing between closely related genotypes.
- **Advantages**: ISSR markers are highly polymorphic and reproducible, combining the benefits of RAPD and SSR markers.
- **Limitations**: ISSR markers can be sensitive to reaction conditions, which may affect their reproducibility.

5. Single Nucleotide Polymorphisms (SNPs)

- **Applications**: SNP markers are becoming increasingly popular in cardamom for high-resolution mapping of traits, association studies, and development of SNP-based genotyping arrays. SNPs are valuable in marker-assisted breeding for traits such as disease resistance and yield.
- **Advantages**: SNPs are abundant across the genome and provide high-resolution genetic information. They are co-dominant markers, allowing for precise genotyping.
- **Limitations**: SNP genotyping requires advanced technology and bioinformatics tools, which can be expensive.

6. Expressed Sequence Tags (ESTs)

- **Applications**: EST markers, derived from expressed genes, have been used in cardamom to identify genes associated with important traits such as flavour, aroma, and disease resistance. ESTs are particularly useful in functional genomics studies.
- **Advantages**: EST markers are linked to functional genes, making them valuable for understanding the genetic basis of important traits and for developing trait-specific markers.
- **Limitations**: The number of available EST markers is limited by the extent of transcriptome data, and they may not cover the entire genome.

7. Sequence-Related Amplified Polymorphism (SRAP)

- **Applications**: SRAP markers are used in cardamom for studying genetic diversity, constructing linkage maps, and identifying markers associated with agronomic traits. They are especially useful for targeting coding sequences.
- **Advantages**: SRAP markers target open reading frames (ORFs) in the genome, making them effective for identifying functional markers associated with traits.

- **Limitations**: SRAP markers are less commonly used than other marker types like SSRs and SNPs, and their application in cardamom is still emerging.

Applications in Cardamom breeding

- **Genetic diversity studies**: Molecular markers are essential for evaluating the genetic diversity of cardamom cultivars and wild relatives, which is crucial for conservation and breeding programs.
- **Marker-assisted selection (MAS)**: Markers linked to important traits such as disease resistance, yield, and quality can be used in MAS to accelerate the development of improved cardamom varieties.
- **Germplasm characterization**: Molecular markers help in the accurate characterization and management of cardamom germplasm collections, ensuring the preservation of genetic diversity and preventing the loss of valuable traits.
- **Phylogenetic studies**: These markers are also used to study the evolutionary relationships among different cardamom species and cultivars, aiding in the understanding of their genetic background and origin.

Molecular markers have significantly enhanced the efficiency and precision of cardamom breeding programs, contributing to the development of superior varieties with improved agronomic traits and resilience to environmental stresses.

2b. Genomics: Genomic studies help in identifying genes responsible for essential oil biosynthesis and stress tolerance.

Genomics in cardamom (*Elettaria cardamomum*) involves the comprehensive study of its genome to understand the genetic basis of key traits such as flavour, aroma, disease resistance, and yield. Advances in genomics have the potential to significantly impact breeding, conservation, and cultivation practices for this valuable spice crop. Here's an overview of the current state of genomics in cardamom:

1. Genome sequencing

- **Whole Genome Sequencing (WGS)**: While complete genome sequencing of cardamom is still emerging, efforts are underway to sequence its genome. The availability of a reference genome would provide a critical resource for understanding gene function, genetic diversity, and the molecular basis of important traits.
- **Draft genomes**: Initial sequencing efforts have led to the development of draft genomes or partial sequences, providing insights into key genes involved in cardamom's unique characteristics, such as volatile oil biosynthesis and disease resistance.

- **Genome size**: Cardamom is known to have a relatively large and complex genome, which poses challenges for sequencing but also offers rich genetic information.

2. Transcriptomics

- **RNA sequencing (RNA-Seq)**: Transcriptomic studies using RNA-Seq have been conducted to analyze gene expression profiles in cardamom, especially under different stress conditions like pathogen attack or abiotic stresses (e.g., drought, salinity). This helps in identifying stress-responsive genes and understanding the regulatory networks involved.
- **Gene expression studies**: Transcriptomics has been used to study the expression of genes involved in the biosynthesis of secondary metabolites such as essential oils, which are critical for the flavour and aroma of cardamom.

3. Genetic mapping and QTL analysis

- **Quantitative Trait Loci (QTL) mapping**: QTL mapping in cardamom is used to link specific genomic regions with important phenotypic traits, such as resistance to diseases like rhizome rot and leaf blight, and agronomic traits like yield and pod size. These maps are essential for identifying marker-trait associations.
- **Linkage maps**: Genetic linkage maps have been constructed using molecular markers (e.g., SSRs, SNPs) to facilitate QTL mapping and marker-assisted selection (MAS). These maps are crucial for breeding programs aiming to improve specific traits in cardamom.

4. Marker-assisted selection (MAS)

- **Trait-specific markers**: Genomic studies have identified molecular markers associated with desirable traits, enabling the use of MAS in cardamom breeding. For instance, markers linked to disease resistance genes are used to select for resistant plants, speeding up the breeding process.
- **Breeding efficiency**: MAS allows for the early selection of desirable traits in seedlings, reducing the time and resources needed to develop new cardamom varieties with improved characteristics.

5. Comparative genomics

- **Comparative analysis with related species**: Comparative genomics involves analyzing the cardamom genome alongside related species, such as ginger or turmeric, to identify conserved genes and pathways. This can help in transferring knowledge from well-studied species to cardamom.

- **Evolutionary insights**: Comparative genomics also aids in understanding the evolutionary history of cardamom, including the divergence of *Elettaria cardamomum* from its wild relatives and other species within the Zingiberaceae family.

6. Functional genomics and gene editing

- **Gene function studies**: Functional genomics involves studying the roles of specific genes in cardamom, particularly those involved in traits like aroma, flavour, and stress tolerance. Techniques such as gene silencing or overexpression are used to validate gene functions.
- **CRISPR-Cas9**: Although still in its early stages for cardamom, gene editing technologies like CRISPR-Cas9 hold potential for targeted modifications of the genome to enhance desirable traits, such as disease resistance or improved yield.

7. Pangenomics

- **Genomic diversity**: Pangenomics involves studying the full range of genetic diversity within cardamom by analyzing the core genome shared by all cultivars and the accessory genome specific to certain varieties. This approach helps identify unique genes and alleles that contribute to important traits.
- **Application in breeding**: Pangenomics can be used to discover and utilize novel genetic variations present in traditional or wild cardamom varieties, potentially introducing new traits into cultivated varieties.

8. Metagenomics

- **Soil and rhizosphere microbiome**: Metagenomic studies have been used to explore the microbial communities associated with cardamom roots and soil. Understanding these communities can help in developing sustainable cultivation practices by leveraging beneficial microbes for plant growth and disease resistance.

Applications of genomics in Cardamom

- **Disease resistance**: Genomics has been pivotal in identifying resistance genes and developing molecular markers for breeding disease-resistant cardamom varieties, particularly against common pathogens like Fusarium, Pythium, and Colletotrichum.
- **Quality improvement**: Genomics helps in understanding and improving the biosynthetic pathways of key secondary metabolites that contribute to cardamom's flavour and aroma, such as terpenoids and cineole.

- **Conservation of genetic resources**: Genomic tools are essential for the conservation of cardamom's genetic diversity, helping to preserve traditional varieties and wild relatives that may harbor useful traits for future breeding efforts.
- **Precision breeding**: The integration of genomics with traditional breeding methods allows for precision breeding, where desirable traits can be introduced into new varieties more efficiently and with greater accuracy.

Genomics in cardamom is still evolving, but it holds great promise for improving the crop's productivity, quality, and resilience to environmental stresses, ensuring that this valuable spice continues to thrive in both traditional and modern agricultural systems.

3. Allspice (*Pimenta dioica*)

3a. Molecular markers: Useful in analyzing genetic variation and phylogenetic relationships.

Molecular markers are essential tools in the genetic study and breeding of allspice (Pimenta dioica), a tropical spice known for its aromatic properties. These markers facilitate the assessment of genetic diversity, identification of cultivars, and selection for desirable traits in breeding programs. Here's an overview of the molecular markers used in allspice:

1. Random Amplified Polymorphic DNA (RAPD)

- **Applications**: RAPD markers have been used to assess genetic diversity among allspice populations and to study the genetic relationships between different genotypes. This technique helps in the identification of genetic variation within and between populations.
- **Advantages**: RAPD is a relatively simple, quick, and cost-effective method that requires only a small amount of DNA.
- **Limitations**: RAPD markers may have lower reproducibility and can be sensitive to changes in experimental conditions, leading to variability in results.

2. Simple Sequence Repeats (SSR) or microsatellites

- **Applications**: SSR markers are highly polymorphic and have been used in genetic diversity studies, germplasm characterization, and the development of genetic linkage maps in allspice. These markers are particularly useful for studying population structure and for use in marker-assisted selection (MAS).
- **Advantages**: SSRs are co-dominant markers, meaning they can detect both alleles in heterozygous individuals, providing detailed and reproducible information on genetic diversity.

- **Limitations**: The development of SSR markers can be labor-intensive and requires an initial investment in sequencing.

3. Amplified Fragment Length Polymorphism (AFLP)

- **Applications**: AFLP markers are used in genetic fingerprinting, assessing genetic diversity, and constructing genetic linkage maps in allspice. They provide a high-resolution method for detecting polymorphisms across the genome.
- **Advantages**: AFLP is a highly sensitive technique capable of generating a large number of markers, making it ideal for comprehensive genetic analysis.
- **Limitations**: AFLP analysis is technically demanding and requires specialized equipment, which may limit its accessibility in some laboratories.

4. Inter Simple Sequence Repeats (ISSR)

- **Applications**: ISSR markers have been used to analyze genetic diversity and phylogenetic relationships among different allspice genotypes. They are effective in distinguishing between closely related genotypes and are valuable in germplasm characterization.
- **Advantages**: ISSR markers are highly polymorphic and reproducible, offering a good balance between simplicity and the ability to detect genetic variation.
- **Limitations**: ISSR analysis can be sensitive to reaction conditions, which may affect reproducibility.

5. Single Nucleotide Polymorphisms (SNPs)

- **Applications**: SNP markers are increasingly used in allspice for high-resolution genetic mapping, association studies, and the development of SNP-based genotyping arrays. They are valuable in marker-assisted breeding for traits such as flavour, aroma, and disease resistance.
- **Advantages**: SNPs are abundant and evenly distributed across the genome, providing high-resolution data. They are co-dominant markers, allowing for precise genotyping and detailed genetic analysis.
- **Limitations**: SNP genotyping requires advanced technology and bioinformatics tools, which can be expensive and require significant expertise.

6. Expressed Sequence Tags (ESTs)

- **Applications**: EST markers, derived from expressed genes, have been used in allspice to identify genes associated with important traits such as essential oil biosynthesis and disease resistance. ESTs are particularly useful in functional genomics studies.

- **Advantages**: EST markers are linked to functional genes, making them valuable for studying the genetic basis of traits and for developing trait-specific markers.
- **Limitations**: The availability of EST markers depends on the extent of transcriptome data, and they may not provide complete coverage of the genome.

7. Sequence-Related Amplified Polymorphism (SRAP)

- **Applications**: SRAP markers are used in allspice to study genetic diversity, construct genetic linkage maps, and identify markers associated with agronomic traits. They are particularly useful for targeting coding sequences.
- **Advantages**: SRAP markers target open reading frames (ORFs) in the genome, making them effective for identifying functional markers associated with specific traits.
- **Limitations**: SRAP markers are less commonly used than other marker types, and their application in allspice is still emerging.

Applications in Allspice breeding

- **Genetic diversity studies**: Molecular markers are critical for evaluating the genetic diversity of allspice populations, which is important for conservation and breeding programs.
- **Marker-assisted selection (MAS)**: Markers linked to important traits, such as disease resistance and essential oil content, can be used in MAS to accelerate the development of improved allspice varieties.
- **Germplasm characterization**: Molecular markers help in accurately characterizing and managing allspice germplasm collections, ensuring the preservation of genetic diversity and the identification of unique genotypes.
- **Phylogenetic studies**: These markers are also used to study the evolutionary relationships among different allspice species and cultivars, aiding in the understanding of their genetic background and origin.

Molecular markers have significantly enhanced the efficiency and precision of allspice breeding programs, contributing to the development of superior varieties with improved traits such as aroma, flavour, and resilience to environmental stresses.

3b. Genomics: Focus on the identification of genes involved in the biosynthesis of aromatic compounds.

Genomics in allspice (Pimenta dioica) is a developing field that provides valuable insights into the genetic basis of key traits such as flavour, aroma, disease resistance, and adaptability to environmental conditions. Advances in genomics are crucial for enhancing breeding programs, conserving genetic diversity, and improving cultivation practices for this economically significant spice. Here's an overview of the current state and potential of genomics in allspice:

1. Genome sequencing

- **Whole Genome Sequencing (WGS)**: Although complete genome sequencing of allspice is still in its early stages, efforts are being made to sequence its genome. A complete reference genome would be a critical resource for understanding the genetic architecture of important traits, identifying gene functions, and developing molecular markers for breeding.
- **Draft genomes**: Initial sequencing efforts have led to the creation of draft genome sequences or partial genomic data, which provide insights into the genes involved in essential oil biosynthesis, disease resistance, and other economically important traits.
- **Genome size**: The genome of allspice is relatively large and complex, similar to other members of the Myrtaceae family, which presents challenges but also opportunities for discovering rich genetic information.

2. Transcriptomics

- **RNA sequencing (RNA-Seq)**: Transcriptomic studies using RNA-Seq have been conducted to analyze gene expression patterns in allspice, particularly in response to different environmental stresses and during various stages of essential oil production. These studies help identify genes that are differentially expressed and provide insights into the molecular pathways involved.
- **Functional genomics**: Transcriptomics is used to study the function of specific genes related to secondary metabolite production, such as the biosynthesis of eugenol, a key compound responsible for allspice's characteristic aroma.

3. Genetic mapping and QTL analysis

- **Quantitative Trait Loci (QTL) mapping**: QTL mapping in allspice aims to identify genomic regions associated with important phenotypic traits, such as oil content, disease resistance, and growth characteristics. These efforts are crucial for understanding the genetic basis of complex traits and for developing marker-assisted selection (MAS) strategies.
- **Linkage maps**: Genetic linkage maps, constructed using molecular markers like SSRs and SNPs, facilitate QTL mapping and the identification of candidate genes for specific traits. These maps are foundational for breeding programs focused on improving allspice.

4. Marker-assisted selection (MAS)

- **Trait-specific markers**: Genomic studies have identified molecular markers linked to desirable traits, such as high essential oil content and disease resistance. These markers are used in MAS to accelerate the selection of superior allspice plants in breeding programs.
- **Breeding efficiency**: MAS enables breeders to screen for desirable traits at the seedling stage, reducing the time and resources required to develop new allspice varieties with improved characteristics.

5. Comparative genomics

- **Comparative analysis with related species**: Comparative genomics involves analyzing the allspice genome alongside those of related species, such as cloves or guava, to identify conserved genes and metabolic pathways. This approach can help in transferring knowledge from well-studied species to allspice and in understanding its evolutionary history.
- **Evolutionary studies**: Comparative genomics also provides insights into the genetic divergence of allspice from other species in the Myrtaceae family, aiding in the study of speciation and genetic diversity within the genus Pimenta.

6. Functional genomics and gene editing

- **Gene function studies**: Functional genomics in allspice focuses on studying the roles of specific genes, particularly those involved in the biosynthesis of essential oils, stress responses, and developmental processes. Techniques such as gene silencing or overexpression are used to validate gene functions.
- **CRISPR-Cas9**: Although gene editing in allspice is still in its infancy, CRISPR-Cas9 technology holds promise for targeted modifications of the genome to enhance traits like flavour, aroma, and disease resistance.

7. Pangenomics

- **Genomic diversity**: Pangenomics involves studying the full range of genetic diversity within allspice by analyzing both the core genome shared by all cultivars and the accessory genome containing unique genes. This approach helps identify genetic variations that contribute to important traits and can be used in breeding programs.
- **Application in breeding**: Pangenomics can help discover novel genetic traits present in different allspice cultivars or wild relatives, providing new opportunities for crop improvement.

8. Metagenomics

- **Soil and rhizosphere microbiome**: Metagenomic studies are used to explore the microbial communities associated with allspice roots and surrounding soil. Understanding these interactions can lead to improved soil health management and enhance plant growth and resistance to diseases.

Applications of genomics in Allspice

- **Improvement of essential oil quality**: Genomics helps in identifying and enhancing the biosynthetic pathways of key compounds like eugenol and myrcene, which determine the flavour and aroma of allspice. Understanding these pathways can lead to the development of varieties with improved oil quality.
- **Disease resistance**: Genomic tools are used to identify resistance genes and develop molecular markers that can be used in breeding programs to produce allspice varieties resistant to common diseases.
- **Conservation of genetic resources**: Genomic studies contribute to the conservation of genetic diversity in allspice, ensuring that valuable traits found in traditional cultivars or wild relatives are preserved for future breeding efforts.
- **Precision breeding**: The integration of genomics with traditional breeding methods allows for precision breeding, where specific desirable traits can be introduced or enhanced in new allspice varieties with greater efficiency.

Genomics in allspice is still evolving but holds great potential for advancing the breeding and cultivation of this important spice. By unlocking the genetic basis of key traits, genomics can help meet the growing demand for high-quality allspice in the global market.

4. Clove (*Syzygium aromaticum*)

4a. Molecular markers: Applied to study genetic diversity and assist in clonal selection.

Molecular markers are critical tools in the genetic study, breeding, and conservation of clove (*Syzygium aromaticum*). These markers help in understanding genetic diversity, identifying cultivars, and selecting for desirable traits such as disease resistance, yield, and oil content in breeding programs. Here's an overview of the molecular markers used in clove:

1. Random Amplified Polymorphic DNA (RAPD)

- **Applications**: RAPD markers have been widely used to assess genetic diversity among clove populations, identify cultivars, and study genetic relationships. They are particularly useful in detecting polymorphisms within and between clove populations.

- **Advantages**: RAPD is a simple, quick, and cost-effective method that requires only a small amount of DNA and does not require prior knowledge of the genome.
- **Limitations**: RAPD markers can be less reproducible and sensitive to experimental conditions, which may lead to variability in results.

2. Simple Sequence Repeats (SSR) or Microsatellites

- **Applications**: SSR markers are highly polymorphic and co-dominant, making them ideal for genetic diversity studies, cultivar identification, and the development of genetic linkage maps in clove. These markers have been used to characterize germplasm collections and to understand population structure.
- **Advantages**: SSRs provide high levels of polymorphism, are highly reproducible, and allow for detailed genetic analysis. They are effective in identifying both alleles in heterozygous individuals, which is valuable for studying genetic diversity.
- **Limitations**: Developing SSR markers can be time-consuming and requires a significant initial investment in sequencing.

3. Amplified Fragment Length Polymorphism (AFLP)

- **Applications**: AFLP markers are used for high-resolution genetic fingerprinting, assessing genetic diversity, and constructing genetic linkage maps in clove. They are effective in detecting a large number of polymorphisms across the genome.
- **Advantages**: AFLP is highly sensitive and can generate a large number of markers, making it suitable for comprehensive genetic analysis and diversity studies.
- **Limitations**: AFLP analysis is technically demanding, requiring specialized equipment and expertise, which may limit its accessibility.

4. Inter Simple Sequence Repeats (ISSR)

- **Applications**: ISSR markers are used to study genetic diversity, phylogenetic relationships, and germplasm characterization in clove. They are effective in detecting polymorphisms and are useful for distinguishing between closely related genotypes.
- **Advantages**: ISSR markers offer high reproducibility and polymorphism, combining the benefits of both RAPD and SSR markers. They are particularly useful in genetic diversity studies.
- **Limitations**: ISSR markers can be sensitive to reaction conditions, which may affect their reproducibility.

5. Single Nucleotide Polymorphisms (SNPs)

- **Applications**: SNP markers are increasingly being used in clove for high-resolution genetic mapping, association studies, and the development of SNP-based genotyping arrays. They are valuable in marker-assisted selection (MAS) for traits such as disease resistance, oil content, and yield.
- **Advantages**: SNPs are abundant, co-dominant, and provide high-resolution genetic data. They are ideal for detailed genotyping and for studying the genetic architecture of complex traits.
- **Limitations**: SNP genotyping requires advanced technology and bioinformatics tools, which can be expensive and require significant expertise.

6. Expressed Sequence Tags (ESTs)

- **Applications**: EST markers, derived from expressed genes, have been used to identify genes associated with important traits such as eugenol biosynthesis, disease resistance, and stress tolerance in clove. ESTs are particularly useful in functional genomics and in the development of trait-specific markers.
- **Advantages**: EST markers are linked to functional genes, making them valuable for studying the genetic basis of economically important traits and for use in MAS.
- **Limitations**: The availability of EST markers is limited by the extent of transcriptome data, and they may not provide comprehensive coverage of the entire genome.

7. Sequence-Related Amplified Polymorphism (SRAP)

- **Applications**: SRAP markers are used in clove for genetic diversity studies, the construction of genetic linkage maps, and identifying markers associated with agronomic traits. They are particularly useful for targeting coding sequences and understanding functional genomics.
- **Advantages**: SRAP markers target open reading frames (ORFs) in the genome, making them effective for identifying functional markers associated with specific traits.
- **Limitations**: SRAP markers are less commonly used than SSRs and SNPs, and their application in clove is still relatively new.

Applications in Clove breeding

- **Genetic diversity studies**: Molecular markers are essential for evaluating the genetic diversity of clove populations, which is critical for conservation and breeding programs. Understanding genetic diversity helps in the selection of parent plants for breeding and in maintaining genetic resources.

- **Marker-assisted selection (MAS)**: Markers linked to important traits, such as eugenol content (a key compound in clove oil), disease resistance, and yield, are used in MAS to accelerate the development of improved clove varieties.
- **Germplasm characterization**: Molecular markers help in accurately characterizing and managing clove germplasm collections, ensuring the preservation of genetic diversity and the identification of unique genotypes for breeding purposes.
- **Phylogenetic studies**: These markers are also used to study the evolutionary relationships among different clove species and cultivars, aiding in the understanding of their genetic background and origin.

Molecular markers have greatly enhanced the precision and efficiency of clove breeding programs, contributing to the development of superior varieties with improved traits such as flavour, aroma, yield, and resistance to environmental stresses.

4b. Genomics: Studies focus on the genomics of essential oil production and disease resistance.

Genomics in clove (*Syzygium aromaticum*) is a field that provides crucial insights into the genetic makeup of this valuable spice, known for its high eugenol content and medicinal properties. Genomics tools help in understanding the genetic basis of key traits such as flavour, aroma, disease resistance, and yield, which are important for breeding programs and the sustainable cultivation of clove. Here's an overview of the state of genomics in clove:

1. Genome Sequencing

- **Whole Genome Sequencing (WGS)**: The complete genome sequencing of clove is still an emerging area, with efforts focusing on sequencing the genome to create a reference. A complete reference genome would significantly enhance the understanding of clove's genetic architecture and facilitate the identification of genes responsible for important traits.
- **Draft genomes**: Initial efforts in sequencing have produced draft genomes or partial sequences, providing a foundation for further research. These sequences offer insights into genes involved in eugenol biosynthesis, disease resistance, and other critical metabolic pathways.
- **Genome size and complexity**: The clove genome is relatively large and complex, similar to other members of the Myrtaceae family. This complexity presents challenges but also offers opportunities to uncover a wealth of genetic information that can be used in breeding and conservation.

2. Transcriptomics

- **RNA sequencing (RNA-Seq)**: Transcriptomic studies using RNA-Seq have been employed to analyze gene expression profiles in clove, particularly under different environmental conditions and during various stages of essential oil production. These studies are essential for identifying genes that are differentially expressed in response to stress and during the biosynthesis of key compounds like eugenol.
- **Functional genomics**: Transcriptomics allows researchers to study the function of specific genes involved in flavour, aroma, and stress responses. This knowledge is critical for understanding the biochemical pathways that produce clove's distinctive aroma and medicinal properties.

3. Genetic Mapping and QTL Analysis

- **Quantitative Trait Loci (QTL) mapping**: QTL mapping in clove is used to link specific genomic regions with phenotypic traits such as oil content, disease resistance, and yield. Identifying QTLs helps in understanding the genetic basis of these traits and provides markers that can be used in marker-assisted selection (MAS).
- **Linkage maps**: Genetic linkage maps constructed using molecular markers (such as SSRs and SNPs) are foundational tools for QTL mapping and gene discovery. These maps help in tracking the inheritance of important traits in breeding programs.

4. Marker-assisted selection (MAS)

- **Trait-specific markers**: Genomic studies have led to the development of molecular markers linked to desirable traits, such as high eugenol content and resistance to diseases like leaf spot and clove bud rot. These markers are used in MAS to select plants with the best genetic potential early in the breeding process.
- **Breeding efficiency**: MAS increases the efficiency of clove breeding programs by allowing the early and accurate selection of plants with desired traits, reducing the time and resources needed to develop new varieties.

5. Comparative Genomics

- **Comparative analysis with related species**: Comparative genomics involves analyzing the clove genome alongside those of related species, such as guava (Psidium guajava) or eucalyptus (Eucalyptus spp.), to identify conserved genes and pathways. This approach can help in transferring knowledge from well-studied species to clove.
- **Evolutionary studies**: Comparative genomics also provides insights into the evolutionary history of clove, helping to understand its genetic

divergence from other members of the Myrtaceae family and the unique traits it has developed.

6. Functional Genomics and Gene Editing

- **Gene function studies**: Functional genomics focuses on understanding the roles of specific genes in clove, particularly those involved in biosynthesis pathways for essential oils, stress responses, and growth regulation. Techniques such as gene silencing, overexpression, or CRISPR-Cas9 gene editing are used to validate the functions of these genes.
- **CRISPR-Cas9**: Although still in its early stages for clove, CRISPR-Cas9 technology holds promise for targeted modifications of the genome to enhance traits like disease resistance, yield, and eugenol content.

7. Pangenomics

- **Genomic diversity**: Pangenomics involves studying the full range of genetic diversity within clove by analyzing the core genome shared by all cultivars and the accessory genome containing unique genes. This approach is valuable for identifying genetic variations that contribute to important traits and can be used to broaden the genetic base of breeding programs.
- **Application in breeding**: Pangenomics can help discover novel genetic traits in different clove cultivars or wild relatives, offering new opportunities for crop improvement and the development of resilient clove varieties.

8. Metagenomics

- **Soil and rhizosphere microbiome**: Metagenomic studies have been used to explore the microbial communities associated with clove roots and surrounding soil. Understanding these interactions can lead to improved soil health management practices and enhance plant growth and resistance to diseases.

Applications of Genomics in Clove

- **Improvement of essential oil quality**: Genomics provides insights into the biosynthetic pathways of eugenol and other compounds that define clove's flavour and aroma. By understanding these pathways, it is possible to breed clove varieties with enhanced oil content and quality.
- **Disease resistance**: Genomic tools are used to identify genes responsible for resistance to major clove diseases, such as leaf spot, and to develop molecular markers for these traits. This helps in breeding clove varieties that are more resilient to diseases.
- **Conservation of genetic resources**: Genomics plays a critical role in the conservation of clove's genetic diversity, helping to preserve traditional

varieties and wild relatives that may harbor useful traits for future breeding efforts.

- **Precision breeding**: The integration of genomics with traditional breeding methods allows for precision breeding, where specific desirable traits can be introduced or enhanced in new clove varieties with greater efficiency.

Future Prospects

- **Advanced breeding techniques**: With further advancements in genomics, there is potential for the development of clove varieties with tailored characteristics, such as specific flavour profiles or enhanced medicinal properties.
- **Sustainable cultivation**: Genomics can contribute to more sustainable clove cultivation practices by enabling the development of varieties that are better adapted to local environmental conditions and resistant to pests and diseases.

Genomics in clove is a rapidly evolving field that promises to enhance the breeding and cultivation of this valuable spice. By unlocking the genetic basis of key traits, genomics can help meet the growing demand for high-quality clove in the global market while ensuring its sustainability and conservation.

5. Cinnamon (*Cinnamomum verum*)

5a. Molecular markers: Help in the identification of superior cultivars with high oil content.

Molecular markers are crucial tools for the genetic study, breeding, and conservation of cinnamon (*Cinnamomum* spp.), a significant spice crop with multiple species contributing to the global market, including *Cinnamomum verum* (true cinnamon) and *Cinnamomum cassia*. These markers assist in understanding genetic diversity, identifying species and cultivars, and selecting desirable traits such as disease resistance, yield, and essential oil content. Here's an overview of the molecular markers used in cinnamon:

1. Random Amplified Polymorphic DNA (RAPD)

- **Applications**: RAPD markers have been widely used to assess genetic diversity within and between cinnamon species and to identify cultivars. They help in understanding the genetic relationships and population structure of different cinnamon species.
- **Advantages**: RAPD is a simple, quick, and cost-effective technique that requires only a small amount of DNA and does not require prior genomic information.

- **Limitations**: RAPD markers may have lower reproducibility and are sensitive to changes in experimental conditions, which can lead to variability in results.

2. Simple Sequence Repeats (SSR) or Microsatellites

- **Applications**: SSR markers are highly polymorphic and have been used for genetic diversity studies, species identification, and the development of genetic linkage maps in cinnamon. They are particularly useful for germplasm characterization and in marker-assisted selection (MAS).
- **Advantages**: SSRs are co-dominant markers, providing detailed genetic information with high reproducibility. They are effective in identifying both alleles in heterozygous individuals, which is valuable for studying genetic diversity and population structure.
- **Limitations**: Developing SSR markers can be time-consuming and requires an initial investment in sequencing.

3. Amplified Fragment Length Polymorphism (AFLP)

- **Applications**: AFLP markers are used for high-resolution genetic fingerprinting, assessing genetic diversity, and constructing genetic linkage maps in cinnamon. They are effective in detecting a large number of polymorphisms across the genome, making them useful for detailed genetic analysis.
- **Advantages**: AFLP is a highly sensitive technique that can generate a large number of markers, making it suitable for comprehensive genetic analysis and diversity studies.
- **Limitations**: AFLP analysis is technically demanding and requires specialized equipment, which may limit its accessibility in some laboratories.

4. Inter Simple Sequence Repeats (ISSR)

- **Applications**: ISSR markers are used to study genetic diversity, phylogenetic relationships, and germplasm characterization in cinnamon. They are effective in distinguishing closely related species and cultivars and are valuable in understanding the evolutionary relationships within the *Cinnamomum* genus.
- **Advantages**: ISSR markers offer high polymorphism and reproducibility, combining the benefits of both RAPD and SSR markers. They are particularly useful for genetic diversity studies in species with limited genomic information.
- **Limitations**: ISSR markers can be sensitive to reaction conditions, which may affect their reproducibility.

5. Single Nucleotide Polymorphisms (SNPs)

- **Applications**: SNP markers are increasingly used in cinnamon for high-resolution genetic mapping, association studies, and the development of SNP-based genotyping arrays. They are valuable in MAS for traits such as oil content, disease resistance, and growth characteristics.
- **Advantages**: SNPs are abundant, co-dominant, and provide high-resolution genetic data. They are ideal for detailed genotyping and for studying the genetic architecture of complex traits in cinnamon.
- **Limitations**: SNP genotyping requires advanced technology and bioinformatics tools, which can be expensive and require significant expertise.

6. Expressed Sequence Tags (ESTs)

- **Applications**: EST markers, derived from expressed genes, have been used to identify genes associated with important traits such as essential oil biosynthesis, disease resistance, and stress tolerance in cinnamon. ESTs are particularly useful in functional genomics and in developing trait-specific markers.
- **Advantages**: EST markers are linked to functional genes, making them valuable for studying the genetic basis of economically important traits and for use in MAS.
- **Limitations**: The availability of EST markers depends on the extent of transcriptome data, and they may not provide comprehensive coverage of the entire genome.

7. Sequence-Related Amplified Polymorphism (SRAP)

- **Applications**: SRAP markers are used in cinnamon for genetic diversity studies, the construction of genetic linkage maps, and identifying markers associated with agronomic traits. They are particularly useful for targeting coding sequences and understanding functional genomics.
- **Advantages**: SRAP markers target open reading frames (ORFs) in the genome, making them effective for identifying functional markers associated with specific traits in cinnamon.
- **Limitations**: SRAP markers are less commonly used than SSRs and SNPs, and their application in cinnamon is still relatively new.

Applications in Cinnamon breeding

- **Genetic diversity studies**: Molecular markers are essential for evaluating the genetic diversity of cinnamon populations, which is critical for conservation and breeding programs. Understanding genetic diversity helps in selecting parent plants for breeding and maintaining genetic resources.

- **Marker-assisted selection (MAS)**: Markers linked to important traits, such as essential oil content and disease resistance, are used in MAS to accelerate the development of improved cinnamon varieties. For instance, markers linked to cinnamaldehyde production (a key compound in cinnamon oil) can be used to select for high-quality oil-producing varieties.
- **Germplasm characterization**: Molecular markers help in accurately characterizing and managing cinnamon germplasm collections, ensuring the preservation of genetic diversity and the identification of unique genotypes for breeding purposes.
- **Phylogenetic studies**: These markers are used to study the evolutionary relationships among different cinnamon species and cultivars, aiding in understanding their genetic background and origin.

Molecular markers have greatly enhanced the precision and efficiency of cinnamon breeding programs, contributing to the development of superior varieties with improved traits such as flavour, aroma, yield, and resilience to environmental stresses.

5b. Genomics: Genomic studies are aimed at understanding the biosynthesis of cinnamaldehyde and other valuable compounds.

Genomics in cinnamon (Cinnamomum spp.) is an emerging field that provides valuable insights into the genetic basis of key traits such as flavour, aroma, disease resistance, and growth characteristics. These insights are crucial for breeding programs, conservation efforts, and the sustainable cultivation of this important spice crop. Here's an overview of the state and potential of genomics in cinnamon:

1. Genome sequencing

- **Whole Genome Sequencing (WGS)**: While complete genome sequencing of cinnamon species like Cinnamomum verum (true cinnamon) and Cinnamomum cassia is still in progress, draft genome sequences and partial genomic data have been generated. These sequences provide a foundation for understanding the genetic architecture of cinnamon and identifying genes involved in essential oil biosynthesis, disease resistance, and other important traits.
- **Genome size and complexity**: The cinnamon genome is relatively large and complex, similar to other members of the Lauraceae family. This complexity poses challenges for sequencing but also offers opportunities for discovering a rich diversity of genes that can be leveraged for crop improvement.

2. Transcriptomics

- **RNA sequencing (RNA-Seq)**: Transcriptomic studies using RNA-Seq have been employed to analyze gene expression patterns in cinnamon, especially under different environmental conditions and during various stages of essential oil production. These studies are essential for identifying genes that are differentially expressed in response to stress, disease, or during the biosynthesis of key compounds like cinnamaldehyde, a major component of cinnamon oil.
- **Functional genomics**: Transcriptomics allows researchers to study the function of specific genes involved in flavour, aroma, disease resistance, and stress responses. This knowledge is critical for understanding the biochemical pathways that produce cinnamon's distinctive aroma and medicinal properties.

3. Genetic mapping and QTL analysis

- **Quantitative Trait Loci (QTL) mapping**: QTL mapping in cinnamon aims to link specific genomic regions with phenotypic traits such as oil content, disease resistance, and growth characteristics. Identifying QTLs helps in understanding the genetic basis of these traits and provides markers that can be used in marker-assisted selection (MAS).
- **Linkage maps**: Genetic linkage maps, constructed using molecular markers (such as SSRs and SNPs), are foundational tools for QTL mapping and gene discovery. These maps help in tracking the inheritance of important traits in breeding programs.

4. Marker-assisted selection (MAS)

- **Trait-specific markers**: Genomic studies have led to the development of molecular markers linked to desirable traits, such as high cinnamaldehyde content and resistance to diseases like leaf spot and root rot. These markers are used in MAS to select plants with the best genetic potential early in the breeding process, increasing the efficiency of breeding programs.
- **Breeding efficiency**: MAS enables breeders to screen for desirable traits at the seedling stage, reducing the time and resources required to develop new cinnamon varieties with improved characteristics.

5. Comparative genomics

- **Comparative analysis with related species**: Comparative genomics involves analyzing the cinnamon genome alongside those of related species, such as other Lauraceae family members like avocado (Persea americana). This approach helps in identifying conserved genes and pathways and can facilitate the transfer of knowledge from well-studied species to cinnamon.

- **Evolutionary studies**: Comparative genomics provides insights into the evolutionary history of cinnamon, helping to understand its genetic divergence from other species and the development of its unique traits.

6. Functional genomics and gene editing

- **Gene function studies**: Functional genomics in cinnamon focuses on understanding the roles of specific genes, particularly those involved in biosynthesis pathways for essential oils, stress responses, and growth regulation. Techniques such as gene silencing, overexpression, or CRISPR-Cas9 gene editing are used to validate the functions of these genes.
- **CRISPR-Cas9**: Although gene editing in cinnamon is still in its infancy, CRISPR-Cas9 technology holds promise for targeted modifications of the genome to enhance traits like disease resistance, yield, and cinnamaldehyde content.

7. Pangenomics

- **Genomic diversity**: Pangenomics involves studying the full range of genetic diversity within cinnamon by analyzing both the core genome shared by all cultivars and the accessory genome containing unique genes. This approach is valuable for identifying genetic variations that contribute to important traits and can be used to broaden the genetic base of breeding programs.
- **Application in breeding**: Pangenomics can help discover novel genetic traits in different cinnamon cultivars or wild relatives, offering new opportunities for crop improvement and the development of resilient cinnamon varieties.

8. Metagenomics

- **Soil and rhizosphere microbiome**: Metagenomic studies explore the microbial communities associated with cinnamon roots and the surrounding soil. Understanding these interactions can lead to improved soil health management practices, enhancing plant growth and resistance to diseases.

Applications of genomics in Cinnamon

- **Improvement of essential oil quality**: Genomics provides insights into the biosynthetic pathways of cinnamaldehyde and other compounds that define cinnamon's flavour and aroma. By understanding these pathways, it is possible to breed cinnamon varieties with enhanced oil content and quality.
- **Disease resistance**: Genomic tools are used to identify genes responsible for resistance to major cinnamon diseases, such as leaf spot and root rot, and to develop molecular markers for these traits. This helps in breeding cinnamon varieties that are more resilient to diseases.

- **Conservation of genetic resources**: Genomics plays a critical role in the conservation of cinnamon's genetic diversity, helping to preserve traditional varieties and wild relatives that may harbor useful traits for future breeding efforts.
- **Precision breeding**: The integration of genomics with traditional breeding methods allows for precision breeding, where specific desirable traits can be introduced or enhanced in new cinnamon varieties with greater efficiency.

Future prospects

- **Advanced breeding techniques**: With further advancements in genomics, there is potential for the development of cinnamon varieties with tailored characteristics, such as specific flavour profiles or enhanced medicinal properties.
- **Sustainable cultivation**: Genomics can contribute to more sustainable cinnamon cultivation practices by enabling the development of varieties that are better adapted to local environmental conditions and resistant to pests and diseases.

Genomics in cinnamon is a rapidly evolving field with the potential to significantly enhance the breeding, conservation, and sustainable cultivation of this valuable spice. By unlocking the genetic basis of key traits, genomics can help meet the growing demand for high-quality cinnamon in the global market while ensuring its sustainability and conservation.

6. Ginger (*Zingiber officinale*)

6a. Molecular markers: Used for genetic mapping and identifying disease-resistant genes.

Molecular markers are essential tools for the genetic study, breeding, and conservation of ginger (Zingiber officinale). These markers help in understanding genetic diversity, identifying cultivars, and selecting desirable traits such as disease resistance, yield, and essential oil content. Here's an overview of the molecular markers used in ginger:

1. Random Amplified Polymorphic DNA (RAPD)

- **Applications**: RAPD markers have been widely used to assess genetic diversity among ginger cultivars, identify genetic relationships, and study population structure. This technique has been particularly useful in distinguishing between different ginger cultivars and studying genetic variation within and between populations.
- **Advantages**: RAPD is a simple, quick, and cost-effective method that requires only a small amount of DNA. It does not require prior knowledge of the genome.

- **Limitations**: RAPD markers can be less reproducible and sensitive to changes in experimental conditions, which may lead to variability in results.

2. Simple Sequence Repeats (SSR) or microsatellites

- **Applications**: SSR markers are highly polymorphic and co-dominant, making them ideal for genetic diversity studies, cultivar identification, and the construction of genetic linkage maps in ginger. These markers are also used in germplasm characterization and marker-assisted selection (MAS).
- **Advantages**: SSRs are highly reproducible and provide detailed genetic information. They are particularly effective for identifying alleles in heterozygous individuals, making them valuable for studying genetic diversity and population structure.
- **Limitations**: Developing SSR markers can be time-consuming and requires an initial investment in sequencing.

3. Amplified Fragment Length Polymorphism (AFLP)

- **Applications**: AFLP markers are used for high-resolution genetic fingerprinting, assessing genetic diversity, and constructing genetic linkage maps in ginger. AFLP is effective in detecting a large number of polymorphisms across the genome, making it useful for detailed genetic analysis.
- **Advantages**: AFLP is highly sensitive and can generate a large number of markers, making it suitable for comprehensive genetic studies and diversity analysis.
- **Limitations**: AFLP analysis is technically demanding, requiring specialized equipment and expertise, which may limit its accessibility.

4. Inter Simple Sequence Repeats (ISSR)

- **Applications**: ISSR markers have been used to study genetic diversity, phylogenetic relationships, and germplasm characterization in ginger. They are particularly useful for distinguishing closely related genotypes and for studying genetic variation in both cultivated and wild ginger species.
- **Advantages**: ISSR markers are highly polymorphic and reproducible, offering a good balance between simplicity and the ability to detect genetic variation. They combine the benefits of both RAPD and SSR markers.
- **Limitations**: ISSR analysis can be sensitive to reaction conditions, which may affect reproducibility.

5. Single Nucleotide Polymorphisms (SNPs)

- **Applications**: SNP markers are increasingly used in ginger for high-resolution genetic mapping, association studies, and the development of SNP-based genotyping arrays. They are valuable in marker-assisted breeding for traits such as disease resistance, oil content, and yield.

- **Advantages**: SNPs are abundant and evenly distributed across the genome, providing high-resolution genetic data. They are co-dominant markers, allowing for precise genotyping and detailed genetic analysis.
- **Limitations**: SNP genotyping requires advanced technology and bioinformatics tools, which can be expensive and require significant expertise.

6. Expressed Sequence Tags (ESTs)

- **Applications**: EST markers, derived from expressed genes, have been used to identify genes associated with important traits such as essential oil biosynthesis, disease resistance, and stress tolerance in ginger. ESTs are particularly useful in functional genomics and in the development of trait-specific markers.
- **Advantages**: EST markers are linked to functional genes, making them valuable for studying the genetic basis of economically important traits and for use in MAS.
- **Limitations**: The availability of EST markers depends on the extent of transcriptome data, and they may not provide comprehensive coverage of the entire genome.

7. Sequence-Related Amplified Polymorphism (SRAP)

- **Applications**: SRAP markers are used in ginger for genetic diversity studies, the construction of genetic linkage maps, and identifying markers associated with agronomic traits. SRAP markers are particularly useful for targeting coding sequences and understanding functional genomics.
- **Advantages**: SRAP markers target open reading frames (ORFs) in the genome, making them effective for identifying functional markers associated with specific traits in ginger.
- **Limitations**: SRAP markers are less commonly used than SSRs and SNPs, and their application in ginger is still relatively new.

Applications in Ginger breeding

- **Genetic diversity studies**: Molecular markers are essential for evaluating the genetic diversity of ginger populations, which is critical for conservation and breeding programs. Understanding genetic diversity helps in selecting parent plants for breeding and maintaining genetic resources.
- **Marker-assisted selection (MAS)**: Markers linked to important traits, such as essential oil content, disease resistance, and yield, are used in MAS to accelerate the development of improved ginger varieties. For instance,

markers linked to compounds that contribute to the medicinal properties of ginger can be used to select high-quality cultivars.

- **Germplasm characterization**: Molecular markers help in accurately characterizing and managing ginger germplasm collections, ensuring the preservation of genetic diversity and the identification of unique genotypes for breeding purposes.
- **Phylogenetic studies**: These markers are also used to study the evolutionary relationships among different ginger species and cultivars, aiding in understanding their genetic background and origin.

Challenges and future prospects

- **Genomic complexity**: The complexity of the ginger genome, with its high level of heterozygosity and polyploidy, presents challenges for genomic studies. However, advances in sequencing technologies and bioinformatics are expected to overcome these challenges, enabling more comprehensive genetic studies.
- **Integration with breeding programs**: The integration of molecular markers into ginger breeding programs is still in its early stages, but it holds great promise for improving the efficiency and precision of breeding efforts.
- **Sustainable cultivation**: Genomic tools can contribute to more sustainable ginger cultivation by enabling the development of varieties that are better adapted to local environmental conditions and resistant to pests and diseases.

Molecular markers have greatly enhanced the precision and efficiency of ginger breeding programs, contributing to the development of superior varieties with improved traits such as flavour, aroma, yield, and resilience to environmental stresses.

6b. Genomics: Studies focus on the metabolic pathways for gingerol and shogaol production.

Genomics in ginger (*Zingiber officinale*) is a field focused on decoding and understanding the genetic makeup of this important medicinal and culinary plant. Ginger is valued not only for its flavour but also for its therapeutic properties, largely attributed to bioactive compounds such as gingerol, shogaol, and zingerone. Genomic research in ginger aims to enhance its agricultural value, improve disease resistance, and optimize the production of these bioactive compounds.

Key areas of genomics research in Ginger

Genome sequencing

- The ginger genome has been sequenced to provide a comprehensive understanding of its genetic structure. This includes identifying the genes involved in growth, development, and the biosynthesis of key secondary metabolites. Genome sequencing lays the foundation for further genetic studies and breeding programs.

Gene identification and functional genomics

- Researchers focus on identifying and characterizing the specific genes responsible for the biosynthesis of medicinal compounds like gingerol and shogaol. Functional genomics involves studying these genes to understand their roles in metabolic pathways and their regulation under various conditions.

Transcriptomics

- Transcriptomics studies in ginger involve analyzing RNA transcripts to understand gene expression patterns across different tissues and developmental stages. This approach helps in identifying genes that are upregulated or downregulated in response to environmental stresses or during the production of bioactive compounds.

Molecular Markers and Marker-assisted selection (MAS)

- Molecular markers, such as SSRs (Simple Sequence Repeats), SNPs (Single Nucleotide Polymorphisms), and AFLPs (Amplified Fragment Length Polymorphisms), are used to identify genetic variations linked to desirable traits. These markers facilitate marker-assisted selection in breeding programs, allowing for the development of ginger varieties with enhanced disease resistance, yield, and quality.

Genetic diversity and population genomics

- Understanding the genetic diversity within ginger populations is crucial for conservation and breeding. Population genomics studies assess the genetic variation among different ginger varieties, providing insights into the evolutionary history and helping to preserve genetic resources for future breeding efforts.

Metabolomics and pathway analysis

- Integrating genomics with metabolomics allows researchers to map the metabolic pathways involved in the production of ginger's bioactive compounds. This helps in identifying key enzymes and regulatory genes that control the biosynthesis of these compounds, which could be targeted for enhancing their production.

CRISPR and gene editing

- Although still in the early stages, CRISPR-Cas9 technology has the potential to be used in ginger to edit specific genes, potentially improving traits such as disease resistance or increasing the production of valuable compounds like gingerol.

Comparative genomics

- Comparative genomics involves comparing the ginger genome with those of other related species, such as turmeric (Curcuma longa) or other members of the Zingiberaceae family. This helps in understanding the genetic basis of key differences between species and in identifying conserved genes that might be important for similar traits across species.

Applications and impact

- **Development of improved varieties**: Genomic insights enable the development of ginger varieties with improved traits, such as higher yield, better quality, and enhanced resistance to diseases like bacterial wilt and rhizome rot.
- **Enhanced medicinal value**: By understanding the genetic basis of bioactive compound production, researchers can develop strategies to increase the levels of these compounds in ginger, thereby enhancing its medicinal properties.
- **Conservation and sustainable use**: Genomic research aids in the conservation of ginger's genetic diversity, which is vital for sustainable cultivation and breeding programs aimed at meeting the increasing global demand for ginger.
- **Disease resistance**: Identifying and manipulating genes linked to disease resistance can lead to the development of ginger varieties that are more resilient to pathogens, reducing the reliance on chemical treatments and increasing crop sustainability.

Challenges and future directions

- **Complex genome**: One of the major challenges in ginger genomics is the complexity of its genome, which can make it difficult to assemble and analyze. However, advances in sequencing technologies and bioinformatics are gradually overcoming these challenges.
- **Limited genomic resources**: Compared to other crops, ginger has fewer genomic resources available. Ongoing research and international collaborations are expected to expand these resources, providing more tools for breeders and researchers.

- **Integration with other 'Omics' technologies**: Future research in ginger genomics will likely involve the integration of genomics with other 'omics' approaches, such as proteomics and metabolomics, to provide a more comprehensive understanding of ginger's biology and its potential applications.

Genomics in ginger holds great promise for improving the cultivation and medicinal value of this important plant, benefiting both farmers and consumers by providing more resilient and high-quality ginger varieties.

7. Turmeric (*Curcuma longa*)

7a. Molecular markers: Employed in selecting high-curcumin varieties and disease resistance.

Molecular markers in turmeric (Curcuma longa) play a crucial role in understanding the genetic diversity, identification, and improvement of this important medicinal plant. Turmeric is widely known for its bioactive compound curcumin, and molecular markers help in the selection and breeding of turmeric varieties with enhanced medicinal properties, yield, and disease resistance.

Key types of molecular markers used in Turmeric

Simple Sequence Repeats (SSRs) or Microsatellites

- SSRs are DNA sequences that consist of repeating units of 1-6 base pairs. They are highly polymorphic, making them useful for genetic diversity studies, cultivar identification, and marker-assisted selection (MAS) in turmeric breeding programs.

Random Amplified Polymorphic DNA (RAPD)

- RAPD markers are generated by amplifying random DNA segments with single primers of arbitrary nucleotide sequence. Although less specific and reproducible than other markers, RAPD has been widely used for genetic diversity analysis and germplasm characterization in turmeric.

Amplified Fragment Length Polymorphism (AFLP)

- AFLP is a technique that combines the reliability of restriction enzyme digestion with the sensitivity of PCR. It is used for assessing genetic diversity, constructing genetic maps, and identifying molecular markers linked to important traits in turmeric.

Inter Simple Sequence Repeats (ISSR)

- ISSR markers amplify the regions between microsatellites. These markers are highly polymorphic and have been used in turmeric for genetic diversity studies, identifying duplicates in germplasm collections, and assessing genetic stability in tissue-cultured plants.

Single Nucleotide Polymorphisms (SNPs)

- SNPs are single base pair changes in the DNA sequence. They are the most abundant type of genetic variation and are used in turmeric for high-resolution mapping of genes, association studies, and for identifying traits related to stress resistance and yield.

Sequence-Related Amplified Polymorphism (SRAP):

- SRAP markers target coding sequences in the genome and are used to study genetic diversity, construct genetic maps, and identify markers linked to specific traits in turmeric.

Applications of molecular markers in Turmeric:

- **Genetic diversity and germplasm characterization**: Molecular markers help in assessing the genetic diversity among different turmeric accessions and cultivars. This is important for conservation, breeding programs, and the identification of superior genotypes.
- **Marker-assisted selection (MAS)**: By linking molecular markers to desirable traits (e.g., high curcumin content, disease resistance), breeders can more effectively select and develop improved turmeric varieties.
- **Phylogenetic studies**: Molecular markers are used to study the evolutionary relationships among different species and varieties of Curcuma, helping to understand the domestication and spread of turmeric.
- **Disease resistance**: Molecular markers linked to disease resistance genes can be used to develop turmeric varieties that are less susceptible to diseases such as leaf blotch and rhizome rot, improving crop yield and quality.
- **Tissue culture and clonal fidelity**: Markers are also used to ensure the genetic stability of tissue-cultured turmeric plants, which is essential for maintaining the quality and consistency of turmeric crops produced via in vitro methods.

The use of molecular markers in turmeric is a powerful tool for advancing breeding programs, conserving genetic resources, and enhancing the medicinal and commercial value of this important spice.

7b. Genomics: Genomic insights focus on curcumin biosynthesis and the genetic basis of rhizome development.

Genomics in turmeric (Curcuma longa) is an expanding field of research aimed at understanding the genetic basis of its valuable traits, such as curcumin production, disease resistance, and adaptability to various environmental conditions. Turmeric is a widely used medicinal plant, primarily for its anti-inflammatory and antioxidant properties, which are largely attributed to the bioactive compound curcumin.

Key areas of genomics research in Turmeric

Genome sequencing

- The turmeric genome has been sequenced to identify and characterize the genes responsible for its unique properties. Genome sequencing provides a comprehensive map of the DNA, helping researchers understand the genetic makeup and how it influences the plant's traits.

Gene identification and functional genomics

- Researchers focus on identifying specific genes involved in the biosynthesis of curcumin and other secondary metabolites. Functional genomics studies aim to understand the role of these genes in the production of medicinal compounds and their regulation under different environmental conditions.

Transcriptomics

- Transcriptomics involves studying the RNA transcripts produced by the turmeric genome under various conditions. This helps in understanding how gene expression is regulated in response to environmental stresses, disease, and during different stages of plant growth.

Molecular Breeding and Marker-assisted selection (MAS)

- Genomic research has identified molecular markers associated with desirable traits, such as high curcumin content, disease resistance, and yield. These markers are used in marker-assisted selection to develop improved turmeric varieties through breeding programs.

Genetic diversity and population genomics

- Understanding the genetic diversity within and between turmeric populations is crucial for conservation and breeding. Population genomics studies analyze the genetic variation across different turmeric accessions, which helps in identifying unique genotypes and understanding the evolutionary history of the species.

Metabolomics and pathway analysis

- By integrating genomics with metabolomics, researchers can map out the metabolic pathways involved in curcumin production. This approach helps in identifying key enzymes and regulatory genes that could be targeted for enhancing the production of specific metabolites through genetic modification or selective breeding.

CRISPR and gene editing

- The application of CRISPR-Cas9 technology in turmeric is still in the early stages, but it holds potential for precisely editing genes to enhance desirable traits. For instance, gene editing could be used to increase curcumin content or improve resistance to diseases.

Comparative genomics

- Comparing the turmeric genome with those of other related species, such as ginger or other Curcuma species, provides insights into the evolution of specific traits and the genetic basis of the differences between these species.

Applications and impact

- **Improved cultivars**: Genomic research helps in developing turmeric varieties with enhanced curcumin content, better resistance to diseases, and higher yields, which are crucial for both medicinal use and commercial production.
- **Conservation of genetic resources**: Genomic studies aid in the conservation of turmeric's genetic diversity, ensuring that valuable traits are preserved for future breeding efforts.
- **Understanding disease resistance**: By identifying genes linked to disease resistance, researchers can develop turmeric varieties that are more resilient to common pests and diseases, reducing the need for chemical treatments.
- **Enhanced medicinal value**: Through genomics, it's possible to manipulate the biosynthetic pathways to increase the production of specific bioactive compounds, thereby enhancing the medicinal value of turmeric.

Challenges and future directions

While significant progress has been made in turmeric genomics, challenges remain, such as the complexity of its genome and the limited genomic resources available. However, ongoing research and technological advancements are expected to overcome these hurdles, leading to more refined genetic tools and improved turmeric cultivars. The integration of genomics with other "omics" approaches, such as proteomics and metabolomics, will further enhance our understanding of turmeric's biology and its potential applications.

8. Black Turmeric (*Curcuma caesia*)

8a. Molecular markers: Utilized for studying genetic diversity and conservation strategies.

Molecular markers in black turmeric (Curcuma caesia) are important tools for understanding the genetic diversity, conservation, and breeding of this unique medicinal plant. Black turmeric is known for its dark bluish-black rhizomes and has various therapeutic properties, making it a valuable plant in traditional medicine.

Key types of molecular markers used in Black Turmeric

Simple Sequence Repeats (SSRs) or microsatellites

- SSRs are short, repetitive DNA sequences that are highly polymorphic and widely distributed across the genome. They are commonly used in black turmeric for genetic diversity studies, population structure analysis, and cultivar identification. SSRs are particularly useful for detecting genetic variations among different accessions of black turmeric.

Random Amplified Polymorphic DNA (RAPD)

- RAPD markers involve the amplification of random DNA segments using arbitrary primers. These markers have been used in black turmeric for genetic diversity assessment, although they have limitations in terms of reproducibility and specificity compared to other marker types.

Inter Simple Sequence Repeats (ISSR)

- ISSR markers amplify regions between microsatellites and are highly polymorphic. In black turmeric, ISSRs are used to study genetic diversity, differentiate between various accessions, and assess genetic stability in tissue culture-derived plants.

Amplified Fragment Length Polymorphism (AFLP)

- AFLP is a technique that combines restriction enzyme digestion with selective PCR amplification. This marker system is used in black turmeric for genetic mapping, diversity analysis, and identifying markers linked to specific traits, such as disease resistance or enhanced medicinal properties.

Single Nucleotide Polymorphisms (SNPs)

- SNPs represent single base pair changes in the DNA sequence and are the most abundant type of genetic variation. SNP markers are used in black turmeric for high-resolution genetic mapping, association studies, and the identification of genetic markers linked to important agronomic traits.

Sequence-Related Amplified Polymorphism (SRAP)

- SRAP markers target coding sequences and are used to analyze genetic diversity, construct genetic maps, and identify markers associated with specific phenotypic traits in black turmeric.

Applications of molecular markers in Black Turmeric

- **Genetic diversity and population structure**: Molecular markers help in assessing the genetic diversity within and among populations of black turmeric. This information is crucial for conservation efforts and breeding programs, ensuring the sustainability and genetic health of the species.
- **Marker-assisted selection (MAS)**: Molecular markers linked to desirable traits, such as high medicinal compound content, disease resistance, or drought tolerance, can be used in marker-assisted selection to develop improved black turmeric varieties.
- **Cultivar identification and authentication**: Molecular markers are essential for identifying and authenticating black turmeric cultivars, which is important for maintaining the purity of genetic lines and preventing adulteration in the market.
- **Conservation of genetic resources**: By understanding the genetic diversity of black turmeric through molecular markers, conservation strategies can be developed to preserve this valuable plant's genetic resources, especially in the face of habitat loss and overharvesting.
- **Breeding and improvement**: Molecular markers are used in breeding programs to select parent plants with desirable traits and to track the inheritance of these traits in offspring. This accelerates the development of improved black turmeric varieties with enhanced medicinal properties and better adaptability to different environmental conditions.

Challenges and future directions:

- **Limited genomic resources**: Like many medicinal plants, black turmeric has limited genomic resources compared to major crops. Expanding these resources through sequencing and marker development is essential for advancing research and breeding efforts.
- **Integration with other 'Omics' approaches**: Future research may involve integrating molecular markers with other 'omics' technologies, such as genomics, transcriptomics, and metabolomics, to gain a more comprehensive understanding of black turmeric's biology and its medicinal properties.
- **Conservation through biotechnology**: The use of molecular markers in combination with biotechnological tools, such as tissue culture and cryopreservation, can enhance the conservation of black turmeric's genetic diversity and support the sustainable use of this valuable medicinal plant.

Overall, molecular markers are a powerful tool for advancing the genetic study and improvement of black turmeric, contributing to its conservation and the development of high-quality, medicinally valuable cultivars.

8b. Genomics: Research is aimed at understanding the unique medicinal properties associated with its phytochemical profile.

Genomics in black turmeric (Curcuma caesia) is an emerging area of research focused on understanding the genetic makeup of this medicinally significant plant. Black turmeric is known for its distinctive bluish-black rhizomes and its use in traditional medicine, particularly in Ayurveda, for its anti-inflammatory, antioxidant, and antimicrobial properties. Genomic research in black turmeric aims to unravel the genetic basis of its unique traits, improve its cultivation, and enhance its medicinal value.

Key aspects of genomics research in Black Turmeric

Genome sequencing

- Although complete genome sequencing of black turmeric is still in the early stages, efforts are underway to sequence its genome to provide a comprehensive understanding of its genetic structure. Genome sequencing is crucial for identifying genes involved in the biosynthesis of its bioactive compounds, such as curcumin and camphor, and understanding the genetic factors underlying its medicinal properties.

Gene identification and functional genomics

- Functional genomics involves studying the genes responsible for important traits in black turmeric, such as the production of medicinal compounds, resistance to diseases, and adaptability to different environmental conditions. Identifying and characterizing these genes helps in understanding how they contribute to the plant's medicinal value and how they can be manipulated for crop improvement.

Transcriptomics

- Transcriptomics studies in black turmeric focus on analyzing gene expression patterns in various tissues, such as rhizomes, leaves, and roots, under different environmental conditions or during different growth stages. This information is essential for understanding how genes are regulated in response to stress, disease, or developmental cues, and how these regulations impact the production of bioactive compounds.

Molecular markers and genetic diversity

- Molecular markers, such as SSRs (Simple Sequence Repeats), SNPs (Single Nucleotide Polymorphisms), and ISSRs (Inter Simple Sequence Repeats), are used to assess the genetic diversity of black turmeric populations. Understanding genetic diversity is important for conservation, breeding, and the development of improved varieties with enhanced traits.

Metabolomics and pathway analysis

- By integrating genomics with metabolomics, researchers can map out the metabolic pathways responsible for the synthesis of key medicinal compounds in black turmeric. This helps identify the enzymes and regulatory genes involved in these pathways, which can be targeted for genetic modification or selective breeding to enhance the production of these compounds.

Comparative genomics

- Comparative genomics involves comparing the genome of black turmeric with those of other Curcuma species, such as Curcuma longa (common turmeric) and Curcuma aromatica. This approach helps identify conserved genes and understand the evolutionary relationships between different species, providing insights into the genetic basis of their distinct characteristics.

CRISPR and Gene editing

- While still in the early stages, CRISPR-Cas9 technology has the potential to be applied in black turmeric for precise gene editing. This could be used to enhance specific traits, such as increasing the concentration of medicinal compounds, improving disease resistance, or modifying growth characteristics to suit different agricultural environments.

Population genomics and conservation

- Genomics research is also important for the conservation of black turmeric. By studying the genetic diversity and population structure, conservation strategies can be developed to protect this species from overharvesting and habitat loss, ensuring its sustainability for future generations.

Applications and impact

- **Development of improved varieties**: Genomic insights enable the development of black turmeric varieties with enhanced medicinal properties, higher yield, and better resistance to diseases and environmental stresses. This is particularly important as the demand for black turmeric increases due to its medicinal uses.
- **Medicinal value enhancement**: By understanding the genetic basis of bioactive compound production, researchers can develop strategies to increase the levels of these compounds in black turmeric, thereby enhancing its medicinal efficacy.
- **Conservation of genetic resources**: Genomic studies help conserve the genetic diversity of black turmeric, which is vital for maintaining its

medicinal properties and ensuring its adaptability to changing environmental conditions.

- **Breeding and crop improvement**: Genomic tools, such as marker-assisted selection, are used to accelerate breeding programs aimed at improving black turmeric for commercial cultivation, focusing on traits such as disease resistance, yield, and quality of medicinal compounds.

Challenges and future directions

- **Limited genomic resources**: One of the major challenges in black turmeric genomics is the limited availability of genomic resources compared to more extensively studied crops. However, ongoing research is gradually expanding these resources, providing a foundation for more advanced studies.
- **Complexity of the genome**: The genome of black turmeric, like other Curcuma species, is likely complex and large, which poses challenges for sequencing and assembly. Advances in sequencing technologies and bioinformatics are helping to overcome these challenges.
- **Integration with other 'Omics' approaches**: Future research in black turmeric will likely involve integrating genomics with proteomics, metabolomics, and other 'omics' technologies to provide a more comprehensive understanding of its biology and to fully exploit its medicinal potential.

Genomics in black turmeric is a promising field that has the potential to unlock new opportunities for enhancing the cultivation, conservation, and medicinal use of this valuable plant. As research progresses, it will likely lead to the development of improved black turmeric varieties that meet the growing demand for natural medicinal products.

9. Mango Ginger (*Curcuma amada*)

9a. Molecular markers: Applied in the differentiation and identification of cultivars.

Molecular markers in mango ginger (*Curcuma amada*) are vital tools for studying the genetic diversity, conservation, and breeding of this unique plant. Mango ginger is known for its distinctive aroma, similar to that of raw mango, and is used in various culinary and medicinal applications. Understanding its genetic makeup through molecular markers helps in the identification, characterization, and improvement of mango ginger varieties.

Key types of molecular markers used in Mango Ginger

Simple Sequence Repeats (SSRs) or Microsatellites

- SSRs are short, repeating DNA sequences that are highly polymorphic and distributed throughout the genome. They are widely used in mango ginger to assess genetic diversity, identify different cultivars, and support marker-assisted selection (MAS) in breeding programs.

Random Amplified Polymorphic DNA (RAPD)

- RAPD markers are generated by amplifying random segments of DNA using single, arbitrary primers. They are used in mango ginger for genetic diversity studies and germplasm characterization, though their reproducibility can be lower compared to other markers.

Inter Simple Sequence Repeats (ISSR)

- ISSR markers amplify regions between microsatellites and are highly polymorphic. In mango ginger, ISSRs are employed to evaluate genetic diversity, distinguish between different accessions, and analyse genetic relationships among populations.

Amplified Fragment Length Polymorphism (AFLP)

- AFLP combines restriction enzyme digestion of DNA with selective PCR amplification. This technique is used in mango ginger for constructing genetic maps, assessing genetic diversity, and identifying markers linked to specific traits, such as disease resistance or quality attributes.

Single Nucleotide Polymorphisms (SNPs)

- SNPs are single base pair variations in the DNA sequence. They are the most abundant type of genetic variation and are used in mango ginger for high-resolution genetic mapping, association studies, and the identification of genes linked to important traits.

Sequence-Related Amplified Polymorphism (SRAP)

- SRAP markers target coding sequences in the genome, making them useful for studying gene expression and identifying markers linked to agronomic traits. In mango ginger, SRAPs can be used to explore genetic diversity and develop improved varieties through breeding.

Applications of molecular markers in Mango Ginger

- **Genetic diversity and germplasm conservation**: Molecular markers help in assessing the genetic diversity within and between mango ginger populations. This information is crucial for conserving genetic resources, especially as mango ginger is often cultivated in specific regions and may face threats from environmental changes and habitat loss.

- **Marker-Assisted Selection (MAS)**: By identifying molecular markers linked to desirable traits, such as disease resistance, yield, or aromatic qualities, breeders can use MAS to develop improved mango ginger varieties more efficiently.
- **Cultivar identification and authentication**: Molecular markers are essential for the accurate identification and authentication of mango ginger cultivars. This helps in maintaining the purity of genetic lines and prevents the misidentification or adulteration of mango ginger products in the market.
- **Breeding and improvement**: Molecular markers facilitate the selection of parent plants with desirable traits, accelerating the breeding process. This can lead to the development of mango ginger varieties with enhanced traits such as higher yield, better disease resistance, and improved flavour or medicinal properties.
- **Phylogenetic studies**: Molecular markers are used to study the evolutionary relationships between mango ginger and other Curcuma species. This helps in understanding the genetic basis of their unique characteristics and can guide breeding efforts to introduce desirable traits from related species.

Challenges and future directions:

- **Limited genomic resources**: Like other minor crops, mango ginger has relatively limited genomic resources. Expanding these resources through the development of more molecular markers and sequencing projects is essential for advancing research and breeding programs.
- **Integration with genomics and other 'Omics'**: Future research will likely involve integrating molecular markers with genomic, transcriptomic, and metabolomic data to gain a more comprehensive understanding of mango ginger's biology and its potential applications. This integrative approach could lead to more targeted breeding strategies and the discovery of novel bioactive compounds.
- **Conservation through biotechnology**: Molecular markers, combined with biotechnological approaches such as tissue culture and cryopreservation, can enhance the conservation of mango ginger's genetic diversity. This is particularly important for preserving this plant in the face of environmental challenges and ensuring its sustainable use.

Molecular markers are powerful tools that significantly contribute to the genetic study, conservation, and improvement of mango ginger. Their application in breeding and conservation efforts will help ensure the sustainability and enhanced utilization of this valuable plant.

9b. Genomics: Focus on understanding the genetic basis for the unique flavour and aroma compounds.

Genomics in mango ginger (*Curcuma amada*) focuses on exploring the genetic makeup of this unique plant, known for its distinct mango-like aroma and its uses in both culinary and medicinal contexts. As a member of the Zingiberaceae family, mango ginger is related to turmeric and ginger, but it possesses unique traits that make it of particular interest in genetic studies.

Key aspects of genomics research in Mango Ginger

Genome sequencing

- Genome sequencing of mango ginger aims to decode its entire genetic blueprint. While the complete genome sequence of mango ginger may still be under development, efforts to sequence its genome would provide critical insights into the genes responsible for its unique aroma, bioactive compounds, and adaptability to various environments.

Gene identification and functional genomics

- Identifying and characterizing specific genes in mango ginger is crucial for understanding the molecular basis of its key traits. Functional genomics studies focus on understanding how these genes contribute to the biosynthesis of essential oils, secondary metabolites, and other bioactive compounds that give mango ginger its medicinal and aromatic properties.

Transcriptomics

- Transcriptomics involves analysing the RNA transcripts produced in mango ginger under various conditions, such as different stages of development or in response to environmental stresses. This helps in identifying which genes are actively expressed and how they contribute to the plant's physiological and biochemical processes.

Molecular markers

- Molecular markers like SSRs (Simple Sequence Repeats), SNPs (Single Nucleotide Polymorphisms), and AFLPs (Amplified Fragment Length Polymorphisms) are used to study genetic diversity, map important traits, and assist in the breeding of improved varieties. These markers are critical for identifying genetic variations and understanding the population structure of mango ginger.

Genetic diversity and conservation

- Understanding the genetic diversity of mango ginger is vital for conservation and breeding purposes. Genomic studies help assess the variability within and between populations of mango ginger, which is crucial for maintaining its genetic health and ensuring its long-term sustainability, especially in the face of environmental changes and habitat loss.

Metabolomics and pathway analysis

- Integrating genomics with metabolomics allows researchers to map out the metabolic pathways involved in the production of mango ginger's bioactive compounds, such as curcumin and essential oils. This can lead to the identification of key enzymes and regulatory genes, which can be targeted for enhancing the production of these compounds through breeding or biotechnological approaches.

Comparative genomics

- Comparative genomics involves comparing the genome of mango ginger with other Curcuma species, such as turmeric (Curcuma longa) and black turmeric (Curcuma caesia). This helps identify conserved and unique genes across species, providing insights into their evolutionary relationships and the genetic basis of their distinct traits.

CRISPR and Gene editing

- Although still at an early stage, CRISPR-Cas9 technology holds potential for application in mango ginger. Gene editing could be used to enhance specific traits, such as improving resistance to diseases, increasing yield, or boosting the production of valuable aromatic compounds.

Applications and impact

- **Development of improved varieties**: Genomic research in mango ginger can lead to the development of improved varieties with enhanced traits such as higher yield, better disease resistance, and increased levels of medicinally valuable compounds. This is particularly important as demand for natural and medicinal plants continues to rise.
- **Enhanced medicinal and aromatic properties**: By understanding the genetic basis of the bioactive compounds in mango ginger, researchers can develop strategies to increase the production of these compounds, thereby enhancing the plant's medicinal and aromatic properties.
- **Conservation of genetic resources**: Genomics plays a crucial role in the conservation of mango ginger's genetic diversity. This is vital for ensuring the sustainability of this plant, especially in regions where it is endemic and may be threatened by environmental changes or overharvesting.
- **Breeding and crop improvement**: Marker-assisted selection and other genomic tools can be used to accelerate breeding programs aimed at

improving mango ginger for commercial cultivation. This could include selecting for traits such as pest and disease resistance, environmental adaptability, and superior flavour or medicinal properties.

Challenges and future directions

- **Limited genomic resources**: Compared to major crops, mango ginger has limited genomic resources. Expanding these resources through further genome sequencing, transcriptome analysis, and marker development is essential for advancing research and breeding efforts.
- **Complexity of the genome**: The genome of mango ginger, like other Curcuma species, may be complex and large, which poses challenges for sequencing and analysis. However, advances in sequencing technologies and bioinformatics are gradually overcoming these hurdles.
- **Integration with other 'Omics' approaches**: Future research will likely involve integrating genomics with other 'omics' technologies, such as proteomics and metabolomics, to provide a more comprehensive understanding of mango ginger's biology and its potential applications in medicine and industry.

Genomics in mango ginger is a promising field that offers new opportunities for enhancing the cultivation, conservation, and utilization of this valuable plant. As research progresses, it will contribute to the development of high-quality mango ginger varieties that meet the growing demand for natural medicinal products and aromatic ingredients.

10. Coriander (*Coriandrum sativum*)

10a. Molecular markers: Used in breeding programs for improving yield and disease resistance.

Molecular markers in coriander (*Coriandrum sativum*) are essential tools used to study the genetic diversity, structure, and improvement of this widely cultivated herb. Coriander, also known as cilantro, is valued for its culinary uses and its seeds' essential oil, which has various medicinal properties. Understanding the genetic makeup of coriander through molecular markers aids in breeding programs, conservation efforts, and the enhancement of desirable traits such as flavour, aroma, disease resistance, and yield.

Key types of molecular markers used in Coriander

Simple Sequence Repeats (SSRs) or Microsatellites

- SSRs are short, repetitive DNA sequences scattered throughout the genome. They are highly polymorphic and are widely used in coriander for assessing genetic diversity, cultivar identification, and population structure analysis.

SSRs are particularly useful for marker-assisted selection (MAS) in breeding programs.

Random Amplified Polymorphic DNA (RAPD)

- RAPD markers are generated by amplifying random segments of DNA with arbitrary primers. In coriander, RAPD has been employed for genetic diversity studies and germplasm characterization. While RAPD is relatively easy to perform, it has limitations in reproducibility and reliability compared to other markers.

Inter Simple Sequence Repeats (ISSR)

- ISSR markers amplify regions between microsatellites and are highly polymorphic. These markers are used in coriander to study genetic diversity, phylogenetic relationships, and to identify and distinguish between different accessions or cultivars.

Amplified Fragment Length Polymorphism (AFLP)

- AFLP combines restriction enzyme digestion of DNA with selective PCR amplification. This technique is utilized in coriander for genetic mapping, diversity studies, and identifying markers linked to important agronomic traits, such as disease resistance or essential oil content.

Single Nucleotide Polymorphisms (SNPs)

- SNPs are single base pair changes in the DNA sequence and are the most common type of genetic variation. SNP markers are increasingly used in coriander for high-resolution genetic mapping, association studies, and the identification of genes linked to traits such as flavour, aroma, and stress tolerance.

Sequence-Related Amplified Polymorphism (SRAP)

- SRAP markers target coding sequences in the genome and are useful for studying gene expression and identifying markers associated with specific phenotypic traits. In coriander, SRAP markers can be used to explore genetic diversity and assist in breeding for improved varieties.

Applications of molecular markers in Coriander

- **Genetic diversity and germplasm conservation**: Molecular markers help assess the genetic diversity of coriander, which is essential for the conservation of genetic resources and the sustainable use of coriander germplasm in breeding programs. This is particularly important for preserving landraces and wild relatives of coriander that may possess valuable traits.

- **Marker-assisted selection (MAS)**: By identifying molecular markers linked to desirable traits, such as high essential oil content, disease resistance, or improved flavour profiles, breeders can use MAS to develop improved coriander varieties more efficiently. This accelerates the breeding process and ensures the selection of superior genotypes.
- **Cultivar identification and authentication**: Molecular markers are vital for the accurate identification and authentication of coriander cultivars. This helps maintain the purity of genetic lines, prevent the misidentification of cultivars, and protect against fraudulent labelling in the market.
- **Breeding and crop improvement**: Molecular markers facilitate the selection of parent plants with desirable traits, aiding in the development of new coriander varieties with enhanced characteristics, such as better yield, improved stress tolerance, and superior quality in terms of flavour and aroma.
- **Phylogenetic studies and evolutionary relationships**: Molecular markers are used to study the phylogenetic relationships between different coriander accessions and related species. This helps in understanding the evolutionary history of coriander and can guide breeding efforts to introduce desirable traits from related species.

Challenges and future directions

- **Limited genomic resources**: Compared to other major crops, coriander has relatively limited genomic resources. Expanding these resources through the development of more molecular markers and sequencing projects is essential for advancing research and breeding efforts.
- **Integration with genomics and other 'Omics'**: Future research will likely involve integrating molecular markers with genomic, transcriptomic, and metabolomic data to gain a more comprehensive understanding of coriander's biology and its potential applications. This integrative approach could lead to more targeted breeding strategies and the discovery of novel bioactive compounds.
- **Conservation through biotechnology**: Molecular markers, combined with biotechnological approaches such as tissue culture and cryopreservation, can enhance the conservation of coriander's genetic diversity. This is particularly important for preserving this plant in the face of environmental challenges and ensuring its sustainable use.

Molecular markers are powerful tools that significantly contribute to the genetic study, conservation, and improvement of coriander. Their application in breeding and conservation efforts will help ensure the sustainability and enhanced utilization of this important herb.

10b. Genomics: Studies explore the genes involved in essential oil biosynthesis and flavour compounds.

Coriander (*Coriandrum sativum*), also known as cilantro, is an important culinary and medicinal herb. Genomic studies in coriander are relatively recent but have become increasingly significant due to the plant's economic and cultural importance.

Key Areas of Genomic Research in Coriander

Genome sequencing

- The genome of coriander has been sequenced to understand its genetic makeup. This helps in identifying genes responsible for important traits like flavour, aroma, disease resistance, and growth characteristics.

Genetic diversity

- Studies have been conducted to explore the genetic diversity within coriander populations. This is crucial for breeding programs aimed at improving traits such as yield, flavour, and resistance to pests and diseases.

Trait mapping

- Genomic tools are used to map traits like oil content, which is a major component of coriander's flavour profile. Understanding the genetic basis of these traits allows for targeted breeding strategies to enhance desired characteristics.

Molecular markers

- Molecular markers have been developed to assist in the selection of specific traits in coriander breeding programs. These markers help in identifying plants that carry desirable genes, making breeding more efficient.

Functional genomics

- Functional genomics studies are focused on understanding the role of specific genes in the development and functioning of coriander plants. This includes research on genes involved in flavour biosynthesis, such as those responsible for the production of linalool, a compound contributing to coriander's characteristic aroma.

Applications of Genomic Research in Coriander

- **Breeding programs**
  - Enhanced breeding programs aimed at improving yield, flavour, and disease resistance through the use of genomic data.

- **Biotechnology**
  - Genetic engineering approaches to enhance specific traits, such as increasing the concentration of beneficial compounds in coriander leaves and seeds.
- **Conservation**
  - Understanding genetic diversity helps in the conservation of coriander germplasm, ensuring that a wide variety of genetic material is available for future breeding and research efforts.

11. Fennel (*Foeniculum vulgare*)

11a. Molecular markers: Applied to understand genetic diversity and improve breeding strategies.

Molecular markers are tools used in plant genetics to identify specific sequences in the DNA that are associated with particular traits. In fennel, which is an important aromatic and medicinal plant, molecular markers are used to study genetic diversity, identify desirable traits, and assist in breeding programs.

Common Types of Molecular Markers Used in Fennel

Simple Sequence Repeats (SSRs)

- Also known as microsatellites, SSRs are short, repeating sequences of DNA that are highly variable among individuals. SSR markers are often used to assess genetic diversity within fennel populations.

Random Amplified Polymorphic DNA (RAPD)

- RAPD markers are used for genetic fingerprinting in fennel. They help in identifying genetic variations without prior knowledge of the DNA sequence, making them useful for studying genetic relationships among different fennel varieties.

Amplified Fragment Length Polymorphism (AFLP)

- AFLP markers are used to generate a large number of markers across the genome, providing detailed information about genetic diversity and relationships. This method is particularly useful in mapping genes linked to important traits in fennel.

Single Nucleotide Polymorphisms (SNPs)

- SNPs are single base-pair changes in the DNA sequence. They are highly abundant across the genome and are used in high-resolution genetic mapping and association studies in fennel.

Applications of Molecular Markers in Fennel

- **Breeding programs**
  - Molecular markers are used to select parent plants with desirable traits, such as disease resistance or enhanced flavour profiles, to produce improved fennel varieties.
- **Genetic diversity studies**
  - By using molecular markers, researchers can assess the genetic diversity within and between fennel populations, which is crucial for conservation and breeding purposes.
- **Trait mapping**
  - Molecular markers help in identifying the genetic loci associated with important traits, such as oil content or resistance to pests, allowing for more targeted breeding efforts.
- **Marker-assisted selection (MAS)**
  - In fennel breeding, MAS is used to track the inheritance of desirable traits through generations, speeding up the breeding process and increasing the efficiency of developing new varieties.

11b. Genomics: Focus on the genomics of essential oil production and stress tolerance.

Fennel is a medicinal and culinary plant with significant economic value. Genomic studies in fennel are essential for understanding its genetic makeup, which can be leveraged to enhance desirable traits such as flavour, medicinal properties, and resistance to pests.

Key areas of genomic research in Fennel

Genome sequencing

- The complete genome sequencing of fennel is an ongoing area of research. Sequencing provides a comprehensive understanding of the genetic blueprint of fennel, which is crucial for identifying genes responsible for key traits.

Transcriptomics

- Transcriptomic studies involve analyzing the expression of genes in fennel. This helps in understanding how different environmental conditions or developmental stages affect gene expression, which in turn influences the plant's growth, flavour, and resistance to stress.

Molecular markers

- Molecular markers such as SSRs (Simple Sequence Repeats) and SNPs (Single Nucleotide Polymorphisms) are developed and used for mapping important traits in fennel. These markers assist in the breeding process by allowing the selection of plants with desirable genetic traits.

Functional genomics

- Functional genomics focuses on the roles of specific genes in fennel, particularly those involved in essential oil biosynthesis, which contributes to the plant's aroma and flavour.

Breeding and genetic improvement

- Genomic data is increasingly used in breeding programs to create fennel varieties with improved traits, such as higher yields, better flavour profiles, and enhanced resistance to diseases.

Applications of genomic research in Fennel

- **Trait improvement:**
  - Genomics aids in the identification and enhancement of traits like oil content, pest resistance, and overall plant vigour.
- **Conservation**
  - Genomic studies help in preserving the genetic diversity of fennel, which is vital for maintaining its adaptability and resilience to environmental changes.
- **Biotechnology**
  - Genetic engineering, guided by genomic insights, allows for the targeted modification of fennel plants to improve specific traits without altering the plant's overall genetic integrity.

12. Fenugreek (*Trigonella foenum-graecum*)

12a. Molecular markers: Used in the study of genetic diversity and to identify beneficial traits.

Fenugreek is an important medicinal and culinary plant known for its seeds and leaves. Molecular markers have become crucial tools in the study of fenugreek genetics, aiding in breeding programs and the conservation of genetic diversity.

Types of molecular markers used in Fenugreek

Simple Sequence Repeats (SSRs)

- SSRs, also known as microsatellites, are DNA sequences that repeat and are highly polymorphic. They are used to assess genetic diversity, which is important for breeding and conservation.

Random Amplified Polymorphic DNA (RAPD)

- RAPD markers are used for genetic fingerprinting in fenugreek. This method helps in identifying genetic variability among different fenugreek cultivars or populations.

Amplified Fragment Length Polymorphism (AFLP)

- AFLP is a technique used to generate a large number of genetic markers, which helps in the study of genetic relationships and diversity within fenugreek populations.

Single Nucleotide Polymorphisms (SNPs)

- SNPs are variations at a single nucleotide position in the DNA sequence. They are highly useful for fine-scale genetic mapping and association studies in fenugreek.

Applications of molecular markers in Fenugreek

- **Breeding programs**
  - Molecular markers are utilized to select fenugreek plants with desirable traits, such as higher yield, better nutritional content, or resistance to diseases. This accelerates the breeding process and improves the efficiency of developing new varieties.
- **Genetic diversity studies**
  - Markers like SSRs and SNPs help in analyzing the genetic diversity within fenugreek populations, which is crucial for maintaining a healthy gene pool and preventing the loss of valuable genetic traits.
- **Marker-assisted selection (MAS)**
  - In fenugreek breeding, MAS involves tracking specific markers associated with desired traits, allowing for the selection of the best candidates for further breeding.

- **Trait mapping**
    - Molecular markers assist in mapping the genes responsible for important traits such as flavour, seed size, and resistance to environmental stresses. This knowledge aids in the targeted breeding of fenugreek with enhanced characteristics.

These markers play a pivotal role in improving fenugreek's agricultural value by enabling more precise and efficient breeding strategies.

12b. Genomics: Research involves identifying genes related to diosgenin production and stress responses.

Fenugreek (*Trigonella foenum-graecum* L.) is an important medicinal plant known for its numerous health benefits, including lowering blood sugar levels, boosting testosterone, and enhancing milk production in breastfeeding women. Genomics in fenugreek involves the study of its genetic makeup to understand its traits, improve its cultivation, and enhance its medicinal properties.

Key areas of genomic research in Fenugreek

Genome sequencing

- **Genome sequencing** of fenugreek involves determining the complete DNA sequence of its genome. This helps in identifying genes responsible for important traits like seed size, oil content, and medicinal properties. Sequencing projects aim to build a comprehensive genetic map of fenugreek.

Genetic diversity studies

- **Genetic diversity studies** use genomics to understand the variation in fenugreek populations. This helps in identifying varieties with desirable traits, which can be used in breeding programs to develop superior fenugreek cultivars.

Marker-assisted selection (MAS)

- **MAS** uses genetic markers linked to important traits to assist in breeding programs. By identifying these markers through genomics, breeders can select plants with desirable traits early in the breeding process, speeding up the development of new varieties.

Functional genomics

- **Functional genomics** studies the function of genes and their roles in fenugreek's growth, development, and response to environmental stresses. Understanding gene function can lead to the development of fenugreek varieties that are more resilient to climate change and diseases.

Metabolomics and transcriptomics

- These approaches study the metabolic and gene expression profiles of fenugreek. They help in understanding how fenugreek produces its bioactive compounds, which are responsible for its medicinal properties. This knowledge can be used to enhance the production of these compounds.

Applications

- **Medicinal enhancement:** Genomics can be used to enhance the production of specific bioactive compounds in fenugreek, making it more effective as a medicinal plant.
- **Agricultural improvement:** Genomics can help develop fenugreek varieties that are more resistant to diseases, pests, and environmental stresses, improving crop yields and sustainability.

13. Cumin (Cuminum cyminum)

13a. Molecular markers: Employed for genetic mapping and breeding for quality traits.

Molecular markers in cumin (Cuminum cyminum L.), an important spice crop, are tools used to study and manipulate the plant's genetic material. These markers play a critical role in plant breeding, genetic diversity studies, and conservation efforts. Here's an overview of molecular markers and their applications in cumin:

Types of molecular markers used in Cumin

RAPD (Random Amplified Polymorphic DNA)

- **RAPD markers** are used to assess genetic diversity and relationships among cumin varieties. They are simple, quick, and cost-effective, though they may have lower reproducibility compared to other markers.

SSR (Simple Sequence Repeats) or microsatellites

- **SSR markers** are highly polymorphic and co-dominant, making them ideal for studying genetic diversity, mapping genomes, and identifying specific traits. They provide precise and reliable results in cumin breeding programs.

AFLP (Amplified Fragment Length Polymorphism)

- **AFLP markers** are used for genome mapping and studying genetic diversity. They are highly sensitive and can detect numerous polymorphisms across the genome.

ISSR (Inter Simple Sequence Repeats)

- **ISSR markers** combine the benefits of SSR and RAPD markers, offering a cost-effective and reliable way to study genetic diversity, population structure, and phylogenetics in cumin.

SNP (Single Nucleotide Polymorphisms)

- **SNP markers** are increasingly used in cumin genomics due to their abundance and stability. They are useful for high-resolution mapping of traits and for genomic selection in breeding programs.

Applications of molecular markers in Cumin

Genetic diversity studies

- Molecular markers help assess the genetic diversity within and between cumin populations. This information is crucial for conserving genetic resources and selecting parent lines for breeding.

Breeding and selection

- Molecular markers assist in marker-assisted selection (MAS), allowing breeders to select cumin plants with desirable traits such as disease resistance, drought tolerance, or high yield early in the breeding process.

Genome mapping

- By using molecular markers, researchers can create linkage maps of the cumin genome. These maps are essential for identifying genes associated with important agronomic traits.

Conservation genetics

- Markers are used to identify and preserve genetic diversity in wild and cultivated cumin populations. This helps in the conservation of rare and endangered varieties.

Hybrid identification

- Molecular markers are useful in identifying hybrids and ensuring the purity of cumin varieties. This is important for maintaining the quality and consistency of cumin as a commercial crop.

Molecular markers are powerful tools in cumin research, aiding in the understanding of its genetic makeup, improving breeding strategies, and conserving its genetic diversity. The use of these markers accelerates the development of superior cumin varieties with improved yield, quality, and stress tolerance, which are essential for meeting the demands of both farmers and consumers.

13b. Genomics: Genomic research focuses on essential oil biosynthesis and drought tolerance.

Genomics in cumin (*Cuminum cyminum* L.) involves the study of its entire genetic material to understand its biology, improve breeding practices, and enhance its agronomic traits. Given the economic and medicinal importance of cumin, genomics research has significant implications for improving yield, disease resistance, and other desirable traits. Here's an overview of the key aspects of genomics in cumin:

Overview of Cumin genomics

Genome sequencing

- The genome of cumin has been sequenced to provide a comprehensive map of its genetic material. This sequencing is crucial for identifying genes responsible for key traits, such as oil content, flavour compounds, and resistance to environmental stresses.
- The availability of a reference genome helps in understanding the molecular basis of various traits and supports advanced breeding techniques.

Transcriptomics

- Transcriptomic studies involve analyzing the expression of genes in cumin under different conditions. This approach helps identify which genes are active during specific stages of growth or in response to environmental stresses, such as drought or pathogens.
- Transcriptomics is particularly useful in understanding the biosynthesis of secondary metabolites, which are responsible for cumin's medicinal properties and distinctive flavour.

Marker-assisted selection (MAS)

- Genomic information is used to identify molecular markers linked to important traits in cumin. These markers can be used in MAS, a breeding technique where plants with desirable genetic markers are selected early in the breeding process.
- MAS accelerates the development of new cumin varieties with improved traits, such as higher yield, disease resistance, and better quality.

Genetic diversity and population genomics

- Studies on the genetic diversity of cumin populations using genomic tools help in understanding the extent of variation within and between populations. This information is critical for the conservation of genetic resources and for designing breeding programs that maintain or enhance genetic diversity.

- Population genomics also aids in identifying unique or rare alleles that could be valuable for breeding programs.

QTL mapping

- Quantitative Trait Loci (QTL) mapping is used to locate the specific regions of the cumin genome that are associated with particular traits, such as seed size, oil content, or drought tolerance.
- QTL mapping provides a more detailed understanding of the genetic architecture of complex traits and can guide the development of improved cumin varieties.

CRISPR and genetic engineering

- The potential application of CRISPR-Cas9 and other genome-editing tools in cumin genomics could allow precise modifications to be made to its DNA. This could lead to the development of cumin varieties with enhanced traits or new properties that are not achievable through traditional breeding methods.
- Genetic engineering, guided by genomic insights, could be used to introduce resistance to specific pests or diseases or to enhance the nutritional content of cumin.

Applications of genomics in Cumin

Improvement of agronomic traits

- Genomic studies can lead to the development of cumin varieties that are more resilient to environmental stresses, have higher yields, or produce seeds with better flavour and aroma profiles.

Disease resistance

- Understanding the genetic basis of disease resistance can help breeders develop cumin varieties that are less susceptible to common pathogens, reducing the need for chemical pesticides.

Nutritional and medicinal enhancement

- Genomics can be used to increase the concentration of bioactive compounds in cumin, enhancing its medicinal properties. This could lead to cumin varieties that are more effective in traditional and modern medicine.

Sustainability and conservation

- Genomic insights help in conserving the genetic diversity of cumin, ensuring that valuable traits are not lost in the face of environmental changes or selective breeding practices.

Genomics in cumin is transforming how this important spice is studied and improved. By providing detailed insights into the genetic makeup of cumin, genomics enables the development of better varieties that meet the needs of farmers, consumers, and the spice industry. The integration of genomic tools in cumin breeding programs promises to enhance the crop's productivity, quality, and resilience, contributing to its sustainable cultivation and use.

14. Ajwain (*Trachyspermum ammi*)

14a. Molecular markers: Useful in studying genetic variability and improving disease resistance.

Molecular markers are powerful tools used in the study and improvement of Ajwain *(Trachyspermum ammi)*, a medicinal and aromatic plant widely used in traditional medicine and as a spice. These markers help in understanding the genetic diversity, identifying specific traits, and assisting in breeding programs. Here's an overview of molecular markers and their applications in Ajwain:

Types of molecular markers used in Ajwain

RAPD (Random Amplified Polymorphic DNA)

- **RAPD markers** are commonly used in Ajwain to assess genetic diversity. They are quick and easy to use, though they may have lower reproducibility compared to more advanced markers.

ISSR (Inter Simple Sequence Repeats)

- **ISSR markers** are used to study genetic diversity and relationships among Ajwain accessions. They offer a balance between the simplicity of RAPD and the specificity of SSR markers.

SSR (Simple Sequence Repeats) or microsatellites

- **SSR markers** are co-dominant and highly polymorphic, making them useful for detailed genetic studies, including diversity analysis, population structure studies, and marker-assisted selection in Ajwain.

AFLP (Amplified Fragment Length Polymorphism)

- **AFLP markers** provide high-resolution data and are used for assessing genetic variation, creating genetic maps, and studying population genetics in Ajwain.

SNP (Single Nucleotide Polymorphisms)

- **SNP markers** are increasingly being used due to their abundance and stability. They provide high-resolution mapping and are useful in genomic selection and association studies in Ajwain.

Applications of molecular markers in Ajwain

Genetic diversity studies

- Molecular markers help assess the genetic diversity of Ajwain populations. This is crucial for the conservation of genetic resources and for identifying diverse genotypes that can be used in breeding programs.

Breeding and selection

- Molecular markers are used in marker-assisted selection (MAS) to identify and select Ajwain plants with desirable traits, such as high essential oil content, disease resistance, or better growth characteristics.

Population structure analysis

- Markers like SSR and AFLP are used to study the population structure of Ajwain, helping researchers understand how different populations are related and how genetic variation is distributed across different regions.

Genetic mapping

- Molecular markers are used to create genetic maps of Ajwain, which are essential for identifying quantitative trait loci (QTL) associated with important agronomic traits. These maps can guide breeding efforts to improve Ajwain varieties.

Hybrid identification

- Molecular markers are also useful for confirming the identity of hybrids in breeding programs, ensuring that the selected plants carry the desired genetic traits.

The use of molecular markers in Ajwain research is essential for advancing our understanding of its genetics and for improving the crop through targeted breeding strategies. By leveraging these markers, scientists and breeders can develop superior Ajwain varieties with enhanced traits, such as increased essential oil yield, better disease resistance, and improved adaptability to different environmental conditions. This not only boosts Ajwain production but also supports its continued use in traditional medicine and culinary applications.

14b. Genomics: Studies explore the genetic basis of thymol production and other aromatic compounds.

Genomics in **ajwain** (*Trachyspermum ammi*), a traditional medicinal plant, involves studying the plant's genetic makeup to understand its properties and potential uses. Ajwain is known for its strong flavour and various health benefits, including antimicrobial and digestive properties.

The focus of genomic studies on ajwain may include

- **Genetic diversity**: Understanding the variations within different ajwain species to improve cultivation practices.
- **Metabolite production**: Identifying genes responsible for the production of bioactive compounds like thymol and carvacrol, which contribute to its medicinal properties.
- **Breeding programs**: Using genomic data to develop strains with enhanced traits, such as higher yield or disease resistance.
- **Conservation**: Protecting genetic diversity in wild populations of ajwain.

This knowledge can help in improving the quality and efficacy of ajwain in traditional medicine and potentially lead to the discovery of new therapeutic compounds.

C. Genetic engineering and CRISPR in spices breeding

Genetic engineering and CRISPR technologies are emerging tools in the breeding and improvement of spice crops. Here's an overview of how they can be applied to the spices you've mentioned:

1. Black Pepper (*Piper nigrum*)

1a. Genetic Engineering in Black Pepper breeding

Efforts include modifying genes related to disease resistance (e.g., resistance to *Phytophthora*) and enhancing flavour compounds. Genetic engineering in black pepper (*Piper nigrum*) breeding is a promising approach to overcome several challenges faced by traditional breeding methods. Here's how genetic engineering is being or could be utilized in black pepper breeding:

i) Disease resistance

- ***Phytophthora* resistance**: Black pepper is highly susceptible to *Phytophthora capsici*, causing root rot and other diseases. Genetic engineering can introduce resistance genes, such as those encoding for pathogenesis-related proteins or other defence-related genes, directly into black pepper plants. This can significantly reduce crop losses due to these diseases.
- **Viral resistance**: Genetic engineering can be used to introduce genes that confer resistance to viral diseases, such as the Piper yellow mottle virus (PYMoV), which affects black pepper productivity.

ii) Pest resistance

- **Nematode resistance**: Nematodes are another major problem in black pepper cultivation. By introducing genes that produce nematode-resistant proteins, black pepper plants can be engineered to withstand nematode attacks, reducing the need for chemical nematicides.

iii) Abiotic stress tolerance

- **Drought and salinity resistance**: Genetic engineering can introduce genes that help black pepper plants tolerate drought and high salinity conditions. For example, genes involved in osmotic adjustment, like those encoding for trehalose or proline synthesis, could be introduced to help the plant cope with water scarcity and saline soils.

iv) Improving yield and quality

- **Enhanced piperine content**: Piperine is the alkaloid responsible for the pungency of black pepper. Genetic engineering can be used to modify the biosynthetic pathway of piperine, potentially increasing its concentration in the peppercorns, thereby enhancing the quality of the spice.
- **Flowering and fruit set**: Genes controlling flowering time and fruit set can be manipulated to increase the yield of black pepper plants. This could involve the introduction of genes that regulate flowering hormones like gibberellins.

v) Nutritional enhancement

- **Biofortification**: Genetic engineering could be used to enhance the nutritional value of black pepper by introducing genes that increase the content of beneficial compounds like vitamins, minerals, or other phytonutrients.

vi) Clonal propagation and uniformity

- **Somatic embryogenesis and cloning**: Genetic engineering techniques can be used to enhance the efficiency of somatic embryogenesis, which is important for clonal propagation of elite black pepper varieties. This ensures uniformity in plant characteristics and maintains the quality of propagated plants.

vii) Biotechnological tools integration

- **CRISPR/Cas9**: The use of CRISPR/Cas9 for targeted gene editing in black pepper is a potential area of research. This technology allows for precise modifications of the black pepper genome, enabling the introduction or modification of specific traits without affecting other characteristics of the plant.

Research and development challenges

While the potential is immense, the application of genetic engineering in black pepper breeding faces challenges such as:

- **Regulatory approvals**: Genetically modified crops must pass through strict regulatory processes before they can be commercialized.
- **Public acceptance**: Consumer acceptance of genetically engineered foods can be a hurdle in certain markets.
- **Technical complexity**: Developing effective transformation protocols and identifying key genes for desired traits are technically demanding tasks.

Overall, genetic engineering offers significant opportunities to improve black pepper breeding by enhancing disease resistance, stress tolerance, and yield, while also enabling the development of plants with superior quality and nutritional profiles.

1b. CRISPR in Black Pepper breeding

Potential applications include precise edits to increase piperine content or to confer resistance to pests and diseases. CRISPR (Clustered Regularly Interspaced Short Palindromic Repeats) technology is a powerful tool for precise genetic modifications and has significant potential in black pepper (Piper nigrum) breeding. Here's how CRISPR can be applied in black pepper breeding:

i) Disease resistance

- **Targeting resistance genes**: CRISPR can be used to edit genes associated with susceptibility to diseases like *Phytophthora capsici* (which causes root rot) and viral pathogens such as the Piper yellow mottle virus (PYMoV). By knocking out or modifying these susceptibility genes, CRISPR can help develop black pepper varieties with enhanced disease resistance.
- **Gene Knock-In**: CRISPR can introduce genes from resistant species directly into black pepper. For example, introducing R genes (resistance genes) from wild relatives that naturally resist *Phytophthora* could be a promising strategy.

ii) Enhancing piperine content

- **Metabolic pathway engineering**: CRISPR can be used to upregulate or modify key enzymes in the piperine biosynthetic pathway. By editing regulatory genes that control the production of piperine, the main alkaloid responsible for black pepper's pungency, the piperine content in the peppercorns can be increased, leading to a higher-quality spice.

iii) Abiotic stress tolerance

- **Drought and salinity tolerance**: CRISPR can target and modify genes involved in the plant's response to abiotic stresses, such as drought and salinity. For instance, genes that regulate water retention or salt exclusion mechanisms can be edited to produce black pepper plants that can thrive in challenging environmental conditions.
- **CRISPR-enhanced root systems**: Modifying genes that influence root architecture could help black pepper plants develop more robust root systems, improving their ability to access water and nutrients in poor soil conditions.

iv) Improving yield and growth characteristics

- **Flowering time regulation**: CRISPR can be used to edit genes that control flowering time and fruit set, which can directly impact yield. By either delaying or advancing flowering, plants can be better synchronized with favorable environmental conditions, leading to improved yield.
- **Growth habit**: CRISPR could be used to modify genes that influence plant architecture, such as those controlling vine length and branching. This could help develop black pepper plants that are easier to cultivate and harvest.

v) Pest resistance

- **Targeting pest susceptibility genes**: CRISPR can knock out or alter genes that make black pepper plants susceptible to common pests like nematodes. This could significantly reduce crop losses and decrease the need for chemical pesticides.
- **Enhancing natural defenses**: By editing genes that regulate the production of secondary metabolites involved in defense, CRISPR could make black pepper more resistant to pest attacks.

vi) Nutritional and medicinal value enhancement

- **Biofortification**: CRISPR can be used to edit genes responsible for the synthesis of beneficial compounds in black pepper. This could involve enhancing the levels of specific antioxidants, vitamins, or other phytonutrients that contribute to the health benefits of black pepper.

vii) Development of CRISPR Knock-In/Out libraries

- **Functional genomics**: Creating a CRISPR knock-in/out library for black pepper could help researchers identify and study the functions of various genes, leading to the discovery of new targets for breeding programs. This library would allow systematic analysis of gene function related to growth, stress responses, and metabolite production.

viii) Precision breeding

- **Trait stacking**: CRISPR allows for the precise stacking of multiple desirable traits within a single cultivar. For example, disease resistance, improved yield, and enhanced piperine content can be introduced simultaneously without the unwanted effects associated with traditional cross-breeding methods.

Challenges and Considerations

- **Efficiency of CRISPR delivery**: One of the key challenges in using CRISPR in black pepper is the efficient delivery of the CRISPR/Cas9 system to the plant cells. Developing reliable transformation protocols is essential.
- **Off-Target effects**: Ensuring the specificity of CRISPR edits to avoid off-target mutations is critical for developing stable and safe black pepper varieties.
- **Regulatory approval**: Like other genetically modified organisms (GMOs), CRISPR-edited plants must undergo regulatory scrutiny before commercialization, which can vary significantly by region.

CRISPR offers a precise, efficient, and versatile tool for black pepper breeding, enabling the development of new varieties with enhanced disease resistance, improved stress tolerance, and better quality traits. As research progresses, CRISPR could become a cornerstone technology in the breeding programs for black pepper, addressing both current challenges and future needs in cultivation.

2. Cardamom (*Elettaria cardamomum*)

2a. Genetic engineering in Cardamom breeding

Focuses on enhancing essential oil content and developing resistance to fungal diseases. Genetic engineering in cardamom (*Elettaria cardamomum*) breeding holds significant potential for improving crop characteristics such as yield, disease resistance, and quality of the spice. Here's an overview of how genetic engineering can be utilized in cardamom breeding:

i) Disease resistance

- **Fungal disease resistance**: Cardamom is vulnerable to several fungal diseases, including *Colletotrichum* blight and *Phytophthora* rot. Genetic engineering can be used to introduce resistance genes from other species that produce antifungal proteins or pathogenesis-related (PR) proteins. For example, overexpressing genes that encode chitinase or glucanase enzymes could enhance the plant's ability to degrade fungal cell walls, providing increased resistance.

- **Viral disease resistance**: Cardamom mosaic virus (CdMV) is a major threat to cardamom production. Genetic engineering can introduce RNA interference (RNAi) constructs that target the viral genome, effectively silencing the virus and preventing its replication within the plant.

ii) Pest resistance

- **Insect resistance**: Genetic engineering can introduce genes encoding proteins like Bt toxin (from *Bacillus thuringiensis*) that are toxic to specific insect pests but safe for humans and non-target organisms. This would reduce the need for chemical pesticides and lower production costs.
- **Nematode resistance**: Cardamom roots are susceptible to nematode infestations. By inserting genes that produce nematode-resistant proteins, such as those encoding cysteine protease inhibitors, cardamom plants could be engineered to resist nematode attacks.

iii) Abiotic stress tolerance

- **Drought and salinity tolerance**: Cardamom is typically grown in regions with abundant rainfall, but climate change is making water availability less predictable. Genetic engineering can be used to introduce genes that help the plant conserve water or tolerate high soil salinity, such as those encoding for osmoprotectants like trehalose or glycine betaine.
- **Temperature tolerance**: With changing climate patterns, temperature extremes are becoming more common. Genetic modifications could help cardamom plants tolerate higher temperatures or prevent heat-induced damage by enhancing the expression of heat shock proteins (HSPs).

iv) Yield improvement

- **Enhanced flowering and fruit set**: By modifying genes that regulate flowering time and fruit set, genetic engineering can help increase the yield of cardamom plants. For instance, altering the expression of genes involved in gibberellin biosynthesis could lead to more prolific flowering and improved fruit set under various environmental conditions.
- **Growth rate enhancement**: Genes that control growth rates, such as those involved in hormone regulation (e.g., auxins and cytokinins), could be engineered to accelerate the growth of cardamom plants, leading to quicker harvesting cycles.

v) Quality improvement

- **Essential oil content**: The quality of cardamom is largely determined by its essential oil content, particularly compounds like cineole and terpenes. Genetic engineering can be used to enhance the biosynthetic pathways responsible for these compounds, either by upregulating the expression of key enzymes or by introducing new metabolic pathways from other species.

- **Flavour profile modification**: Through genetic engineering, specific genes involved in the production of flavour compounds can be targeted to enhance or modify the flavour profile of cardamom, making it more desirable for specific markets or culinary applications.

vi) Nutritional and medicinal enhancement

- **Biofortification**: Cardamom can be engineered to enhance its nutritional value by increasing the content of vitamins, antioxidants, or other beneficial compounds. This could involve the introduction of genes that boost the biosynthesis of compounds like flavonoids, which have known health benefits.

vii) Clonal propagation and uniformity

- **Somatic embryogenesis**: Genetic engineering can be used to improve the efficiency of somatic embryogenesis, a key technique for the clonal propagation of elite cardamom varieties. This ensures uniformity in the propagated plants and maintains the desired traits in large-scale production.

Challenges and considerations

- **Genetic transformation efficiency**: Developing effective protocols for the stable transformation of cardamom plants is a major challenge, given the species' recalcitrance to genetic modification.
- **Regulatory hurdles**: Like all genetically modified crops, genetically engineered cardamom varieties would need to pass through stringent regulatory processes, which can vary by region and may involve significant time and cost.
- **Public perception**: The acceptance of genetically modified cardamom by consumers and farmers is crucial for the successful commercialization of such varieties.

Genetic engineering offers a suite of tools that can be harnessed to address the challenges faced in cardamom cultivation, such as disease pressure, environmental stresses, and the need for higher yields and better quality. While still in the early stages of application in cardamom, the potential benefits make it a promising avenue for future research and development in cardamom breeding programs.

2b. CRISPR in Cardamom breeding

CRISPR could be used to target genes involved in oil biosynthesis pathways or to enhance stress tolerance. CRISPR (Clustered Regularly Interspaced Short Palindromic Repeats) is a cutting-edge tool that allows for precise genome editing and has significant potential in cardamom (*Elettaria cardamomum*) breeding. Here's how CRISPR technology could be applied to improve cardamom breeding:

i) Enhancing disease resistance

- **Targeting susceptibility genes**: CRISPR can be used to knock out or modify genes that make cardamom susceptible to diseases such as *Cardamom mosaic virus* (CdMV), *Phytophthora* rot, and fungal infections like *Colletotrichum* blight. For instance, by disrupting susceptibility (S) genes, CRISPR can enhance the plant's natural defences against these pathogens.
- **Gene insertion for resistance**: CRISPR can facilitate the insertion of genes from other species or even wild relatives that confer resistance to specific pathogens. For example, introducing genes encoding pathogenesis-related (PR) proteins can improve resistance to various fungal and bacterial pathogens.

ii) Improving pest resistance

- **Editing genes for insect resistance**: CRISPR can be used to introduce or modify genes that produce insecticidal proteins, such as those derived from *Bacillus thuringiensis* (Bt). This can make cardamom plants resistant to common pests like borers and caterpillars.
- **Nematode resistance**: By editing genes involved in the plant's response to nematode infestation, CRISPR can help develop nematode-resistant cardamom varieties. This might involve enhancing the expression of natural nematicidal compounds within the plant.

iii) Enhancing abiotic stress tolerance

- **Drought and salinity tolerance**: CRISPR can target genes that regulate water use efficiency, root architecture, and salt tolerance. By editing these genes, cardamom plants could be engineered to better withstand drought and saline soil conditions, making them more resilient to climate change.
- **Heat tolerance**: CRISPR can be employed to enhance the expression of heat shock proteins (HSPs) or other genes that help the plant cope with high-temperature stress, which is increasingly important as global temperatures rise.

iv) Increasing yield and growth rate

- **Manipulating flowering genes**: CRISPR can be used to edit genes that control flowering time, enabling synchronization with optimal environmental conditions. This could lead to better fruit set and increased yields.

- **Growth regulation**: Editing genes involved in hormone regulation, such as those controlling gibberellins, can lead to faster growth and shorter times to harvest, improving overall productivity.

v) Enhancing essential oil content

- **Modifying biosynthetic pathways**: The quality of cardamom is largely dependent on its essential oil content, particularly compounds like cineole and terpenes. CRISPR can be used to upregulate or modify genes in the biosynthetic pathways of these essential oils, increasing their concentration and improving the spice's market value.
- **Flavour and aroma enhancement**: CRISPR can precisely edit genes responsible for flavour and aroma compounds, allowing for the development of cardamom varieties with enhanced or modified flavour profiles tailored to specific consumer preferences.

vi) Improving nutritional and medicinal properties

- **Biofortification**: CRISPR can be used to enhance the nutritional value of cardamom by increasing the levels of beneficial compounds such as antioxidants, vitamins, or other phytonutrients. For example, editing genes involved in the synthesis of flavonoids or other health-promoting compounds could lead to cardamom varieties with enhanced medicinal properties.

vii) Developing CRISPR Knock-In/Knock-Out libraries

- **Functional genomics**: Establishing a CRISPR-based knock-in/knock-out library in cardamom would allow researchers to systematically study gene functions related to key traits such as disease resistance, stress tolerance, and essential oil production. This could lead to the identification of novel targets for future breeding programs.

viii) Precision breeding for multiple traits

- **Trait stacking**: CRISPR enables the simultaneous modification of multiple genes, allowing for the stacking of desirable traits in a single cardamom variety. For instance, a CRISPR-engineered cardamom plant could be designed to have enhanced resistance to diseases, improved yield, and superior essential oil content, all within one generation of breeding.

Challenges and Considerations

- **Delivery of CRISPR components**: Effective delivery of CRISPR/Cas9 components into cardamom cells is a challenge that requires efficient transformation methods, which can be a limiting factor.
- **Off-target effects**: Ensuring that CRISPR edits are specific and do not cause unintended mutations in other parts of the genome is critical for the stability and safety of the modified plants.

- **Regulatory and public acceptance**: Like other genetically modified organisms, CRISPR-edited cardamom would need to navigate regulatory approvals and public perception, which can vary widely depending on the region.

CRISPR technology offers a powerful tool for advancing cardamom breeding by enabling precise genetic modifications that can enhance disease resistance, improve stress tolerance, increase yield, and refine the quality of the spice. As research progresses, CRISPR is likely to become an integral part of modern cardamom breeding programs, helping to address both current challenges and future demands in the spice industry.

3. Allspice (Pimenta dioica)

3a. Genetic engineering in Allspice breeding:

Modifications might aim at improving the yield of essential oils or resistance to environmental stresses. Genetic engineering offers significant potential for improving allspice (Pimenta dioica) breeding, addressing challenges such as disease resistance, yield, quality, and environmental stress tolerance. Here's how genetic engineering can be applied in allspice breeding:

i) Enhancing disease resistance

- **Fungal and bacterial resistance**: Allspice is susceptible to various fungal and bacterial pathogens. Genetic engineering can introduce resistance genes that encode antifungal proteins, such as chitinases or glucanases, and antibacterial peptides. This could significantly reduce crop losses due to diseases like root rot and leaf spot.
- **Viral resistance**: Although less common in allspice, viral infections can still pose a threat. Genetic engineering can employ RNA interference (RNAi) to silence viral genes, preventing their replication and spread within the plant.

ii) Improving pest resistance

- **Insect resistance**: Allspice is sometimes attacked by pests like borers and leaf miners. Genetic engineering can introduce genes from *Bacillus thuringiensis* (Bt), which produce proteins toxic to these insects but safe for humans and non-target organisms. This would reduce the reliance on chemical insecticides and lower production costs.
- **Nematode resistance**: Nematodes can affect the root systems of allspice, leading to reduced growth and yield. Genetic engineering can introduce nematode-resistant genes, such as those encoding for cysteine protease inhibitors, to make the plant resistant to these pests.

iii) Abiotic stress tolerance

- **Drought and salinity tolerance**: Allspice cultivation can be impacted by water scarcity and high soil salinity, especially in regions prone to drought or poor soil quality. Genetic engineering can enhance the plant's tolerance to these conditions by introducing genes that improve water use efficiency or salt tolerance, such as those involved in the synthesis of osmoprotectants like proline or trehalose.
- **Temperature tolerance**: With climate change leading to more extreme weather conditions, enhancing allspice's ability to withstand high temperatures through the introduction of heat shock proteins (HSPs) or other stress-related genes can improve its resilience.

iv) Improving yield and growth rate

- **Manipulating growth hormones**: Genetic engineering can alter the expression of genes related to plant growth hormones such as gibberellins, cytokinins, and auxins, potentially leading to faster growth rates and higher yields. This could result in quicker maturity and more frequent harvesting cycles.
- **Flowering and fruit set**: Modifying genes that regulate flowering time and fruit development can help synchronize allspice production with optimal environmental conditions, enhancing overall yield.

v) Enhancing essential oil content

- **Increasing key aromatic compounds**: The quality and market value of allspice are largely determined by its essential oil content, particularly eugenol, which gives it its characteristic flavour and aroma. Genetic engineering can be used to upregulate the biosynthetic pathways responsible for eugenol production by increasing the expression of specific enzymes involved in this pathway.
- **Modifying flavour profiles**: Genetic engineering can also be used to modify other flavour compounds, potentially enhancing or diversifying the flavour profile of allspice to meet specific market demands or culinary preferences.

vi) Nutritional and medicinal enhancement

- **Biofortification**: Allspice has medicinal properties due to its antioxidant content and other bioactive compounds. Genetic engineering can enhance the production of these compounds, such as flavonoids or phenolic acids, making the spice even more beneficial for health. This could involve the introduction or overexpression of genes involved in the biosynthesis of these compounds.

vii) Clonal propagation and uniformity

- **Somatic embryogenesis**: To ensure uniformity in plant characteristics and maintain the quality of propagated plants, genetic engineering can be used to improve the efficiency of somatic embryogenesis in allspice. This would facilitate large-scale clonal propagation of elite varieties, ensuring consistency in commercial production.

Challenges and considerations

- **Genetic transformation protocols**: Developing effective transformation methods for allspice is a challenge, as the species may have specific biological characteristics that make genetic engineering difficult. Establishing reliable protocols for gene insertion and expression is crucial.
- **Regulatory approvals**: Like all genetically modified organisms (GMOs), genetically engineered allspice must undergo rigorous regulatory review before commercialization, which can be a lengthy and costly process.
- **Public acceptance**: Consumer and producer acceptance of genetically engineered allspice will be critical for the success of such initiatives, especially in markets where there is resistance to GMOs.

Genetic engineering presents a valuable tool for advancing allspice breeding by enabling targeted improvements in disease resistance, yield, stress tolerance, and essential oil content. While challenges remain in the application of this technology to allspice, the potential benefits make it a promising area for future research and development in spice breeding programs.

3b. CRISPR in Allspice breeding

Applications could include precise editing to enhance the production of eugenol or other aromatic compounds. CRISPR (Clustered Regularly Interspaced Short Palindromic Repeats) technology has revolutionized genetic engineering by enabling precise editing of DNA sequences. Its application in allspice breeding, though not widespread yet, holds significant potential for improving various traits in the plant. Here's how CRISPR could be used in allspice breeding:

- **Trait improvement**: CRISPR can be employed to enhance desirable traits in allspice, such as aroma, flavour, and yield. By targeting specific genes responsible for these traits, CRISPR can introduce or remove mutations that optimize these characteristics.
- **Disease resistance**: Allspice is vulnerable to various diseases that can affect its growth and productivity. CRISPR can be used to edit genes associated with disease resistance, making the plants more resilient to pathogens.
- **Growth and development**: Modifying genes involved in growth and development could help produce allspice plants that are better suited to

different environments, have shorter maturation times, or are more robust under stressful conditions.

- **Biotic and abiotic stress tolerance**: CRISPR could also be used to enhance the tolerance of allspice plants to biotic stresses (like pests) and abiotic stresses (such as drought or salinity). This would involve editing genes that confer resistance or tolerance to these conditions.
- **Metabolic pathways**: Allspice contains various bioactive compounds, and CRISPR could be used to enhance or modify the production of these compounds by editing the genes involved in their biosynthetic pathways.

The integration of CRISPR in allspice breeding could lead to significant advancements in the quality, resilience, and productivity of allspice, making it a valuable tool for future agricultural practices. However, further research and development are necessary to fully realize its potential in this specific plant species.

4. Clove (Syzygium aromaticum)

4a. Genetic engineering in clove breeding:

Targeted genetic changes to improve clove oil content or resistance to pests.

To improve clove oil content or resistance to pests through genetic engineering, targeted genetic changes can be applied using modern biotechnological tools. Here are some specific strategies:

1. Enhancing Clove Oil Content

Clove oil, primarily composed of **eugenol**, is highly valued for its medicinal and aromatic properties. Genetic engineering can increase the synthesis of eugenol or other beneficial compounds in clove oil by targeting specific genes involved in its biosynthesis.

- **Targeting the eugenol synthase gene**: Eugenol synthase is a key enzyme in the biosynthesis of eugenol. By upregulating this gene or introducing optimized variants through gene editing (e.g., **CRISPR-Cas9**), it is possible to boost eugenol production.
- **Manipulating terpenoid pathways**: Terpenoids are precursors of many essential oils. Genetic modifications in the **mevalonate pathway** or **methylerythritol phosphate (MEP) pathway** can enhance the flow of precursors towards the biosynthesis of eugenol and other valuable terpenoids.
- **Gene overexpression**: Overexpression of genes involved in oil production, such as **Eugenol O-methyltransferase (EOMT)**, can lead to increased oil yield or alteration of its composition to enhance specific desired compounds.

2. Improving Pest Resistance

Clove trees are susceptible to various pests, such as **clove borers** and **leaf-eating insects**. Genetic engineering can introduce traits for pest resistance, reducing crop losses and dependency on chemical pesticides.

- **Bt genes**: One common approach is the introduction of **Bacillus thuringiensis (Bt)** genes, which produce proteins toxic to specific insect pests. This technique has been widely used in crops like cotton and corn, and can be applied to cloves to provide resistance against leaf-eating pests.
- **RNA Interference (RNAi)**: This technology can be used to silence essential genes in pests that attack clove plants. By delivering RNA molecules that interfere with pest gene expression (e.g., targeting genes essential for insect survival), RNAi can offer a highly specific form of pest control.
- **Enhancing natural defense mechanisms**: Clove plants produce various secondary metabolites that deter pests. Genetic modification to upregulate these **defensive compounds**, such as phenolics or terpenoids, could enhance the plant's natural resistance. The expression of genes from other species that produce insecticidal compounds can also be explored.

3. CRISPR-Cas9 for Precision Editing

Using **CRISPR-Cas9**, scientists can precisely edit genes related to clove oil biosynthesis or pest resistance without introducing foreign DNA, which could be more acceptable in regulatory environments. This method allows for fine-tuning specific metabolic pathways, potentially leading to more robust, high-oil-yielding, and pest-resistant clove varieties.

- **Oil content**: Target the **eugenol synthase** and **terpenoid pathways** to increase oil content.
- **Pest resistance**: Introduce **Bt genes**, use **RNAi** for pest-specific gene silencing, or enhance natural defense compounds.
- **CRISPR-Cas9**: Use this tool for precise genetic modifications, improving traits without foreign gene introduction.

These targeted genetic modifications provide a sustainable approach to improving clove cultivation, enhancing both yield and resilience.

4b. CRISPR in Clove breeding

Potential for editing genes involved in the biosynthesis of eugenol and other bioactive compounds. **CRISPR-Cas9** holds significant potential in clove breeding, particularly for editing genes involved in the biosynthesis of **eugenol** and other bioactive compounds. Eugenol, a major component of clove oil, and various other terpenoids and phenolics contribute to the medicinal and aromatic properties of cloves. By using CRISPR, targeted genetic changes can be made to optimize the production of these valuable compounds.

Key advantages of CRISPR in Clove breeding

1. **Precision**: CRISPR allows for precise editing of specific genes, enabling fine-tuning of metabolic pathways involved in eugenol production without introducing foreign DNA.
2. **Reduced breeding time**: Traditional clove breeding can take several years due to the plant's long reproductive cycle. CRISPR accelerates trait improvements by directly targeting genes of interest.
3. **Customizing bioactive compounds**: CRISPR can be used to increase the concentration of specific bioactive compounds in clove oil, enhancing its medicinal and commercial value.

Potential targets for CRISPR in Eugenol and Bioactive compound biosynthesis

1. **Eugenol Synthase Gene (EGS)**: This is the key enzyme responsible for the conversion of coniferyl acetate into eugenol. CRISPR can be used to upregulate or optimize the activity of **EGS** to enhance eugenol production.
   - **Gene editing strategy**: By increasing the expression of **EGS** or introducing mutations that improve enzyme efficiency, the overall yield of eugenol can be significantly boosted.
2. **Biosynthesis Pathway Regulation**:
   - **MEP pathway (Methylerythritol Phosphate Pathway)**: This pathway leads to the production of **isoprenoids**, which are precursors of eugenol and other terpenoids. CRISPR can be used to enhance flux through the MEP pathway by editing key regulatory genes, increasing the availability of precursors for eugenol biosynthesis.
   - **Mevalonate pathway**: An alternative to the MEP pathway, this pathway can also be targeted to increase the production of terpenoids, contributing to higher eugenol and other essential oil yields.
3. **Modifying competing pathways**: In many plants, there are competing pathways that may divert precursors away from eugenol biosynthesis. CRISPR can deactivate or downregulate genes in these pathways to ensure more metabolic precursors are directed towards eugenol production.
4. **Overexpression of transcription factors**: Certain transcription factors control the expression of multiple genes in the eugenol biosynthesis pathway. CRISPR can be used to activate or enhance transcription factors that upregulate the entire biosynthetic pathway, leading to a coordinated increase in eugenol and other bioactive compounds.

CRISPR for Enhancing Other Bioactive Compounds in Clove

1. **Flavonoids and phenolics**: These compounds have antioxidant and antimicrobial properties. Genes involved in the **phenylpropanoid pathway**, which produces these compounds, can be edited to enhance the concentration of health-promoting phenolics like **quercetin** and **kaempferol**.
    - **Editing Strategy**: Knockout or overexpression of key enzymes, such as **phenylalanine ammonia-lyase (PAL)**, which is involved in the early steps of phenolic biosynthesis, could boost overall phenolic content.
2. **Terpenoid modification**: Terpenoids are responsible for the aroma and therapeutic properties of clove oil. CRISPR can target specific genes to alter the composition of terpenoids, improving the commercial and medicinal value of clove oil. This could involve increasing the levels of **beta-caryophyllene** or other desired terpenes.
3. **Resistance to environmental stress**: CRISPR could also target genes related to stress responses, ensuring that plants maintain high levels of bioactive compounds even under adverse environmental conditions, such as drought or pathogen attack.

Challenges and Future Outlook:

- **Off-Target Effects**: While CRISPR is precise, off-target mutations are a potential concern. Advanced versions of CRISPR (e.g., **CRISPR-Cas12** or **base editing**) could minimize these risks.
- **Regulatory hurdles**: Genetically modified organisms (GMOs), including CRISPR-edited plants, often face regulatory scrutiny. However, plants edited without introducing foreign DNA might have an easier path to approval.

CRISPR offers a powerful tool for enhancing the **eugenol content** and **bioactive properties** of cloves by precisely editing genes in key biosynthetic pathways. By targeting eugenol synthase, terpenoid pathways, and transcriptional regulators, CRISPR can significantly boost the production of valuable compounds while maintaining or even enhancing the plant's resilience to environmental stresses. This opens up new possibilities for producing higher-quality clove oil with tailored chemical profiles for medicinal, aromatic, and commercial applications.

5. Cinnamon (*Cinnamomum verum*)

5a. Genetic engineering in Cinnamon breeding

Enhancements could focus on increasing cinnamaldehyde content or resistance to pathogens.

Genetic engineering in **cinnamon breeding** can focus on two major areas: enhancing **cinnamaldehyde content** and improving **resistance to pathogens**. Cinnamaldehyde is the primary compound responsible for cinnamon's flavor and aroma, making it a valuable target for genetic improvements. At the same time, breeding for pathogen resistance ensures healthier plants and better yields.

1. Increasing Cinnamaldehyde Content through Genetic Engineering

Cinnamaldehyde is synthesized via the **phenylpropanoid pathway**, which is responsible for the production of several important plant metabolites, including lignins, flavonoids, and aromatic compounds.

Potential Genetic Engineering Strategies

- **Upregulating key enzymes in the phenylpropanoid pathway**:
    - The production of cinnamaldehyde is catalyzed by **cinnamyl alcohol dehydrogenase (CAD)** and **cinnamoyl-CoA reductase (CCR)**, key enzymes that convert phenylpropanoid intermediates into cinnamaldehyde precursors.
    - **CRISPR-Cas9** or **gene overexpression** techniques can be used to enhance the activity of these enzymes, directing more metabolic precursors towards cinnamaldehyde production.
- **Editing Regulatory Genes for Enhanced Production**:
    - Transcription factors that regulate the phenylpropanoid pathway, such as **MYB** and **bHLH** genes, could be overexpressed or edited using CRISPR to activate the genes involved in cinnamaldehyde biosynthesis.
    - **RNA interference (RNAi)** could be used to downregulate competing pathways that siphon off precursors, ensuring more resources are available for cinnamaldehyde production.
- **Metabolic Pathway Optimization:**
    - By using **metabolic engineering**, the entire pathway can be optimized for higher flux towards cinnamaldehyde. For example, altering the **ferulic acid** and **coumaric acid** pools can increase the supply of precursors, boosting cinnamaldehyde yields.

2. Improving Pathogen Resistance

Cinnamon is vulnerable to several pathogens, including **fungal diseases** like **leaf spot**, **root rot**, and **stem blight**. Pathogen resistance can be enhanced by introducing or editing genes related to plant immunity.

Genetic Engineering Approaches for Pathogen Resistance

- **CRISPR for targeted resistance gene activation**
    - **CRISPR-Cas9** can be employed to edit genes involved in the plant's natural immune response, such as those encoding for **R proteins** (resistance proteins) that detect pathogen invasion and trigger defense mechanisms.
    - Editing **PAMP-triggered immunity (PTI)** or **effector-triggered immunity (ETI)** pathways could enhance the plant's ability to recognize and fend off invading pathogens.
- **Transgenic Resistance via Antimicrobial Peptides**
    - Introducing **antimicrobial peptide (AMP) genes** from other species can provide cinnamon with a direct line of defense against fungal and bacterial pathogens. AMPs disrupt the membranes of pathogens, preventing their spread.
    - **Bt (Bacillus thuringiensis)** genes, which have been used successfully in other crops, can provide resistance to specific pests or pathogens.
- **RNA Interference (RNAi) for pathogen gene silencing**:
    - RNAi can be used to silence pathogen virulence genes. This technique involves creating double-stranded RNA that targets essential genes in the pathogen, preventing it from reproducing or attacking the plant.
    - For fungal pathogens, RNAi can be used to target genes involved in spore formation or infection mechanisms, reducing their ability to infect cinnamon plants.
- **Engineering for Systemic Acquired Resistance (SAR)**
    - **Systemic acquired resistance** is a "whole-plant" immune response that is activated when a pathogen infects a single part of the plant. By overexpressing SAR-related genes, such as those encoding for **salicylic acid** or **pathogenesis-related (PR) proteins**, cinnamon plants can become more resistant to a wide range of pathogens.

Combining Cinnamaldehyde Enhancement and Pathogen Resistance

Through genetic engineering, it is possible to combine traits for both enhanced cinnamaldehyde content and improved pathogen resistance, resulting in cinnamon plants that are more productive and resilient.

- **Gene Stacking**: Multiple genes related to cinnamaldehyde biosynthesis and pathogen resistance can be stacked in the same plant through transgenic methods or CRISPR to achieve both objectives simultaneously.

Challenges and Opportunities

- **Off-Target effects**: As with all gene-editing techniques, minimizing off-target mutations is critical, especially when using CRISPR. Advanced CRISPR versions (such as **CRISPR-Cas12** or **base editing**) could provide more precise edits.
- **Regulatory approval**: Engineered plants, particularly those modified through transgenic methods, may face regulatory hurdles, though gene-edited plants (without foreign DNA) could potentially face fewer obstacles in some regions.
- **Consumer acceptance**: Public perception of genetically modified plants can influence the adoption of engineered cinnamon, though the absence of foreign DNA through CRISPR might improve consumer acceptance.

Genetic engineering offers a powerful toolkit to enhance **cinnamaldehyde content** and **pathogen resistance** in cinnamon plants. By targeting the phenylpropanoid pathway for improved cinnamaldehyde production and employing advanced techniques such as **CRISPR** and **RNAi** for pathogen resistance, breeders can develop more resilient and higher-yielding cinnamon varieties, addressing both economic and agricultural challenges in cinnamon production.

5b. CRISPR in Cinnamon breeding

Could be used for precise editing to boost essential oil yield or to confer resistance to diseases like leaf blight. **CRISPR-Cas9** technology has great potential for improving cinnamon breeding, particularly to boost **essential oil yield** (rich in **cinnamaldehyde**) and enhance resistance to **leaf blast disease**. Below are the specific ways CRISPR can be applied to these goals:

1. Boosting Essential Oil Yield

Cinnamon oil is valued for its high content of **cinnamaldehyde**, which gives it its characteristic aroma and medicinal properties. By using CRISPR to target genes involved in the **biosynthesis of cinnamaldehyde** and other components of cinnamon oil, the overall yield of essential oils can be increased.

CRISPR Strategies for Oil Yield Enhancement

- **Targeting the phenylpropanoid pathway**
    - **Cinnamaldehyde** is produced through the **phenylpropanoid pathway**, which involves key enzymes like **cinnamyl alcohol dehydrogenase (CAD)** and **cinnamoyl-CoA reductase (CCR)**.
    - CRISPR can be used to **upregulate genes** encoding these enzymes, leading to increased production of precursors and, consequently, a higher yield of cinnamaldehyde in the essential oil.

- **Optimizing Metabolic Pathways**
  - CRISPR can be employed to increase the **flux** through the **mevalonate pathway (MVA)** or **methylerythritol phosphate (MEP)** pathway, which supply precursors for terpenoid and cinnamaldehyde biosynthesis. By editing these upstream pathways, the metabolic resources available for oil production can be enhanced.
- **Gene Silencing of Competing Pathways**
  - Some pathways might divert resources away from cinnamaldehyde production. CRISPR could be used to knock out or silence genes in these competing pathways, ensuring that more metabolic precursors are directed towards cinnamaldehyde and other valuable terpenes in cinnamon oil.
- **Overexpression of Regulatory Genes**
  - CRISPR can target **transcription factors** that regulate multiple genes in the oil biosynthesis pathway. By activating or boosting the expression of key transcription factors, the entire pathway for essential oil production can be upregulated.

2. Enhancing Resistance to Leaf Blast Disease

Leaf blast disease, caused by the fungus **pyricularia grisea**, can severely affect cinnamon trees by damaging leaves and reducing plant vigor. CRISPR offers precise and effective strategies to improve cinnamon's resistance to this pathogen.

CRISPR Strategies for Leaf Blast Disease Resistance

- **Editing Disease Resistance (R) Genes**
  - Plants use **resistance (R) genes** to detect pathogen attack and trigger immune responses. CRISPR can be used to enhance or introduce **R genes** that specifically recognize leaf blast pathogen effectors, enabling the plant to mount a stronger and faster defense response.
- **Improving Pathogen Recognition Pathways**
  - **Pattern Recognition Receptors (PRRs)** in plants detect pathogen-associated molecular patterns (PAMPs). CRISPR can be used to improve the efficiency of these **PRRs**, enhancing the plant's ability to recognize and respond to fungal infections. This would activate the **PAMP-triggered immunity (PTI)**, which provides a broad-spectrum defense against pathogens like **Pyricularia grisea**.

- **CRISPR for Systemic Acquired Resistance (SAR)**
  - CRISPR can edit genes involved in **Systemic Acquired Resistance (SAR)**, a plant-wide immune response that provides long-lasting protection against a range of pathogens. SAR-related genes, such as those encoding for **salicylic acid (SA)** or **pathogenesis-related (PR) proteins**, can be upregulated to boost the plant's overall resistance to leaf blast disease.
- **Silencing Pathogen Virulence Factors:**
  - Using CRISPR, it is possible to target and silence **pathogen virulence genes** via **Host-Induced Gene Silencing (HIGS)**. By modifying the plant to produce small interfering RNAs (siRNAs) that target essential fungal genes, the pathogen's ability to infect the plant can be greatly reduced.
- **Strengthening Plant Cell Walls:**
  - Strengthening plant cell walls can inhibit pathogen penetration. CRISPR can target genes that enhance the production of **lignin** or other structural components, making it more difficult for **pyricularia grisea** to invade plant tissues. For instance, upregulating **lignin biosynthesis genes** could reinforce the plant's physical barriers against fungal attack.

Combined Approach: Enhancing Both Oil Yield and Disease Resistance

- **Simultaneous editing**: Using CRISPR, multiple genes can be edited simultaneously (a process known as **multiplex CRISPR**), allowing breeders to enhance **both cinnamaldehyde production** and **disease resistance** in a single breeding cycle. This can create cinnamon varieties that not only produce higher-quality oil but are also more resilient to diseases like leaf blast.

CRISPR technology offers precise, efficient, and potentially rapid solutions to improve both the **oil yield** and **disease resistance** of cinnamon plants. By targeting specific genes involved in **cinnamaldehyde biosynthesis** and **pathogen resistance mechanisms**, breeders can develop superior cinnamon varieties that are more productive and resilient to diseases like **leaf blast**, ensuring both higher economic returns and sustainable cultivation practices.

6. Ginger (Zingiber officinale)

6a. Genetic engineering in Ginger breeding

Aimed at improving rhizome yield, enhancing gingerol content, or increasing resistance to diseases like soft rot. Genetic engineering can be applied to ginger (**Zingiber officinale**) to address key challenges and improve traits such as **rhizome yield, gingerol content**, and **resistance to diseases** like **soft rot**. By employing advanced techniques such as **CRISPR-Cas9, RNA interference (RNAi)**, and **transgenic approaches**, these improvements can be targeted with precision.

1. Improving Rhizome Yield

The rhizome, or underground stem, is the most commercially valuable part of ginger, used for its culinary and medicinal properties. Enhancing its yield through genetic engineering can increase productivity.

Genetic Strategies for Enhancing Rhizome Yield

- **Manipulating growth-related hormones**
    - The **rhizome yield** is influenced by growth hormones such as **auxins, cytokinins**, and **gibberellins**. By using CRISPR or gene overexpression, the genes involved in **hormone biosynthesis** and **signal transduction** (e.g., **GA3** for gibberellin) can be regulated to promote rhizome enlargement and faster growth.
- **Photosynthesis Efficiency**
    - Improving photosynthetic efficiency can enhance biomass accumulation, which directly affects rhizome yield. Genetic modifications aimed at optimizing **chlorophyll content**, **carbon fixation enzymes** (such as **ribulose-1,5-bisphosphate carboxylase/oxygenase (RuBisCO)**), and **light-harvesting complexes** can boost the overall plant growth, leading to larger rhizomes.
- **Modifying Rhizome-Specific Growth Genes**
    - Genes that control rhizome formation and growth, such as those involved in **cell division** and **expansion** (e.g., **cyclin-dependent kinases**), can be edited to enhance the size and number of rhizomes. Overexpression of these genes could lead to increased rhizome mass.

2. Enhancing Gingerol Content

Gingerols are the bioactive compounds responsible for the spicy flavor and health benefits of ginger. Enhancing gingerol content can increase the medicinal and market value of ginger.

Genetic Strategies for Increasing Gingerol Content

- **Upregulating gingerol biosynthesis pathway**
    - **Gingerol** is synthesized via the **phenylpropanoid pathway**, followed by **polyketide synthase** activity. Key enzymes such as **phenylalanine ammonia-lyase (PAL), cinnamate-4-hydroxylase (C4H)**, and **polyketide synthases (PKS)** play a crucial role in this process.

    - CRISPR-Cas9 can be used to **upregulate key genes** in the gingerol biosynthesis pathway, increasing the production of **precursors** and ensuring higher gingerol content in the rhizomes.
- **Suppressing Competing Pathways**
    - Competing metabolic pathways may divert precursors away from gingerol biosynthesis. By using **RNAi** or CRISPR, genes involved in these competing pathways can be **knocked out or silenced**, ensuring that more metabolic resources are channeled into gingerol production.
- **Increasing Precursor Availability**
    - Overexpressing genes that produce **shikimic acid**, a precursor for many secondary metabolites like gingerol, can enhance the supply of precursors for gingerol biosynthesis. Genetic engineering can also increase the availability of **acetyl-CoA** and other building blocks required for gingerol production.

3. Increasing Resistance to Diseases like Soft Rot

Soft rot, caused by **Pythium spp.** and other pathogens, is one of the most damaging diseases in ginger production, leading to significant yield losses. Genetic engineering can help enhance ginger's resistance to soft rot and other pathogens.

Genetic Strategies for Disease Resistance

- **Introducing pathogen-resistant genes**
    - By introducing **R-genes** (resistance genes) from other species, ginger can be made more resistant to soft rot pathogens. These R-genes encode proteins that recognize pathogen effectors and trigger immune responses.
    - CRISPR can be used to **edit ginger's innate immune receptors**, improving their ability to detect and respond to the soft rot pathogen, thus activating **pattern-triggered immunity (PTI)** or **effector-triggered immunity (ETI)**.
- **Enhancing Cell Wall Integrity**
    - Soft rot pathogens attack the ginger rhizome by degrading cell walls. By editing genes involved in **cell wall biosynthesis**, such as those responsible for **lignin** or **cellulose production**, genetic engineering can strengthen the rhizome's defenses against pathogen invasion.
- **Host-Induced Gene Silencing (HIGS)**
    - **RNA interference (RNAi)** can be used to silence essential genes in the soft rot pathogen. Ginger plants can be engineered to produce **small interfering RNAs (siRNAs)** that target and silence genes critical for the pathogen's growth or virulence, such as enzymes involved in cell wall degradation.

- **Expression of Antimicrobial Peptides (AMPs)**
  - Transgenic approaches can introduce genes encoding **antimicrobial peptides** that directly inhibit the growth of pathogens. These peptides disrupt the cell membranes of bacteria and fungi, reducing infection rates.
- **Systemic Acquired Resistance (SAR)**
  - Engineering ginger plants to enhance **systemic acquired resistance (SAR)** could provide long-lasting immunity against a wide range of pathogens, including those causing soft rot. SAR-related genes, such as those involved in **salicylic acid** or **jasmonic acid** pathways, can be targeted to enhance the plant's defense mechanisms.

Summary of Genetic Engineering in Ginger

- **Rhizome yield**: Target growth hormones, photosynthesis efficiency, and rhizome-specific genes for larger and more productive rhizomes.
- **Gingerol content**: Upregulate key enzymes in the gingerol biosynthesis pathway and suppress competing metabolic pathways to increase gingerol levels.
- **Disease resistance**: Enhance resistance to soft rot through R-genes, RNAi (HIGS), and antimicrobial peptides, or by strengthening cell wall integrity.

By leveraging genetic engineering techniques like CRISPR and RNAi, ginger breeding can be enhanced to improve **rhizome yield**, **gingerol content**, and **disease resistance**. These advancements can help ginger producers meet market demands for higher-quality products while reducing losses from diseases such as **soft rot**, ultimately contributing to more sustainable and profitable ginger cultivation.

6b. CRISPR in Ginger breeding

Targeting genes involved in secondary metabolite production (e.g., gingerol biosynthesis) and disease resistance.

Using **CRISPR-Cas9** technology, specific genes involved in **gingerol biosynthesis, soft rot resistance**, and **bacterial wilt resistance** in ginger can be targeted to improve its medicinal and agricultural traits. Here's how CRISPR can be applied to each of these aspects:

1. Targeting Genes for Gingerol Biosynthesis

Gingerol, the primary bioactive compound in ginger, is synthesized through the **phenylpropanoid pathway** and subsequent modifications. Genetic engineering using CRISPR can directly influence the expression of key enzymes in this pathway to enhance gingerol production.

Key Genes Involved in Gingerol Biosynthesis

- **Phenylalanine Ammonia-Lyase (PAL)**
  - PAL catalyzes the conversion of phenylalanine into cinnamic acid, which is the first step in the phenylpropanoid pathway.
  - **CRISPR strategy**: Upregulate or overexpress the **PAL** gene to boost the production of precursors for gingerol.
- **Cinnamate-4-Hydroxylase (C4H)**
  - C4H converts cinnamic acid to p-coumaric acid, an important intermediate in gingerol biosynthesis.
  - **CRISPR strategy**: Enhance C4H expression to increase the flow of intermediates toward gingerol production.
- **4-Coumarate-CoA Ligase (4CL)**
  - This enzyme converts p-coumaric acid to p-coumaroyl-CoA, a critical step in the formation of gingerol.
  - **CRISPR strategy**: Upregulate **4CL** to ensure that more precursor molecules are available for gingerol biosynthesis.
- **Polyketide Synthases (PKS)**
  - PKS enzymes are involved in synthesizing the polyketide backbone of gingerol.
  - **CRISPR strategy**: Edit PKS genes to increase the rate of gingerol synthesis, resulting in higher concentrations of gingerol in the rhizomes.
- **Shikimate Pathway Genes**
  - Since gingerol biosynthesis begins with the shikimate pathway, increasing the supply of **shikimic acid** precursors through CRISPR editing could enhance overall gingerol production.

2. Targeting Genes for Soft Rot Resistance

Soft rot, caused by **Pythium spp.**, is a major disease affecting ginger rhizomes. CRISPR can be used to enhance ginger's resistance by targeting genes involved in pathogen recognition and immune response.

Key Genes for Soft Rot Resistance

- **R Genes (Resistance Genes)**
  - **R genes** encode proteins that recognize pathogen attack and activate the plant's defense responses.

- **CRISPR strategy**: Edit or introduce R genes that specifically detect **Pythium spp.** effector proteins, boosting the plant's resistance. Enhancing R-gene expression would improve the ability of ginger to mount a rapid immune response.

- **Pattern Recognition Receptors (PRRs)**:
  - These receptors recognize **pathogen-associated molecular patterns (PAMPs)** and trigger **PAMP-triggered immunity (PTI)**.
  - **CRISPR Strategy**: Modify PRR genes to increase their sensitivity to **Pythium** pathogens, enabling the plant to activate PTI more effectively. This would result in earlier and stronger immune responses to soft rot pathogens.
- **Pathogenesis-Related (PR) Proteins**
  - PR proteins are involved in defending plants against pathogens by degrading pathogen cell walls or inhibiting their growth.
  - **CRISPR strategy**: Overexpress PR protein genes to increase the amount of these defensive proteins, making ginger more resistant to soft rot infections.
- **Lignin Biosynthesis Genes**:
  - Strengthening cell walls by increasing **lignin** production can help prevent soft rot pathogens from penetrating and infecting the rhizome.
  - **CRISPR strategy**: Upregulate genes involved in **lignin biosynthesis** to reinforce cell walls, reducing susceptibility to pathogen invasion.

3. Targeting Genes for Bacterial Wilt Resistance

Bacterial wilt, caused by ***Ralstonia solanacearum***, is another devastating disease affecting ginger. CRISPR can help develop resistance by targeting genes involved in bacterial recognition and the plant's defence mechanisms.

Key Genes for Bacterial Wilt Resistance

- **Salicylic Acid (SA) pathway genes**
  - The **SA pathway** is crucial for activating **systemic acquired resistance (SAR)**, which provides long-term resistance to bacterial pathogens.
  - **CRISPR strategy**: Enhance the expression of genes involved in the **salicylic acid biosynthesis** pathway, promoting SAR and increasing resistance to ***Ralstonia solanacearum***.
- **NPR1 (Nonexpressor of Pathogenesis-Related Genes 1)**
  - **NPR1** is a key regulator of SAR and mediates the expression of defence-related genes.

  - **CRISPR Strategy**: Upregulate **NPR1** to boost the plant's systemic defences against bacterial pathogens like ***Ralstonia***.
- **Reactive Oxygen Species (ROS) defence mechanisms**
  - ROS are produced in response to pathogen attack and can kill invading bacteria.
  - **CRISPR strategy**: Enhance genes responsible for **ROS production** or regulation, allowing ginger to produce more ROS at infection sites, improving its resistance to bacterial wilt.
- **Cell Wall Reinforcement**
  - ***Ralstonia*** bacteria weaken plant cell walls during infection. Strengthening cell walls by increasing **cellulose** and **hemicellulose** production could make it harder for the pathogen to spread.
  - **CRISPR strategy**: Upregulate genes involved in **cell wall reinforcement** to limit pathogen invasion.

Combined Approach: Multiplex CRISPR Editing

- **Multiplex CRISPR**: By simultaneously editing multiple genes involved in gingerol biosynthesis, soft rot resistance, and bacterial wilt resistance, it is possible to enhance several important traits in ginger in a single breeding cycle.
  - **Example**: Targeting **PAL** for gingerol production, **R genes** for soft rot resistance, and **NPR1** for bacterial wilt resistance at the same time can result in ginger varieties with improved bioactive content and disease resistance.

CRISPR offers precise and powerful tools for targeting **key genes** involved in **gingerol biosynthesis**, **soft rot resistance**, and **bacterial wilt resistance** in ginger. By editing and upregulating specific genes, ginger varieties can be developed that not only produce higher levels of valuable gingerol but also exhibit increased resilience to devastating diseases like soft rot and bacterial wilt, leading to more productive and sustainable cultivation practices.

7. Turmeric (*Curcuma longa*)

7a. Genetic engineering in turmeric breeding

Focused on increasing curcumin content and developing drought and pest-resistant varieties.Genetic engineering holds great potential for improving key traits in **turmeric** (***Curcuma longa***), such as **drought resistance**, **curcumin content**, **high yield**, and **disease resistance**. Using tools like **CRISPR-Cas9**, **transgenic approaches**, and **RNA interference (RNAi)**, these traits can be enhanced to meet growing agricultural and market demands. Here's a breakdown of how these traits can be targeted:

1. Enhancing Drought Resistance

Drought stress is a major issue in turmeric cultivation, as it affects growth, yield, and the quality of rhizomes. Genetic engineering can improve the plant's ability to tolerate water scarcity by targeting genes involved in **water retention**, **stress responses**, and **root architecture**.

Genetic Engineering Strategies for Drought Resistance

- **Abscisic Acid (ABA) pathway**
  - ABA is a critical hormone in plants' response to drought, regulating **stomatal closure** to reduce water loss.
  - **CRISPR Strategy**: Upregulate genes in the **ABA biosynthesis pathway**, such as **NCED (9-cis-epoxycarotenoid dioxygenase)**, to increase ABA production, resulting in enhanced drought tolerance.
- **Drought-Responsive Genes**
  - **DREB (Drought-Responsive Element Binding)** transcription factors regulate the expression of genes that protect cells during drought.
  - **CRISPR or overexpression strategy**: Enhance the expression of **DREB** or other **AP2/ERF family** genes to activate stress response pathways, improving the plant's ability to cope with drought.
- **Root Architecture Modifications**:
  - Genes that control root growth, such as **root hair formation genes** or **auxin transporters**, can be edited to increase root depth or surface area, improving water uptake during drought.
  - **CRISPR Strategy**: Edit genes like **PIN-FORMED (PIN)** to promote deeper root systems, enabling better access to water in dry soils.

2. Increasing Curcumin Content

Curcumin, the primary bioactive compound in turmeric, is responsible for its medicinal and therapeutic properties. Genetic engineering can be used to increase curcumin production by targeting its biosynthetic pathway.

Genetic Engineering Strategies for Curcumin Enhancement

- **Targeting curcumin biosynthesis pathway**
  - **Curcumin synthase (CURS)** enzymes are key in converting feruloyl-CoA and coumaroyl-CoA into curcumin.
  - **CRISPR strategy**: Upregulate **CURS1**, **CURS2**, and **CURS3** genes to enhance the conversion of precursors into curcumin, increasing overall curcumin content.

- **Increasing Precursor Availability**
  - Curcumin biosynthesis begins with the **phenylpropanoid pathway**, which produces the necessary precursors (ferulic acid and coumaric acid).
  - **CRISPR Strategy**: Upregulate upstream genes like **phenylalanine ammonia-lyase (PAL)** and **cinnamate-4-hydroxylase (C4H)** to boost the availability of precursors for curcumin synthesis.
- **Suppressing Competing Pathways**
  - By knocking out or silencing genes in competing pathways that divert precursors away from curcumin production, more resources can be directed toward curcumin biosynthesis.
  - RNAi strategy: Use RNA interference to silence genes in the lignin biosynthesis pathway, redirecting metabolic flux towards curcumin production.

3. Enhancing Yield

Improving turmeric yield involves increasing biomass accumulation, rhizome development, and overall plant vigor. Genetic modifications targeting hormonal regulation, photosynthesis efficiency, and nutrient uptake can lead to higher yields.

Genetic Engineering Strategies for High Yield

- **Growth hormones (auxins, cytokinins, and gibberellins)**
  - Auxins and gibberellins play essential roles in cell division, elongation, and rhizome development.
  - **CRISPR strategy**: Modify genes controlling hormone levels, such as **GA3** (gibberellin biosynthesis) or **PIN1** (auxin transport), to promote larger rhizomes and overall plant growth.
- **Photosynthesis Efficiency**
  - Improving the efficiency of **photosynthesis** can lead to better biomass accumulation, which translates into larger rhizomes.
  - **CRISPR strategy**: Edit genes involved in the **Calvin cycle** and **RuBisCO activity** to optimize carbon fixation, improving growth and yield.
- **Nutrient Uptake Genes**
  - Genes related to nutrient uptake, especially **nitrogen** and **phosphorus transporters**, can be edited to improve nutrient efficiency, which supports higher yields.
  - **CRISPR Strategy**: Enhance the expression of genes like **NRT (Nitrate transporter)** and **PHO1 (Phosphate transporter)** to boost nutrient uptake and utilization.

4. Increasing Disease Resistance

Turmeric is susceptible to several diseases, including **rhizome rot**, **leaf spot**, and **bacterial wilt**. Enhancing disease resistance through genetic engineering can help reduce yield losses and dependency on chemical treatments.

Genetic Engineering Strategies for Disease Resistance

- **R Genes (*Resistance Genes*)**
    - **R genes** encode proteins that detect pathogen effectors and activate immune responses.
    - **CRISPR Strategy**: Edit or introduce **R genes** specific to pathogens affecting turmeric, such as those responsible for **rhizome rot** caused by **Pythium spp.** or **bacterial wilt** caused by **Ralstonia solanacearum**.
- **Pattern Recognition Receptors (PRRs)**
    - **PRRs** detect pathogen-associated molecular patterns (PAMPs) and trigger **PAMP-triggered immunity (PTI)**.
    - **CRISPR Strategy**: Enhance **PRR gene expression** to improve the plant's recognition of fungal and bacterial pathogens, activating a strong defense response.
- **Antimicrobial Peptides (AMPs)**
    - AMPs can be introduced to directly inhibit pathogen growth by disrupting microbial cell membranes.
    - **Transgenic approach**: Introduce genes encoding **antimicrobial peptides** from other plants or organisms to provide broad-spectrum resistance against bacterial and fungal pathogens.
- **Cell Wall Reinforcement**
    - Strengthening the plant's cell walls can prevent pathogens from invading.
    - **CRISPR strategy**: Edit genes involved in **lignin** and **cellulose biosynthesis** to reinforce cell walls, making it harder for pathogens to penetrate.

Summary of Genetic Engineering Strategies in Turmeric

- **Drought resistance**: Target ABA signaling and root architecture genes to improve water-use efficiency and enhance drought tolerance.
- **Curcumin content**: Upregulate curcumin biosynthesis genes and suppress competing pathways to boost curcumin production.
- **Yield enhancement**: Manipulate hormonal regulation, photosynthesis efficiency, and nutrient uptake to increase biomass and rhizome yield.
- **Disease resistance**: Introduce or edit R genes, PRRs, AMPs, and cell wall-related genes to improve resistance to major pathogens.

By leveraging genetic engineering techniques like **CRISPR-Cas9, transgenic approaches**, and **RNA interference**, turmeric can be bred to possess **higher yields, increased curcumin content, enhanced drought tolerance**, and **stronger disease resistance**. These advancements can significantly improve the profitability and sustainability of turmeric cultivation, meeting both market demands and agricultural challenges.

7b. CRISPR in Turmeric breeding

Precise gene editing in turmeric using **CRISPR-Cas9** can target key pathways to enhance **curcumin biosynthesis** and improve **resistance to biotic and abiotic stresses**. These modifications can lead to higher curcumin content, increased resistance to pathogens, and improved resilience to environmental stresses like drought.

1. Precise Editing of Genes for Curcumin Biosynthesis

Curcumin, a polyphenolic compound found in turmeric, is synthesized through a series of enzymatic reactions. By targeting specific genes involved in its biosynthesis, CRISPR can enhance curcumin production.

Key genes in the curcumin biosynthesis pathway

- **Curcumin synthase (CURS) Genes**
    - Curcumin is primarily synthesized by **curcumin synthase enzymes (CURS1, CURS2, and CURS3)**.
    - **CRISPR strategy**: Upregulate or optimize the **CURS1, CURS2, and CURS3** genes to increase the conversion of feruloyl-CoA and coumaroyl-CoA into curcumin. Enhancing their expression can directly boost curcumin yield.
- **Phenylalanine Ammonia-Lyase (PAL)**
    - **PAL** catalyzes the first step in the **phenylpropanoid pathway**, converting phenylalanine to cinnamic acid, a precursor to curcumin.
    - **CRISPR Strategy**: Overexpress **PAL** to increase the availability of precursors for curcumin biosynthesis, leading to higher curcumin production.
- **Cinnamate-4-Hydroxylase (C4H)**
    - C4H is responsible for converting cinnamic acid into p-coumaric acid, which is an intermediate in the curcumin biosynthesis pathway.
    - **CRISPR strategy**: Enhance the activity of **C4H** to ensure a steady flow of intermediates for curcumin biosynthesis.

- **4-Coumarate-CoA Ligase (4CL)**
  - **4CL** activates p-coumaric acid by converting it into p-coumaroyl-CoA, which is essential for curcumin biosynthesis.
  - **CRISPR strategy**: Upregulate **4CL** to increase the availability of activated intermediates required for curcumin formation.
- **Feruloyl-CoA Synthase**
  - This enzyme produces **feruloyl-CoA**, one of the key precursors for curcumin synthesis.
  - **CRISPR strategy**: Increase the expression of **feruloyl-CoA synthase** to ensure a higher supply of this precursor, leading to elevated curcumin levels.

2. Precise Editing of Genes for Biotic Stress Resistance

Biotic stress refers to damage caused by living organisms such as pathogens (fungi, bacteria, viruses) and pests. By using CRISPR, turmeric can be genetically enhanced to resist common diseases and pests.

Key Genes for Biotic Stress Resistance

- **R Genes (Resistance Genes)**:
  - **R genes** encode proteins that recognize pathogen attack and activate plant immune responses.
  - **CRISPR strategy**: Introduce or edit **R genes** that confer resistance to specific pathogens affecting turmeric, such as **Pythium spp.** (causing rhizome rot) or **Ralstonia solanacearum** (causing bacterial wilt). Enhancing R gene expression improves pathogen recognition and immune activation.
- **Pattern Recognition Receptors (PRRs)**:
  - PRRs detect **pathogen-associated molecular patterns (PAMPs)** and trigger **PAMP-triggered immunity (PTI)**.
  - **CRISPR strategy**: Enhance PRR genes to improve the detection of microbial invaders, leading to faster and stronger activation of immune responses. This reduces susceptibility to fungal, bacterial, and viral infections.
- **Pathogenesis-Related (PR) Proteins**:
  - PR proteins are defensive proteins that play a role in inhibiting pathogen spread by degrading cell walls or other vital structures of pathogens.
  - **CRISPR strategy**: Upregulate **PR genes** to boost the production of antimicrobial proteins, offering broad-spectrum resistance against pathogens.

- **Host-Induced Gene Silencing (HIGS)**
    - HIGS involves using **RNA interference (RNAi)** to silence essential genes in pathogens or pests.
    - **CRISPR strategy**: Engineer turmeric to produce RNAi molecules that specifically target and silence critical genes in invading pathogens, rendering them less virulent or unable to infect the plant. This strategy can be particularly effective against fungi and insects.

3. Precise Editing of Genes for Abiotic Stress Resistance

Abiotic stress, including drought, salinity, and temperature extremes, can severely affect turmeric yields. Genetic engineering can enhance the plant's ability to withstand these environmental stresses.

Key Genes for Abiotic Stress Resistance

- **DREB (Drought-Responsive Element Binding) Genes**
    - **DREB** transcription factors regulate a wide range of genes involved in stress tolerance, particularly in response to drought and high salinity.
    - **CRISPR Strategy**: Upregulate **DREB** genes to enhance the plant's response to drought and salt stress. This can improve water retention, osmotic balance, and survival under harsh environmental conditions.
- **Abscisic Acid (ABA) biosynthesis pathway**
    - **ABA** is a plant hormone critical for **stomatal closure**, reducing water loss during drought.
    - **CRISPR strategy**: Enhance ABA biosynthesis by editing key genes in the **ABA pathway** (e.g., **NCED**), which leads to increased ABA levels and improved drought tolerance through reduced transpiration.
- **Heat Shock Proteins (HSPs)**
    - HSPs are produced in response to heat stress and help in protecting cellular proteins from denaturation.
    - **CRISPR strategy**: Upregulate **HSP genes** to enhance tolerance to high temperatures, ensuring the plant can survive and maintain productivity under heat stress.
- **Dehydration-Responsive Proteins**
    - These proteins help plants retain water and protect cells during drought conditions.
    - **CRISPR strategy**: Increase the expression of **dehydration-responsive proteins** to improve water-use efficiency and reduce cell damage under water-limited conditions.

- **Aquaporins (Water Channels)**
  - Aquaporins regulate water movement within plant tissues, crucial during periods of drought or high salinity.
  - **CRISPR strategy**: Enhance the expression of **aquaporin genes** to improve water uptake and retention during abiotic stress, thereby increasing drought resilience.

Through precise gene editing with **CRISPR-Cas9**, key genes in turmeric can be modified to enhance **curcumin biosynthesis**, improve **biotic stress resistance** (against pathogens like fungi and bacteria), and strengthen **abiotic stress tolerance** (to drought, heat, and salinity). This approach enables the development of superior turmeric varieties with enhanced medicinal value, higher yields, and greater environmental resilience.

By targeting genes like **CURS1** for curcumin production, **R genes** for disease resistance, and **DREB** for drought tolerance, genetic engineering can unlock the full potential of turmeric cultivation in a range of environments.

8. Black Turmeric (Curcuma caesia)

8a. Genetic engineering in Black Turmeric

Modifications may enhance the medicinal properties by increasing bioactive compound concentrations.

Genetic engineering can be applied to the improvement of black turmeric (Curcuma caesia) for various purposes, including enhancing its bioactive compounds, increasing yield, improving disease resistance, and adapting to different environmental conditions. Here are some key areas where genetic engineering can be used for the improvement of black turmeric:

1. Enhancement of Bioactive Compounds

Black turmeric is valued for its medicinal properties due to compounds like curcumin, essential oils, flavonoids, and terpenoids. Genetic engineering can:

- **Target Genes**: Identify genes responsible for the biosynthesis of key bioactive compounds. This would involve genes involved in curcumin synthesis, terpenoid biosynthesis pathways, and the enzymes that modulate their activity.
- **Gene Editing Tools**: CRISPR-Cas9 or RNA interference (RNAi) can be employed to enhance or suppress specific genes. By targeting these pathways, the concentration of desirable compounds like curcumin could be increased, or undesirable compounds might be reduced.

- **Introduction of Novel Pathways**: Foreign genes can be introduced to develop new pathways for creating novel bioactive compounds. This can be done by transferring genes from other plants that produce similar or complementary compounds.
- **Promoter Selection**: Select promoters that ensure high expression of the desired biosynthetic genes in the plant tissue where bioactive compound production is most active.
- **Testing and Optimization**: The modified plants would need to be analyzed for changes in bioactive compound profiles using techniques such as HPLC (High-Performance Liquid Chromatography) to confirm successful modification.

2. Improved Yield and Growth Characteristics

- **Enhanced growth rate**: Genetic modifications can be made to improve the plant's growth rate by introducing genes related to hormone regulation (such as gibberellins or auxins) that control cell division and elongation.
- **Improved biomass production**: Genes related to biomass accumulation can be manipulated to increase the overall yield of the plant, making cultivation more efficient for commercial purposes.

3. Disease and Pest Resistance

- **Increased resistance to pathogens**: Genetic engineering can introduce genes that make black turmeric more resistant to fungal, bacterial, or viral pathogens. For example, resistance genes from other plants can be introduced to activate defense mechanisms.
- **Insect Resistance**: By incorporating genes that produce insecticidal proteins (like Bt genes), the plant can be made more resistant to insect pests, reducing the need for chemical pesticides.

4. Environmental Adaptation and Stress Tolerance

- **Drought and salinity tolerance**: Genetic modifications can be made to introduce or enhance stress-related genes, improving the plant's ability to survive in adverse environmental conditions such as drought or high salinity.
- **Temperature Tolerance**: Genes involved in cold or heat tolerance can be incorporated to allow black turmeric to be grown in a wider range of climates.

5. Improvement of Reproductive Efficiency

- **Faster propagation**: Genetic engineering can improve the plant's reproductive processes, including the production of viable seeds or enhanced vegetative propagation, which is important for scaling up production.

- **Flowering time regulation**: By altering flowering-related genes, black turmeric can be made to flower earlier or later, depending on the needs of the cultivation environment.

6. Introduction of Novel Traits

- **Nutritional enhancement**: Genetic engineering can also be used to fortify black turmeric with additional nutrients or beneficial compounds that are not naturally present in the plant.
- **Metabolic engineering for novel products**: New metabolic pathways can be introduced to produce novel compounds with potential pharmaceutical applications, thus expanding the commercial use of black turmeric.

By applying genetic engineering, black turmeric can be improved in numerous ways to enhance its medicinal value, agronomic traits, and adaptability to different environments, making it more commercially viable and therapeutically potent.

8b. CRISPR in Black Turmeric breeding

Could be applied to edit genes responsible for unique phytochemical production, improving both yield and quality.

Gene editing in black turmeric (Curcuma caesia) using **CRISPR-Cas9** offers a powerful method to improve both the yield and quality of the plant by targeting genes responsible for phytochemical production and agronomic traits. CRISPR is an efficient tool to precisely alter specific genes, allowing scientists to manipulate pathways involved in the synthesis of bioactive compounds, growth, and resilience. Here's how CRISPR can be applied to black turmeric for enhanced phytochemical production, yield, and quality:

1. Targeting Phytochemical Biosynthesis Pathways

Black turmeric contains a variety of bioactive compounds, including **curcumin**, **terpenoids**, **essential oils**, and **flavonoids**. CRISPR can enhance the production of these compounds by:

- **Activating Key Enzymes**: CRISPR can be used to upregulate genes encoding enzymes in the curcumin biosynthetic pathway, such as *curcumin synthase* and *phenylalanine ammonia-lyase*, to increase curcumin content, a compound known for its anti-inflammatory and antioxidant properties.
- **Enhancing terpenoid production**: Genes involved in terpenoid synthesis (e.g., *geranyl diphosphate synthase*) can be edited to produce higher levels of terpenoids, which contribute to the plant's antimicrobial and anti-cancer properties.
- **Creating novel phytochemicals**: CRISPR can introduce new biosynthetic pathways by editing or adding genes from other plants. This could lead to

the production of novel bioactive compounds not naturally present in black turmeric, potentially increasing its medicinal and commercial value.

2. Improving Yield through Genetic Modifications

Enhancing the overall productivity of black turmeric can be achieved through CRISPR-mediated edits in genes that regulate plant growth and development:

- **Increasing biomass**: Targeting genes that regulate biomass accumulation, such as those involved in **cell division** and **photosynthesis efficiency**, can lead to increased plant mass, improving yield.
- **Growth hormone regulation**: Editing genes associated with **plant hormones** like gibberellins or auxins could enhance shoot and root development, leading to faster growth and more robust plants.
- **Optimizing nutrient uptake**: CRISPR could be used to modify genes involved in nutrient uptake, improving the plant's ability to absorb essential minerals from the soil, thereby boosting growth and yield.

3. Enhancing Quality through Gene Editing

In addition to improving yield, CRISPR can also enhance the quality of black turmeric by altering genes that influence its phytochemical profile:

- **Tailoring flavonoid content**: By editing genes involved in the flavonoid biosynthesis pathway (such as *chalcone synthase* or *flavonoid 3'-hydroxylase*), it's possible to increase the concentration of specific flavonoids that have higher antioxidant activity, improving the plant's medicinal efficacy.
- **Balancing essential oils**: CRISPR can be used to regulate the synthesis of essential oils by editing genes responsible for the production of monoterpenes and sesquiterpenes, optimizing the plant's aroma and therapeutic properties.
- **Silencing undesirable genes**: CRISPR can knock out genes responsible for producing compounds that may reduce the plant's quality or shelf life, ensuring that only the most beneficial phytochemicals are produced.

4. Improving Environmental Tolerance and Stress Resistance

Yield and quality in black turmeric can also be improved by enhancing the plant's resistance to environmental stresses, which can be achieved through CRISPR-mediated edits:

- **Drought and salinity tolerance**: CRISPR can be used to edit stress-responsive genes, such as those related to **abscisic acid signaling** or **osmoprotectants** (e.g., *proline biosynthesis* genes), allowing the plant to thrive in suboptimal conditions like drought or saline soils.

- **Disease resistance**: By targeting genes involved in the plant's immune response (e.g., *R-genes*), CRISPR can make black turmeric more resistant to common pests and pathogens, reducing losses and ensuring consistent yield.
- **Cold and heat tolerance**: Editing temperature-sensitive genes can help black turmeric adapt to a wider range of growing environments, ensuring better yield and quality in different climates.

5. Customizing Metabolism for Novel Compound Production

- **Metabolic engineering for unique compounds**: CRISPR can be used to introduce new metabolic pathways by editing multiple genes simultaneously, allowing black turmeric to produce completely new compounds that have pharmaceutical or nutraceutical applications. For instance, introducing genes for rare curcuminoids or other terpenoids could provide unique therapeutic properties.

CRISPR Strategy for Black Turmeric Improvement

1. **Gene Identification**: Identify and characterize key genes involved in curcumin, terpenoid, flavonoid, and essential oil biosynthesis.
2. **Guide RNA (gRNA) Design**: Design specific guide RNAs that target these genes for knockout or activation using CRISPR-Cas9.
3. **Gene editing and validation**: Apply CRISPR to the selected genes and validate the edits by sequencing and analyzing the plant's phytochemical profile using techniques like High-Performance Liquid Chromatography (HPLC) or mass spectrometry.
4. **Field trials**: Test the edited plants in controlled environments to assess improvements in yield, phytochemical content, stress tolerance, and overall quality.

By leveraging CRISPR-Cas9 for precise gene edits, black turmeric can be transformed into a more potent medicinal plant with higher yields, improved quality, and greater resilience, offering significant benefits for both growers and consumers.

9. Mango Ginger (*Curcuma amada*)

9a. Genetic engineering in Mango Ginger breeding

Improving the flavor profile and increasing disease resistance in **mango ginger** (Curcuma amada) through genetic engineering offers a promising approach to enhance its commercial and agricultural value. Mango ginger, known for its distinct mango-like flavor and medicinal properties, can benefit from genetic interventions to enhance its taste, aroma, and resistance to pests and diseases.

1. Improving flavor profile through genetic engineering

Mango ginger's flavor profile is largely influenced by its **volatile oils, terpenes,** and **phenolic compounds**. By using genetic engineering tools such as **CRISPR-Cas9, RNA interference (RNAi)**, or **gene overexpression**, these pathways can be modified to enhance desirable flavor characteristics.

a. Increasing essential oils and volatile compounds

- **Terpene Biosynthesis**: Terpenes are responsible for the aromatic profile of mango ginger. By enhancing the expression of genes involved in terpene biosynthesis (e.g., *geranyl diphosphate synthase* or *limonene synthase*), the overall aroma and flavor can be intensified. Terpenes like **limonene** and **pinene** contribute to citrusy and piney notes, enhancing the mango-like flavor.
- **Volatile phenolic compounds**: Modifying genes related to phenylpropanoid metabolism, such as *phenylalanine ammonia-lyase (PAL)*, can increase the production of volatile phenolic compounds. This can enhance the spicy and fruity undertones characteristic of mango ginger.

b. Enhancing Sugar Content

- **Sucrose and starch metabolism**: Genes that regulate sugar biosynthesis and accumulation can be edited to increase the sweetness of mango ginger. For instance, overexpressing enzymes like **sucrose synthase** or **invertase** could boost sugar levels, resulting in a more palatable flavor.
- **Regulation of organic acids**: Modifying genes involved in organic acid metabolism (such as *malate dehydrogenase*) can help balance acidity, which is crucial for enhancing the overall flavor profile. Reducing excessive acidity can make the flavor smoother and more appealing.

c. Modifying Secondary Metabolites

- **Curcuminoid pathway modulation**: Curcuminoids contribute to both flavor and medicinal properties in many members of the Curcuma genus. Modifying the curcumin synthesis pathway (e.g., *curcumin synthase*) can enhance or reduce curcuminoid content, depending on the desired flavor intensity.
- **Flavonoid enhancement**: By editing flavonoid biosynthetic genes, the production of compounds responsible for bitterness or astringency can be controlled, leading to a smoother flavor. This can be done by downregulating specific enzymes such as *chalcone synthase* or *flavonoid 3'-hydroxylase*.

2. Increasing disease resistance through genetic engineering

Mango ginger is susceptible to various diseases caused by fungi, bacteria, and pests. Genetic engineering can enhance the plant's resistance to these challenges by introducing or modifying genes responsible for immune responses and defense mechanisms.

a. Fungal and Bacterial Resistance

- **Pathogenesis-Related (PR) proteins**: These proteins play a crucial role in plant defense against pathogens. Overexpression of genes encoding **PR proteins** (e.g., *PR-1*, *PR-2*, *PR-5*) can enhance the plant's resistance to fungal and bacterial infections. This would make mango ginger more resistant to common pathogens like *Fusarium* or *Pythium*, which cause root rot and wilt diseases.
- **Chitinase and glucanase**: These enzymes degrade the cell walls of fungal pathogens. Introducing or upregulating genes for **chitinase** or **β-glucanase** can boost the plant's ability to fight off fungal infections, improving overall disease resistance.

b. **Insect and Pest Resistance**

- **Bt (Bacillus thuringiensis) toxin genes**: The introduction of **Bt genes**, which produce proteins toxic to certain insect pests, can provide resistance to common pests like caterpillars or beetles. This reduces the need for chemical pesticides, promoting more sustainable agriculture.
- **Protease Inhibitors**: Introducing genes that encode **protease inhibitors** can interfere with the digestive processes of insects, deterring pests from feeding on the plant. This strategy has been used in other crops to provide broad-spectrum insect resistance.

c. **Resistance to Viral Pathogens**

- **RNA Interference (RNAi)**: RNAi technology can be used to silence viral genes by introducing small RNA molecules that target viral RNA. This can prevent viruses from replicating in the plant cells. RNAi has been effective in controlling viral diseases in various crops and could be adapted for use in mango ginger to prevent viral infections.

d. **Broad-Spectrum Disease Resistance via R-genes**

- **R-gene overexpression**: R-genes (resistance genes) recognize pathogen attack and trigger immune responses. By identifying and overexpressing key R-genes in mango ginger, the plant's resistance to a wide range of pathogens can be enhanced. R-genes can activate **hypersensitive response (HR)**, which leads to localized cell death at the infection site, preventing the spread of pathogens.

e. **Salicylic Acid Pathway Activation**

- The **salicylic acid (SA)** pathway is crucial for systemic acquired resistance (SAR), which provides long-term protection against a broad range of pathogens. Overexpression of genes involved in SA biosynthesis or signaling, such as *NPR1* (Non-expressor of PR genes), can enhance the plant's innate immunity, making it more resistant to disease outbreaks.

Genetic engineering techniques for Mango Ginger

1. **CRISPR-Cas9**: This precise gene-editing tool can knock out undesirable genes (e.g., genes that make the plant susceptible to diseases) or activate genes involved in flavor enhancement or disease resistance.
2. **RNAi**: RNA interference can silence specific genes that contribute to vulnerability to pathogens or poor flavor traits.
3. **Agrobacterium-mediated transformation**: This method can be used to introduce foreign genes, such as **Bt genes** for insect resistance or genes from other plants that enhance disease resistance.

Potential strategy for genetic engineering in Mango Ginger

1. **Gene Discovery**: Identify the genes responsible for flavor compounds, disease resistance, and growth traits in mango ginger. This could involve transcriptome analysis and genome sequencing.
2. **CRISPR/Cas9 Editing**: Design guide RNAs to target and modify the desired genes. For flavor, target genes involved in terpene and sugar metabolism. For disease resistance, focus on PR proteins, R-genes, and chitinases.
3. **Transformation and regeneration**: Use **Agrobacterium tumefaciens** or other transformation methods to introduce the CRISPR-Cas9 machinery or RNAi constructs into mango ginger cells.
4. **Validation and testing**: Grow the genetically modified plants and test for enhanced flavor compounds using **gas chromatography-mass spectrometry (GC-MS)** and for disease resistance through pathogen inoculation trials.

Conclusion

Through genetic engineering, mango ginger can be improved to produce more intense and desirable flavors while also becoming more resistant to diseases. By targeting the biosynthesis pathways of key flavor compounds and enhancing the plant's natural defense mechanisms, it is possible to create a more robust and commercially viable crop.

9b. CRISPR in Mango Ginger breeding

Enhancing the aroma and medicinal properties of **mango ginger** (Curcuma amada) through gene editing using **CRISPR-Cas9** involves precise modifications to the biosynthetic pathways responsible for volatile aroma compounds, essential oils, and bioactive metabolites that contribute to its therapeutic value. Below is an approach that combines both aroma enhancement and the improvement of medicinal properties using CRISPR technology.

1. Enhancing aroma through gene editing

The distinctive aroma of mango ginger is largely due to its **terpenes** and **volatile oils**. Key genes in the biosynthesis of these compounds can be targeted using CRISPR to increase the production of specific aromatic compounds.

a. Terpene Biosynthesis Pathway

- **Key terpene synthase genes**: Terpenes like **limonene**, **α-pinene**, and **β-caryophyllene** contribute to mango ginger's citrusy, piney, and woody scents. By editing or overexpressing **terpene synthase** genes (e.g., *limonene synthase*, *α-pinene synthase*), CRISPR can enhance the production of these terpenes, resulting in a more intense and desirable aroma.
- **Regulation of geranyl diphosphate (GPP) production**: GPP is a key precursor in the biosynthesis of many terpenes. Editing genes that regulate GPP synthesis, such as **geranyl diphosphate synthase (GPPS)**, can increase the availability of precursors, boosting overall terpene production.

b. Volatile Oils and Phenolic Compounds

- **Phenylpropanoid pathway enhancement**: Volatile phenolic compounds contribute significantly to the aroma profile. CRISPR can be used to upregulate genes in the **phenylpropanoid pathway**, such as *phenylalanine ammonia-lyase (PAL)* and *cinnamate-4-hydroxylase (C4H)*. This can increase the levels of **volatile phenolics** like **eugenol** and **methyl cinnamate**, both known for their strong aromatic properties.
- **Modification of methylation pathways**: Volatile esters such as **methyl salicylate** contribute to fruity and floral aromas. By editing genes involved in the methylation of phenolic compounds (e.g., *salicylic acid methyltransferase*), CRISPR can boost the production of these esters, enhancing the aroma profile.

2. Enhancing medicinal properties through CRISPR

Mango ginger has a range of medicinal properties, including anti-inflammatory, antimicrobial, and antioxidant activities, which are linked to its bioactive compounds such as **curcuminoids**, **terpenoids**, **flavonoids**, and **phenolics**.

CRISPR can be used to edit genes involved in the production of these bioactives to enhance their concentration and efficacy.

a. Curcuminoid Production

- **Curcumin synthase gene activation**: Curcuminoids, particularly **curcumin**, have potent anti-inflammatory and antioxidant effects. By targeting and activating curcumin biosynthesis genes such as **curcumin synthase (CURS)**, CRISPR can increase the curcuminoid content in mango ginger, enhancing its medicinal properties.
- **Modulation of precursor pathways**: Curcumin is synthesized from **p-coumaric acid** and **ferulic acid**. Genes involved in the **shikimate pathway** and **phenylpropanoid pathway** that supply these precursors, such as *phenylalanine ammonia-lyase (PAL)* and *4-coumarate ligase (4CL)*, can be upregulated to boost curcumin production.

b. Increasing Flavonoids for Antioxidant Activity

- **Upregulation of flavonoid biosynthesis genes**: Flavonoids, like **quercetin** and **kaempferol**, are powerful antioxidants. Editing genes in the flavonoid biosynthesis pathway, such as *chalcone synthase (CHS)* and *flavonoid 3'-hydroxylase (F3'H)*, can enhance the production of these compounds. This will not only improve the plant's antioxidant properties but also contribute to its overall therapeutic value.

c. Boosting Terpenoid and Essential Oil Content

- **Editing terpene synthase genes for medicinal terpenoids**: In addition to their role in aroma, terpenoids such as **β-caryophyllene** and **α-pinene** have antimicrobial and anti-inflammatory properties. By editing specific **terpene synthase** genes, CRISPR can increase the production of these medicinal terpenoids. For example, increasing **β-caryophyllene** can enhance the anti-inflammatory potential of mango ginger.
- **Enhancing antimicrobial volatile oils**: Essential oils like **eugenol** and **methyl eugenol** exhibit strong antimicrobial properties. By editing enzymes involved in their biosynthesis, such as **eugenol synthase**, CRISPR can increase their levels, improving the plant's capacity to fight bacterial and fungal infections.

d. Enhancing Alkaloid and Phenolic Content

- **Alkaloid biosynthesis**: Alkaloids often contribute to medicinal properties such as analgesic and anti-cancer effects. CRISPR can be used to upregulate or introduce genes involved in alkaloid biosynthesis, potentially boosting these compounds in mango ginger.

- **Phenolic compounds for enhanced anti-inflammatory effects**: Phenolic acids, such as **ferulic acid** and **p-coumaric acid**, have known anti-inflammatory properties. By editing genes like *cinnamate-4-hydroxylase (C4H)* and *ferulate 5-hydroxylase (F5H)*, CRISPR can boost the production of these phenolics, enhancing the medicinal value of mango ginger.

3. Approach to Gene Editing Using CRISPR

a. Gene Identification and Targeting

- **Genome sequencing and gene mapping**: The first step is to identify the genes involved in the biosynthesis of terpenes, curcuminoids, flavonoids, and other medicinal compounds through genome sequencing and transcriptome analysis.
- **CRISPR guide RNA design**: Once the key genes are identified, guide RNAs (gRNAs) can be designed to target specific regions in these genes for editing.

b. CRISPR-Cas9 Delivery Methods

- **Agrobacterium-mediated Transformation**: This is a common method to deliver CRISPR components into plant cells, where the desired gene edits can be made.
- **Particle Bombardment**: Alternatively, CRISPR-Cas9 DNA constructs can be delivered into mango ginger cells through particle bombardment or biolistics, which is useful for plants that are difficult to transform.

c. Validation of Gene Edits

- **Molecular analysis**: The gene-edited plants are analyzed using **PCR** and **sequencing** to confirm successful edits in the target genes.
- **Phytochemical profiling**: Techniques like **gas chromatography-mass spectrometry (GC-MS)** and **high-performance liquid chromatography (HPLC)** are used to measure changes in the levels of terpenes, curcuminoids, and other phytochemicals.
- **Bioassays for medicinal properties**: The medicinal efficacy of the edited plants can be evaluated through **antioxidant assays** (e.g., DPPH assay) and **anti-inflammatory tests** to confirm the enhancement of medicinal properties.

4. Expected Outcomes

By employing CRISPR for gene editing in mango ginger, the following enhancements can be achieved:

- **Stronger aroma**: Increased levels of terpenes and volatile oils will result in a more intense and appealing mango-like aroma.
- **Improved medicinal efficacy**: Higher concentrations of curcuminoids, flavonoids, and terpenoids will improve the plant's anti-inflammatory, antimicrobial, and antioxidant properties.
- **Balanced flavor profile**: Adjusting the levels of volatile phenolic compounds and terpenes will also enhance the taste, making mango ginger more suitable for culinary and medicinal applications.

CRISPR-Cas9 offers a precise and powerful method to enhance both the **aroma** and **medicinal properties** of mango ginger by targeting specific genes involved in the production of key bioactive compounds. This approach not only improves the sensory appeal of the plant but also boosts its therapeutic potential, making it more valuable for both medicinal and commercial purposes.

10. Coriander (Coriandrum sativum)

10a. Genetic engineering in Coriander breeding

Increasing essential oil content and improving resistance to pests and environmental stresses in **coriander** (Coriandrum sativum) through genetic engineering can significantly enhance its commercial value, flavor, and resilience. Genetic engineering tools, particularly **CRISPR-Cas9**, can be employed to target genes responsible for essential oil biosynthesis, pest resistance, and tolerance to environmental stresses like drought, salinity, and heat. Below is an approach to achieve these improvements in coriander.

1. Increasing Essential Oil Content through Genetic Engineering

The essential oil in coriander is mainly composed of **linalool**, **geraniol**, and **terpinene**, which contribute to its characteristic aroma and flavor. Genetic modifications can boost the production of these oils by targeting the metabolic pathways involved in **terpene biosynthesis**.

a. Upregulating Terpene Synthase Genes

- **Terpene synthase genes (TPSs)**: Terpene synthase enzymes catalyze the formation of key compounds in essential oil production. Genes like **linalool synthase** (for linalool), **geranyl diphosphate synthase (GPPS)**, and **geraniol synthase** (for geraniol) can be overexpressed using CRISPR to enhance terpene synthesis, increasing the essential oil content in coriander.
- **Enhanced Geranyl Diphosphate (GPP) Production**: GPP is the precursor for the biosynthesis of monoterpenes such as linalool and geraniol. Editing genes that regulate the GPP pathway, such as **geranyl diphosphate synthase (GPPS)**, can increase the availability of this precursor, leading to higher production of essential oils.

b. Improving Flux through the Methylerythritol Phosphate (MEP) Pathway

- **MEP pathway enzyme activation**: The MEP pathway produces precursors for monoterpenes, diterpenes, and other essential oil components. Genes like **1-deoxy-D-xylulose-5-phosphate synthase (DXS)** and **2-C-methyl-D-erythritol 4-phosphate synthase (MCS)** can be targeted to increase the flux of metabolites through this pathway, enhancing the overall production of essential oils.

c. Reducing Competing Pathways

- **Knocking out competing pathways**: Some metabolic pathways might divert resources away from essential oil production. For example, using CRISPR to knock out genes involved in **carbohydrate metabolism** or other non-essential pathways can redirect precursors and energy toward terpene biosynthesis, increasing essential oil output.

2. Improving Resistance to Pests through Genetic Engineering

Pests like aphids, mites, and fungal pathogens can reduce coriander yield and quality. Genetic engineering can enhance coriander's resistance to pests by introducing or modifying genes related to defense mechanisms.

a. Incorporating Bt Genes for Insect Resistance

- **Bt (bacillus thuringiensis) toxin genes**: Bt genes encode proteins that are toxic to certain insect pests like caterpillars and aphids. Introducing **Bt genes** into coriander can provide resistance against a range of insect pests, reducing the need for chemical pesticides and enhancing crop yield.

b. Enhancing Production of Secondary Metabolites with Insecticidal Properties

- **Terpenoid production for pest defense**: Terpenoids, in addition to being aromatic compounds, have insecticidal properties. Increasing the production of terpenoids like **linalool** and **α-terpineol** through the upregulation of **TPS genes** can not only improve the essential oil content but also make the plant less attractive to insect pests.
- **Phenylpropanoid pathway activation**: Flavonoids and phenolic compounds are known for their role in plant defense. By enhancing genes in the **phenylpropanoid pathway**, such as **phenylalanine ammonia-lyase (PAL)** and **chalcone synthase (CHS)**, coriander can increase the production of secondary metabolites that deter insect pests and protect against pathogens.

c. RNA Interference (RNAi) for Pest Resistance

- **RNAi technology**: RNA interference can be used to silence essential genes in pests that feed on coriander. For example, genes involved in the digestive systems of aphids or mites can be targeted, leading to their inability to process plant material effectively, thus reducing pest populations.

3. Improving Environmental Stress Tolerance through Genetic Engineering

Coriander is often exposed to environmental stresses like **drought**, **heat**, and **salinity**, which can negatively affect its growth and yield. Genetic engineering can enhance the plant's tolerance to these stresses by targeting genes involved in stress responses.

a. Drought Tolerance

- **Overexpression of Drought-Tolerant Genes**: Genes like **DREB (Dehydration Responsive Element Binding proteins)** and **AREB (ABA Responsive Element Binding proteins)** are crucial in regulating drought tolerance in plants. CRISPR can be used to overexpress these genes, enabling coriander to better manage water stress by activating **abscisic acid (ABA)**-dependent and **ABA-independent** pathways, which improve water retention and reduce transpiration.
- **Aquaporin regulation**: **Aquaporins** are membrane proteins that regulate water movement within plant cells. By editing genes that regulate aquaporins, coriander's ability to maintain water balance during drought can be enhanced, improving overall resilience to dry conditions.

b. Salinity Tolerance

- **Sodium transporter genes**: Salinity stress affects coriander by causing ionic imbalances, particularly sodium toxicity. Overexpression of **NHX (Na+/H+ antiporter)** genes can improve the plant's ability to sequester sodium ions in vacuoles, thus reducing their toxic effects and enhancing growth under saline conditions.
- **Proline biosynthesis**: Proline accumulation helps plants mitigate the effects of osmotic stress caused by salinity. By editing **proline biosynthesis genes** (such as *P5CS*), CRISPR can increase proline levels in coriander, improving its ability to cope with saline environments.

c. Heat Tolerance

- **Heat shock proteins (HSPs)**: HSPs help plants survive high temperatures by protecting proteins from denaturation and assisting in protein refolding. Overexpressing **HSP genes** like **HSP70** and **HSP90** through CRISPR can help coriander plants tolerate heat stress more effectively.

- **Heat tolerance transcription factors**: Genes like **HSFA1** and **HSFA3**, which are transcription factors regulating the heat stress response, can be edited to boost the plant's ability to activate protective mechanisms when exposed to high temperatures.

4. Approach to Gene Editing Using CRISPR

a. Gene Identification and Targeting

- **Identify key genes**: The first step is to identify genes involved in essential oil biosynthesis, pest resistance, and stress tolerance through genome analysis and transcriptome studies.
- **CRISPR guide RNA design**: Once the key genes are identified, guide RNAs (gRNAs) are designed to target specific regions of the genome for knockout, upregulation, or precise edits.

b. CRISPR-Cas9 Delivery Methods

- **Agrobacterium-mediated transformation**: A common method for delivering the CRISPR-Cas9 machinery into plant cells, suitable for transforming coriander.
- **Particle bombardment**: For coriander cells that are difficult to transform with Agrobacterium, particle bombardment can be used to deliver CRISPR constructs into the plant cells.

c. Validation and Testing

- **Molecular confirmation**: DNA sequencing and PCR are used to confirm the successful edits in coriander's genome.
- **Phytochemical profiling**: Essential oil content is analyzed using **gas chromatography-mass spectrometry (GC-MS)** to measure increases in terpene levels, while pest resistance is evaluated through bioassays.
- **Stress Tolerance assays**: Coriander plants are subjected to drought, salinity, and heat stress to assess improvements in stress resilience.

5. Expected Outcomes

By using CRISPR-Cas9 to edit genes responsible for essential oil production, pest resistance, and stress tolerance, coriander can be improved as follows:

- **Increased essential oil content**: Higher levels of linalool, geraniol, and other terpenes will enhance coriander's flavor and aroma, making it more appealing for culinary and industrial uses.
- **Enhanced pest resistance**: Incorporating Bt genes and upregulating secondary metabolite pathways will reduce crop losses from insect pests and pathogens, improving overall yield.

- **Improved stress tolerance**: By editing genes related to drought, salinity, and heat tolerance, coriander will be able to thrive in more challenging environmental conditions, ensuring better crop stability and productivity.

Genetic engineering, particularly CRISPR, offers a powerful tool for enhancing both the **essential oil content** and **resilience** of coriander. By targeting the key pathways involved in terpene biosynthesis, pest defense, and stress response, coriander can be improved for greater productivity, higher-quality oils, and improved resistance to environmental challenges and pests.

10b. CRISPR in Coriander breeding

CRISPR-Cas9 can be used for precise gene editing in coriander (**Coriandrum sativum**) to enhance both **flavor compound production** and **disease resistance**. By targeting specific genes involved in essential oil biosynthesis, flavor compounds like **linalool** and **geraniol** can be enhanced. Similarly, genes associated with defense mechanisms can be edited to improve resistance to pathogens such as fungi, bacteria, and insects. Here's how gene editing can be applied in coriander for these improvements.

1. Precise Gene Editing for Flavor Compound Production

Flavor compounds in coriander are largely determined by the **biosynthesis of terpenes**, **phenolic compounds**, and **volatile oils**. These compounds, such as **linalool**, **geraniol**, and **α-pinene**, define the plant's characteristic aroma and taste.

a. Enhancing linalool production

- **Target Gene: Linalool Synthase (LIS)**: Linalool is a major monoterpene contributing to coriander's distinctive aroma. By using CRISPR to upregulate the **LIS gene**, which catalyzes the conversion of **geranyl diphosphate (GPP)** to linalool, the production of this flavor compound can be significantly increased.
- **Gene editing strategy**: The **promoter region** of the LIS gene can be edited to enhance its expression. Alternatively, a **knock-in** strategy could be used to introduce stronger promoters, resulting in higher levels of linalool synthesis.

b. Boosting Geraniol Production

- **Target Gene: Geraniol Synthase (GES)**: Geraniol contributes to the sweet, floral aroma in coriander. CRISPR can be used to overexpress **GES**, increasing the production of geraniol from **GPP**.
- **Pathway optimization**: To increase the availability of **GPP**, CRISPR can also be used to enhance the expression of **geranyl diphosphate synthase (GPPS)**, which catalyzes the production of GPP, a precursor for both linalool and geraniol.

c. Balancing Other Terpenes

- **Target Gene: Myrcene Synthase (MS)**: **Myrcene** contributes to the earthy and spicy notes in coriander. By using CRISPR to modulate the **MS gene**, the production of myrcene can be adjusted. If a more floral and sweet aroma is desired, the MS gene could be downregulated, shifting the balance toward compounds like linalool and geraniol.
- **Gene editing strategy**: **Knockout** of myrcene synthase through CRISPR-Cas9 can reduce the levels of myrcene, thereby enhancing the relative concentrations of linalool and geraniol.

d. Modifying the Phenylpropanoid Pathway

- **Target Genes: PAL (Phenylalanine Ammonia-Lyase)** and **C4H (Cinnamate-4-Hydroxylase)**: These genes are involved in the production of **volatile phenolics** and **flavonoids**, which influence both aroma and flavor. By editing these genes, the production of key phenolic compounds can be enhanced to improve coriander's flavor profile.
- **Gene Editing Strategy**: Upregulation of PAL and C4H via promoter editing could lead to an increase in phenolic compounds, enhancing the spicy and aromatic notes of coriander.

2. Precise Gene Editing for Disease Resistance

Coriander is susceptible to various pests and diseases, including fungal infections (e.g., **Fusarium** and **Pythium**), bacterial infections, and insect attacks. CRISPR can be used to edit genes that improve the plant's ability to resist these diseases.

a. Enhancing Fungal Resistance

- **Target gene: Chitinase (CHI)**: **Chitinase** enzymes degrade the chitin in fungal cell walls, providing resistance to fungal pathogens. By overexpressing the CHI gene using CRISPR, coriander can be made more resistant to fungal diseases such as **Fusarium wilt**.
- **Gene editing strategy**: CRISPR can be used to insert additional copies of the CHI gene or to enhance its expression by editing regulatory sequences, leading to increased production of chitinase enzymes.
- **Target gene: Pathogenesis-Related (PR) Proteins**: **PR proteins**, such as **PR-1** and **PR-5**, play a role in the plant's defense against fungal and bacterial pathogens. Enhancing the expression of these genes using CRISPR can make coriander more resistant to infections like **root rot** caused by **pythium**.
- **Gene editing strategy**: **Promoter editing** to boost PR protein levels or insertion of additional resistance genes can be achieved using CRISPR.

b. Improving Resistance to Bacterial Pathogens

- **Target gene: R-Genes (Resistance Genes)**: **R-genes** recognize pathogen attack and trigger immune responses. CRISPR can be used to introduce or modify specific **R-genes** that enhance resistance to bacterial pathogens.
- **Gene editing strategy**: Introducing **broad-spectrum R-genes**, such as **NBS-LRR** genes (Nucleotide-Binding Site Leucine-Rich Repeat), can improve the plant's ability to detect and respond to bacterial infections.

c. Enhancing Insect Resistance

- **Target Gene: Bt (Bacillus thuringiensis) Toxin Genes**: Introducing **Bt genes** into coriander through CRISPR provides resistance to insect pests such as caterpillars and beetles. Bt proteins are toxic to many insect larvae but safe for humans, making it an environmentally friendly solution.
- **Gene editing strategy**: The **Bt gene** can be introduced into the coriander genome using CRISPR to provide continuous protection against pests without the need for chemical pesticides.
- **Target Gene: protease inhibitors**: **Protease inhibitors** interfere with the digestive enzymes of insects, making it harder for them to feed on the plant. By overexpressing protease inhibitor genes, coriander can become less attractive to insect pests.
- **Gene editing strategy**: CRISPR can be used to enhance the expression of protease inhibitor genes, such as **cystatin** or **serpin**, to create natural insect resistance.

d. Improving Resistance to Viral Pathogens

- **Target gene: RNA Interference (RNAi)**: CRISPR can be used to introduce RNAi constructs that target viral RNA, silencing viral genes and preventing infections like those caused by **cucumber mosaic virus (CMV)**, which can affect coriander.
- **Gene Editing Strategy**: RNAi constructs targeting viral replication proteins can be introduced via CRISPR to prevent the virus from multiplying in plant cells.

3. CRISPR Gene Editing Approach in Coriander

a. Gene identification and targeting

- **Gene selection**: First, genes responsible for flavor production (e.g., **terpene synthases**) and disease resistance (e.g., **PR proteins** and **R-genes**) need to be identified. This can be done through genomic studies and transcriptomic analysis of coriander.

- **Guide RNA (gRNA) design**: Once target genes are identified, gRNAs specific to these genes are designed to direct CRISPR-Cas9 to the precise locations in the coriander genome for editing.

b. CRISPR-Cas9 Delivery Methods

- **Agrobacterium-mediated transformation**: One of the most common methods for gene delivery in plants, Agrobacterium can be used to transfer the CRISPR-Cas9 system into coriander cells.
- **Particle bombardment (Biolistics)**: If Agrobacterium transformation is not efficient, biolistics can be used to deliver CRISPR-Cas9 plasmids directly into coriander cells.

c. Validation and analysis

- **Molecular Analysis**: After editing, molecular tools like **PCR**, **DNA sequencing**, and **qRT-PCR** are used to confirm the presence of desired edits in the coriander genome.
- **Phytochemical profiling**: To evaluate the effect of gene editing on flavor compound production, techniques like **gas chromatography-mass spectrometry (GC-MS)** and **high-performance liquid chromatography (HPLC)** are used to quantify changes in essential oil and terpene content.
- **Disease resistance testing**: Edited plants are exposed to fungal, bacterial, and insect pests in controlled trials to assess improvements in resistance.

4. Expected Outcomes of CRISPR Gene Editing in Coriander

- **Enhanced flavor profile**: Increased production of key flavor compounds like linalool, geraniol, and phenolics will result in a more intense and desirable aroma and taste.
- **Improved disease resistance**: Edited coriander plants will be more resistant to common pathogens such as **Fusarium**, **Pythium**, and **bacterial blight**, reducing losses due to disease.
- **Increased pest resistance**: By introducing **Bt genes** and upregulating secondary metabolites, coriander will become less susceptible to insect pests, reducing the need for pesticide use.
- **Better crop yield and quality**: With improved resilience to pests, diseases, and enhanced flavor, coriander's overall yield and market value will increase.

CRISPR-Cas9 allows precise editing of genes involved in **flavor compound production** and **disease resistance** in coriander, offering a promising approach to enhancing its culinary and agricultural value. Through targeted edits in terpene biosynthesis and defense pathways, coriander can be optimized for both **higher essential oil content** and **greater resistance to pests and diseases**, resulting in a more robust and valuable crop.

11. Fennel (Foeniculum vulgare)

11a. Genetic engineering in Fennel breeding

Increasing the **yield of essential oils** and improving **drought tolerance** in fennel (**Foeniculum vulgare**) through genetic engineering can significantly enhance its economic value and adaptability to varying environmental conditions. Fennel is known for its aromatic properties due to compounds like **anethole**, **fenchone**, and **limonene**, which are key components of its essential oils. Genetic engineering, particularly using **CRISPR-Cas9** or **Agrobacterium-mediated transformation**, can be employed to modify biosynthetic pathways responsible for essential oil production and to enhance the plant's resilience to drought stress.

1. Increasing Essential Oil Yield in Fennel through Genetic Engineering

Fennel's essential oils are composed of various terpenes and phenylpropanoids, with **anethole** being the most abundant compound. Enhancing the biosynthesis of these compounds involves targeting key genes in the **terpene biosynthesis pathway** and the **phenylpropanoid pathway**.

a. Upregulating Anethole and Phenylpropanoid Biosynthesis

- **Target gene: Eugenol Synthase (EGS)** and **Anethole Synthase**: **Anethole** is a derivative of **phenylpropanoids**, and genes like **eugenol synthase (EGS)** and other enzymes in the phenylpropanoid pathway can be upregulated to enhance anethole production.
- **Gene editing strategy**: Using **CRISPR-Cas9**, the **promoter regions** of anethole synthase and EGS can be modified to boost their expression, leading to increased conversion of **cinnamic acid** into anethole. Additionally, knocking out competing pathways can redirect more precursors into anethole biosynthesis.

b. Increasing the Production of Terpenes (Limonene, Fenchone)

- **Target gene: Limonene Synthase (LIS)** and **Fenchone Synthase**: Terpenes such as **limonene** and **fenchone** are important for fennel's essential oil profile. By overexpressing **limonene synthase** and **fenchone synthase**, the levels of these compounds can be enhanced.
- **Pathway optimization**: **Geranyl diphosphate synthase (GPPS)**, which provides the precursor for monoterpenes like limonene, can be upregulated to increase the flux toward terpene production, resulting in higher yields of essential oils.

c. Optimizing the Methylerythritol Phosphate (MEP) Pathway

- **Target gene: 1-Deoxy-D-Xylulose-5-Phosphate Synthase (DXS)**: The MEP pathway provides precursors for the biosynthesis of monoterpenes and sesquiterpenes. By overexpressing **DXS**, a key enzyme in the MEP pathway, more precursors can be made available for essential oil biosynthesis.
- **Gene editing strategy**: CRISPR can be used to enhance **DXS** activity by modifying its regulatory elements, increasing the overall production of essential oil components like anethole and fenchone.

2. Improving Drought Tolerance in Fennel through Genetic Engineering

Drought stress is a major limiting factor in fennel cultivation, as it affects plant growth, essential oil yield, and quality. Genetic engineering can improve fennel's drought tolerance by modifying genes involved in **water retention**, **osmotic adjustment**, and **stress response pathways.**

a. Enhancing Abscisic Acid (ABA) Signaling for Drought Response

- **Target gene: DREB (Dehydration-responsive element-binding proteins)**: **DREB** transcription factors play a key role in activating drought-responsive genes, including those involved in water conservation and stress adaptation. Overexpressing **DREB1A** or **DREB2** using CRISPR can improve fennel's drought tolerance by activating the ABA pathway.
- **Gene editing strategy**: CRISPR can be used to enhance DREB expression, leading to increased expression of drought-response genes that improve water-use efficiency and promote survival under water-limited conditions.

b. Regulating Aquaporins for Water Retention

- **Target gene: Aquaporins**: Aquaporins are membrane proteins that facilitate water movement across cell membranes, playing a crucial role in maintaining water balance under drought stress. Overexpression of **PIP1 (Plasma Membrane Intrinsic Proteins)** can help improve the plant's water retention and transport under drought conditions.
- **Gene editing strategy**: CRISPR can be employed to upregulate aquaporin genes, enhancing the plant's ability to absorb and retain water, thus improving its drought tolerance.

c. Boosting Osmoprotectant Production

- **Target gene: Proline biosynthesis (P5CS)**: **Proline** acts as an osmoprotectant, helping plants to maintain cell turgor and enzyme function under stress. Overexpression of **P5CS (Pyrroline-5-Carboxylate Synthetase)**, a key enzyme in proline biosynthesis, can increase proline accumulation, enhancing drought tolerance.

- **Gene editing strategy**: CRISPR can be used to modify the **P5CS** gene or its regulatory elements, leading to higher proline levels, which will help fennel withstand drought by stabilizing cellular structures and maintaining osmotic balance.

d. Reducing Water Loss via Stomatal Regulation

- **Target gene: Guard Cell ABA-signaling components (OST1)**: Stomatal closure is a critical response to drought that reduces water loss. The **OST1 (Open Stomata 1)** gene plays a key role in ABA-mediated stomatal closure. By enhancing the expression of **OST1**, fennel can reduce transpiration under drought conditions.
- **Gene editing strategy**: CRISPR-Cas9 can be used to edit the regulatory regions of the **OST1** gene, enhancing the plant's ability to close stomata more efficiently during drought, thereby minimizing water loss.

e. Improving Root Architecture for Better Water Uptake

- **Target Gene: Root Growth-Related Genes (EXPA1)**: Genes that control root growth, such as **EXPA1 (Expansin A1)**, can be modified to enhance root system development. A more extensive root system allows better water uptake, improving the plant's ability to survive in dry conditions.
- **Gene editing strategy**: CRISPR can be used to modify root growth-related genes, enhancing root depth and lateral root branching to improve water uptake during drought.

3. Approach to Gene Editing Using CRISPR in Fennel

a. Gene Identification and Selection

- **Gene discovery**: Identify genes involved in essential oil biosynthesis (e.g., **DXS, LIS, anethole synthase**) and drought tolerance (e.g., **DREB, aquaporins, P5CS**). This can be done through genomic analysis, transcriptomics, and pathway mapping in fennel.
- **gRNA design**: Design **guide RNAs (gRNAs)** that target the specific genes for editing. CRISPR-Cas9 can then be used to either **knock out**, **knock in**, or **upregulate** these genes.

b. CRISPR-Cas9 Delivery Methods

- **Agrobacterium-mediated transformation**: Agrobacterium can be used to introduce CRISPR-Cas9 constructs into fennel cells, enabling precise editing of target genes.
- **Particle Bombardment (Biolistics)**: For plants where Agrobacterium transformation is inefficient, particle bombardment can be used to deliver CRISPR constructs into fennel cells.

c. Validation of Gene Editing

- **Molecular analysis**: After editing, molecular techniques like **PCR**, **qRT-PCR**, and **DNA sequencing** are used to confirm successful gene edits.
- **Phytochemical profiling**: Techniques like **GC-MS** and **HPLC** are used to measure changes in essential oil composition and yield. Testing for increased levels of anethole, limonene, and fenchone will confirm the success of essential oil enhancement.
- **Drought tolerance testing**: Edited plants are subjected to drought conditions to evaluate improvements in drought tolerance, including water retention, root development, and proline accumulation.

4. Expected Outcomes

a. Increased essential oil yield

- **Higher anethole and terpene levels**: With enhanced expression of **anethole synthase**, **limonene synthase**, and other related genes, fennel plants will produce higher yields of essential oils, improving their commercial value for flavoring, fragrance, and medicinal uses.
- **Optimized terpene profile**: By targeting the terpenoid biosynthesis pathways, fennel's essential oil profile can be tailored to increase the concentrations of valuable compounds like **anethole**, **limonene**, and **fenchone**.

b. Improved Drought Tolerance

- **Increased water use efficiency**: Enhanced expression of drought-related genes like **DREB**, **aquaporins**, and **P5CS** will improve fennel's ability to survive and grow in water-limited environments, ensuring stable yields even under drought stress.
- **Better osmotic adjustment**: By increasing proline levels and improving stomatal regulation, fennel plants will be better equipped to maintain cellular function and reduce water loss during periods of drought.

11b. CRISPR in Fennel breeding

Anethole is a key flavor component in fennel (**Foeniculum vulgare**), responsible for its sweet, licorice-like aroma. Enhancing anethole production through **gene editing**, particularly using **CRISPR-Cas9**, involves targeting the **phenylpropanoid pathway**, which is responsible for the biosynthesis of **anethole** from precursors such as **phenylalanine**. Here's a detailed approach for increasing anethole production in fennel through gene editing.

1. Biosynthetic Pathway of Anethole

Anethole is synthesized via the **phenylpropanoid pathway**, which converts **phenylalanine** into a series of intermediates, ultimately producing **anethole**. The key steps in this pathway involve:

- **Phenylalanine Ammonia-Lyase (PAL)**: Converts phenylalanine into **cinnamic acid**.
- **Cinnamate-4-Hydroxylase (C4H)**: Converts cinnamic acid into **p-coumaric acid**.
- **Eugenol synthase (EGS)**: Converts **coniferyl alcohol** into **eugenol**, a precursor to anethole.
- **Anethole synthase (AS)**: Converts **eugenol** into **anethole**.

2. Key genes to target for anethole production

To increase anethole production in fennel, gene editing can be used to upregulate or enhance the activity of genes responsible for its biosynthesis.

a. Phenylalanine ammonia-Lyase (PAL)

- **Role**: PAL is the entry-point enzyme in the phenylpropanoid pathway. It converts phenylalanine into cinnamic acid, which is an important precursor for anethole.
- **CRISPR Strategy**: Use **CRISPR-Cas9** to enhance the expression of the **PAL** gene by editing its promoter region or introducing stronger promoters, leading to higher levels of phenylalanine conversion into cinnamic acid. This provides more substrate for downstream production of anethole.

b. Cinnamate-4-Hydroxylase (C4H)

- **Role**: C4H converts cinnamic acid into **p-coumaric acid**, another important intermediate in the phenylpropanoid pathway.
- **CRISPR strategy**: Overexpressing the **C4H** gene via CRISPR can increase the flux of intermediates toward anethole synthesis. This can be achieved by editing the regulatory elements of the C4H gene, boosting the production of p-coumaric acid.

c. Eugenol synthase (EGS)

- **Role**: EGS catalyzes the conversion of **coniferyl alcohol** to **eugenol**, a direct precursor to anethole.
- **CRISPR strategy**: Upregulating **EGS** using CRISPR can boost eugenol production, which is then converted into anethole. By editing the **EGS gene** or its regulatory regions, more eugenol is produced, enhancing the availability of substrate for anethole synthesis.

d. Anethole synthase (AS)

- **Role**: AS is the enzyme that directly converts **eugenol** to **anethole**, the final step in the biosynthetic pathway.
- **CRISPR strategy**: Overexpression of **AS** is critical for increasing anethole production. CRISPR can be used to introduce multiple copies of the **AS gene** or to enhance its expression by modifying the promoter region. This will ensure higher conversion rates of eugenol into anethole.

3. CRISPR-Cas9 Strategy for Enhancing Anethole Production

a. Gene identification and gRNA design

- **Gene Identification**: The first step is to sequence the fennel genome and identify the genes responsible for **PAL, C4H, EGS**, and **AS**. These genes are key players in the phenylpropanoid and anethole biosynthesis pathways.
- **gRNA design**: Once the target genes are identified, specific **guide RNAs (gRNAs)** are designed to target the regulatory regions or coding sequences of these genes for editing. The aim is to increase the expression of these genes by either modifying promoters or inserting additional copies.

b. CRISPR Delivery into Fennel

- **Agrobacterium-mediated transformation**: One of the most efficient methods for delivering CRISPR-Cas9 constructs into fennel cells is through **Agrobacterium tumefaciens**. This bacterium can transfer the CRISPR-Cas9 system into the plant genome, enabling targeted gene editing.
- **Particle bombardment (Biolistics)**: For fennel plants that are difficult to transform using Agrobacterium, **particle bombardment** can be used to introduce the CRISPR-Cas9 constructs into the cells.

c. Editing Gene Promoters for Upregulation

- **Promoter editing**: The promoter regions of **PAL, C4H, EGS,** and **AS** can be edited to enhance their expression. CRISPR can be used to knock out repressive elements in these promoters or to introduce synthetic promoters that drive higher expression of these genes, leading to an increased flux through the anethole biosynthesis pathway.

d. Validation of Gene Edits

- **PCR and sequencing**: After the transformation, the gene edits can be validated using **PCR** and **DNA sequencing** to confirm that the target genes have been successfully edited.
- **Gene Expression Analysis**: **qRT-PCR** can be used to measure the expression levels of the edited genes, ensuring that the edits lead to higher transcription of **PAL, C4H, EGS,** and **AS**.

- **Essential oil profiling**: Techniques such as **gas chromatography-mass spectrometry (GC-MS)** can be used to quantify the levels of anethole and other related compounds in the essential oils of the edited fennel plants.

4. Expected Outcomes

a. Increased anethole production

- By upregulating the key enzymes involved in the **phenylpropanoid pathway**, particularly **anethole synthase**, fennel plants will produce higher levels of anethole. This will lead to an enhancement of the sweet, licorice-like aroma, improving the plant's marketability for culinary and industrial purposes.

b. Improved essential oil yield

- Alongside an increase in anethole, the overall essential oil yield in fennel is likely to increase due to the higher production of other terpenes and phenylpropanoids. This can make fennel more valuable for the essential oil industry.

c. Optimized Flavour Profile

- Targeted edits to specific terpene and phenylpropanoid pathway genes will also influence the production of other flavor compounds, potentially optimizing fennel's aroma and flavor profile for specific commercial needs.

Using **CRISPR-Cas9** to enhance **anethole** production in fennel is a precise and effective strategy for improving the plant's flavor and essential oil content. By targeting key genes in the **phenylpropanoid biosynthetic pathway**—including **PAL**, **C4H**, **EGS**, and **AS**—the yield of anethole can be significantly increased. This gene-editing approach not only boosts the production of this important flavor compound but also enhances the overall essential oil profile of fennel, making it more valuable for both culinary and industrial applications.

12. Fenugreek (*Trigonella foenum-graecum*)

12a. Genetic engineering in Fenugreek breeding

To increase diosgenin content, yield, and disease resistance in fenugreek (*Trigonella foenum-graecum*) through genetic engineering, several strategies can be employed:

1. Enhancing Diosgenin Biosynthesis

Diosgenin is a steroidal sapogenin, and its production can be boosted by manipulating the genes involved in its biosynthetic pathway. Techniques include:

- **Overexpression of key enzymes**: Genes encoding enzymes such as cycloartenol synthase, cytochrome P450 enzymes, and glycosyltransferases, which are involved in steroidal saponin biosynthesis, can be overexpressed to increase diosgenin content.
- **Gene editing (CRISPR/Cas9)**: CRISPR/Cas9 can be used to knock out or modify negative regulators of the diosgenin biosynthesis pathway, thereby enhancing its production.

2. Improving Yield

Genetic engineering approaches to increase overall biomass and seed yield include:

- **Modifying plant growth regulators**: Genes involved in the synthesis or signaling of plant hormones such as gibberellins, auxins, and cytokinins can be manipulated to promote growth and development.
- **Improving photosynthesis**: Overexpression of genes related to photosynthesis, such as RuBisCO activase or carbonic anhydrase, can improve photosynthetic efficiency, leading to better growth and yield.
- **Stress tolerance genes**: Introducing genes that confer tolerance to abiotic stresses like drought and salinity (e.g., DREB1 or HSPs) could help fenugreek thrive in suboptimal conditions, enhancing yield.

3. Enhancing Disease Resistance

Disease resistance in fenugreek can be increased by incorporating:

- **Pathogen resistance (R) genes**: Transgenic approaches can introduce *R* genes that recognize specific pathogens and trigger plant defense mechanisms. This can be done by transferring resistance genes from related species or other plants.
- **Antimicrobial peptide genes**: Introducing genes that encode antimicrobial peptides like defensins or thionins can provide broad-spectrum resistance to fungal, bacterial, and viral pathogens.
- **RNA interference (RNAi)**: RNAi technology can be used to silence essential genes in pathogens, thereby preventing their growth and infection in fenugreek.

4. Metabolic Pathway Engineering

By targeting the metabolic flux of diosgenin precursors, metabolic engineering strategies could ensure more precursor metabolites are directed toward diosgenin production, instead of competing pathways. This involves:

- **Overexpressing precursor supply genes**: Increasing the availability of precursors like cholesterol by enhancing the expression of upstream genes in the mevalonate pathway.

- **Reducing competing pathways**: Silencing or knocking out enzymes that divert precursors away from diosgenin biosynthesis.

5. Plant Transformation Techniques

- **Agrobacterium-mediated transformation**: This is commonly used for fenugreek and involves the introduction of the desired gene constructs into fenugreek's genome through *Agrobacterium tumefaciens.*
- **CRISPR/Cas9-based editing**: This can be used to precisely modify or edit genes involved in diosgenin biosynthesis and disease resistance, providing a precise and efficient way to improve traits.

By combining these genetic engineering strategies, fenugreek plants can be engineered to have higher diosgenin content, improved yield, and enhanced resistance to diseases.

12b. CRISPR in Fenugreek breeding

Modifying genes related to the biosynthesis of bioactive compounds and stress tolerance in fenugreek using CRISPR/Cas9 can be a powerful approach to improving the plant's health benefits and resilience. Here's a step-by-step breakdown of how CRISPR can be applied for both:

1. Targeting Bioactive Compound Biosynthesis

Fenugreek is known for producing bioactive compounds such as diosgenin, flavonoids, and other saponins. To enhance the biosynthesis of these compounds using CRISPR/Cas9, we can:

a. Identify and target key genes in diosgenin and saponin biosynthesis

- **Cycloartenol synthase (CAS)** and **Cytochrome P450 enzymes** are key in the production of steroidal saponins like diosgenin. You can:
  - **Upregulate genes**: Use CRISPR activation (CRISPRa) to enhance the expression of genes that catalyze key steps in diosgenin biosynthesis.
  - **Knockout suppressor genes**: Inactivate genes that negatively regulate diosgenin production using CRISPR/Cas9 knockouts to prevent the diversion of precursors away from saponin biosynthesis.
  - **Pathway optimization**: Ensure higher flux towards diosgenin by knocking out competing metabolic pathways (e.g., by targeting genes involved in the synthesis of other triterpenoids).

b. Editing Regulatory Genes to Enhance Production

- **Transcription factors** involved in regulating bioactive compound biosynthesis can be edited. Examples include WRKY, MYB, or bHLH transcription factors that control the expression of secondary metabolite-related genes. CRISPR/Cas9 can knock out negative regulators or enhance positive ones to increase overall bioactive compound production.

2. Improving Stress Tolerance through CRISPR/Cas9

To enhance fenugreek's resistance to abiotic stresses (such as drought, salinity, and temperature extremes), genes that regulate the plant's stress response can be modified. CRISPR can be used to modify the following:

a. DREB (Dehydration Responsive Element Binding) Genes

- **DREB transcription factors** are key players in the plant's response to drought and cold. Using CRISPR/Cas9, you can:
  - **Overexpress DREB genes** through CRISPRa to increase the plant's tolerance to drought and salinity.
  - **Mutate repressor genes** that limit DREB activity, ensuring the plant's stress response is activated more quickly or strongly.

b. Heat Shock Proteins (HSPs)

- **HSP genes** help the plant cope with heat stress. By targeting genes like **HSP70** or **HSP90**, CRISPR can be used to:
  - **Increase expression**: Upregulate these genes to provide enhanced protection against heat.
  - **Knock out repressors**: Silence genes that inhibit HSP production under stress conditions.

c. ABA (Abscisic Acid) Signaling Pathway

- **Abscisic acid** is crucial in drought and salt stress responses. You can use CRISPR to:
  - **Knock out ABA-inactivating enzymes**: This will maintain higher levels of ABA during stress.
  - **Target ABA receptor genes (like PYR/PYL)** to modify ABA sensitivity, making fenugreek more responsive to environmental stresses.

d. ROS (Reactive Oxygen Species) Scavenging Pathways

- **Reactive oxygen species** accumulate under stress, leading to cellular damage. Genes like **superoxide dismutase (SOD)** or **catalase** are important for detoxifying ROS.
  - **Upregulate SOD or catalase** through CRISPRa to increase the plant's ability to detoxify harmful ROS.
  - **Edit ROS-related transcription factors**, like NAC or WRKY, which control the expression of multiple antioxidant enzymes.

3. CRISPR Tools and Methods for Fenugreek

- **CRISPR/Cas9** is the primary tool, where the Cas9 protein introduces double-strand breaks at the target gene site, followed by error-prone repair or insertion of new sequences via homology-directed repair (HDR).
- **CRISPR/Cas12a (Cpf1)**: An alternative to Cas9, which creates staggered cuts and may offer more precise gene edits.
- **CRISPRa and CRISPRi**: CRISPR activation (a) and interference (i) allow modulation of gene expression without cutting DNA, making them useful for upregulating or downregulating target genes.
- **Delivery**: Fenugreek cells can be transformed via **Agrobacterium-mediated transformation** or **gene gun** technology to deliver the CRISPR/Cas9 components.

4. Example Targets for Fenugreek Using CRISPR

Gene	Function	CRISPR Modification	Outcome
CAS (Cycloartenol Synthase)	Steroidal saponin biosynthesis	Knock-in (upregulate) or CRISPRa	Increased diosgenin content
DREB	Stress tolerance (drought/salt)	Overexpression using CRISPRa	Enhanced drought and salt tolerance
SOD (Superoxide Dismutase)	ROS detoxification	CRISPRa for increased expression	Improved stress resilience (oxidative stress)
HSP70	Heat shock response	Upregulation through CRISPRa	Increased tolerance to heat stress
ABA-inactivating enzyme	ABA degradation (stress response)	Knockout via CRISPR/Cas9	Enhanced drought response due to higher ABA levels

5. Considerations and Challenges

- **Gene Off-target effects**: It's crucial to ensure that the CRISPR system only edits the intended target genes without causing unintended mutations elsewhere in the genome.
- **Regulatory pathways**: Careful attention must be paid to the plant's regulatory networks to avoid negative trade-offs, such as growth reduction due to excessive activation of stress pathways.

By using CRISPR/Cas9 to modify genes related to bioactive compound biosynthesis and stress tolerance, fenugreek can be engineered to produce more valuable compounds like diosgenin while becoming more resilient to environmental stresses, leading to both agricultural and medicinal benefits.

13. Cumin (*Cuminum cyminum*)

Genetic engineering & CRISPER gene editing in Cumin breeding

Enhancing the **essential oil content**, **flavor compounds**, **yield**, and **resistance to environmental stresses** in cumin (*Cuminum cyminum*) through genetic engineering involves targeted modifications to key metabolic and regulatory pathways. Here's a breakdown of how this can be achieved:

1. Increasing essential oil content and flavor compounds

The essential oils in cumin are responsible for its characteristic flavor and aroma, mainly composed of **monoterpenes** like **cuminaldehyde**, **β-pinene**, **γ-terpinene**, and **limonene**. To boost the production of these flavor compounds, genetic engineering strategies can target the biosynthetic pathways involved in monoterpene production.

a. Key Enzyme Overexpression

Monoterpenes are synthesized via the **methylerythritol phosphate (MEP)** pathway. Enhancing the expression of key enzymes in this pathway could increase essential oil production:

- **1-Deoxy-D-xylulose-5-phosphate synthase (DXS)** and **1-Deoxy-D-xylulose-5-phosphate reductoisomerase (DXR)**: Overexpression of these enzymes can channel more carbon into the MEP pathway, leading to higher monoterpene output.
- **Monoterpene synthases (e.g., limonene synthase, α-pinene synthase)**: Overexpressing these enzymes could increase the biosynthesis of specific terpenes like **limonene** and **α-pinene**, key components of cumin's flavor.

b. Transcription Factor Activation

Transcription factors such as **MYB** or **bHLH** regulate the expression of genes involved in secondary metabolite biosynthesis. Overexpressing or activating these transcription factors using CRISPR activation (CRISPRa) or other methods can boost the production of terpenes and essential oils in cumin.

c. CRISPR/Cas9 to Knock Out Competing Pathways

CRISPR/Cas9 can be used to knock out genes in competing biosynthetic pathways that might divert precursors away from essential oil production. For instance:

- Target genes in the **sesquiterpene pathway**, if it competes for precursor availability with monoterpenes.

2. Enhancing Yield

Enhancing cumin yield through genetic engineering involves improving both **biomass production** and **seed yield**. This can be achieved by:

a. Improving photosynthesis efficiency

- **RuBisCO Activase (RCA)**: Overexpressing genes like **RCA** can improve the efficiency of photosynthesis, leading to greater biomass accumulation and higher yields.
- **Carbonic anhydrase**: Overexpression of carbonic anhydrase can enhance CO□ fixation, which supports more efficient photosynthesis and increases plant growth.

b. Modulating growth hormone pathways

- **Gibberellin (GA) pathway**: Engineering genes involved in gibberellin biosynthesis can promote stem elongation, seed production, and overall growth. This can be achieved through either overexpression of GA biosynthesis genes or the silencing of GA-degrading enzymes.
- **Cytokinin and auxin regulation**: Cytokinins promote cell division, and auxins regulate growth. Genetic modification to balance the levels of these hormones can lead to better vegetative growth and seed yield.

c. Introducing stress-tolerant growth promoters

- Introducing genes from model plants (e.g., DREB transcription factors or HSPs) that enhance growth under suboptimal environmental conditions can boost overall biomass and yield in cumin.

3. Enhancing Resistance to Environmental Stresses

Cumin is often exposed to **drought, salinity, heat**, and **pathogen attacks**. Genetic engineering can increase its resilience by:

a. Introducing Stress-Responsive Genes

- **DREB (Dehydration Responsive Element Binding) transcription factors**: These play a major role in the plant's response to abiotic stresses like drought and high salinity. Overexpressing **DREB1A** can enhance the plant's ability to tolerate water-deficit conditions.
- **LEA (Late Embryogenesis Abundant) proteins**: LEA proteins protect cells under stress by stabilizing proteins and membranes. Introducing LEA genes can improve drought and salinity tolerance.

b. Improving Reactive Oxygen Species (ROS) Scavenging

- **Superoxide dismutase (SOD)** and **catalase (CAT)**: These are important antioxidant enzymes that help the plant cope with oxidative stress, which is a common response to environmental stresses like drought and heat. Overexpression of these genes can enhance cumin's resistance to oxidative damage caused by stresses.

c. Modulating Abscisic Acid (ABA) Signaling

Abscisic acid (ABA) is crucial for stress response, especially drought tolerance. Genetic modifications in the ABA pathway can:

- **Increase ABA levels** during stress by knocking out ABA-inactivating enzymes (e.g., **ABA 8'-hydroxylase**) using CRISPR/Cas9, which leads to a stronger and faster stress response.
- **Enhance ABA sensitivity** by upregulating ABA receptors, such as **PYR/PYL** receptors, which leads to quicker stomatal closure, reducing water loss during drought.

d. Heat Shock Proteins (HSPs)

HSPs help plants survive under heat stress by preventing protein denaturation. Introducing or overexpressing **HSP70** or **HSP90** can help cumin tolerate higher temperatures.

4. Optimizing Terpene Production for Flavor and Essential Oil Enhancement

- **Terpene synthase Genes**: Monoterpene and sesquiterpene biosynthesis genes, like **terpinene synthase**, **pinene synthase**, and **limonene synthase**, can be overexpressed or edited to produce higher levels of specific flavor compounds.
- **Co-expression of enzyme pathways**: Genes like **cytochrome P450 monooxygenases** and **methyltransferases** can be co-expressed to modify terpenes into flavor-enhancing derivatives.

5. Gene Editing Techniques for Cumin

- **CRISPR/Cas9**: Use CRISPR/Cas9 to knock out unwanted pathways, silence negative regulators, or precisely edit genes involved in terpene biosynthesis and stress tolerance.
- **CRISPRa (activation)**: This can be used to upregulate beneficial genes that improve essential oil biosynthesis, yield, or stress tolerance.
- **CRISPR/Cas12a (Cpf1)**: Another gene-editing tool, which offers more precise and flexible editing, suitable for targeting multiple genes simultaneously in complex pathways like terpene biosynthesis.

6. Example Targets for Genetic Engineering in Cumin

Gene	Function	Modification	Expected Outcome
DXS, DXR	Monoterpene biosynthesis (MEP pathway)	Overexpression or CRISPRa	Increased production of essential oils (monoterpenes)
DREB1A	Drought and salt stress response	Overexpression through CRISPRa	Improved tolerance to drought and salinity
RuBisCO Activase	Photosynthetic efficiency	Overexpression	Increased biomass and seed yield
SOD, Catalase	ROS scavenging under stress	Overexpression	Enhanced tolerance to oxidative stress
Terpene Synthases (e.g., Limonene Synthase)	Flavor compound biosynthesis	Overexpression or CRISPRa	Increased levels of limonene and other flavor compounds
HSP70, HSP90	Heat stress response	Overexpression	Increased tolerance to high temperatures
PYR/PYL (ABA receptor)	Drought stress response	Overexpression	Enhanced drought tolerance through improved ABA sensitivity

By targeting specific biosynthetic pathways and stress response genes through **genetic engineering** and **CRISPR/Cas9** technology, cumin can be modified to:

- Produce higher levels of **essential oils** and **flavor compounds** like monoterpenes.
- **Increase yield** by enhancing photosynthesis, growth hormone regulation, and stress resilience.
- **Enhance stress tolerance** to drought, salinity, and heat by modulating stress-responsive pathways and antioxidant systems.

These modifications can result in cumin plants with greater economic value, improved flavor, and higher resilience to changing environmental conditions.

14. Ajwain (*Trachyspermum ammi*)

Genetic engineering and CRISPER gene editing technology in Ajwain breeding

Increasing the yield, concentration of **thymol** and other essential oils in **ajwain** (*Trachyspermum ammi*) through genetic engineering requires targeting key biosynthetic pathways and regulatory mechanisms. Below is a strategic approach to achieve these goals:

1. Enhancing Thymol and Other Essential Oil Production

Thymol is the primary bioactive compound in ajwain, contributing to its essential oil content. It is a **monoterpene** produced through the **methylerythritol phosphate (MEP)** pathway. Genetic modifications can focus on boosting the production of thymol and other related essential oils (e.g., carvacrol, γ-terpinene) by targeting key enzymes and transcription factors.

a. Overexpressing Key Enzymes in Thymol Biosynthesis

- **Geranyl diphosphate synthase (GPPS)**: This enzyme produces **geranyl diphosphate (GPP)**, a precursor for monoterpenes like thymol. Overexpressing the **GPPS** gene can enhance the supply of GPP, increasing thymol and other monoterpenes.
- **Thymol synthase**: Specific enzymes that convert GPP into thymol are key to its production. Overexpressing **thymol synthase** can directly increase thymol biosynthesis.
- **γ-terpinene synthase**: γ-terpinene is a precursor to thymol. Enhancing the expression of **γ-terpinene synthase** can ensure higher production of thymol.

b. Optimizing the MEP Pathway

The **MEP pathway** is responsible for the biosynthesis of monoterpenes. Enhancing this pathway can result in a higher flux of metabolites towards essential oil production:

- **1-deoxy-D-xylulose-5-phosphate synthase (DXS)** and **1-deoxy-D-xylulose-5-phosphate reductoisomerase (DXR)** are critical enzymes at the start of the MEP pathway. Overexpressing these genes can boost the supply of monoterpene precursors, leading to increased thymol and essential oil content.

c. CRISPR/Cas9 for Pathway Optimization

CRISPR can be used to

- **Knock out competing pathways** that consume intermediates needed for thymol biosynthesis. For instance, targeting sesquiterpene biosynthesis pathways could redirect precursors like GPP towards thymol and other monoterpenes.
- **Edit transcription factors** such as **MYB** or **WRKY**, which regulate the expression of essential oil biosynthesis genes. Overexpressing or activating these transcription factors could upregulate the entire pathway.

2. Increasing Yield

Yield in ajwain can be improved by enhancing biomass production, seed yield, and the plant's ability to thrive under various conditions. Genetic engineering can target several aspects to improve yield:

a. Boosting Photosynthesis Efficiency

- **RuBisCO Activase (RCA)**: Overexpressing the gene encoding RuBisCO activase can improve the efficiency of the photosynthesis process, increasing biomass and seed yield.
- **Carbonic anhydrase**: Overexpression of carbonic anhydrase can enhance carbon dioxide fixation, leading to better growth and increased yield.

b. Modifying Plant Growth Regulators

- **Gibberellin (GA) pathway**: Gibberellins play a crucial role in plant height, seed development, and overall growth. Overexpression of GA biosynthesis genes or downregulation of GA-degrading enzymes (such as **GA2-oxidase**) can promote more robust growth and higher yield.
- **Auxin regulation**: Auxins control cell elongation and division. By modulating genes involved in auxin transport and signaling (such as **PIN-FORMED (PIN)** proteins or **auxin response factors (ARFs)**), you can enhance vegetative growth and seed yield.

c. Introducing Stress-Resistant Traits

- **Drought and salinity tolerance**: Introducing stress-responsive genes, like **DREB1A** (Dehydration Responsive Element Binding protein), can enhance ajwain's ability to grow in suboptimal conditions, which indirectly increases yield.
- **Heat tolerance**: Overexpression of **Heat Shock Proteins (HSPs)**, particularly **HSP70**, can improve ajwain's resilience to high temperatures, contributing to stable yields under environmental stress.

3. Enhancing Resistance to Environmental Stresses

Ajwain often faces stresses like drought, salinity, and heat. Genetic engineering can be used to enhance its resilience to these stresses, leading to better growth and higher yield.

a. Improving antioxidant defense

- **Superoxide Dismutase (SOD)** and **Catalase (CAT)**: Overexpression of these antioxidant enzymes helps the plant detoxify reactive oxygen species (ROS) that accumulate under stress, preventing cellular damage and improving plant health.

b. Enhancing water use efficiency

- **Aquaporins**: These proteins regulate water transport in cells. Overexpressing **aquaporin genes** can improve water uptake and retention, especially under drought conditions.

c. Modulating ABA (Abscisic Acid) pathway

- ABA is crucial for drought and salt tolerance. By editing genes involved in ABA synthesis or signaling, such as **ABA receptors (PYR/PYL)** or **ABA catabolism genes**, the plant can better manage water loss under drought conditions, leading to increased survival and yield.

4. Optimizing Secondary Metabolism and Essential Oil Composition

Secondary metabolites like thymol and other essential oils can be influenced by the availability of precursors and the activity of biosynthetic enzymes. Genetic engineering can:

- **Increase precursor availability** by targeting key enzymes in the MEP pathway (e.g., DXS, DXR, GPPS).
- **Enhance terpene synthase activity**: By overexpressing thymol synthase and γ-terpinene synthase, more precursors are directed towards the production of flavor and aromatic compounds.

5. Gene Editing Approaches for Ajwain

Using advanced genetic tools like CRISPR/Cas9 and CRISPRa (CRISPR activation), several modifications can be made to ajwain's genome:

- **CRISPR/Cas9 knockouts**: These can silence competing pathways or negative regulators of essential oil biosynthesis.
- **CRISPRa (activation)**: CRISPRa can be used to upregulate genes involved in thymol biosynthesis and yield improvement without disrupting the genome.
- **Multiplex CRISPR**: Simultaneous editing of multiple genes involved in thymol biosynthesis and stress tolerance can be achieved using multiplex CRISPR, improving the efficiency of genetic modifications.

6. Example Gene Targets for Ajwain Enhancement

Gene	Function	Genetic Modification	Outcome
GPPS (Geranyl diphosphate synthase)	Precursor to monoterpenes	Overexpression	Increased thymol and essential oil production
Thymol synthase	Converts precursors into thymol	Overexpression	Higher thymol content
γ-terpinene synthase	Precursor for thymol and carvacrol	Overexpression	Enhanced essential oil production
DXS, DXR	MEP pathway (monoterpene biosynthesis)	Overexpression	Increased overall essential oil content
DREB1A	Drought and salinity tolerance	Overexpression through CRISPRa	Improved resistance to drought and salinity
SOD, Catalase	Antioxidant enzymes	Overexpression	Better stress tolerance, improving yield
RuBisCO Activase (RCA)	Photosynthetic efficiency	Overexpression	Increased biomass and seed yield
HSP70	Heat tolerance	Overexpression	Enhanced tolerance to high temperatures

By focusing on **thymol biosynthesis** and related pathways, as well as **enhancing plant growth and stress resistance**, ajwain can be genetically modified to produce higher levels of essential oils, including thymol, and achieve higher overall yields. Employing gene-editing technologies like **CRISPR/Cas9** allows precise manipulation of these pathways, offering a robust approach to improving the economic and medicinal value of ajwain.

Marker assisted breeding in Black Pepper

Marker-assisted breeding (MAB) in black pepper involves using molecular markers to select plants with desirable traits, improving the efficiency and precision of breeding programs. The process typically includes identifying genetic markers linked to important traits such as disease resistance, yield, or quality. These markers are then used to screen plants at the seedling stage, allowing breeders to select only those with the desired traits for further development.

MAB in black pepper can significantly speed up the breeding process, as it allows for early selection and reduces the need for extensive field trials. Additionally, it can help in combining multiple desirable traits into a single plant, leading to the development of superior black pepper varieties.

MAB for Phytophthora resistance varieties in Black Pepper

Marker-assisted breeding (MAB) for Phytophthora resistance in black pepper focuses on identifying and utilizing genetic markers linked to resistance against *Phytophthora capsici*, a pathogen responsible for devastating diseases like foot rot. Through MAB, breeders can develop black pepper varieties that are more resistant to this pathogen by selecting plants that carry specific resistance genes.

The process generally involves screening the black pepper germplasm to identify resistant genotypes. Once the markers associated with resistance are identified, these markers are used to screen and select breeding lines or populations, allowing breeders to combine resistance with other desirable traits, such as high yield or quality. This approach accelerates the breeding process and increases the efficiency of developing resistant varieties.

MAB for high piperin and high yield in Black Pepper

Marker-assisted breeding (MAB) for high piperine content and high yield in black pepper involves identifying and utilizing molecular markers linked to these desirable traits. The process typically starts with identifying genetic markers associated with increased piperine levels, which is the primary alkaloid responsible for black pepper's pungency, and markers linked to traits that contribute to higher yields, such as plant vigor and fruit set.

Once these markers are identified, they are used to screen breeding populations. Plants that exhibit both high piperine content and high yield potential, as indicated by the presence of these markers, are selected for further breeding. This approach accelerates the development of superior black pepper varieties that combine both high piperine content and high yield, making the breeding process more efficient and targeted.

Markers used for variety identification in Black Pepper

In black pepper, markers used for variety identification typically involve molecular markers such as SSRs (Simple Sequence Repeats), RAPDs (Random Amplified Polymorphic DNA), and AFLPs (Amplified Fragment Length Polymorphism). These markers are employed to differentiate between various black pepper cultivars, aiding in the accurate identification of varieties. The use of these markers ensures consistency in quality and helps in the conservation and management of genetic resources.

SSR markers in Black Pepper

The document does not provide specific information about SSR markers in black pepper. However, SSR (Simple Sequence Repeat) markers are commonly used in genetic studies and breeding programs for black pepper. These markers are valuable for assessing genetic diversity, variety identification, and marker-assisted selection.

To obtain specific SSR markers, including their sequences and details about their availability, you would typically need to consult specialized genomic databases like NCBI's GenBank or look at peer-reviewed studies focusing on black pepper genomics. These resources often provide detailed information on the SSR markers used in black pepper research, including their sequences, associated traits, and potential applications in breeding programs.

RAPD markers in Black Pepper

RAPD (Random Amplified Polymorphic DNA) markers have been extensively used in black pepper for genetic diversity studies, cultivar identification, and breeding programs. These markers are advantageous because they do not require prior knowledge of the genome, making them useful for assessing genetic variation and relationships among black pepper varieties.

RAPD markers work by amplifying random segments of the genome, leading to the generation of polymorphic patterns that can be compared across different black pepper accessions. This helps in identifying unique genetic fingerprints associated with specific traits or cultivars, aiding in the selection and breeding of desirable varieties.

AFLP markers in Black Pepper

AFLP (Amplified Fragment Length Polymorphism) markers have been used in black pepper for assessing genetic diversity, cultivar identification, and in breeding programs. AFLP markers are highly polymorphic and do not require prior knowledge of the genome, making them particularly useful in species like black pepper, where extensive genomic resources might not be available. In black pepper, AFLP markers have been employed to evaluate genetic relationships among different cultivars and to map important traits such as disease resistance and yield. These markers help in identifying distinct genetic profiles, which can then be used to guide breeding programs aimed at improving specific traits in black pepper.

Marker assisted selection in Cardamom

Marker-assisted selection (MAS) in cardamom involves using molecular markers to assist in the selection of desirable traits during the breeding process. This technique allows breeders to identify plants with specific genetic traits early in the breeding cycle, speeding up the development of new varieties with improved characteristics.

Key Aspects of Marker-Assisted Selection (MAS) in Cardamom

1. Identification of molecular markers

- Specific DNA sequences (markers) are associated with traits of interest, such as disease resistance, yield, or quality traits like aroma and flavor. These markers are identified through genetic mapping and other molecular biology techniques.

2. Screening and selection

- Cardamom plants are screened for the presence of these markers. Plants that carry the desired markers are selected for further breeding. This is faster and more precise than traditional breeding methods, which rely on phenotype (observable traits) alone.

3. Improvement of desirable traits

- MAS can be used to improve traits such as:
  - **Disease resistance**: Selecting plants that are resistant to pests and diseases like *Katte* or *Kokke kandu.*
  - **Yield**: Enhancing the productivity of the plants.
  - **Quality traits**: Improving the aroma, flavor, and size of cardamom seeds.

4. Application in Breeding Programs

- MAS is integrated into cardamom breeding programs to develop new varieties that meet market demands and are better adapted to environmental conditions or farming practices.

5. Environmental and Economic Benefits

- By accelerating the breeding process, MAS reduces the time and resources needed to develop new varieties, making cardamom cultivation more sustainable and profitable.

Overall, MAS represents a significant advancement in cardamom breeding, allowing for more efficient and targeted improvement of this valuable spice crop.

Genomic selection in Cardamom

Genomic selection (GS) in cardamom is an advanced breeding technique that uses genomic information to predict the performance of plants and select individuals with the best genetic potential for desired traits. This method involves the use of dense genetic markers across the entire genome to make predictions about the breeding value of plants, even before phenotypic traits are fully expressed.

Key Aspects of Genomic Selection (GS) in Cardamom

1. Whole-genome markers

- Unlike marker-assisted selection (MAS), which uses a few markers linked to specific traits, genomic selection utilizes a large number of markers spread across the entire genome. These markers collectively capture the genetic variation influencing complex traits.

2. Training population

- A population of cardamom plants with known genetic markers and phenotypic data (such as yield, disease resistance, or quality traits) is used to develop a statistical model. This model links the genetic markers with the observed traits.

3. Prediction model

- The statistical model developed from the training population is used to predict the genetic value of other plants in the breeding program. These predictions help in selecting the best individuals for breeding, even if they have not yet expressed the desired traits.

4. Accelerated breeding

- GS allows breeders to make selection decisions early in the plant's life cycle, which significantly reduces the time required to develop new cardamom varieties with improved traits. This is especially valuable for perennial crops like cardamom, where traditional breeding cycles can be long.

5. Improvement of complex traits

- Genomic selection is particularly useful for improving complex traits that are controlled by many genes, such as yield, drought tolerance, or resistance to multiple diseases. It provides a more comprehensive approach compared to traditional methods, which may miss some genetic variations.

6. Increased Genetic gain

- By selecting the best individuals based on their genomic estimated breeding values (GEBVs), genomic selection can lead to higher genetic gain per breeding cycle, making the breeding process more efficient and effective.

7. Application in Cardamom breeding

- Genomic selection in cardamom can be used to develop new varieties that are more productive, disease-resistant, and better suited to changing environmental conditions. This is increasingly important as the demand for high-quality cardamom continues to grow in global markets.

Benefits of Genomic selection

- **Efficiency**: Speeds up the breeding process by allowing earlier and more accurate selection.
- **Precision**: Improves the accuracy of selecting plants with the best genetic potential.
- **Cost-effectiveness**: Reduces the costs associated with field trials and long breeding cycles.

In summary, genomic selection represents a powerful tool in the improvement of cardamom, enabling the development of superior varieties that meet the needs of farmers and markets more quickly and efficiently than traditional methods.

MAS in coriander breeding

Marker-assisted selection (MAS) in coriander breeding is a powerful tool used to enhance the efficiency and precision of developing new coriander varieties with desired traits. Here's an overview of how MAS is applied in coriander breeding:

1. Trait Identification and marker development

- **Trait of Interest**: Breeders first identify specific traits that are desirable in coriander, such as disease resistance, yield, flavor, and essential oil content.
- **Marker Identification**: Molecular markers (such as SSRs, SNPs) linked to these traits are identified using genetic mapping techniques. These markers serve as proxies to track the presence of the desired traits in coriander plants.

2. Marker-assisted selection process

- **Crossbreeding**: Coriander plants with desirable traits are crossbred. The offspring are then screened using molecular markers to identify those carrying the desired genes.
- **Selection**: Only the plants that have the molecular markers associated with the desired traits are selected for further breeding. This speeds up the process compared to traditional selection methods, which rely solely on phenotype observation.

3. Application in disease resistance

- **Disease resistance**: MAS is particularly useful in breeding coriander varieties resistant to common diseases such as Fusarium wilt or downy mildew. By using markers linked to resistance genes, breeders can efficiently develop resistant varieties.

4. Enhancing Essential Oil Content

- **Oil quality and yield**: MAS is used to select coriander plants with high essential oil content and specific oil profiles, which are important for both culinary and industrial uses.

5. Integration with other breeding techniques

- MAS is often integrated with traditional breeding methods and other modern techniques like genomic selection and CRISPR to enhance overall breeding efficiency and effectiveness.

6. Advantages of MAS in coriander breeding

- **Precision**: MAS allows breeders to target specific traits with greater accuracy.
- **Speed**: It significantly reduces the breeding cycle by enabling early selection of desirable traits.
- **Cost-Effectiveness**: Although initial setup costs for MAS can be high, it ultimately reduces costs associated with extensive field trials and phenotypic selection.

7. Challenges

- **Marker development**: Developing reliable markers for all desired traits can be challenging and resource-intensive.
- **Genetic diversity**: The effectiveness of MAS can be limited by the genetic diversity within the coriander breeding population.

In summary, MAS in coriander breeding is a critical tool that enhances the development of superior coriander varieties by enabling precise selection for traits like disease resistance, yield, and essential oil content. It represents a significant advancement over traditional breeding methods, contributing to more efficient and targeted breeding programs.

Genomic selection in coriander breeding

Genomic selection (GS) is an advanced breeding technique that leverages the entire genome of a plant to predict the performance of breeding candidates, making it a powerful tool in coriander breeding. Here's an overview of how GS is applied in coriander breeding:

1. Introduction to genomic selection

- **Definition**: Genomic selection involves using genome-wide markers to predict the genetic potential or breeding value of individual plants. Unlike MAS, which focuses on a few specific markers, GS uses data from across the entire genome.

- **Purpose**: The primary goal of GS is to improve complex traits controlled by many genes, such as yield, quality traits (e.g., essential oil content in coriander), and resistance to diseases.

2. Steps in genomic selection

- **Reference population**: The process starts with a reference population that has been genotyped (using markers like SNPs) and phenotyped (measured for traits of interest).
- **Model Training**: A statistical model is trained to associate the genomic data with the phenotypic data. This model estimates the effects of the genetic markers on the traits.
- **Prediction**: The trained model is then used to predict the performance of new breeding candidates based on their genomic data alone, without the need for extensive phenotyping.

3. Application in Coriander breeding

- **Trait Improvement**: GS can be used to improve complex traits in coriander, such as:
  - **Yield**: Enhancing overall productivity by selecting plants with the best genetic makeup for yield.
  - **essential oil content and composition**: Targeting higher levels and desired profiles of essential oils.
  - **Disease resistance**: Improving resistance to multiple diseases simultaneously by selecting plants with favorable genomic profiles.
- **Breeding efficiency**: GS allows for the early selection of superior plants before they reach maturity, accelerating the breeding cycle and reducing costs associated with field trials.

4. Advantages of genomic selection in Coriander

- **Accuracy**: By considering the entire genome, GS provides a more accurate prediction of complex traits than traditional selection methods or MAS.
- **Speed**: GS can significantly reduce the time required to develop new coriander varieties.
- **Cost-effectiveness**: While initial genotyping can be expensive, the overall process reduces costs by minimizing the need for large-scale field trials.

5. Challenges in implementing genomic selection

- **High initial costs**: The need for comprehensive genotyping and the development of prediction models requires significant upfront investment.
- **Data management**: Managing and analyzing the large datasets generated in GS can be complex.

- **Model accuracy**: The accuracy of predictions depends on the quality of the reference population and the models used, which must be regularly updated to reflect new data.

6. Integration with other breeding techniques

- **Combined Approaches**: Genomic selection is often used in combination with traditional breeding, MAS, and newer techniques like CRISPR to optimize outcomes.
- **Continual improvement**: As more genomic data becomes available, the models used in GS can be refined to improve prediction accuracy further.

7. Future prospects

- **Increased adoption**: As the cost of genotyping decreases and computational tools improve, GS is likely to become more widely adopted in coriander breeding programs.
- **Precision breeding**: The use of GS, particularly when integrated with other technologies, will enhance the precision and efficiency of coriander breeding, leading to the development of superior varieties that meet specific market demands.

In summary, genomic selection represents a significant advancement in coriander breeding, enabling the efficient selection of plants with the best genetic potential for complex traits. By utilizing the entire genome, GS offers greater accuracy, faster breeding cycles, and the potential for significant improvements in yield, quality, and disease resistance in coriander.

MAS in Fenugreek breeding

Marker-assisted selection (MAS) in fenugreek breeding is a technique used to enhance the development of fenugreek varieties with desirable traits by using molecular markers linked to those traits. Fenugreek (Trigonella foenum-graecum), known for its medicinal and nutritional value, can benefit significantly from MAS in breeding programs. Here's how MAS is applied in fenugreek breeding:

1. Target traits in Fenugreek

- **Disease resistance**: Resistance to common diseases such as powdery mildew and downy mildew is a key target for fenugreek breeding.
- **Yield and biomass**: Improving the overall yield and biomass production is another important goal.
- **Nutritional content**: Enhancing the content of bioactive compounds like diosgenin, saponins, and fiber.
- **Abiotic stress tolerance**: Developing varieties that can withstand drought, salinity, and other environmental stresses.

2. Marker development and validation

- **Trait identification**: Breeders start by identifying key traits of interest. For example, identifying disease resistance genes or genes associated with high diosgenin content.
- **Marker discovery**: Through techniques like QTL mapping and genome-wide association studies (GWAS), molecular markers linked to these traits are discovered. These markers could be SSRs (simple sequence repeats), SNPs (single nucleotide polymorphisms), or other types of markers.
- **Validation**: The identified markers are validated in different fenugreek populations to ensure they are reliably linked to the desired traits.

3. Application of MAS in Fenugreek breeding

- **Breeding process**: Once markers are identified and validated, they are used to screen breeding populations. Plants carrying the favorable markers are selected early in the breeding cycle.
- **Speeding up selection**: By focusing on marker profiles rather than waiting for phenotypic expression (which can take longer), breeders can accelerate the selection process.
- **Pyramiding traits**: MAS allows for the combination (pyramiding) of multiple traits, such as disease resistance and high nutritional content, into a single variety more efficiently than conventional breeding.

4. Examples of MAS in Fenugreek

- **Disease resistance**: MAS has been used to develop fenugreek varieties resistant to powdery mildew by selecting for markers associated with resistance genes.
- **Nutritional enhancement**: Markers linked to high diosgenin content have been used to develop fenugreek varieties with improved medicinal properties.

5. Advantages of MAS in Fenugreek breeding

- **Precision**: MAS provides a high level of precision, ensuring that only plants with the desired genetic traits are selected.
- **Efficiency**: The selection process is faster, as it doesn't require the plants to be grown to maturity before making selections.
- **Cost-effective**: Over time, MAS can reduce the costs associated with field trials and labor-intensive phenotyping.

6. Challenges and limitations

- **Marker availability**: The success of MAS depends on the availability of well-validated markers linked to important traits. In crops like fenugreek, where genetic research may be less developed, finding such markers can be challenging.
- **Genetic diversity**: Limited genetic diversity in fenugreek populations may reduce the effectiveness of MAS.
- **Resource intensive**: Initial development of markers and setting up MAS infrastructure can be resource-intensive.

7. Future Prospects

- **Integration with genomic selection**: Combining MAS with genomic selection (GS) could further improve the efficiency of fenugreek breeding programs by leveraging genome-wide information.
- **CRISPR and molecular breeding**: Advances in gene editing technologies like CRISPR could complement MAS, enabling even more precise modifications in fenugreek.

MAS in fenugreek breeding is a valuable tool that enhances the efficiency of developing improved fenugreek varieties. By leveraging molecular markers, breeders can achieve faster and more accurate selection for traits such as disease resistance, nutritional content, and stress tolerance, ultimately leading to better-performing fenugreek varieties.

Genomic selection in Fenugreek

Genomic selection (GS) in fenugreek breeding is an advanced breeding strategy that uses genome-wide marker data to predict the breeding value of individual plants, allowing for the selection of superior genotypes early in the breeding cycle. This technique is particularly valuable for improving complex traits that are controlled by multiple genes, such as yield, nutritional content, and stress resistance. Here's how genomic selection is applied in fenugreek breeding:

1. Overview of genomic selection

- **Definition**: Genomic selection involves using statistical models that incorporate data from across the entire genome to predict the performance of plants for various traits. This contrasts with marker-assisted selection (MAS), which typically focuses on a few specific markers linked to particular traits.
- **Purpose**: The goal of GS is to increase the efficiency and accuracy of breeding programs by enabling the selection of plants based on their genetic potential rather than relying solely on observed traits.

2. Steps in implementing genomic selection

- **Reference population development**: A reference population is first developed and both genotyped (using markers such as SNPs) and phenotyped (measuring traits of interest). This population serves as the basis for developing predictive models.
- **Model training**: Statistical models, often based on methods like Best Linear Unbiased Prediction (BLUP) or machine learning algorithms, are trained using the genotypic and phenotypic data from the reference population. These models learn to associate genetic markers with trait performance.
- **Prediction of breeding values**: The trained models are then used to predict the genetic potential or genomic estimated breeding values (GEBVs) of new fenugreek breeding lines based on their genomic data alone, without the need for phenotyping.
- **Selection**: Plants with the highest predicted GEBVs are selected for further breeding or cultivation, allowing breeders to make informed decisions early in the breeding process.

3. Application in Fenugreek breeding

- **Yield improvement**: GS can be used to select fenugreek plants with higher predicted yield, which is a complex trait influenced by many genes.
- **Nutritional quality**: Traits such as the concentration of bioactive compounds (e.g., diosgenin, saponins) and protein content, which are crucial for fenugreek's medicinal and nutritional value, can be enhanced using GS.
- **Disease resistance**: Resistance to diseases like powdery mildew can be improved by predicting and selecting plants with genomic profiles associated with resistance.
- **Abiotic stress tolerance**: GS can help in developing fenugreek varieties that are better adapted to abiotic stresses like drought or salinity by selecting for stress-tolerance traits.

4. Advantages of genomic selection in Fenugreek

- **Increased accuracy**: By considering the entire genome, GS provides more accurate predictions for complex traits compared to traditional selection methods.
- **Shorter breeding cycles**: GS allows for the early selection of superior genotypes, reducing the time needed to develop new varieties.
- **Cost-efficiency**: Although genotyping is costly, GS reduces the need for extensive field trials and phenotyping, ultimately lowering overall breeding costs.

5. Challenges in implementing genomic selection

- **High initial costs**: The setup for GS requires significant investment in genotyping and computational resources.
- **Data management**: Handling and analyzing the large volumes of genomic data can be complex and requires specialized software and expertise.
- **Model reliability**: The accuracy of predictions depends on the quality of the reference population and the robustness of the statistical models used.

6. Integration with other breeding techniques

- **Combined approaches**: GS can be combined with marker-assisted selection (MAS) and traditional breeding methods to optimize the breeding process.
- **CRISPR and gene editing**: Advances in gene editing, such as CRISPR, can complement GS by allowing precise modifications of target genes identified through genomic selection.

7. Future Prospects

- **Wider adoption**: As genotyping costs decrease and computational methods improve, GS is likely to become a standard tool in fenugreek breeding programs.
- **Precision breeding**: GS will enable more precise breeding strategies, leading to the development of fenugreek varieties that meet specific agronomic, nutritional, and medicinal needs.

Genomic selection represents a significant advancement in fenugreek breeding, offering a powerful tool for improving complex traits such as yield, nutritional quality, and stress resistance. By utilizing genome-wide data, GS enhances the efficiency, accuracy, and speed of breeding programs, ultimately leading to the development of superior fenugreek varieties that can meet the demands of both farmers and consumers.

Genomic selection in Cardamom

Genomic selection (GS) in cardamom is an advanced breeding technique that uses genomic information to predict the performance of plants and select individuals with the best genetic potential for desired traits. This method involves the use of dense genetic markers across the entire genome to make predictions about the breeding value of plants, even before phenotypic traits are fully expressed.

Key aspects of genomic selection (GS) in Cardamom

8. Whole-genome markers

- Unlike marker-assisted selection (MAS), which uses a few markers linked to specific traits, genomic selection utilizes a large number of markers spread across the entire genome. These markers collectively capture the genetic variation influencing complex traits.

9. Training population

- A population of cardamom plants with known genetic markers and phenotypic data (such as yield, disease resistance, or quality traits) is used to develop a statistical model. This model links the genetic markers with the observed traits.

10. Prediction model

- The statistical model developed from the training population is used to predict the genetic value of other plants in the breeding program. These predictions help in selecting the best individuals for breeding, even if they have not yet expressed the desired traits.

11. Accelerated breeding

- GS allows breeders to make selection decisions early in the plant's life cycle, which significantly reduces the time required to develop new cardamom varieties with improved traits. This is especially valuable for perennial crops like cardamom, where traditional breeding cycles can be long.

12. Improvement of complex traits

- Genomic selection is particularly useful for improving complex traits that are controlled by many genes, such as yield, drought tolerance, or resistance to multiple diseases. It provides a more comprehensive approach compared to traditional methods, which may miss some genetic variations.

13. Increased genetic gain

- By selecting the best individuals based on their genomic estimated breeding values (GEBVs), genomic selection can lead to higher genetic gain per breeding cycle, making the breeding process more efficient and effective.

14. Application in Cardamom breeding

- Genomic selection in cardamom can be used to develop new varieties that are more productive, disease-resistant, and better suited to changing environmental conditions. This is increasingly important as the demand for high-quality cardamom continues to grow in global markets.

Benefits of genomic selection

- **Efficiency**: Speeds up the breeding process by allowing earlier and more accurate selection.
- **Precision**: Improves the accuracy of selecting plants with the best genetic potential.
- **Cost-effectiveness**: Reduces the costs associated with field trials and long breeding cycles.

In summary, genomic selection represents a powerful tool in the improvement of cardamom, enabling the development of superior varieties that meet the needs of farmers and markets more quickly and efficiently than traditional methods.

MAS in coriander breeding

Marker-assisted selection (MAS) in coriander breeding is a powerful tool used to enhance the efficiency and precision of developing new coriander varieties with desired traits. Here's an overview of how MAS is applied in coriander breeding:

1. Trait Identification and marker development

- **Trait of interest**: Breeders first identify specific traits that are desirable in coriander, such as disease resistance, yield, flavour, and essential oil content.
- **Marker identification**: Molecular markers (such as SSRs, SNPs) linked to these traits are identified using genetic mapping techniques. These markers serve as proxies to track the presence of the desired traits in coriander plants.

2. Marker-assisted selection process

- **Crossbreeding**: Coriander plants with desirable traits are crossbred. The offspring are then screened using molecular markers to identify those carrying the desired genes.
- **Selection**: Only the plants that have the molecular markers associated with the desired traits are selected for further breeding. This speeds up the process compared to traditional selection methods, which rely solely on phenotype observation.

3. Application in disease resistance

- **Disease resistance**: MAS is particularly useful in breeding coriander varieties resistant to common diseases such as Fusarium wilt or downy mildew. By using markers linked to resistance genes, breeders can efficiently develop resistant varieties.

4. Enhancing essential oil content

- **Oil quality and yield**: MAS is used to select coriander plants with high essential oil content and specific oil profiles, which are important for both culinary and industrial uses.

5. Integration with other breeding techniques

- MAS is often integrated with traditional breeding methods and other modern techniques like genomic selection and CRISPR to enhance overall breeding efficiency and effectiveness.

6. Advantages of MAS in Coriander breeding

- **Precision**: MAS allows breeders to target specific traits with greater accuracy.
- **Speed**: It significantly reduces the breeding cycle by enabling early selection of desirable traits.
- **Cost-effectiveness**: Although initial setup costs for MAS can be high, it ultimately reduces costs associated with extensive field trials and phenotypic selection.

7. Challenges

- **Marker development**: Developing reliable markers for all desired traits can be challenging and resource-intensive.
- **Genetic diversity**: The effectiveness of MAS can be limited by the genetic diversity within the coriander breeding population.

In summary, MAS in coriander breeding is a critical tool that enhances the development of superior coriander varieties by enabling precise selection for traits like disease resistance, yield, and essential oil content. It represents a significant advancement over traditional breeding methods, contributing to more efficient and targeted breeding programs.

Genomic selection in coriander breeding

Genomic selection (GS) is an advanced breeding technique that leverages the entire genome of a plant to predict the performance of breeding candidates, making it a powerful tool in coriander breeding. Here's an overview of how GS is applied in coriander breeding:

1. Introduction to genomic selection

- **Definition**: Genomic selection involves using genome-wide markers to predict the genetic potential or breeding value of individual plants. Unlike MAS, which focuses on a few specific markers, GS uses data from across the entire genome.

- **Purpose**: The primary goal of GS is to improve complex traits controlled by many genes, such as yield, quality traits (e.g., essential oil content in coriander), and resistance to diseases.

2. Steps in genomic selection

- **Reference population**: The process starts with a reference population that has been genotyped (using markers like SNPs) and phenotyped (measured for traits of interest).
- **Model training**: A statistical model is trained to associate the genomic data with the phenotypic data. This model estimates the effects of the genetic markers on the traits.
- **Prediction**: The trained model is then used to predict the performance of new breeding candidates based on their genomic data alone, without the need for extensive phenotyping.

3. Application in Coriander breeding

- **Trait improvement**: GS can be used to improve complex traits in coriander, such as:
  - **Yield**: Enhancing overall productivity by selecting plants with the best genetic makeup for yield.
  - **Essential oil content and composition**: Targeting higher levels and desired profiles of essential oils.
  - **Disease resistance**: Improving resistance to multiple diseases simultaneously by selecting plants with favorable genomic profiles.
- **Breeding efficiency**: GS allows for the early selection of superior plants before they reach maturity, accelerating the breeding cycle and reducing costs associated with field trials.

4. Advantages of genomic selection in Coriander

- **Accuracy**: By considering the entire genome, GS provides a more accurate prediction of complex traits than traditional selection methods or MAS.
- **Speed**: GS can significantly reduce the time required to develop new coriander varieties.
- **Cost-effectiveness**: While initial genotyping can be expensive, the overall process reduces costs by minimizing the need for large-scale field trials.

5. Challenges in implementing genomic selection

- **High initial costs**: The need for comprehensive genotyping and the development of prediction models requires significant upfront investment.

- **Data management**: Managing and analyzing the large datasets generated in GS can be complex.
- **Model accuracy**: The accuracy of predictions depends on the quality of the reference population and the models used, which must be regularly updated to reflect new data.

6. Integration with other breeding techniques

- **Combined approaches**: Genomic selection is often used in combination with traditional breeding, MAS, and newer techniques like CRISPR to optimize outcomes.
- **Continual improvement**: As more genomic data becomes available, the models used in GS can be refined to improve prediction accuracy further.

7. Future prospects

- **Increased adoption**: As the cost of genotyping decreases and computational tools improve, GS is likely to become more widely adopted in coriander breeding programs.
- **Precision breeding**: The use of GS, particularly when integrated with other technologies, will enhance the precision and efficiency of coriander breeding, leading to the development of superior varieties that meet specific market demands.

In summary, genomic selection represents a significant advancement in coriander breeding, enabling the efficient selection of plants with the best genetic potential for complex traits. By utilizing the entire genome, GS offers greater accuracy, faster breeding cycles, and the potential for significant improvements in yield, quality, and disease resistance in coriander.

MAS in Fenugreek breeding

Marker-assisted selection (MAS) in fenugreek breeding is a technique used to enhance the development of fenugreek varieties with desirable traits by using molecular markers linked to those traits. Fenugreek (Trigonella foenum-graecum), known for its medicinal and nutritional value, can benefit significantly from MAS in breeding programs. Here's how MAS is applied in fenugreek breeding:

1. Target traits in fenugreek

- **Disease resistance**: Resistance to common diseases such as powdery mildew and downy mildew is a key target for fenugreek breeding.
- **Yield and biomass**: Improving the overall yield and biomass production is another important goal.
- **Nutritional content**: Enhancing the content of bioactive compounds like diosgenin, saponins, and fiber.

- **Abiotic stress tolerance**: Developing varieties that can withstand drought, salinity, and other environmental stresses.

2. Marker development and validation

- **Trait identification**: Breeders start by identifying key traits of interest. For example, identifying disease resistance genes or genes associated with high diosgenin content.
- **Marker discovery**: Through techniques like QTL mapping and genome-wide association studies (GWAS), molecular markers linked to these traits are discovered. These markers could be SSRs (simple sequence repeats), SNPs (single nucleotide polymorphisms), or other types of markers.
- **Validation**: The identified markers are validated in different fenugreek populations to ensure they are reliably linked to the desired traits.

3. Application of MAS in Fenugreek breeding

- **Breeding process**: Once markers are identified and validated, they are used to screen breeding populations. Plants carrying the favorable markers are selected early in the breeding cycle.
- **Speeding Up selection**: By focusing on marker profiles rather than waiting for phenotypic expression (which can take longer), breeders can accelerate the selection process.
- **Pyramiding traits**: MAS allows for the combination (pyramiding) of multiple traits, such as disease resistance and high nutritional content, into a single variety more efficiently than conventional breeding.

4. Examples of MAS in Fenugreek

- **Disease resistance**: MAS has been used to develop fenugreek varieties resistant to powdery mildew by selecting for markers associated with resistance genes.
- **Nutritional enhancement**: Markers linked to high diosgenin content have been used to develop fenugreek varieties with improved medicinal properties.

5. Advantages of MAS in Fenugreek breeding

- **Precision**: MAS provides a high level of precision, ensuring that only plants with the desired genetic traits are selected.
- **Efficiency**: The selection process is faster, as it doesn't require the plants to be grown to maturity before making selections.
- **Cost-effective**: Over time, MAS can reduce the costs associated with field trials and labor-intensive phenotyping.

6. Challenges and limitations

- **Marker availability**: The success of MAS depends on the availability of well-validated markers linked to important traits. In crops like fenugreek, where genetic research may be less developed, finding such markers can be challenging.
- **Genetic diversity**: Limited genetic diversity in fenugreek populations may reduce the effectiveness of MAS.
- **Resource intensive**: Initial development of markers and setting up MAS infrastructure can be resource-intensive.

7. Future prospects

- **Integration with genomic selection**: Combining MAS with genomic selection (GS) could further improve the efficiency of fenugreek breeding programs by leveraging genome-wide information.
- **CRISPR and molecular breeding**: Advances in gene editing technologies like CRISPR could complement MAS, enabling even more precise modifications in fenugreek.

MAS in fenugreek breeding is a valuable tool that enhances the efficiency of developing improved fenugreek varieties. By leveraging molecular markers, breeders can achieve faster and more accurate selection for traits such as disease resistance, nutritional content, and stress tolerance, ultimately leading to better-performing fenugreek varieties.

Genomic selection in Fenugreek

Genomic selection (GS) in fenugreek breeding is an advanced breeding strategy that uses genome-wide marker data to predict the breeding value of individual plants, allowing for the selection of superior genotypes early in the breeding cycle. This technique is particularly valuable for improving complex traits that are controlled by multiple genes, such as yield, nutritional content, and stress resistance. Here's how genomic selection is applied in fenugreek breeding:

1. Overview of genomic selection

- **Definition**: Genomic selection involves using statistical models that incorporate data from across the entire genome to predict the performance of plants for various traits. This contrasts with marker-assisted selection (MAS), which typically focuses on a few specific markers linked to particular traits.
- **Purpose**: The goal of GS is to increase the efficiency and accuracy of breeding programs by enabling the selection of plants based on their genetic potential rather than relying solely on observed traits.

2. Steps in implementing genomic selection

- **Reference population development**: A reference population is first developed and both genotyped (using markers such as SNPs) and phenotyped (measuring traits of interest). This population serves as the basis for developing predictive models.
- **Model training**: Statistical models, often based on methods like Best Linear Unbiased Prediction (BLUP) or machine learning algorithms, are trained using the genotypic and phenotypic data from the reference population. These models learn to associate genetic markers with trait performance.
- **Prediction of breeding values**: The trained models are then used to predict the genetic potential or genomic estimated breeding values (GEBVs) of new fenugreek breeding lines based on their genomic data alone, without the need for phenotyping.
- **Selection**: Plants with the highest predicted GEBVs are selected for further breeding or cultivation, allowing breeders to make informed decisions early in the breeding process.

3. Application in Fenugreek breeding

- **Yield improvement**: GS can be used to select fenugreek plants with higher predicted yield, which is a complex trait influenced by many genes.
- **Nutritional quality**: Traits such as the concentration of bioactive compounds (e.g., diosgenin, saponins) and protein content, which are crucial for fenugreek's medicinal and nutritional value, can be enhanced using GS.
- **Disease resistance**: Resistance to diseases like powdery mildew can be improved by predicting and selecting plants with genomic profiles associated with resistance.
- **Abiotic stress tolerance**: GS can help in developing fenugreek varieties that are better adapted to abiotic stresses like drought or salinity by selecting for stress-tolerance traits.

4. Advantages of genomic selection in Fenugreek

- **Increased accuracy**: By considering the entire genome, GS provides more accurate predictions for complex traits compared to traditional selection methods.
- **Shorter breeding cycles**: GS allows for the early selection of superior genotypes, reducing the time needed to develop new varieties.
- **Cost-efficiency**: Although genotyping is costly, GS reduces the need for extensive field trials and phenotyping, ultimately lowering overall breeding costs.

5. Challenges in implementing genomic selection

- **High initial costs**: The setup for GS requires significant investment in genotyping and computational resources.
- **Data management**: Handling and analyzing the large volumes of genomic data can be complex and requires specialized software and expertise.
- **Model reliability**: The accuracy of predictions depends on the quality of the reference population and the robustness of the statistical models used.

6. Integration with other breeding techniques

- **Combined approaches**: GS can be combined with marker-assisted selection (MAS) and traditional breeding methods to optimize the breeding process.
- **CRISPR and gene editing**: Advances in gene editing, such as CRISPR, can complement GS by allowing precise modifications of target genes identified through genomic selection.

7. Future prospects

- **Wider adoption**: As genotyping costs decrease and computational methods improve, GS is likely to become a standard tool in fenugreek breeding programs.
- **Precision breeding**: GS will enable more precise breeding strategies, leading to the development of fenugreek varieties that meet specific agronomic, nutritional, and medicinal needs.

Genomic selection represents a significant advancement in fenugreek breeding, offering a powerful tool for improving complex traits such as yield, nutritional quality, and stress resistance. By utilizing genome-wide data, GS enhances the efficiency, accuracy, and speed of breeding programs, ultimately leading to the development of superior fenugreek varieties that can meet the demands of both farmers and consumers.

8

Seed Production and Certification in Spices

Seed production in spices varies depending on the specific crop and its biological characteristics. Below is an overview of the seed production methods for each of the spices you've mentioned:

1. Ginger (*Zingiber officinale*)

- **Propagation method**: Ginger is propagated through rhizomes rather than true seeds. Selected healthy rhizomes are cut into pieces, each with at least one viable bud (eye).
- **Planting**: The rhizome pieces are planted in well-drained soil with high organic content.
- **Maintenance**: Regular watering and mulching are essential to maintain soil moisture. Weeding and pest management are also crucial.
- **Harvesting**: The rhizomes are harvested 8-10 months after planting. Some are reserved for the next planting season.

2. Turmeric (*Curcuma longa*)

- **Propagation method**: Similar to ginger, turmeric is propagated using rhizomes.
- **Planting**: Rhizomes are cut into pieces with at least one bud each and planted in a prepared field.
- **Maintenance**: Requires consistent soil moisture, so mulching and irrigation are essential. Regular weeding and protection against pests and diseases are necessary.
- **Harvesting**: Rhizomes are harvested 7-9 months after planting. Selected rhizomes are stored for the next planting season.

3. Mango Ginger (*Curcuma amada*)

- **Propagation method**: Like ginger and turmeric, mango ginger is propagated through rhizomes.
- **Planting**: Rhizome pieces with buds are planted in well-prepared soil.

- **Maintenance**: Requires regular irrigation and mulching to retain soil moisture. Weed control and pest management are essential.
- **Harvesting**: The rhizomes are harvested 7-8 months after planting.

4. Black Turmeric (*Curcuma caesia*)

- **Propagation method**: Propagation is done through rhizomes, similar to other Curcuma species.
- **Planting**: The rhizome pieces are planted in well-drained soil with high organic matter.
- **Maintenance**: Regular watering, mulching, and weeding are required to ensure proper growth.
- **Harvesting**: Rhizomes are harvested 8-10 months after planting, with a portion kept aside for future planting.

5. Coriander (*Coriandrum sativum*)

- **Propagation method**: Coriander is grown from seeds.
- **Planting**: Seeds are directly sown into the field, with the option of transplanting seedlings in some cases.
- **Maintenance**: Requires regular watering and weed control. Thinning of plants may be necessary to ensure proper growth.
- **Harvesting**: Seed heads are harvested when they turn brown and seeds mature. Seeds are then dried and cleaned.

6. Cumin (*Cuminum cyminum*)

- **Propagation method**: Cumin is propagated through seeds.
- **Planting**: Seeds are sown directly into the field, usually in rows.
- **Maintenance**: Regular weeding, thinning, and pest control are important. Irrigation should be minimized as cumin is drought-tolerant.
- **Harvesting**: Seeds are harvested when they turn brown and the plant is mature. The plants are cut and dried, and seeds are then threshed out.

7. Fennel (*Foeniculum vulgare*)

- **Propagation method**: Fennel is propagated through seeds.
- **Planting**: Seeds are sown directly in rows or in nursery beds for later transplantation.
- **Maintenance**: Requires regular irrigation, weeding, and thinning. Pest and disease control are also necessary.
- **Harvesting**: Seeds are harvested when the seed heads turn brown. Plants are cut, dried, and seeds are threshed.

8. Fenugreek (*Trigonella foenum-graecum*)

- **Propagation method**: Fenugreek is grown from seeds.
- **Planting**: Seeds are directly sown into the field in rows.
- **Maintenance**: Regular weeding, thinning, and irrigation are needed, especially during dry periods.
- **Harvesting**: Seeds are harvested when pods turn yellow and start to dry. The plants are cut, dried, and seeds are threshed.

9. Ajwain (*Trachyspermum ammi*)

- **Propagation method**: Ajwain is propagated through seeds.
- **Planting**: Seeds are directly sown in well-prepared soil in rows.
- **Maintenance**: Regular irrigation and weeding are important for optimal growth. Thinning of plants may be required.
- **Harvesting**: Seeds are harvested when the seed heads turn brown. Plants are cut, dried, and seeds are threshed.

Each of these spices requires careful attention to soil conditions, moisture, and pest management to ensure successful seed production. Proper harvesting and post-harvest handling are also critical to maintain the quality of the seeds.

Seed certification in spices

Seed certification is a process that ensures the quality of seeds used for cultivation, verifying that they meet certain standards of purity, viability, and freedom from disease. In spices like ginger, turmeric, mango ginger, black turmeric, coriander, cumin, fennel, fenugreek, and ajwain, seed certification ensures that the planting material is true to type and free from pests and diseases. Below is a general overview of seed certification processes for these spices.

1. Ginger (*Zingiber officinale*)

- **Certification focus**: Since ginger is propagated through rhizomes, the certification process focuses on the quality and health of the rhizomes rather than seeds.
- **Standards**: Certified ginger rhizomes should be free from diseases like bacterial wilt, nematodes, and other pathogens. The rhizomes should be of uniform size and have a high sprouting rate.
- **Inspection**: Field inspections ensure that the plants are healthy, and rhizomes are inspected before certification.

2. Turmeric (*Curcuma longa*)

- **Certification focus**: Like ginger, turmeric is propagated through rhizomes, and certification focuses on the health and quality of these rhizomes.

- **Standards**: Certified turmeric rhizomes must be free from diseases, particularly rhizome rot, and should have uniformity in size and good sprouting ability.
- **Inspection**: Regular inspections are conducted to ensure disease-free plants, and rhizomes are checked before certification.

3. Mango Ginger (*Curcuma amada*)

- **Certification focus**: Rhizomes are the primary focus for certification, ensuring they are disease-free and of high quality.
- **Standards**: Certified rhizomes must be free from viral and bacterial diseases and show good vigor and sprouting capability.
- **Inspection**: Similar to ginger and turmeric, field inspections are conducted, and rhizomes are evaluated for certification.

4. Black Turmeric (*Curcuma caesia*)

- **Certification focus**: The certification process centers on the health of the rhizomes.
- **Standards**: Rhizomes must be free from common diseases, particularly fungal infections, and should have uniformity and good viability.
- **Inspection**: Field and rhizome inspections are conducted before certification.

5. Coriander (*Coriandrum sativum*)

- **Certification focus**: Certification for coriander seeds involves ensuring purity, germination rate, and freedom from pests and diseases.
- **Standards**: Certified seeds must have a high germination rate, be free from admixtures, and have uniform size and colour.
- **Inspection**: Field inspections monitor the crop's health, and seed samples are tested for purity and germination before certification.

6. Cumin (*Cuminum cyminum*)

- **Certification focus**: Certification focuses on seed purity, viability, and disease freedom.
- **Standards**: Seeds must have high germination potential, be free from contaminants, and be uniform in size and shape.
- **Inspection**: Regular inspections are carried out to ensure crop health, and seed testing is done before certification.

7. Fennel (*Foeniculum vulgare*)

- **Certification focus**: Certification ensures that fennel seeds are of high quality, with good germination rates and free from diseases.

- **Standards**: Seeds must be uniform in size, free from weed seeds, and exhibit high germination potential.
- **Inspection**: Field inspections check for crop health, and seeds are tested for germination and purity.

8. Fenugreek (*Trigonella foenum-graecum*)

- **Certification focus**: Certification ensures seed purity, high germination rates, and freedom from diseases.
- **Standards**: Seeds must have uniformity in size, high viability, and be free from weed seeds and contaminants.
- **Inspection**: Field inspections ensure crop health, and seed testing is done for certification.

9. Ajwain (*Trachyspermum ammi*)

- **Certification focus**: The certification process ensures seed quality, including purity and germination.
- **Standards**: Seeds should be free from pests, diseases, and contaminants, with high germination rates.
- **Inspection**: Crop inspections and seed testing are performed before certification.

General seed certification process

1. **Seed selection**: High-quality seeds or planting material (rhizomes in the case of ginger, turmeric, etc.) are selected from healthy, disease-free plants.
2. **Field inspection**: Several inspections are conducted during the growing season to ensure the crop is free from pests, diseases, and off-types.
3. **Seed processing**: After harvesting, seeds or rhizomes are cleaned, dried, and treated (if necessary) to meet certification standards.
4. **Laboratory testing**: Seeds are tested in laboratories for germination rates, purity, and freedom from diseases and pests.
5. **Certification tagging**: Once the seeds meet all the criteria, they are certified and tagged with appropriate certification labels for distribution.

Seed certification plays a crucial role in maintaining the quality of planting material, ensuring higher yields, disease resistance, and overall crop health.

Seed standards of spices for certification

The certification of seeds for spices like ginger, turmeric, mango ginger, black turmeric, coriander, cumin, fennel, fenugreek, and ajwain requires adherence to specific seed standards. These standards ensure that the seeds are of high quality, genetically pure, and free from diseases and pests. Below are the general seed standards required for certification of these spices:

1. Ginger (*Zingiber officinale*)

- **Propagation**: Ginger is propagated through rhizomes, not true seeds.
- **Standards for rhizomes**:
  - **Purity**: Rhizomes should be true to type, without admixtures of other varieties.
  - **Health**: Rhizomes must be free from diseases such as bacterial wilt, soft rot, and nematodes.
  - **Size and weight**: Rhizomes should be uniform in size, with a minimum weight (typically 20-25 grams per piece).
  - **Viability**: Rhizomes must show a high sprouting percentage, usually over 80%.
  - **Appearance**: Rhizomes should be healthy, with no mechanical injuries or signs of pest damage.

2. Turmeric (*Curcuma longa*)

- **Propagation**: Like ginger, turmeric is propagated through rhizomes.
- **Standards for rhizomes**:
  - **Purity**: Rhizomes must be pure, with no admixture of other varieties.
  - **Health**: Rhizomes should be free from diseases like rhizome rot and pest infestations.
  - **Size and weight**: Rhizomes should have uniform size, typically with a minimum weight of 20 grams.
  - **Viability**: High sprouting percentage is required, generally over 85%.
  - **Appearance**: Rhizomes should be clean, free from mechanical damage, and have a smooth surface.

3. Mango Ginger (*Curcuma amada*)

- **Propagation**: Mango ginger is also propagated through rhizomes.
- **Standards for rhizomes**:
  - **Purity**: Rhizomes should be true to the specific variety of mango ginger.
  - **Health**: Free from fungal infections, rot, and pest damage.
  - **Size and weight**: Rhizomes should be of uniform size, with a minimum weight similar to other ginger species.
  - **Viability**: Rhizomes should exhibit high sprouting potential, typically above 80%.
  - **Appearance**: Rhizomes should be clean, with no discolouration or damage.

4. Black Turmeric (*Curcuma caesia*)

- **Propagation**: Propagated through rhizomes.
- **Standards for rhizomes**:
  - **Purity**: Rhizomes should be pure, with no admixture of other turmeric varieties.
  - **Health**: Rhizomes must be free from common diseases and pests.
  - **Size and weight**: Uniformity in size and a minimum weight requirement, typically around 20 grams per piece.
  - **Viability**: High sprouting percentage, generally over 80%.
  - **Appearance**: Rhizomes should be dark in colour (typical of black turmeric), free from damage and pests.

5. Coriander (*Coriandrum sativum*)

- **Propagation**: Coriander is propagated through true seeds.
- **Seed standards**:
  - **Genetic purity**: Seeds should have a genetic purity of at least 98%.
  - **Physical purity**: Seeds must have at least 97% physical purity.
  - **Germination rate**: A minimum germination rate of 70% is required.
  - **Moisture content**: Seeds should have a moisture content of 8-10% to ensure good storage life.
  - **Freedom from pests and diseases**: Seeds must be free from visible mold, fungal infections, and pests.
  - **Size and uniformity**: Seeds should be uniform in size and free from broken or damaged seeds.

6. Cumin (*Cuminum cyminum*)

- **Propagation**: Cumin is propagated through seeds.
- **Seed standards**:
  - **Genetic purity**: Seeds should have a genetic purity of at least 98%.
  - **Physical purity**: Seeds should have a physical purity of at least 97%.
  - **Germination rate**: Minimum germination rate should be 70%.
  - **Moisture content**: Seeds should have a moisture content between 8-10%.
  - **Freedom from pests and diseases**: Seeds must be free from pests, diseases, and weed seeds.
  - **Size and uniformity**: Seeds should be uniform, plump, and free from shriveled or immature seeds.

7. Fennel (*Foeniculum vulgare*)

- **Propagation**: Fennel is propagated through seeds.
- **Seed standards**:
  - **Genetic purity**: Seeds should have a genetic purity of at least 98%.
  - **Physical purity**: Seeds must have a physical purity of at least 97%.
  - **Germination rate**: Minimum germination rate should be 70%.
  - **Moisture content**: Seeds should have a moisture content of 8-10%.
  - **Freedom from pests and diseases**: Seeds should be free from any pest infestations and diseases.
  - **Size and uniformity**: Seeds should be uniform, clean, and free from foreign material.

8. Fenugreek (*Trigonella foenum-graecum*)

- **Propagation**: Fenugreek is propagated through seeds.
- **Seed standards**:
  - **Genetic purity**: Seeds should have a genetic purity of at least 98%.
  - **Physical purity**: Seeds should have a physical purity of at least 97%.
  - **Germination rate**: Minimum germination rate should be 75%.
  - **Moisture content**: Seeds should have a moisture content of 8-10%.
  - **Freedom from pests and diseases**: Seeds must be free from pests and disease.
  - **Size and uniformity**: Seeds should be uniform, healthy, and free from damaged seeds.

9. Ajwain (*Trachyspermum ammi*)

- **Propagation**: Ajwain is propagated through seeds.
  - **Seed standards**:
  - **Genetic purity**: Seeds should have a genetic purity of at least 98%.
  - **Physical purity**: Seeds should have a physical purity of at least 97%.
  - **Germination rate**: Minimum germination rate should be 70%.
  - **Moisture content**: Seeds should have a moisture content of 8-10%.
  - **Freedom from pests and diseases**: Seeds should be free from visible signs of mold, fungal infections, and pests.
  - **Size and uniformity**: Seeds should be uniform in size, free from foreign matter, and fully mature.

General certification requirements

- **Labeling**: Certified seeds must be labeled with information on the variety, purity, germination rate, and date of certification.
- **Sampling and testing**: Seeds must undergo rigorous sampling and testing to ensure they meet the specified standards.
- **Field inspection**: Regular inspections during the growing season to ensure that the seed production fields are free from diseases, pests, and off-types.
- **Documentation**: Detailed records of the seed production process, including source of seed, field history, and pest and disease management practices, must be maintained.

These standards help ensure that the seeds and planting materials used in the cultivation of these spices are of the highest quality, leading to better yields, disease resistance, and overall crop performance.

Post-harvest handling in spices for seed certification

Post-harvest handling is a critical stage in the production of high-quality seeds for certification, especially for spices like ginger, turmeric, mango ginger, black turmeric, coriander, cumin, fennel, fenugreek, and ajwain. Proper post-harvest practices ensure that the seeds or rhizomes maintain their viability, purity, and health, which are essential for meeting certification standards. Below are the key steps involved in the post-harvest handling of these spices for seed certification:

1. Ginger (*Zingiber officinale*)

- **Harvesting**: Ginger rhizomes are harvested when they reach maturity, usually 8-10 months after planting.
- **Cleaning**: After harvest, rhizomes should be cleaned thoroughly to remove soil and debris. Washing should be done with clean water, followed by air drying in the shade to prevent fungal growth.
- **Curing**: Rhizomes can be cured by sun drying for a short period to toughen the skin. This helps to reduce moisture content and prolong storage life.
- **Sorting and grading**: Rhizomes are sorted and graded based on size, weight, and the absence of damage or disease.
- **Storage**: Store the cleaned, cured, and graded rhizomes in well-ventilated storage facilities with controlled temperature and humidity. Rhizomes should be stored in layers, separated by sand or sawdust, to prevent sprouting and decay.

2. Turmeric (*Curcuma longa*)

- **Harvesting**: Turmeric rhizomes are harvested when the leaves start to yellow and wither, typically 7-9 months after planting.

- **Cleaning**: Rhizomes are washed to remove soil and dirt. They should be dried in the shade to remove surface moisture.
- **Boiling and drying**: For seed purposes, boiling may be skipped or minimized. If done, boiling should be brief to avoid damaging the rhizomes. Afterward, rhizomes should be dried thoroughly in the shade.
- **Sorting and grading**: Rhizomes are sorted to remove any that are damaged or diseased. Grading is done based on size and quality.
- **Storage**: Store the rhizomes in cool, dry, and well-ventilated conditions. Regular checks should be conducted to remove any rhizomes showing signs of spoilage or sprouting.

3. Mango Ginger (*Curcuma amada*)

- **Harvesting**: Rhizomes are harvested when the plant leaves start to dry, typically 7-8 months after planting.
- **Cleaning**: Remove soil and debris by washing the rhizomes. Air-dry them in the shade to reduce moisture content.
- **Curing**: Cure the rhizomes by sun drying briefly to harden the skin.
- **Sorting and grading**: Select rhizomes that are uniform in size, free from damage, and healthy for storage.
- **Storage**: Store the rhizomes in a cool, dry place with good air circulation. Use materials like sawdust or sand to prevent direct contact between rhizomes.

4. Black Turmeric (*Curcuma caesia*)

- **Harvesting**: Harvest black turmeric rhizomes when the plant leaves dry up, around 8-10 months after planting.
- **Cleaning**: Wash the rhizomes thoroughly to remove all soil and debris, followed by shade drying.
- **Curing**: Like other turmeric varieties, brief sun drying helps in curing and reducing moisture.
- **Sorting and grading**: Sort rhizomes to remove any that are diseased or damaged, and grade them based on size and quality.
- **Storage**: Store in a cool, dry environment with good ventilation, possibly with layers of sand or sawdust to prevent sprouting.

5. Coriander (*Coriandrum sativum*)

- **Harvesting**: Seeds are harvested when the seed heads turn brown, and the seeds are fully mature.
- **Cleaning**: Thresh the seeds to separate them from the chaff, and then clean them by winnowing or using mechanical seed cleaners.

- **Drying**: Dry the seeds to a moisture content of around 8-10% to prevent fungal growth and maintain viability.
- **Sorting and grading**: Remove any impurities, undersized, or damaged seeds. Seeds should be graded to ensure uniformity in size and quality.
- **Storage**: Store the seeds in airtight containers or bags in a cool, dry place. Seeds should be protected from pests and moisture during storage.

6. Cumin (*Cuminum cyminum*)

- **Harvesting**: Harvest when the plants turn yellow, and the seeds have matured but before they start shattering.
- **Cleaning**: Thresh the plants to separate the seeds, followed by cleaning to remove chaff and other debris.
- **Drying**: Seeds should be dried to a moisture content of 8-10% to ensure longevity and prevent mold.
- **Sorting and grading**: Sort out any broken or immature seeds and grade them by size and quality.
- **Storage**: Store in moisture-proof containers in a cool, dry environment to maintain seed quality and viability.

7. Fennel (*Foeniculum vulgare*)

- **Harvesting**: Harvest the seed heads when they turn brown and the seeds are fully mature.
- **Cleaning**: Thresh to separate the seeds, then clean to remove any foreign material.
- **Drying**: Dry seeds to a moisture content of 8-10% to avoid spoilage.
- **Sorting and grading**: Sort out impurities, damaged seeds, and grade the seeds to ensure uniformity.
- **Storage**: Store in dry, cool conditions, preferably in airtight containers, to preserve seed viability.

8. Fenugreek (*Trigonella foenum-graecum*)

- **Harvesting**: Harvest when the pods turn yellow and begin to dry, but before they shatter.
- **Cleaning**: Thresh the seeds and clean them thoroughly to remove chaff and other impurities.
- **Drying**: Ensure seeds are dried to a moisture content of 8-10%.
- **Sorting and grading**: Remove any immature or damaged seeds and grade the remaining seeds for uniformity.
- **Storage**: Store in airtight containers in a cool, dry place to maintain seed quality.

9. Ajwain (*Trachyspermum ammi*)

- **Harvesting**: Harvest when the seed heads turn brown, and the seeds are fully mature.
- **Cleaning**: Thresh the seeds to separate them from the plant material, then clean them thoroughly.
- **Drying**: Dry the seeds to a moisture content of around 8-10%.
- **Sorting and grading**: Sort out any broken or immature seeds and ensure uniformity in size.
- **Storage**: Store in a cool, dry environment in moisture-proof containers to maintain seed viability and prevent contamination.

General post-harvest considerations for seed certification:

- **Hygiene**: Ensure all handling and processing equipment are clean to avoid contamination.
- **Pest management**: Store seeds or rhizomes in pest-free environments, using appropriate pest control measures if necessary.
- **Moisture control**: Maintain optimal moisture levels during drying and storage to prevent mold growth and maintain seed viability.
- **Regular monitoring**: Periodically check stored seeds or rhizomes for signs of spoilage, pest infestation, or sprouting, and take corrective measures as needed.
- **Labelling and documentation**: Properly label all batches with relevant information (variety, harvest date, etc.), and keep detailed records to ensure traceability and compliance with certification standards.

Adhering to these post-harvest handling practices helps ensure that the seeds or planting materials are of high quality, viable, and meet the standards required for certification.

QPM production standards in tree spices

Quality Planting Material (QPM) production in tree spices like black pepper, small cardamom, long cardamom, clove, allspice, and cinnamon is essential to ensure healthy and productive plants. The methods for QPM production in these spices involve careful selection, propagation, and management practices. Here's a summary of QPM production methods for these spices:

1. Black Pepper (*Piper nigrum*)

- **Propagation method**: Black pepper is primarily propagated through vegetative methods, such as cuttings.

- **Selection**: Select healthy, disease-free mother plants with high yield potential.
- **Nursery production**: Stem cuttings, particularly from lateral branches (orthotropic shoots), are preferred. The cuttings are treated with rooting hormones and planted in polybags or raised beds.
- **Maintenance**: Regular watering, shading, and protection from pests and diseases are necessary in the nursery.
- **Field planting**: Once rooted, the cuttings are transplanted into the field. Proper care during the establishment phase is critical.

2. Small Cardamom (*Elettaria cardamomum*)

- **Propagation method**: Small cardamom is propagated through suckers (vegetative method) or seeds.
- **Selection**: Select high-yielding, disease-resistant mother plants.
- **Nursery production**: Suckers are separated from the mother plant and planted in nurseries. Seed propagation involves sowing seeds in nursery beds under shade.
- **Maintenance**: Suckers or seedlings are maintained with adequate watering, shading, and protection from pests.
- **Field planting**: Transplant the suckers or seedlings to the field once they are well-established. Ensure proper spacing and shade.

3. Long Cardamom (*Amomum subulatum*)

- **Propagation method**: Similar to small cardamom, long cardamom is propagated through vegetative methods (suckers) or seeds.
- **Selection**: High-yielding and disease-resistant mother plants are selected.
- **Nursery production**: Suckers are carefully separated and planted in nursery beds or polybags. Seed propagation involves raising seedlings in shaded nursery beds.
- **Maintenance**: Regular watering, shading, and pest management are necessary to maintain healthy nursery plants.
- **Field planting**: Transplant well-established suckers or seedlings to the field, ensuring adequate shade and moisture.

4. Clove (*Syzygium aromaticum*)

- **Propagation method**: Clove is typically propagated through seeds, although vegetative propagation (grafting or cuttings) is also possible.
- **Selection**: Select seeds from healthy, high-yielding trees. Only fully mature and ripened fruits should be harvested for seeds.

- **Nursery production**: Seeds are sown in nursery beds under shade, or young seedlings are grown in polybags. Clove seeds have a short viability period, so they must be sown soon after harvesting.
- **Maintenance**: Seedlings require regular watering, shading, and protection from pests and diseases.
- **Field planting**: Once the seedlings are robust enough, they are transplanted into the field. Proper care is necessary during the initial years to establish the plants.

5. Allspice (*Pimenta dioica*)

- **Propagation method**: Allspice can be propagated by seeds or vegetatively through cuttings or grafting.
- **Selection**: Select seeds from high-quality fruits or take cuttings from healthy, high-yielding trees.
- **Nursery production**: Seeds are sown in nursery beds or polybags under shade. Cuttings can be treated with rooting hormones and planted in suitable nursery media.
- **Maintenance**: Ensure regular watering, shading, and pest control in the nursery.
- **Field planting**: Transplant seedlings or rooted cuttings to the field, ensuring proper spacing and care during establishment.

6. Cinnamon (*Cinnamomum verum*)

- **Propagation method**: Cinnamon is propagated by seeds, cuttings, or layering.
- **Selection**: Select seeds from high-quality, mature fruits or take cuttings from healthy branches.
- **Nursery production**: Seeds are sown in nursery beds under shade. Cuttings are treated with rooting hormones and planted in polybags.
- **Maintenance**: Nursery plants require regular watering, shading, and pest management.
- **Field planting**: Transplant well-established seedlings or rooted cuttings to the field. Cinnamon plants need good care during the initial years to establish a strong root system.

General considerations for QPM production

1. **Selection of mother plants**: Only healthy, disease-free, and high-yielding plants should be selected as mother plants for producing QPM.
2. **Nursery management**: The nursery environment must be carefully controlled, with appropriate shading, watering, and pest/disease management practices in place.

3. **Propagation techniques**: The choice of propagation method (seed, cutting, grafting, layering) depends on the specific spice and its biological characteristics.
4. **Disease and pest control**: Regular monitoring and control of pests and diseases are crucial in both nursery and field environments to maintain the quality of the planting material.
5. **Certification and quality control**: Quality control measures should be in place to ensure that the QPM meets the required standards before being distributed to farmers.

Producing high-quality planting material is key to ensuring the long-term productivity and health of these tree spices. Proper management practices at every stage—from selection and propagation to nursery management and field establishment—are critical to achieving this goal.

Certification in tree spices QPM

Certification of Quality Planting Material (QPM) in tree spices such as black pepper, small cardamom, long cardamom, clove, allspice, and cinnamon is a crucial process that ensures the propagation materials meet specific quality standards. This certification helps in maintaining the genetic purity, health, and overall quality of the planting material, which is vital for ensuring high yields and disease resistance in the crops.

QPM certification steps for tree spices

1. Selection of mother plants

- **Criteria**: Only plants that are healthy, high-yielding, and free from diseases and pests are selected as mother plants. These plants should also exhibit desirable traits like resistance to local pests and diseases and adaptation to the local climate.
- **Inspection**: The mother plants undergo rigorous inspections to ensure they meet the necessary quality criteria. These inspections are usually carried out by certification agencies or designated agricultural officers.

2. Propagation and nursery management

- **Propagation methods**: Depending on the spice, propagation may involve seeds, cuttings, grafting, or other vegetative methods. The propagation method used must be appropriate for the spice and ensure that the genetic characteristics of the mother plant are preserved.
- **Nursery standards**: The nursery where the QPM is grown must meet specific standards, including proper spacing, shading, watering, and pest control. The nursery environment must be conducive to producing healthy and robust planting material.

3. **Field inspections**

- **Monitoring**: Regular field inspections are conducted throughout the growth period of the nursery plants. Inspectors look for signs of disease, pest infestation, and overall plant health.
- **Compliance**: The nursery must comply with the guidelines and standards set by the certification body. Non-compliance may lead to the rejection of the planting material for certification.

4. **Testing for purity and viability**

- **Genetic purity**: The planting material is tested to ensure that it is genetically pure and true to the type of the mother plant. This is especially important for maintaining the uniformity and quality of the crop in the field.
- **Viability**: The material is also tested for viability, ensuring that the seeds or vegetative cuttings have a high germination or rooting rate.

5. **Pest and disease control**

- **Phytosanitary measures**: The planting material must be free from any pests and diseases. This involves stringent phytosanitary measures, including the treatment of planting material to eliminate potential threats.
- **Certification testing**: Samples may be taken for laboratory testing to check for the presence of specific pathogens or pests that could affect the quality of the planting material.

6. **Certification and labelling**

- **Certification process**: Once the planting material passes all inspections and tests, it is certified by the relevant agricultural authority or certification body. This certification indicates that the material meets the required standards for quality, purity, and health.
- **Labelling**: Certified planting materials are labeled with a certification tag that includes information such as the certification number, the name of the certifying body, and details about the planting material (e.g., variety, date of certification).
- **Documentation**: All certification processes are documented, and records are maintained to ensure traceability and compliance with certification standards.

Specific considerations for each spice

- **Black Pepper**: Certification focuses on ensuring that cuttings are taken from disease-free mother plants and are free from pests like nematodes and fungal infections.

- **Small and Long Cardamom**: Certification includes ensuring that suckers or seedlings are free from viral diseases and have high vigor and productivity potential.
- **Clove**: Given that clove is often propagated by seeds, certification ensures that the seeds are from high-yielding, disease-free mother plants and have high germination rates.
- **Allspice**: Certification ensures that vegetative cuttings or seeds are free from diseases, particularly fungal pathogens, and exhibit good growth characteristics.
- **Cinnamon**: Certification focuses on ensuring that the planting material is from high-quality, disease-free mother plants, with specific attention to the prevention of diseases like leaf spot and bark canker.

Benefits of certification

- **Assurance of quality**: Certification provides assurance to farmers that the planting material they are using is of high quality and will perform well under local conditions.
- **Higher yields**: Certified QPM is more likely to produce higher yields due to its genetic purity and resistance to diseases.
- **Market access**: Certified planting materials often have better market access, as they meet the standards required by many buyers and regulatory bodies.
- **Traceability**: Certification ensures that the source of the planting material can be traced back to the nursery, providing transparency and accountability.

Certification of QPM in tree spices is a comprehensive process that involves multiple stages of inspection, testing, and approval to ensure that the planting material meets the highest standards of quality. This process is crucial for sustaining the productivity and health of spice crops.

Certification standards Spices QPM nurseries

Certification standards for Quality Planting Material (QPM) nurseries in tree spices like black pepper, small cardamom, long cardamom, clove, allspice, and cinnamon are established to ensure that nurseries produce high-quality, disease-free, and genetically pure planting materials. These standards are essential for maintaining the health and productivity of spice crops. Below are the key certification standards typically required for QPM nurseries:

1. Black Pepper (*Piper nigrum*)

- **Mother plant selection:**
  - Only use high-yielding, disease-free, and true-to-type mother plants.

- Regularly inspect mother plants for pests, diseases, and physiological disorders.

- **Propagation standards**
    - Use lateral branch cuttings (orthotropic shoots) from healthy plants.
    - Ensure proper treatment of cuttings with fungicides and rooting hormones.
- **Nursery management**
    - Maintain adequate spacing, shading, and humidity in the nursery.
    - Ensure regular watering and proper drainage to prevent waterlogging.
    - Implement strict phytosanitary measures to prevent the introduction of pests and diseases.
- **Quality control**
    - Conduct regular inspections and remove any diseased or weak plants.
    - Test soil and water used in the nursery for contaminants.

2. Small Cardamom (*Elettaria cardamomum*)

- **Mother plant selection**
    - Select mother plants based on high yield, resistance to diseases like capsule rot and mosaic virus, and quality of the cardamom pods.
- **Propagation standards**
    - Propagate through healthy suckers or seeds from certified sources.
    - Ensure that suckers are disease-free and have a good root system.
- **Nursery management**
    - Provide adequate shading, watering, and pest control.
    - Use well-drained, fertile soil with a good mix of organic matter.
    - Maintain proper hygiene to prevent the spread of diseases.
- **Quality control**
    - Regularly inspect nurseries for pests and diseases, and quarantine affected plants.
    - Ensure seedlings are robust and free from viral and fungal infections before certification.

3. Long Cardamom (*Amomum subulatum*)

- **Mother plant selection**
    - Choose mother plants with high productivity and resistance to common diseases such as leaf blight and anthracnose.

- **Propagation standards**
  - Propagate through disease-free suckers or certified seeds.
  - Suckers should be healthy, with a well-developed root system and no signs of disease.
- **Nursery management**
  - Maintain high humidity levels, especially during the early growth stages.
  - Implement integrated pest management (IPM) strategies to keep pests under control.
  - Use organic fertilizers and well-composted organic matter.
- **Quality control**
  - Conduct regular inspections for signs of diseases and pests.
  - Ensure that the nursery environment is conducive to the production of healthy, vigorous plants.

4. Clove (*Syzygium aromaticum*)

- **Mother plant selection**
  - Use seeds or cuttings from trees with high yield, good quality clove buds, and resistance to diseases like clove canker and leaf spot.
- **Propagation standards**
  - For seeds, ensure they are from mature, healthy fruits, and sown immediately due to their short viability.
  - For vegetative propagation, use healthy cuttings treated with rooting hormones.
- **Nursery management**
  - Maintain optimal temperature and moisture levels in the nursery.
  - Provide adequate shading to young seedlings and ensure proper ventilation.
  - Regularly monitor and control fungal infections and pests.
- **Quality control**
  - Inspect seedlings for vigor, disease-free status, and true-to-type characteristics.
  - Ensure proper hardening of seedlings before field planting.

5. Allspice (*Pimenta dioica*)

- **Mother plant selection**
  - Select mother plants known for their aromatic quality, high yield, and disease resistance.

- **Propagation standards**
  - Propagate through seeds or vegetative cuttings from healthy plants.
  - Ensure seeds are viable and cuttings are treated to promote rooting and prevent diseases.
- **Nursery management**:
  - Maintain appropriate shading, watering, and humidity levels.
  - Implement pest and disease management practices, including regular inspections.
  - Use well-draining soil with adequate organic matter.
- **Quality control**
  - Ensure that allspice plants are true-to-type and free from pests and diseases before certification.
  - Test the nursery stock periodically for genetic purity and viability.

6. Cinnamon (*Cinnamomum verum*)

- **Mother plant selection**
  - Use plants with high bark yield, good aroma, and resistance to diseases like leaf spot and dieback.
- **Propagation standards**
  - Propagate through seeds, cuttings, or air layering, ensuring that the material is from disease-free plants.
  - Treat cuttings and air-layered plants with rooting hormones to ensure good establishment.
- **Nursery management**
  - Provide shade, regular watering, and maintain soil fertility.
  - Control pests like leaf miners and diseases with appropriate IPM techniques.
- **Quality control**
  - Inspect plants regularly to ensure they are healthy, disease-free, and true-to-type.
  - Ensure that cinnamon plants are well-hardened before transplanting.

General Certification Requirements

1. **Record keeping**: Nurseries must maintain detailed records of mother plants, propagation methods, pest and disease control measures, and inspection reports.
2. **Sanitation**: Nurseries must adhere to strict sanitation protocols to prevent the spread of diseases and pests.

3. **Traceability**: The origin of all planting material must be traceable, from mother plants to the final QPM.
4. **Inspection and testing**: Regular inspections by certification authorities are required, along with laboratory testing for diseases and genetic purity.
5. **Labeling and documentation**: Certified QPM must be properly labeled with certification tags that include information on the species, variety, date of certification, and certifying authority.
6. **Compliance with regulations**: Nurseries must comply with local and national regulations governing the production and certification of QPM.

These standards help ensure that the QPM produced in nurseries is of the highest quality, which is essential for maintaining healthy, productive tree spice plantations.

Isolation distance for seed spices crops

Seed spices	Foundation seed (m)	Certified seed(m)
Fenugreek (highly self-pollinated)	50	25
Coriander (Cross-pollinated)	800	400
Cumin (Cross pollinated)	800	400
Fennel (Cross pollinated)	1000	800
Ajowain (Cross pollinated)	1000	800
Dill (Cross pollinated)	1000	800

Ginger: For ginger (**Zingiber officinale**) seed production, isolation distances are generally shorter compared to those required for crops that rely heavily on cross-pollination. Since ginger is primarily propagated through rhizomes and does not produce viable seeds in the same way as sexually reproducing plants, the main concern with isolation is more about preventing the spread of pests and diseases rather than cross-pollination. However, if ginger is being cultivated for true seed production (which is rare), the recommended isolation distance would typically be around **10-20 meters** to prevent any potential contamination from different varieties, though this is not a common practice due to ginger's method of propagation. For maintaining genetic purity or for specific breeding programs, you would follow these isolation distances more strictly.

Turmeric: For **turmeric** (*Curcuma longa*), which is primarily propagated through rhizomes similar to ginger, the isolation distance is generally not as critical as it is for seed-propagated crops. Turmeric does not usually produce viable seeds under normal cultivation conditions, so the primary concern is again about preventing the spread of pests and diseases rather than ensuring genetic purity through cross-pollination. However, when considering genetic purity in rare cases of seed production or specific breeding programs, an isolation distance of around **50-100 meters** is often recommended. This helps to prevent any potential cross-contamination with different varieties, although such practices are not common in standard turmeric cultivation.

Cardamom: For **cardamom** (*Elettaria cardamomum*), which is a cross-pollinated crop primarily pollinated by bees, maintaining proper isolation distances is crucial to prevent cross-pollination between different varieties. The recommended isolation distance for cardamom seed production is typically **400 to 500 meters**. This distance helps ensure the purity of the seed by minimizing the risk of cross-pollination from other cardamom plants, which is particularly important for maintaining varietal characteristics in commercial seed production.

Black Pepper: For **black pepper** (*Piper nigrum*), which is primarily propagated through vegetative cuttings rather than seeds, isolation distance is less of a concern in typical cultivation. However, in cases where black pepper is grown from seeds or for breeding purposes, maintaining genetic purity might require attention to isolation distances. If isolation is necessary, such as in a breeding program or when growing distinct varieties close to each other, a distance of about **100 to 200 meters** is recommended to prevent any potential cross-pollination, though cross-pollination in black pepper is not as common as in some other crops. The main focus typically remains on preventing the spread of diseases rather than on cross-pollination.

Clove: For **clove** (*Syzygium aromaticum*), which is primarily propagated through seeds or vegetative means, maintaining proper isolation distance is important when the focus is on seed production, especially to ensure varietal purity.Clove is cross-pollinated by insects, so to prevent cross-pollination between different clove varieties, an isolation distance of **at least 300 meters** is recommended. This distance helps ensure that the seeds produced remain true to the desired variety by minimizing the chances of cross-pollination from other nearby clove plants.

Allspice: For **allspice** (*Pimenta dioica*), which is typically cross-pollinated by insects, maintaining an appropriate isolation distance is important to preserve the genetic purity of the seeds when producing allspice plants. The recommended isolation distance for allspice is generally around **400 meters**. This distance helps prevent cross-pollination between different varieties or unwanted genetic mixing from nearby plants, ensuring that the seeds collected are true to the desired variety.

Nutmeg: For **nutmeg** (*Myristica fragrans*), which is a dioecious plant (having separate male and female trees), cross-pollination is necessary for fruit production. However, in seed production or breeding programs, maintaining genetic purity might require consideration of isolation distances.

The recommended isolation distance for nutmeg is typically around **300 to 400 meters**. This distance helps prevent unwanted cross-pollination between different varieties or types of nutmeg, ensuring the purity of seeds or maintaining specific breeding lines. This is particularly important when distinct varieties are grown in proximity or when maintaining the genetic integrity of a specific nutmeg variety is desired.

Cinnamon

1. Selection of Mother Plants

- **High-Quality Selection:** Choose healthy, disease-free, and high-yielding mother plants with desirable traits such as robust growth and aromatic bark.
- **Genetic Diversity:** Ensure genetic diversity by selecting a variety of mother plants to maintain the resilience and adaptability of the seed population.

2. Pollination

- **Natural pollination:** Cinnamon plants are typically self-pollinating, but cross-pollination can be encouraged to enhance genetic variation.
- **Manual pollination:** In controlled environments, manual pollination might be performed to ensure seed production, especially if natural pollinators are scarce.

3. Seed Collection

- **Timing:** Collect seeds when the fruits (berries) turn a dark purple or black color, indicating full ripeness.
- **Harvesting:** Gently pluck the ripe fruits by hand to avoid damaging the seeds.

4. Seed Processing

- **Pulp Removal:** Remove the pulp surrounding the seeds by soaking the fruits in water for a few days, then rubbing them to release the seeds.
- **Drying:** Dry the cleaned seeds in a shaded area to reduce moisture content, which helps prevent fungal infections.

5. Seed Storage

- **Temperature and humidity:** Store seeds in a cool, dry place. Seeds can be kept in airtight containers with moisture-absorbing materials to extend viability.
- **Viability duration:** Cinnamon seeds are recalcitrant, meaning they lose viability quickly and should ideally be sown soon after harvest.

6. Sowing and germination

- **Soil preparation:** Prepare a well-drained, fertile seedbed with organic matter to provide a suitable environment for germination.
- **Sowing depth:** Sow the seeds at a shallow depth, covering them lightly with soil.
- **Watering:** Keep the seedbed moist but not waterlogged to promote germination, which typically occurs within 2 to 4 weeks.

7. Transplanting

- **Seedling selection:** Once seedlings have developed a few true leaves, select the healthiest ones for transplanting.
- **Transplanting:** Carefully transplant the seedlings to their permanent location, ensuring they are spaced appropriately to allow for growth.

Field inspection

1. Pre-Inspection preparation

- **Familiarization:** Inspectors should familiarize themselves with the crop, its growth stages, and the specific varietal characteristics of the seed spice being produced.
- **Record review:** Examine the planting records, seed source, and field history to identify any potential issues before inspection.

2. Inspection stages

- **Initial inspection:** Conduct an early inspection shortly after the crop has been established to ensure proper plant spacing, emergence, and overall plant health.
- **Mid-Season inspection:** Inspect during the vegetative stage to check for any off-type plants, pests, diseases, and weeds. This is also the time to assess the uniformity of the crop.
- **Pre-Harvest inspection:** Inspect just before harvesting to ensure that the crop has developed correctly, focusing on seed maturity, purity, and the absence of contaminants.

3. Key Inspection criteria

- **Plant uniformity:** Ensure uniformity in plant height, leaf shape, and color, which indicates genetic purity. Off-type plants should be removed.
- **Pest and disease monitoring:** Inspect for signs of pest infestation or disease. Infected plants should be treated or removed to prevent contamination of the seed lot.
- **Weed control:** Check for the presence of weeds, which can compete with the crop and contaminate the seed. Effective weed control measures should be in place.
- **Isolation distance:** Ensure that the field meets the required isolation distance from other crops of the same species to prevent cross-pollination and maintain varietal purity.
- **Seed setting:** Inspect for proper seed setting, ensuring that the plants are producing a good quantity of viable seeds.

- **Roguing:** Identify and remove any off-type plants, diseased plants, or those that do not conform to the varietal characteristics.

4. Documentation and reporting

- **Field Inspection reports:** Document each inspection in detail, noting any issues, actions taken, and recommendations for future inspections or improvements.
- **Photographic evidence:** Capture images of any problematic areas or issues to provide visual evidence in the inspection report.
- **Follow-up inspections:** Schedule follow-up inspections if necessary to ensure that corrective actions have been effective.

5. Final inspection and certification

- **Final assessment:** Conduct a final inspection when the seeds are ready for harvest to verify that all standards have been met.
- **Certification:** If the seed crop passes all inspections, it can be certified as meeting the quality standards for seed production.

6. Post-harvest inspection

- **Seed cleaning and processing:** Inspect the seed cleaning and processing facilities to ensure that the seeds are properly handled and contaminants are removed.
- **Storage conditions:** Verify that seeds are stored in conditions that prevent moisture, pests, and disease from compromising seed quality.

1. Pre-inspection preparation

- **Understanding the Crop:** Inspectors should be familiar with the growth stages, characteristic features, and specific varietal traits of ginger, turmeric, black turmeric, and mango ginger.
- **Review records:** Check planting records, seed material sources, and field history to anticipate any potential problems.

2. Inspection stages

• Initial inspection (Establishment Stage)

- **Planting material:** Ensure that the rhizomes used are healthy, disease-free, and true to type.
- **Emergence and spacing:** Inspect for uniform emergence and proper spacing between plants to ensure good air circulation and growth.

- **Mid-season inspection (Vegetative stage)**
  - **Plant health:** Assess the plants for uniformity in growth, leaf color, and shape. Look for any signs of disease, pests, or nutrient deficiencies.
  - **Weed control:** Ensure effective weed management to avoid competition and contamination.
  - **Irrigation and drainage:** Check for proper irrigation practices and drainage to prevent waterlogging, which can lead to rhizome rot.
- **Pre-Harvest inspection (Rhizome maturity)**
  - **Rhizome development:** Inspect the rhizomes for size, uniformity, and health. Ensure they are free from rot, pests, and diseases.
  - **Isolation:** Confirm that the field is isolated adequately from other varieties or species to maintain genetic purity.
  - **Off-type plants:** Identify and rogue out any off-type plants that do not conform to the varietal characteristics.

3. Key inspection criteria

- **Uniformity:** Ensure that plants are uniform in growth and development, indicating that the planting material is genetically true to type.
- **Disease and pest monitoring**
  - **Common diseases:** Look for symptoms of common diseases such as soft rot, bacterial wilt, and leaf spot in ginger and turmeric.
  - **Pest damage:** Inspect for signs of pest infestation, particularly root-knot nematodes and shoot borers, which can affect rhizome quality.
- **Weed control:** Confirm that weeds are managed effectively to prevent them from competing with the crops and contaminating the rhizomes.

4. Roguing

- **Off-Type removal:** Remove any plants that do not exhibit the desired varietal characteristics, such as differences in leaf shape, color, or rhizome appearance.
- **Disease and pest infected plants:** Rogue out and destroy any plants showing symptoms of disease or severe pest infestation to prevent the spread to healthy plants.

5. Documentation and reporting

- **Inspection records:** Keep detailed records of each inspection, including observations, actions taken, and any recommendations for the grower.
- **Photographs:** Document issues with photographs to provide evidence in reports.

- **Follow-up inspections:** Schedule additional inspections if needed to ensure that corrective actions have been successful.

6. Final inspection and certification

- **Final check:** Before harvesting, conduct a final inspection to ensure that the rhizomes meet all quality standards and are ready for certification as planting material.
- **Harvesting:** Oversee the harvest to ensure that rhizomes are handled properly to avoid damage, which can lead to post-harvest diseases.

7. Post-harvest inspection

- **Cleaning and curing:** Inspect the cleaning and curing processes of the rhizomes to ensure that they are free of soil, pests, and diseases.
- **Storage:** Check storage conditions to ensure that rhizomes are kept in a cool, dry place to prevent spoilage.

Pre-inspection preparation

- **Crop familiarization:** Understand the growth stages, morphological traits, and specific varietal characteristics of each crop.
- **Review Planting Records:** Check the source of planting material, field history, and previous inspections to identify any potential issues.

Inspection Stages

Initial Inspection (Establishment Stage)

- **Planting Material:** Verify that seeds or vegetative planting materials (like cuttings or seedlings) are healthy, disease-free, and true to type.
- **Germination and establishment:** Ensure that germination or sprouting is uniform and that plants are establishing well in the field.

Mid-Season inspection (Vegetative to flowering stage)

- **Plant health:** Inspect for uniform growth in height, leaf size, and color, which indicates genetic purity.
- **Pest and disease monitoring:** Look for signs of common pests and diseases that affect these crops:
  - **Black Pepper:** Inspect for pests like aphids, mealybugs, and diseases like Phytophthora foot rot.
  - **Cardamom:** Check for thrips, borers, and diseases like rhizome rot and mosaic virus.
  - **Cinnamon:** Inspect for leaf spot, bark borer, and other fungal infections.

    - **Allspice, Clove, Nutmeg:** Monitor for fungal diseases like leaf blight, and check for pests like scales and beetles.
- **Weed control:** Ensure that weeds are effectively managed to prevent competition and contamination.

Pre-Harvest inspection (Maturity stage):

- **Flowering and Fruiting:** Inspect for proper flowering and fruit setting. Ensure that the fruits or seeds are developing uniformly and healthily.
- **Isolation Distance:** Confirm that the crop is sufficiently isolated from other varieties or species to prevent cross-pollination and maintain varietal purity.
- **Off-Type Plants:** Identify and remove any off-type plants that do not conform to the expected varietal characteristics.

Key inspection criteria:

- **Genetic purity:** Ensure uniformity in plant characteristics like height, leaf structure, and fruit/seed morphology.
- **Disease and pest control**
    - **Monitoring:** Regularly check for pest infestations and disease outbreaks. Quick intervention is necessary to protect seed quality.
    - **Treatment:** Verify that any chemical or biological treatments used are appropriate and effective.
- **Crop maintenance:** Assess irrigation practices, nutrient management, and overall crop care to ensure optimal growth conditions.

Roguing

- **Off-type removal:** Remove any plants that deviate from the desired variety characteristics, such as differences in plant height, leaf shape, or fruit/seed characteristics.
- **Disease and pest management:** Rogue out and destroy any plants severely affected by pests or diseases to prevent spread.

Documentation and reporting

- **Inspection logs:** Maintain detailed records of each inspection, noting any irregularities, actions taken, and recommendations.
- **Photographic documentation:** Take photos of any significant issues or deviations to include in the inspection report.
- **Follow-up inspections:** Schedule additional inspections as needed to verify that corrective actions are effective.

Final inspection and certification

- **Pre-harvest inspection:** Conduct a final inspection before harvest to ensure that the crop meets all required standards for seed production.
- **Harvest monitoring:** Oversee the harvest to ensure that fruits/seeds are harvested at the correct maturity stage and handled properly to prevent damage or contamination.

Post-harvest inspection

- **Seed processing:** Inspect seed cleaning, drying, and storage processes to ensure that seeds are handled correctly to maintain viability and purity.
- **Storage conditions:** Verify that seeds are stored under optimal conditions to prevent deterioration from moisture, pests, or diseases.

Crop-specific notes

- **Black Pepper:** Ensure that peppercorns are fully mature before harvest for maximum viability.
- **Cardamom:** Harvest capsules when they turn greenish-yellow, ensuring proper drying to prevent mold.
- **Cinnamon:** Focus on the quality of bark for propagation rather than seeds.
- **Allspice, Clove, Nutmeg:** Ensure that the fruits or seeds are fully mature and handled with care during harvesting and processing.

These inspection steps are crucial to maintaining the quality, purity, and viability of the seeds or planting material for these spice crops, ensuring successful propagation and high-quality yield in future generations.

Objectionable weeds in seed spices

Fenugreek: Melilotus (senji)

Cumin: *Plantago pumila* (Jiri)

Ajowain : *Ammi majus*

Nigella : *Nigella damescena*

Designated seed borne and other diseases in seed spices

Fenugreek: Root rot (*Rhizoctonia solani*), downy mildew (*Peronospora trigonella*), leaf spot (*Cercospora traversions*), powdery mildew (*Erisiphae* spp.)

Coriander: Fusarium wilt (*Fusarium oxysporum* f.sp. *coriandrii*), stem gall (*Protomyces macrosporus*), downy mildew, powdery mildew (*Erisiphae* spp.)

Cumin: Fusarium wilt (*Fusarium oxysporum* f.sp. *cuminaii*), Cumin blight (*Alternaria cuminaii*), powdery mildew (*Erisiphae* spp.)

Fennel: Alternaria blight (*Alternaria tenius*) and Ramularia blight (*Ramularia foeniculaii*)

Ajowain : Collar and root rot (*Scleroticum rolfsii*), Blight (*Alternaria* spp.)

Dill: Root rot (*Fusarium oxysporum*)

Nigella: Root rot (*Fusarium oxysporum*)

Anise: Alternaria blight, powdery mildew

Celery: Yellow virus

Caraway: Alternaria blight, powdery mildew

Specific seed requirement for seed spices crops

S.N.	Name of Crops	Maximum permitted limits (%)					
		Foundation seed			Certified seed		
		Off type	Objectionable weed	Diseased plants	Off type	Objectionable weed	Diseased plants
1	Fenugreek	0.10	0.01	0.10	0.20	0.02	0.50
2	Coriander	0.10	-	0.10	0.50	-	0.50
3	Cumin	0.10	0.01	0.10	0.20	0.02	0.50
4	Fennel	0.10	-	0.10	0.20	-	0.50
5	Ajowain	0.10	0.01	0.10	0.20	0.02	0.50
6	Dill	0.10	0.01	0.10	0.20	0.02	0.50
7	Anise	0.10	0.01	0.10	0.20	-	0.50
8	Celery	0.10	0.01	0.10	0.20	-	0.50
9	Caraway	0.10	0.01	0.10	0.20	0.02	0.50
10	Nigella	0.10	0.01	0.10	0.50	0.02	0.50

Seed standards for seed spices crops

S.N.	Name of Crops	Seed Standard (Percent)											
		Pure seed (minimum)		Inert matter (Maximum)		Other crop seed (Maximum)		Total weed seed (Maximum)		Germination (%)		Moisture (%)	
		F	C	F	C	F	C	F	C	F	C	F	C
1	Fenugreek	98	98	2	2	0.10	0.20	0.10	0.20	70	70	8	8
2	Coriander	98	97	2	3	0.10	0.20	0.10	0.20	65	65	10	10
3	Cumin	95	95	5	5	0.05	0.10	0.10	0.20	65	65	8	8
4	Fennel	95	95	5	5	0.05	0.10	0.10	0.20	70	70	8	8
5	Ajowain	95	95	5	5	0.05	0.10	0.10	0.20	65	65	8	8
6	Dill	95	95	5	5	0.05	0.10	0.10	0.20	70	70	8	8
7	Anise	95	95	2	3	0.10	0.20	0.10	0.20	65	65	10	10
8	Celery	95	95	5	5	0.05	0.10	0.10	0.20	65	65	8	8
9	Caraway	95	95	5	5	0.05	0.10	0.10	0.20	70	70	8	8
10	Nigella	95	95	5	5	0.05	0.10	0.10	0.20	65	65	8	8

Specific seed quality of rhizomatous spices

1. Ginger (*Zingiber officinale*)

- **Seed Material:** Rhizomes
- **Size:** Each seed rhizome should weigh between 20-25 grams, containing at least 2-3 healthy buds (eyes).
- **Quality:** The rhizomes should be free from diseases, pests, and physical damage. They should be well-matured and not sprouted excessively before planting.
- **Variety:** Use recommended high-yielding varieties that are suitable for the local growing conditions.
- **Quantity:** Approximately 1,500 to 2,000 kg of seed rhizomes per hectare.

2. Turmeric (*Curcuma longa*)

- **Seed Material:** Rhizomes
- **Size:** Rhizomes should be about 30-40 grams each, with 2-3 buds.
- **Quality:** Select well-developed, healthy rhizomes that are free from pests and diseases, particularly free from rhizome rot.
- **Variety:** Use disease-resistant and high-yielding varieties suitable for the region.
- **Quantity:** Around 2,000 to 2,500 kg of seed rhizomes per hectare.

3. Black Turmeric (Curcuma caesia)

- **Seed Material:** Rhizomes
- **Size:** Rhizomes should weigh about 25-30 grams each, with 2-3 active buds.
- **Quality:** Ensure rhizomes are healthy, free from pests, diseases, and physical injuries, with a characteristic dark blue to black inner flesh.
- **Variety:** Use genuine black turmeric rhizomes as it's less common and often confused with other Curcuma species.
- **Quantity:** Approximately 1,800 to 2,000 kg of seed rhizomes per hectare.

4. Mango Ginger (*Curcuma amada*)

- **Seed Material:** Rhizomes
- **Size:** Rhizomes should be around 20-25 grams each, with 2-3 viable buds.
- **Quality:** Rhizomes should be fresh, healthy, and free from any signs of rot, pests, or diseases, with a characteristic mango-like aroma.
- **Variety:** Use high-yielding varieties adapted to local conditions.
- **Quantity:** About 1,500 to 2,000 kg of seed rhizomes per hectare.

General Guidelines for All spices Crops

- **Selection:** Choose rhizomes from healthy, disease-free mother plants that are true to type and show vigorous growth.
- **Pre-planting treatment:** Treat the rhizomes with appropriate fungicides or bio-control agents to prevent soil-borne diseases.
- **Storage:** Store seed rhizomes in a cool, dry place with good ventilation to prevent sprouting and rot before planting.

Ensuring these specific seed requirements will help in establishing a healthy crop, leading to higher yields and better-quality produce.

Specific requirement for QPM production in tree spices

1. Black Pepper (*Piper nigrum*)

- **Propagation Material**
    - **Cuttings:** Select healthy, disease-free cuttings with 3-5 nodes from high-yielding mother plants.
    - **Rooting media:** Use a well-drained, fertile mixture of sand, soil, and organic matter for rooting the cuttings.
- **Nursery management**
    - Maintain shade (50% shade) and adequate humidity in the nursery.
    - Regularly water and apply fungicide to prevent fungal infections.
- **Selection:** Choose plants that are true to type, showing uniform growth and vigor.
- **Variety:** Use certified and recommended high-yielding varieties.

2. Cardamom (*Elettaria cardamomum*)

- **Propagation material**
    - **Suckers:** Use healthy, disease-free suckers with well-developed roots from high-yielding mother clumps.
    - **Seedlings:** If using seeds, ensure they are collected from superior, high-yielding mother plants.
- **Nursery management**
    - Grow suckers in a shaded, well-drained area.
    - Regular watering, and pest and disease management are essential.
- **Selection:** Ensure uniformity in size, vigor, and disease resistance in the selected planting material.
- **Variety:** Use elite clones or certified high-yielding varieties.

3. Cinnamon (*Cinnamomum verum*)

- **Propagation material:**
  - **Seeds:** Fresh, viable seeds should be used, sown immediately after collection as they lose viability quickly.
  - **Cuttings:** Semi-hardwood cuttings from healthy, high-yielding mother plants can also be used.
- **Nursery management:**
  - Use a well-drained, fertile seedbed or rooting medium.
  - Provide partial shade and maintain consistent moisture levels.
- **Selection:** Select vigorous, disease-free seedlings or rooted cuttings.
- **Variety:** Use proven high-quality clones or selected varieties with desirable bark characteristics.

4. Clove (*Syzygium aromaticum*)

- **Propagation material**
  - **Seeds:** Fresh, fully ripe seeds (drupe stage) from healthy, high-yielding trees should be sown immediately as clove seeds lose viability quickly.
- **Nursery managemen**
  - Seeds should be sown in a well-drained seedbed with partial shade.
  - Regular watering is essential to prevent drying out.
- **Selection:** Select uniform, healthy seedlings with a strong root system.
- **Variety:** Use certified or locally adapted high-yielding varieties.

5. Allspice (*Pimenta dioica*)

- **Propagation material**
  - **Seeds:** Use fresh seeds from mature, healthy, high-yielding trees.
  - **Cuttings:** Vegetative propagation through cuttings or air layering can also be used.
- **Nursery management**
  - Sow seeds in a well-drained seedbed under shade.
  - Regular watering and protection from pests are essential.
- **Selection:** Choose vigorous, uniform seedlings or rooted cuttings.
- **Variety:** Utilize varieties known for their superior spice quality.

6. Vanilla (*Vanilla planifolia*)

- **Propagation material**
  - **Cuttings:** Select healthy vine cuttings with 6-8 nodes from high-yielding, disease-free mother plants.
- **Nursery management**
  - Maintain high humidity and partial shade in the nursery.
  - Rooting medium should be well-drained, with regular watering.
- **Selection:** Ensure uniform, disease-free rooted cuttings with strong root development.
- **Variety:** Use high-yielding, disease-resistant vanilla varieties.

7. Nutmeg (*Myristica fragrans*)

- **Propagation material**
  - **Seeds:** Fresh seeds from fully ripe fruits, preferably from high-yielding, disease-free mother trees.
  - **Grafting:** Grafting of selected scion wood onto seedling rootstocks can also be practiced.
- **Nursery management**
  - Seeds should be sown immediately after collection in a well-drained seedbed.
  - Provide partial shade and maintain consistent moisture.
- **Selection:** Select seedlings or grafted plants that show uniform growth and vigor.
- **Variety:** Use certified varieties or selections known for high yield and quality of mace and nutmeg.

General requirements across Allspices crops

- **Disease-free material:** Ensure that all propagation material is free from diseases and pests, with regular inspections during nursery stages.
- **Nursery conditions:** Provide optimal conditions such as appropriate shade, humidity, and drainage in the nursery.
- **Certification:** Use certified or locally recommended varieties to ensure genetic purity and high yield potential.
- **Regular monitoring:** Continuously monitor the health and growth of the planting material, with prompt action taken against any signs of disease or pest infestation.

By adhering to these specific requirements, high-quality QPM can be produced, ensuring successful cultivation and productivity of these valuable spice crops.

9

Intellectual Property Rights (IPR) in Spices Breeding

Intellectual Property Rights (IPR)

Intellectual Property Rights refer to the legal protections given to the creators of original works, including inventions, literary and artistic works, designs, symbols, names, and images. These rights allow creators to control the use of their creations, preventing unauthorized use by others.

Types of intellectual property rights

1. **Patents**: Protect inventions, giving the patent holder the exclusive right to make, use, sell, and distribute the invention for a certain period.
2. **Copyrights**: Protect original works of authorship, such as books, music, and art. Copyright gives the creator exclusive rights to reproduce, distribute, and perform their work.
3. **Trademarks**: Protect symbols, names, and slogans used to identify goods or services. Trademarks prevent others from using similar marks that could confuse consumers.
4. **Trade secrets**: Protect confidential business information, such as formulas, processes, or methods, that give a business a competitive edge.

Importance of IPR

- **Incentive for innovation**: IPR provides financial incentives for creators by ensuring they can profit from their work.
- **Economic growth**: By protecting inventions and creations, IPR encourages investment in research and development, leading to economic advancement.
- **Consumer protection**: Trademarks help consumers identify the source of goods and services, ensuring they get what they expect in terms of quality.

Need of IPR

The need for Intellectual Property Rights (IPR) is crucial for several reasons, spanning economic, legal, and social dimensions. Here's why IPR is important:

1. Encouragement of innovation and creativity

- **Incentive for creation**: IPR provides creators and inventors with exclusive rights to their work, which serves as a motivation to develop new ideas, products, and technologies. Knowing that they can control and potentially profit from their inventions encourages individuals and companies to invest time and resources in innovation.

2. Economic growth

- **Attracting investment**: Strong IPR protection attracts investment in research and development, as businesses are more likely to invest in areas where their innovations are protected from being copied by competitors.
- **Market expansion**: IPR helps companies to protect their products and expand into new markets, knowing that their intellectual property is safeguarded, thus fostering economic development.

3. Legal protection and commercialization

- **Monetization of IP**: IPR allows creators to license, sell, or commercially exploit their intellectual property, providing a legal framework for monetizing their innovations.
- **Prevention of unauthorized use**: IPR enables creators to take legal action against unauthorized use or infringement of their work, thereby protecting their rights and ensuring fair competition.

4. Consumer protection

- **Assurance of quality and authenticity**: Trademarks and brand protection ensure that consumers can distinguish between genuine and counterfeit products, safeguarding them against low-quality or unsafe goods.

5. Cultural and educational growth

- **Preservation of cultural heritage**: Copyright laws help protect and preserve cultural works, ensuring that they are properly credited and respected, contributing to cultural continuity and growth.
- **Access to information**: By regulating the use of intellectual property, IPR ensures that creators are credited for their work, but it also facilitates access to information, knowledge, and culture in a controlled manner.

6. Global trade

- **Standardization in international markets**: IPR harmonizes the protection of intellectual property across borders, which is essential in today's globalized economy. This helps in creating a stable environment for international trade.

7. Ethical considerations

- **Recognition and respect for creators**: IPR ensures that creators and innovators receive recognition for their contributions, fostering a culture of respect and ethical practices in business and society.

In summary, Intellectual Property Rights are essential for fostering innovation, protecting creators, ensuring fair competition, and contributing to overall economic and cultural development.

IPR in spices breeding

Intellectual Property Rights (IPR) in the context of spices breeding is a critical aspect of agricultural innovation and protection. Here's how IPR applies to spices breeding:

1. Plant Breeders' Rights (PBR)

- **Protection of new varieties**: Plant Breeders' Rights (PBR) are a form of IPR that gives breeders exclusive control over the propagation of new plant varieties, including spices. This means breeders can control the sale, reproduction, and distribution of their newly developed spice varieties.
- **Encouragement of breeding innovation**: PBR incentivizes the development of new, improved varieties of spices by ensuring that breeders can benefit financially from their work. This includes varieties that might have enhanced yield, disease resistance, or specific flavour profiles.

2. Patents

- **Protection of biotechnological inventions**: In some jurisdictions, specific biotechnological methods used in spice breeding (such as genetic modification or specific hybridization techniques) can be patented. A patent on these methods ensures that others cannot use the same techniques without permission, protecting the breeder's innovation.
- **Limited scope for natural varieties**: It's important to note that naturally occurring spice varieties cannot be patented, but the processes or innovations related to their breeding can be.

3. Geographical Indications (GI)

- **Protection of regional spice varieties**: Certain spices, like saffron from Kashmir or black pepper from Malabar, can be protected under Geographical Indications. GI recognizes the unique qualities of spices linked to a specific region and prevents unauthorized use of the name by producers outside the region.

- **Market differentiation**: GI protection helps in distinguishing authentic products in the market, adding value to the spices and ensuring that the traditional knowledge and practices of local communities are recognized.

4. Trademarks

- **Brand protection**: Breeders or companies developing new spice varieties can use trademarks to protect the names or logos under which their spices are sold. This ensures brand recognition and protects against counterfeit products in the market.

5. Trade secrets

- **Protection of breeding techniques**: Some breeding methods or processes, particularly those that involve proprietary knowledge or trade secrets, can be protected under trade secret laws. This form of IPR is useful when the breeding process involves a unique, non-patentable method.

6. Biodiversity and traditional knowledge

- **Respect for indigenous knowledge**: Spices breeding often involves traditional knowledge from local communities. IPR in spices breeding must also consider biodiversity laws and regulations that protect the rights of these communities, ensuring they benefit from the use of their knowledge and genetic resources.
- **Access and benefit-sharing (ABS)**: Under the Convention on Biological Diversity (CBD) and the Nagoya Protocol, there are mechanisms to ensure that benefits arising from the use of genetic resources and traditional knowledge (such as breeding new varieties of spices) are shared fairly with the communities that provide them.

7. Challenges and considerations

- **Balancing innovation and access**: While IPR encourages innovation by protecting breeders, it also poses challenges, such as ensuring that farmers and small-scale producers have access to new varieties and that the genetic diversity of spices is maintained.
- **Ethical considerations**: The commercialization of spice varieties raises ethical concerns, particularly in relation to biopiracy (unauthorized use of genetic resources) and the rights of indigenous communities.

In summary, IPR in spices breeding plays a crucial role in promoting innovation, protecting breeders' interests, and ensuring the sustainable use of genetic resources. However, it also requires careful balancing to protect the rights of farmers, indigenous communities, and ensure equitable benefit-sharing.

PBR in spices breeding

Plant Breeders' Rights (PBR) are a specific type of Intellectual Property Right that grants plant breeders exclusive control over the propagation of their new plant varieties, including those of spices. Here's how PBR applies to spices breeding:

1. Protection of new varieties

- **Exclusive rights**: PBR gives breeders the exclusive right to produce, sell, and market the seeds or planting material of their new spice variety. This means that others cannot legally propagate or sell the protected variety without the breeder's permission.
- **Duration of protection**: Typically, PBR protection lasts for 20 to 25 years, depending on the country and the type of plant. This period allows breeders to recoup their investment and profit from their innovation.

2. Criteria for protection

To be eligible for PBR, a new spice variety must meet specific criteria:

- **Novelty**: The variety must be new, meaning it has not been sold or otherwise disposed of with the breeder's consent before the application for PBR.
- **Distinctiveness**: The variety must be clearly distinguishable from any other known variety. This could involve differences in taste, aroma, yield, disease resistance, or other agronomic traits.
- **Uniformity**: The variety must be sufficiently uniform in its characteristics, meaning that plants of the same variety should be similar in traits like growth habit, leaf shape, or colour.
- **Stability**: The variety must remain consistent in its essential characteristics after repeated propagation.

3. Breeder's rights

- **Control over propagation**: The breeder has the right to authorize or prevent others from producing, selling, exporting, or importing the protected variety.
- **Breeding exemption**: PBR typically includes a "breeder's exemption," allowing other breeders to use a protected variety to develop new varieties without needing permission from the original breeder. However, the new variety must be distinct from the original.
- **Farmers' privilege**: In many jurisdictions, small-scale farmers are allowed to save and replant seeds from protected varieties on their own farms, although they may not sell the seeds without authorization.

4. Application in spices breeding

- **Development of new varieties**: PBR is particularly relevant in spices breeding, where breeders might develop new varieties with improved traits, such as higher yield, better disease resistance, or specific flavour profiles. These new varieties can be protected under PBR, ensuring that the breeder has control over their commercial use.
- **Commercialization**: PBR enables breeders to commercialize new spice varieties, either by selling the seeds or licensing the rights to others. This can lead to greater availability of improved spice varieties for farmers and consumers.

5. Challenges and considerations

- **Access to genetic resources**: The use of genetic resources from traditional varieties in breeding new spice varieties must comply with access and benefit-sharing regulations, ensuring that the original custodians of these resources are fairly compensated.
- **Ethical and legal concerns**: In some cases, the protection of a spice variety through PBR might raise concerns about biopiracy, where traditional knowledge and genetic resources are used without adequate compensation or recognition to the communities that developed them.

6. Global framework and national laws

- **UPOV convention**: The International Union for the Protection of New Varieties of Plants (UPOV) sets the framework for PBR globally. Countries that are members of UPOV implement its standards through their national laws.
- **National implementation**: Each country has its own system for granting PBR, and breeders must apply in each country where they seek protection. This can be costly but is essential for protecting the variety in different markets.

7. Impact on the spice industry

- **Innovation and competition**: PBR encourages innovation in spice breeding, leading to the development of superior varieties that can enhance production, improve quality, and increase market competitiveness.
- **Economic benefits**: The protection of new varieties through PBR can provide significant economic benefitsto breeders, allowing them to capitalize on their innovations and contribute to the growth of the spice industry. This, in turn, can lead to the availability of better-quality spices for consumers and increased profitability for farmers who adopt these new varieties.

In summary, Plant Breeders' Rights in spices breeding provide a vital mechanism for encouraging innovation and protecting the intellectual property of breeders. By securing exclusive rights over new spice varieties, breeders can ensure that their investments are rewarded, while also contributing to the advancement of agricultural practices and the spice industry as a whole.

Patents in spices breeding

Patents in spices breeding represent a specialized area of intellectual property that provides protection for specific innovations related to the breeding and development of new spice varieties. Here's how patents apply in this context:

1. Scope of patents in spices breeding

- **Biotechnological innovations**: Patents can protect biotechnological processes used in the breeding of spices, such as genetic modification, marker-assisted selection, or other innovative techniques that result in the development of new or improved spice varieties.
- **Genetically modified organisms (GMOs)**: If a spice variety has been genetically modified to exhibit new traits (such as enhanced flavour, increased yield, or pest resistance), the specific genetic modification process or the resultant variety can be patented in many jurisdictions.
- **Breeding methods**: Specific, novel methods of breeding spices that are not obvious and provide a significant technical advancement can be patented. This includes methods that result in a new spice variety with unique characteristics.

2. Criteria for patentability

For an invention in spices breeding to be patentable, it must meet the following criteria:

- **Novelty**: The invention must be new, meaning it has not been previously disclosed or available to the public.
- **Inventive step (Non-Obviousness)**: The invention must not be obvious to someone with knowledge and experience in the field. It should involve an inventive step that distinguishes it from existing knowledge or practices.
- **Industrial applicability**: The invention must be capable of being used or applied in an industry, including agriculture.

3. Patentable innovations in spices breeding

- **Genetic traits**: Patents can be granted for specific genetic traits introduced into a spice plant, such as disease resistance or drought tolerance.

- **Hybrid varieties**: While the specific hybrid plant itself may be protected under Plant Breeders' Rights, the method of producing a particular hybrid or the use of a specific genetic marker in the breeding process may be patented.
- **Processing techniques**: Innovations in processing spices post-harvest, such as methods to enhance flavour, preserve freshness, or improve storage life, can also be patented.

4. Impact on the spice industry

- **Encouragement of research and development**: Patents incentivize companies and individuals to invest in research and development of new breeding techniques and spice varieties, knowing they can protect and potentially monetize their inventions.
- **Commercial advantage**: A patented breeding technique or genetically modified spice variety can give a company a competitive edge in the market, allowing them to control the use of their innovation and potentially charge higher prices.

5. Challenges and considerations

- **Access to genetic resources**: The use of naturally occurring genetic material in creating patented spice varieties raises ethical and legal questions about access and benefit-sharing, particularly regarding traditional knowledge and biodiversity.
- **Biopiracy concerns**: There are concerns that patents on spices or their breeding methods could lead to biopiracy, where companies patent genetic material or traditional knowledge without adequately compensating or acknowledging the indigenous communities that developed or conserved these resources.
- **Balancing innovation and access**: While patents encourage innovation, they can also restrict access to new technologies or varieties, particularly for small-scale farmers who may not afford patented seeds or breeding methods.

6. Global framework and national laws

- **International treaties**: The Agreement on Trade-Related Aspects of Intellectual Property Rights (TRIPS) sets minimum standards for patent protection globally. Under TRIPS, countries can determine the extent of patent protection for plant varieties, often balancing patents with other forms of protection like Plant Breeders' Rights.
- **Country-specific laws**: Different countries have varying approaches to patenting plant varieties and breeding methods. For example, the United

States allows for broad patent protection on plants and biotechnological processes, while some other countries might have more restrictive policies.

7. Ethical and legal considerations

- **Benefit-sharing**: Ensuring that patents on spices breeding respect international agreements like the Convention on Biological Diversity (CBD) and the Nagoya Protocol, which emphasize fair and equitable sharing of benefits arising from the use of genetic resources.
- **Public access vs. Private rights**: The patent system must balance private rights with public interests, ensuring that patents do not unduly restrict access to new varieties or technologies that could benefit society.

In summary, patents in spices breeding provide a powerful tool for protecting and commercializing innovative breeding methods and genetic modifications. However, their application requires careful consideration of ethical, legal, and social implications to ensure that innovation is encouraged while also respecting the rights and needs of all stakeholders involved, especially in relation to biodiversity and traditional knowledge.

Designs in spices breeding

In the context of spices breeding, "designs" typically refers to the concept of protecting the aesthetic or functional aspects of plant-related products rather than the plants themselves. However, design rights can intersect with spices breeding in several ways, especially when it involves the commercial presentation, packaging, or branding of the new varieties. Here's how design rights might apply:

1. Packaging and branding designs

- **Aesthetic appeal**: The design of packaging for spices, including the visual appearance, shape, colour schemes, and branding elements, can be protected under design rights. This is important for differentiating products in the marketplace.
- **Innovative packaging**: If a breeder or company develops unique packaging that enhances the usability or preservation of spices, such as specialized containers that maintain freshness or extend shelf life, these designs can be protected.

2. Designs of cultivation and processing equipment

- **Innovative tools**: If the breeding process involves the use of specially designed tools or equipment (for instance, tools that help in planting or harvesting specific varieties of spices more efficiently), the design of these tools can be protected.

- **Processing equipment**: The design of machinery or equipment used in the post-harvest processing of spices, such as drying, grinding, or packaging machines, can also be protected if they have unique design elements.

3. **Unique plant structures**

- **Aesthetic plant features**: Although plant varieties themselves are typically protected under Plant Breeders' Rights, certain aesthetic features (like the distinct appearance of ornamental plants) could be considered under design protection in some jurisdictions if they are unique and non-functional.
- **Decorative aspects**: If a new spice plant variety has been bred for decorative purposes, such as being grown and sold as an ornamental plant with unique leaves or flowers, the visual aspects might be eligible for design protection, though this is more common with ornamental plants than spices.

4. **Design rights in agricultural products**

- **Overall presentation**: The overall design of how the spice product is presented to the consumer, including the arrangement of the product in the packaging, can be protected. This could include innovative display methods that make the spice more appealing or easier to use.
- **Marketing materials**: The design of marketing materials, logos, or other visual elements used to promote new spice varieties can be protected to ensure brand consistency and recognition.

5. **Legal protection and enforcement**

- **Design registration**: To protect designs in the context of spices breeding, the design must be registered with the relevant intellectual property office. The design must be new, original, and have an individual character.
- **Enforcement**: Once a design is registered, the holder has the exclusive right to use the design and prevent others from copying or using it without permission.

6. **Importance in the market**

- **Market differentiation**: Design protection helps companies differentiate their products in the marketplace, making them more attractive to consumers. In the highly competitive spices market, this can be a significant advantage.
- **Brand identity**: Protecting the design elements associated with a spice variety or its packaging helps build and maintain brand identity, which is crucial for consumer recognition and loyalty.

7. Challenges and considerations

- **Overlap with Other IP Rights**: There is often an overlap between design rights and other forms of intellectual property, such as trademarks and copyrights. Managing these overlapping rights can be complex but is important for comprehensive protection.
- **Limited scope**: Design rights protect only the aesthetic aspects, not the functional or technical aspects of a product. This means that while the appearance of packaging or tools can be protected, the underlying technology or process often requires different forms of protection, such as patents.

In summary, while design rights in spices breeding are more about the protection of the aesthetic and presentation aspects related to the breeding and marketing of spices, they play a crucial role in ensuring that breeders and companies can protect and capitalize on their innovations in how these products are presented and sold.

Geographical Indications (GI) in spices breeding

Geographical Indications (GI) play a significant role in spices breeding by linking specific spice varieties to particular geographic regions known for their unique qualities.

Definition: A Geographical Indication (GI) is a type of intellectual property right that identifies a product as originating from a specific place, where a particular quality, reputation, or characteristic of the product is essentially attributable to its geographical origin.

Some Examples include "Malabar Pepper" from India, "Ceylon Cinnamon" from Sri Lanka, or "Kashmir Saffron" from India. These spices are recognized globally for their unique characteristics tied to their geographical origins.

Importance of GI in spices breeding

- **Quality assurance**: GIs help maintain and ensure the quality of spices by linking them to specific regions that produce them under particular conditions, often involving traditional methods of cultivation and processing.
- **Preservation of traditional knowledge**: GIs protect traditional breeding practices and methods that have been developed over centuries in specific regions. This helps in preserving cultural heritage and traditional knowledge.
- **Market differentiation**: GIs provide a marketing advantage by differentiating a product in the global market. Consumers often associate GIs with quality and authenticity, which can lead to premium pricing.

Process of obtaining GI for spices

- **Application**: To obtain GI status, a group of producers, an organization, or a government entity must apply to the relevant authorities, providing evidence that the spice's unique qualities are directly linked to its geographic origin.
- **Documentation**: The application must include details on the geographical area, the specific qualities of the spice, the traditional methods used in its production, and the historical connection between the spice and the region.
- **Approval and registration**: Once approved, the GI is registered, and only those producers within the designated area who adhere to the specific methods can use the GI label.

Benefits of GI protection in spices breeding

- **Economic benefits**: GIs can enhance the market value of spices, allowing producers to charge higher prices for products that carry a recognized GI label. This can lead to increased income for farmers and communities in the region.
- **Protection against misuse**: GIs protect against the misuse of the geographical name by ensuring that only spices genuinely originating from the region and produced under specific conditions can use the GI label. This prevents counterfeiting and preserves the reputation of the product.
- **Promotion of regional development**: GIs can boost local economies by promoting tourism and creating a brand identity for the region, which can lead to the development of related industries.

Challenges and considerations

- **Compliance and Monitoring**: Ensuring that all producers within the GI region adhere to the specific standards required for GI protection can be challenging. This includes monitoring production practices to maintain the quality and reputation of the GI.
- **Global recognition**: While GIs are recognized under international agreements like the TRIPS Agreement (Trade-Related Aspects of Intellectual Property Rights), obtaining and enforcing GI protection in foreign markets can be complex and costly.
- **Balancing tradition and innovation**: While GIs help preserve traditional methods, they may also limit innovation in breeding new varieties or adopting modern cultivation practices if these are perceived as deviating from traditional methods.

Impact on the Spice Industry

- **Consumer trust**: GIs build consumer trust by assuring them of the authenticity and quality of the spice. This can lead to brand loyalty and increased demand for GI-labeled spices.
- **Encouragement of sustainable practices**: GIs often encourage sustainable agricultural practices, as they emphasize traditional methods and the preservation of the environment in the region of origin.
- **Global market access**: GIs can open up new markets for spices by creating a recognized brand that appeals to consumers looking for high-quality, authentic products.

Notable GI-Labeled spices

- **Tellicherry Pepper**: From the Malabar coast of India, known for its large, bold peppercorns with a robust flavour.
- **Ceylon Cinnamon**: From Sri Lanka, distinguished by its delicate and aromatic flavour, different from the more common cassia.
- **Kashmir saffron**: Renowned for its deep colour and strong aroma, grown in the high-altitude regions of Kashmir.

In summary, Geographical Indications in spices breeding are crucial for protecting the unique qualities and reputations of spices tied to specific regions. They provide economic benefits to local producers, preserve traditional practices, and offer consumers assurance of quality and authenticity. GIs are a powerful tool in promoting and sustaining the cultural and economic value of spice-producing regions.

Spices GIs in India

India is home to a rich diversity of spices, many of which have been granted Geographical Indication (GI) status due to their unique qualities tied to specific regions. Here are some notable spices from India that have received GI status:

1. Malabar Pepper

- **Region**: Malabar Coast, Kerala
- **Features**: Malabar Pepper, also known as Tellicherry pepper, is renowned for its bold, pungent flavour and large, uniform peppercorns. The humid climate and unique soil conditions of the Malabar Coast contribute to its distinctive taste.
- **GI Status**: Recognized for its quality and historical significance, Malabar Pepper is one of the most famous pepper varieties in the world.

2. Kashmir Saffron

- **Region**: Kashmir Valley, Jammu and Kashmir
- **Features**: Known for its deep red colour and strong aroma, Kashmir Saffron is considered one of the finest saffron varieties globally. It is cultivated in the high-altitude fields of Kashmir, where the climate conditions enhance its quality.
- **GI Status**: This GI recognition helps protect the authenticity of Kashmir Saffron against lower-quality imitations.

3. Coorg Green Cardamom

- **Region**: Coorg (Kodagu), Karnataka
- **Features**: Coorg Green Cardamom is valued for its large size, green colour, and rich aroma. The hilly terrain and tropical climate of Coorg make it ideal for cultivating high-quality cardamom.
- **GI status**: The GI status helps preserve the traditional cultivation methods and supports the local economy.

4. Byadgi Chilli

- **Region**: Byadgi, Karnataka
- **Features**: Byadgi Chilli is known for its bright red colour and mild pungency, making it a popular choice for adding colour to dishes without overwhelming heat. It is extensively used in South Indian cuisine.
- **GI status**: The GI status helps ensure the quality and authenticity of Byadgi Chilli in both domestic and international markets.

5. Sichuan Pepper (*Zanthoxylum rhetsa*)

- **Region**: North eastern states (Arunachal Pradesh, Sikkim, Nagaland)
- **Features**: Often referred to as Indian Sichuan pepper, this spice has a unique numbing effect on the palate and is used in both local cuisine and traditional medicine.
- **GI status**: The GI recognition helps promote and protect this spice, which is gaining popularity beyond its traditional regions.

6. Alleppey green Cardamom

- **Region**: Alleppey, Kerala
- **Features**: Known for its vibrant green colour and intense aroma, Alleppey Green Cardamom is highly sought after in international markets. It is considered one of the finest cardamom varieties, used extensively in both culinary and medicinal applications.

- **GI status**: The GI status ensures the continued recognition and authenticity of this premium spice.

7. Naga Chilli (Bhut Jolokia)

- **Region**: Nagaland, Assam, and Manipur
- **Features**: Known as one of the hottest chillies in the world, Bhut Jolokia or Naga Chilli is famous for its extreme spiciness and is used in various culinary and medicinal applications.
- **GI status**: The GI status helps protect this chilli's unique identity and promotes its use in specialty products globally.

8. Salem Turmeric GI

- **Region**: Salem, a district in Tamil Nadu, India.
- **Characteristics**: Salem turmeric is known for its high curcumin content, which gives it a deep yellow colour and potent medicinal properties. It is valued for its strong aroma and flavour, making it a preferred choice in both culinary and medicinal uses.
- **GI status**: The GI tag helps protect the unique qualities of Salem turmeric and prevents unauthorized use of the name.

9. Kandhamal Turmeric GI

- **Region**: Kandhamal district in Odisha, India.
- **Characteristics**: Kandhamal turmeric is organic and known for its high curcumin content as well, but it is distinct due to its cultivation in the hilly terrains of Kandhamal, where the climate and soil contribute to its unique flavour and colour. This turmeric is prized for its organic production methods, often grown by tribal communities in the region.
- **GI status**: The GI tag for Kandhamal turmeric not only protects its name but also supports the livelihoods of the tribal farmers who cultivate it, ensuring they get fair prices for their product.

Both of these turmeric varieties are significant in the global spice market and are recognized for their quality and medicinal properties. The GI tags help preserve their authenticity and support the local economies where they are grown.

Impact of GI on Spices in India

- **Economic benefits**: The GI status helps these regions secure a premium price for their products, benefiting local farmers and communities by boosting income and promoting sustainable agriculture.
- **Cultural preservation**: GI recognition preserves traditional farming practices and supports the cultural heritage associated with spice cultivation in specific regions.

- **Global recognition**: GIs enhance the international reputation of Indian spices, making them more competitive in global markets and helping to combat counterfeit products.

In summary, Geographical Indications in India are vital in protecting and promoting the country's rich heritage of spice

Intellectual property right includes

1. **Plant Variety Protection (PVP):** This grants breeders exclusive control over the propagation of a new plant variety for a certain period, usually up to 25 years. The variety must be novel, distinct, uniform, and stable to qualify for protection.
2. **Patents:** In some jurisdictions, it may be possible to patent a new plant variety if it meets the criteria for patentability, such as being novel, non-obvious, and useful.
3. **Trademarks:** Breeders may also use trademarks to protect the branding of a specific spice variety.
4. **Trade secrets:** Breeders might protect their methods of breeding and cultivation as trade secrets if they do not want to disclose the specifics publicly, which is a requirement for patents.
5. **Copyrights:** Copyright in spice breeding typically involves the protection of intellectual property rights related to the development and creation of new spice varieties. Copyright primarily covers artistic works, literature, music, and other creative outputs rather than functional innovations like plant varieties.

Trademark in spices breeding

In the context of spice breeding, trademarks play a crucial role in protecting the branding and identity of a specific spice variety or a brand associated with it. Here's how trademarks are relevant in this field:

1. **Brand identity:** A trademark can protect the name, logo, or slogan associated with a specific spice variety or a product line of spices. This ensures that consumers can identify the source of the product, distinguishing it from competitors.
2. **Market differentiation:** Trademarks help differentiate one breeder's spice variety from others in the market. For example, if a breeder develops a unique type of chili pepper, they can trademark the name under which it is sold, ensuring that no other company can use the same name for a different product.
3. **Consumer trust:** Trademarks build consumer trust by signaling consistency and quality. When consumers see a trademarked name, they associate it with the particular characteristics and quality of the spice that they expect.

4. **Legal protcction:** A trademark provides legal protection against infringement. If another company tries to sell a similar product using a name or logo that is confusingly similar to a registered trademark, the trademark owner can take legal action to protect their brand.
5. **Global branding:** For spice breeders operating internationally, trademarks help in maintaining a consistent brand identity across different markets, which is crucial for global recognition and consumer loyalty.

For example, a breeder who has developed a unique variety of cinnamon might trademark the name "Golden Spice Cinnamon," ensuring that only their company can market cinnamon under that name.

Trade secrets in spices breeding

In spice breeding, trade secrets are an important form of intellectual property protection, particularly for methods, processes, and knowledge that are valuable but not publicly disclosed. Here's how trade secrets are used in spice breeding:

1. Protection of breeding techniques

- **Breeding methods:** Specific methods used to breed new spice varieties, such as crossbreeding techniques, genetic modifications, or selection processes, can be kept as trade secrets. This includes knowledge about which specific plants to cross, how to achieve desired traits, and the conditions required for optimal breeding outcomes.
- **Cultivation practices:** Details about soil preparation, watering schedules, fertilization, and other cultivation techniques that optimize the growth and quality of a particular spice can also be protected as trade secrets.

2. Preservation of genetic material

- **Proprietary strains:** If a breeder develops a unique strain of a spice plant, the genetic makeup of that strain can be kept confidential. This ensures that competitors do not gain access to the exact genetic material that gives the spice its unique qualities.
- **Seed handling:** Methods for storing and handling seeds to preserve their viability and enhance their traits over time can be kept as trade secrets.

3. Confidential business information

- **Supply chain details:** Information related to sourcing, supplier agreements, and distribution strategies for the bred spice varieties can be protected as trade secrets.
- **Market research:** Insights into market demands, consumer preferences, and potential new markets for specific spice varieties can be valuable and kept confidential.

4. Advantages of trade secrets in spice breeding

- **No expiration:** Unlike patents or plant variety protections that have a fixed term, trade secrets can last indefinitely as long as they remain confidential.
- **Cost-effective:** Maintaining a trade secret does not require the registration fees and legal processes associated with patents.
- **Broad coverage:** Trade secrets can cover a wide range of information, from technical processes to business strategies, which might not be patentable.

5. Challenges

- **Risk of disclosure:** The main challenge with trade secrets is that once the secret is disclosed, either intentionally or unintentionally, it loses its protection.
- **No legal protection against independent discovery:** If another breeder independently develops the same breeding technique or method, they can use it freely, as trade secrets do not provide protection against independent discovery.

In summary, trade secrets allow spice breeders to protect valuable knowledge that gives them a competitive advantage without the need for public disclosure, as required by patents or plant variety protections.

Legal and ethical considerations surrounding Intellectual Property Rights (IPR) in spice breeding

The legal and ethical considerations surrounding Intellectual Property Rights (IPR) in spice breeding are complex, involving issues of ownership, access, and benefit-sharing. Here's an overview:

Legal Considerations

1. Compliance with national and international laws

- **Patent law:** While some countries allow patents on plant varieties, others do not. Breeders must understand and comply with the relevant laws in each jurisdiction where they seek protection.
- **Plant Variety Protection (PVP):** Many countries offer specific PVP laws that grant breeders exclusive rights to their new varieties. Compliance with the UPOV (International Union for the Protection of New Varieties of Plants) Convention is essential in countries that are members.
- **Trade secrets law:** Breeders using trade secrets must ensure proper legal safeguards, such as non-disclosure agreements (NDAs), to protect confidential information.
- **Trademark law:** Proper registration and enforcement of trademarks are necessary to protect brand identity.

2. Biodiversity and access laws

- **Convention on Biological Diversity (CBD):** The CBD emphasizes the need for access to genetic resources and fair benefit-sharing. Breeders must ensure that they comply with national laws on biodiversity, which often require permission and benefit-sharing agreements when using genetic resources.
- **Nagoya protocol:** This supplementary agreement to the CBD focuses on Access and Benefit-Sharing (ABS) related to genetic resources, requiring breeders to share benefits with the countries of origin of the genetic material.

3. IPR enforcement

- **Infringement and disputes:** Breeders must be prepared to enforce their IPR through legal action if necessary, and they should be aware of the potential for disputes over ownership and rights.
- **Cross-border enforcement:** Enforcing IPR across borders can be challenging, requiring a solid understanding of international law and cooperation with foreign legal systems.

Ethical considerations

1. Benefit-sharing

- **Equitable sharing:** Ethical considerations dictate that breeders should share the benefits derived from spice breeding, particularly with communities and countries that provide the genetic material. This includes monetary compensation, technology transfer, or development projects.
- **Respecting indigenous knowledge:** Many spice varieties have been cultivated and improved by indigenous communities for generations. Ethical breeding practices require acknowledging and compensating these contributions, respecting their rights over traditional knowledge.

2. Biopiracy

- **Avoiding exploitation:** Biopiracy refers to the unethical appropriation of genetic resources or traditional knowledge without proper authorization or compensation. Breeders must avoid such practices and ensure that their actions do not exploit the resources of developing countries or indigenous peoples.
- **Informed consent:** Before accessing genetic resources or traditional knowledge, breeders should obtain informed consent from the local communities or authorities, ensuring that they are aware of and agree to the intended use.

3. Sustainability and biodiversity

- **Environmental impact:** Breeders should consider the environmental impact of their activities, ensuring that their breeding practices do not harm biodiversity or lead to the depletion of genetic resources.
- **Sustainable use:** Ethical breeding involves promoting the sustainable use of spice varieties, ensuring that they can be preserved and utilized by future generations.

4. Access to seeds and plant varieties

- **Farmers' Rights:** Ethical IPR practices should respect farmers' rights to save, use, exchange, and sell seeds. Breeders should avoid restrictive practices that limit farmers' access to seeds or impose high costs.
- **Affordability and Accessibility:** New spice varieties developed through breeding should be accessible to farmers, especially in developing countries. High licensing fees or restrictive patents can hinder access to these varieties, raising ethical concerns about equity and food security.

Balancing legal and ethical considerations

Balancing the legal rights of breeders with ethical responsibilities toward communities, the environment, and future generations is critical in spice breeding. Breeders should strive to protect their innovations while promoting fairness, sustainability, and respect for all stakeholders involved.

Impact of IPR on spice breeding programs

Intellectual Property Rights (IPR) have a significant impact on spice breeding programs, influencing the development, commercialization, and accessibility of new spice varieties. The effects can be both positive and negative, depending on how IPR is applied and managed. Here's an analysis of the impact:

Positive Impacts

1. Incentivizing innovation

- **Encouraging research and development:** IPR, such as patents and Plant Variety Protection (PVP), provide breeders with exclusive rights to their innovations. This exclusivity incentivizes investment in research and development, leading to the creation of new and improved spice varieties.
- **Economic returns:** The ability to secure IPR allows breeders to recoup their investment through licensing fees or direct sales, making it financially viable to undertake long-term breeding programs.

2. Enhancing quality and variety

- **Development of superior varieties:** With the protection of IPR, breeders can focus on developing spice varieties with improved traits, such as higher yields, better resistance to pests and diseases, and enhanced flavours or aromas.
- **Market differentiation:** IPR, particularly trademarks, help in establishing unique brands for new spice varieties, enabling breeders to differentiate their products in the market and cater to specific consumer preferences.

3. Global collaboration and trade

- **International breeding programs:** IPR facilitates cross-border collaborations by providing a legal framework to protect and share breeding innovations. This can lead to the development of new varieties that are suitable for diverse climatic and geographic conditions.
- **Trade and export opportunities:** Protected varieties can be marketed globally, opening up new trade opportunities and enhancing the global competitiveness of breeding programs.

Negative Impacts

1. Restricted access to genetic resources

- **Limited availability for farmers:** Strong IPR protections, such as patents, can restrict farmers' access to new spice varieties, particularly if they are unable to afford the seeds or plants due to high costs. This can lead to reduced biodiversity and limit the availability of diverse crops.
- **Impact on traditional practices:** IPR can negatively impact traditional farming practices, where farmers save and exchange seeds. Restrictions on these activities due to IPR can undermine traditional knowledge and practices.

2. Potential for biopiracy

- **Exploitation of indigenous knowledge:** IPR systems, especially patents, can sometimes lead to biopiracy, where companies or individuals patent genetic resources or traditional knowledge without proper consent or benefit-sharing with indigenous communities.
- **Ethical concerns:** The commercialization of traditionally bred or naturally occurring spice varieties under IPR can raise ethical issues, particularly when the benefits do not flow back to the original custodians of the resources.

3. Barriers to innovation

- **IPR as a barrier for small breeders:** Small-scale breeders and farmers may find it challenging to navigate the complexities and costs associated with obtaining IPR, limiting their ability to protect and commercialize their innovations.
- **Concentration of power:** IPR can lead to the concentration of breeding knowledge and genetic resources in the hands of a few large companies, reducing competition and diversity in the breeding sector.

4. Impact on biodiversity

- **Narrowing genetic diversity:** The focus on a few highly profitable varieties protected by IPR may reduce the genetic diversity of spices, as traditional or less commercially viable varieties may be neglected.
- **Long-term sustainability:** Over-reliance on patented or protected varieties might undermine the long-term sustainability of spice agriculture, especially if these varieties are not adapted to changing environmental conditions or diverse ecosystems.

Balancing IPR with broader goals

To ensure that IPR supports rather than hinders spice breeding programs, it's essential to strike a balance between protecting innovations and ensuring access, equity, and sustainability. This includes promoting benefit-sharing arrangements, safeguarding farmers' rights, and ensuring that IPR systems do not unduly restrict the use of genetic resources or traditional knowledge. By aligning IPR with ethical and sustainable practices, breeding programs can contribute positively to agricultural diversity and food security while fostering innovation in the spice industry.

10

Future Prospects and Trends in Spices Breeding

1. Genomic selection and CRISPR technology

- The adoption of genomic selection (GS) and CRISPR-Cas9 technology in spice breeding is likely to accelerate. These technologies can significantly improve the precision of breeding programs, enabling the development of varieties with enhanced traits like disease resistance, yield, and quality.

2. Climate-resilient varieties

- With climate change impacting agriculture globally, there is a growing trend towards breeding spices that are resilient to extreme weather conditions. This includes the development of drought-resistant, heat-tolerant, and flood-tolerant varieties.

3. Biofortification

- Breeding programs are increasingly focusing on biofortification, which involves increasing the nutritional value of spices. This could lead to spices that are richer in essential nutrients like vitamins, minerals, and antioxidants.

4. Digital phenotyping and AI

- The integration of digital phenotyping tools and artificial intelligence (AI) is expected to revolutionize spice breeding. These technologies can enhance the efficiency of breeding programs by providing real-time data analysis and predictive modeling.

5. Sustainable breeding practices

- There is a rising emphasis on sustainability in agriculture. Breeding programs are likely to incorporate practices that reduce environmental impact, such as using less water, reducing pesticide usage, and developing organic farming-compatible varieties.

6. Precision agriculture integration

- Precision agriculture techniques, supported by IoT devices, drones, and satellite imagery, are expected to play a crucial role in monitoring and managing spice crops more effectively. This integration can optimize inputs and improve the quality and yield of spice crops.

7. Consumer-driven breeding

- Consumer preferences are shifting towards spices with specific flavors, health benefits, and organic origins. Breeding programs are increasingly tailored to meet these demands, resulting in spices with unique properties that cater to niche markets.

These trends indicate a future where spice breeding is more precise, sustainable, and aligned with both environmental challenges and consumer needs.

A) Innovations and emerging technologies

Innovations and emerging technologies are transforming spice breeding, enabling the development of improved varieties with desirable traits such as enhanced flavor, disease resistance, and adaptability to climate change. Here are some of the key innovations and technologies in this field:

1. Genomic Selection

- **Genomic Selection (GS)** is a cutting-edge breeding technique that uses DNA markers across the genome to predict the performance of plants. In spices, GS accelerates the selection process by identifying plants with desirable traits such as high yield, disease resistance, or specific flavour profiles early in the breeding cycle.

2. CRISPR-Cas9 Gene Editing

- **CRISPR-Cas9** is a powerful tool for precise gene editing. In spice breeding, it allows for the targeted modification of genes responsible for traits like disease resistance, pungency, or aroma. This technology can create new spice varieties faster and more efficiently than traditional breeding methods.

3. Marker-assisted selection (MAS)

- **Marker-assisted selection** involves using molecular markers linked to desirable traits to guide breeding decisions. In spices, MAS helps breeders focus on plants with specific characteristics, such as resistance to pests or improved nutritional content, reducing the time and cost required to develop new varieties.

4. Genomic and transcriptomic studies

- Advances in **genomics** and **transcriptomics** provide deep insights into the genetic makeup of spice plants. These studies identify the genes and pathways involved in key traits, enabling breeders to develop more targeted breeding strategies and create varieties with enhanced flavours, better yield, or improved stress tolerance.

5. High-throughput phenotyping

- **High-throughput phenotyping** uses advanced imaging techniques and sensors to quickly and accurately measure plant traits. In spices, this technology allows for the rapid screening of large numbers of plants, identifying those with superior qualities like size, colour, or flavour, and speeding up the breeding process.

6. Bioinformatics and data analytics

- **Bioinformatics** tools and **data analytics** play a crucial role in managing the vast amounts of genetic data generated in spice breeding programs. These technologies help in the identification of genetic markers, prediction of breeding outcomes, and optimization of breeding strategies, leading to more efficient and effective breeding programs.

7. Tissue culture and micropropagation

- **Tissue culture** and **micropropagation** techniques are used to rapidly multiply spice plants with desirable traits. These methods ensure the production of large quantities of genetically uniform and disease-free plants, which is particularly important for maintaining the quality and consistency of spice crops.

8. Digital breeding platforms

It integrates various data sources, such as genomic information, phenotypic data, and environmental conditions, to support decision-making in spice breeding. These platforms use machine learning and AI to predict the best breeding combinations and optimize breeding cycles, reducing time and costs.

9. Precision agriculture integration

Precision agriculture tools, such as drones, satellite imagery, and IoT sensors, are increasingly used in spice breeding to monitor crop health, soil conditions, and environmental factors. This data-driven approach allows breeders to make informed decisions about selecting and breeding plants that are best suited to specific growing conditions.

10. Biofortification and nutritional enhancement

Biofortification involves breeding spices with enhanced nutritional content, such as increased levels of essential oils, vitamins, or minerals. This approach not only improves the health benefits of spices but also meets consumer demand for functional foods.

11. Sustainable breeding practices

Innovations in sustainable breeding practices focus on developing spice varieties that require fewer inputs (such as water and fertilizers), are resistant to pests and diseases, and are adaptable to changing climatic conditions. This helps reduce the environmental impact of spice farming while maintaining high yields and quality.

These emerging technologies and innovations are making spice breeding more efficient, sustainable, and responsive to the needs of farmers and consumers, paving the way for the development of superior spice varieties that meet the demands of the future.

B) Global trends in spice consumption and production

Spices breeding is closely linked to global trends in spice consumption and production, which are influenced by factors such as consumer preferences, health consciousness, climate change, and agricultural practices. Here's how global trends are shaping spices breeding:

1. Increasing demand for exotic and authentic flavors

- **Consumer Preferences**: There is a growing global appetite for exotic and authentic flavors, driven by culinary trends and the increasing popularity of international cuisines. Breeding programs are focused on enhancing the unique flavor profiles of spices like chili, pepper, turmeric, and cardamom to meet this demand.
- **Breeding Focus**: Breeding efforts are directed towards developing varieties with intense, unique flavours and aromas. For instance, breeders are selecting for specific essential oil compositions that enhance the flavor intensity of spices like cinnamon and nutmeg.

2. Health and wellness trends

- **Functional Foods**: Spices are increasingly valued for their health benefits, such as anti-inflammatory, antioxidant, and antimicrobial properties. The trend towards functional foods is driving the breeding of spices with enhanced bioactive compounds.
- **Breeding Focus**: Breeding programs are emphasizing the development of spice varieties with higher concentrations of beneficial compounds, such as curcumin in turmeric and capsaicin in chili peppers, to cater to health-conscious consumers.

3. Organic and sustainable agriculture

- **Sustainability Concerns**: With the rise in demand for organic and sustainably produced spices, there is a push towards breeding varieties that are well-suited to organic farming practices, including pest resistance and reduced reliance on chemical inputs.

- **Breeding Focus**: Breeders are working on developing disease-resistant and pest-tolerant spice varieties that can thrive in organic systems. This includes varieties that require less water, are resilient to climate variability, and have a lower environmental footprint.

4. Climate change and crop resilience

- **Impact of climate change**: Global climate change is affecting spice production, with changing temperatures, precipitation patterns, and increased incidences of pests and diseases. This has made the development of climate-resilient spice varieties a priority.
- **Breeding focus**: Breeding programs are increasingly focused on developing heat-tolerant, drought-resistant, and disease-resistant spice varieties. This includes creating spices that can adapt to new growing regions as traditional spice-growing areas become less viable due to climate change.

5. Food Safety and quality standards

- **Global standards**: As international trade in spices increases, there is a stronger emphasis on meeting global food safety and quality standards. This includes ensuring spices are free from contaminants like aflatoxins, pesticides, and heavy metals.
- **Breeding focus**: Breeders are developing spice varieties that are less susceptible to contamination and can be grown with minimal pesticide use. This includes selecting for natural resistance to fungal infections that produce harmful mycotoxins.

6. Technological advancements in agriculture

- **Precision agriculture**: The adoption of precision agriculture technologies is transforming spice production. These technologies allow for more efficient management of spice crops, optimizing yield and quality.
- **Breeding focus**: Spice breeding programs are integrating digital tools and precision agriculture techniques to select and develop varieties that are highly productive under specific environmental conditions. This approach ensures that the spices meet the exacting standards required for global markets.

7. Global supply chain dynamics

- **Supply chain disruptions**: The COVID-19 pandemic and geopolitical tensions have highlighted vulnerabilities in global spice supply chains, leading to a renewed focus on regional production and the development of locally adapted varieties.

- **Breeding focus**: Breeding efforts are now prioritizing the development of spice varieties that can be grown in non-traditional regions, ensuring a stable and diversified supply chain. This includes creating varieties that are adaptable to different climates and soil conditions.

8. Traceability and certification

- **Consumer awareness**: There is increasing consumer demand for traceability and certification in the spice market, with a preference for fair trade, organic, and sustainably sourced products.
- **Breeding focus**: Breeding programs are not only focusing on the agronomic traits of spices but also on how these traits can support certification processes. For example, developing varieties that are more suited to organic farming practices helps farmers achieve organic certification more easily.

These global trends in spice consumption and production are driving innovation in spice breeding, ensuring that new varieties are better suited to meet the challenges of modern agriculture, consumer demands, and global market requirements.

C) Sustainability and organic breeding approaches

Sustainability and organic breeding approaches in spices are increasingly important as the global demand for environmentally friendly and health-conscious products grows. These approaches focus on developing spice varieties that are well-suited to organic farming systems and that contribute to more sustainable agricultural practices. Here's how sustainability and organic breeding are being applied in spice cultivation:

1. Breeding for disease and pest resistance

- **Challenges**: Organic farming restricts the use of synthetic pesticides and herbicides, making crops more vulnerable to diseases and pests.
- **Breeding focus**: Sustainable and organic breeding programs emphasize the development of spice varieties with natural resistance to common diseases and pests. This includes using traditional breeding methods as well as modern techniques like Marker-Assisted Selection (MAS) to identify and propagate resistant traits.
- **Examples**: Varieties of chili peppers that are resistant to viral infections, or turmeric with resistance to rhizome rot, help reduce the reliance on chemical treatments.

2. Adaptation to low input conditions

- **Challenges**: Organic farming relies on natural inputs, which can be less potent than synthetic fertilizers and pesticides, leading to challenges in maintaining high yields.

- **Breeding Focus**: Breeding programs are selecting for traits that enable spices to thrive in low-input environments typical of organic systems. This includes improving nutrient-use efficiency, where plants can make better use of organic fertilizers, and enhancing water-use efficiency.
- **Examples**: Breeding cumin or coriander varieties that perform well in nutrient-poor soils or with minimal irrigation.

3. Enhancing soil health and biodiversity

- **Challenges**: Maintaining soil health is a key principle of sustainable and organic farming, where chemical inputs are minimized.
- **Breeding focus**: Sustainable breeding programs focus on developing spice varieties that contribute to soil health. This might include selecting plants with deeper root systems that improve soil structure, or varieties that are compatible with intercropping systems that enhance biodiversity.
- **Examples**: Varieties of ginger or turmeric that are suited to polycultures, where they grow alongside nitrogen-fixing plants, can help improve soil fertility naturally.

4. Non-GMO breeding techniques

- **Challenges**: Organic certification standards often prohibit the use of genetically modified organisms (GMOs), which limits the tools available for breeding.
- **Breeding focus**: Organic breeding relies on non-GMO techniques, such as traditional cross-breeding, selection, and MAS. These techniques are used to introduce desirable traits without genetic modification.
- **Examples**: Development of organically certified pepper varieties through selective breeding that do not involve gene editing technologies like CRISPR.

5. Resilience to climate change

- **Challenges**: Climate change presents a significant challenge to spice production, with increasing temperatures, changing precipitation patterns, and more frequent extreme weather events.
- **Breeding focus**: Breeding for climate resilience involves developing spice varieties that can withstand stresses like drought, heat, and salinity. These traits are crucial for maintaining productivity in organic systems, which may have less access to resources to mitigate these stresses.
- **Examples**: Drought-resistant varieties of black pepper or heat-tolerant cardamom can help ensure stable yields in changing climates.

6. Reduction of environmental impact

- **Challenges**: Conventional agriculture can lead to environmental degradation, including soil erosion, water pollution, and loss of biodiversity.
- **Breeding focus**: Sustainable spice breeding aims to minimize the environmental footprint of cultivation. This includes developing varieties that require less water, are more efficient in nutrient uptake, and have a lower risk of contaminating the environment.
- **Examples**: Varieties of spices like saffron or vanilla that can be grown with minimal water and in marginal soils, reducing the pressure on natural water resources.

7. Supporting organic certification

- **Challenges**: To meet organic certification standards, spices must be grown without synthetic inputs and often require specific varieties that are compatible with organic practices.
- **Breeding focus**: Breeding programs support organic certification by developing varieties that are specifically suited for organic farming, ensuring that they can produce high yields and quality crops under organic conditions.
- **Examples**: Developing organically certified varieties of basil or oregano that are known for their high essential oil content, which is a key factor in organic spice markets.

8. Consumer and market demand

- **Challenges**: There is a growing consumer demand for spices that are organic, sustainably produced, and free from chemical residues.
- **Breeding focus**: Breeders are increasingly focused on developing spice varieties that not only meet these demands but also possess superior flavor, aroma, and nutritional properties. This enhances the marketability of sustainably and organically produced spices.
- **Examples**: High-curcumin turmeric or high-piperine black pepper varieties are developed to cater to the organic health food market.

These sustainability and organic breeding approaches are critical for developing spice varieties that align with global trends toward more environmentally responsible and health-conscious agriculture. As consumer demand for organic products continues to grow, these breeding strategies will play a crucial role in ensuring that spice production can meet both market needs and environmental challenges.

11

Practical Applications and Field Techniques

Practical applications and field techniques in spices breeding are essential for developing new varieties that meet the needs of farmers, processors, and consumers. These techniques involve a combination of traditional methods and modern technologies to select, test, and propagate spice plants with desirable traits. Here's an overview of the key practical applications and field techniques used in spices breeding:

1. Selection and hybridization

- **Mass selection**: This traditional technique involves selecting the best-performing plants from a large population based on desirable traits such as yield, flavor, and disease resistance. The selected plants are then propagated to create the next generation.
- **Hybridization**: Cross-breeding different varieties or species of spices to combine desirable traits from both parents. This technique is used to create hybrids with improved characteristics, such as higher yield, better flavor, or enhanced resistance to pests and diseases.
- **Example**: Cross-pollinating different varieties of chili peppers to produce hybrids with higher capsaicin content for spicier flavors.

2. Marker-assisted selection (MAS)

- **Application**: MAS uses molecular markers linked to specific traits to identify and select plants that carry desirable genes. This technique speeds up the breeding process by allowing breeders to screen plants at the seedling stage rather than waiting for mature traits to develop.
- **Field technique**: DNA is extracted from plant tissues (such as leaves) and analyzed in the laboratory to identify the presence of markers associated with traits like disease resistance or yield.
- **Example**: Using MAS to select turmeric plants with higher curcumin content, a key compound with medicinal properties.

3. Genetic diversity and germplasm collection

- **Application**: Collecting and maintaining a wide range of genetic material (germplasm) from different regions is crucial for spice breeding. This genetic diversity provides the raw material for developing new varieties with unique traits.
- **Field technique**: Germplasm collection involves traveling to different spice-growing regions to collect seeds, rhizomes, or cuttings. These samples are then conserved in gene banks or field collections for future breeding work.
- **Example**: Collecting wild relatives of black pepper to incorporate drought resistance traits into cultivated varieties.

4. Tissue culture and micropropagation

- **Application**: Tissue culture techniques are used to propagate large numbers of genetically identical plants from a small amount of plant tissue. This method is particularly useful for multiplying disease-free plants and preserving elite breeding lines.
- **Field technique**: Small tissue samples (explants) are taken from the parent plant and grown in sterile, nutrient-rich media under controlled conditions. The resulting plantlets are then acclimatized and transplanted to the field.
- **Example**: Micropropagation of vanilla plants to rapidly multiply disease-resistant and high-yielding varieties.

5. Field trials and evaluation

- **Application**: Field trials are conducted to evaluate the performance of new spice varieties under real-world growing conditions. These trials assess traits like yield, flavor, pest resistance, and adaptability to different environments.
- **Field technique**: New varieties are planted in experimental plots alongside standard varieties for comparison. Data on growth, yield, disease incidence, and other traits are collected throughout the growing season.
- **Example**: Conducting multi-location trials of new ginger varieties to assess their performance across different climates and soil types.

6. Disease and pest resistance screening

- **Application**: Screening for resistance to diseases and pests is a critical step in breeding programs. Identifying resistant plants helps reduce the need for chemical controls, making spice production more sustainable.
- **Field technique**: Plants are exposed to pathogens or pests either in controlled environments (greenhouses) or in the field where the pests are naturally present. Resistant individuals are selected for further breeding.

- **Example**: Screening chili pepper varieties for resistancc to root-knot nematodes in infested soil conditions.

7. Participatory breeding

- **Application**: Involving farmers in the breeding process ensures that new varieties meet the practical needs and preferences of end-users. This approach increases the adoption and success of new varieties.
- **Field technique**: Farmers are involved in selecting and evaluating new varieties in their own fields, providing feedback on traits such as taste, yield, and ease of cultivation.
- **Example**: Collaborating with farmers in India to develop turmeric varieties that are easy to harvest and have high market value.

8. Mutation breeding

- **Application**: Mutation breeding involves exposing seeds or plant tissues to chemicals or radiation to induce mutations. Some of these mutations may result in desirable traits, which are then selected for breeding.
- **Field technique**: Seeds are treated with mutagens and then grown in the field. Plants that exhibit beneficial mutations, such as improved yield or disease resistance, are selected for further breeding.
- **Example**: Developing new cardamom varieties with improved pod size and flavor through induced mutations.

9. High-throughput phenotyping

- **Application**: High-throughput phenotyping uses advanced imaging and sensor technologies to rapidly measure plant traits on a large scale. This method allows breeders to evaluate many plants quickly and accurately.
- **Field technique**: Drones, multispectral cameras, and ground-based sensors are used to collect data on traits like plant height, leaf area, and chlorophyll content across large breeding populations.
- **Example**: Using drones to assess the growth and health of pepper plants in a breeding trial.

10. Cross-pollination techniques

- **Application**: Cross-pollination is used to combine desirable traits from two different spice plants. Controlled pollination ensures that only the selected parent plants contribute to the next generation.
- **Field technique**: Pollen is manually transferred from the male flower of one plant to the female flower of another under controlled conditions. This technique is often used in crops where natural pollination does not reliably produce hybrids.

- **Example**: Hand-pollinating saffron flowers to create hybrids with enhanced flavor and color intensity.

These practical applications and field techniques are crucial for advancing spice breeding, helping to develop new varieties that are more productive, resilient, and better suited to the needs of farmers and consumers.

A) Floral Biology and crossing techniques in spices

1. Floral biology of Black pepper

Floral biology in black pepper (*Piper nigrum*) is an important area of study for understanding its reproductive biology, which is crucial for breeding and improving cultivation practices. Here's an overview of the key aspects of floral biology in black pepper:

1. Flower structure and morphology

- Flower type: Black pepper plants produce small, inconspicuous flowers that are unisexual, meaning they are either male or female.
- Inflorescence: The flowers are arranged in spikes or clusters known as spikes or racemes. These inflorescences are axillary, arising from the leaf axils.
- Flower parts: Each flower consists of:
  - Sepals: Usually small and not very noticeable.
  - Petals: Also small and not prominent.
  - Stamens: Male flowers have four stamens. Each stamen consists of a filament and an anther that produces pollen.
  - Pistil: Female flowers have a pistil that includes the ovary, style, and stigma. The ovary contains the ovules, which develop into fruits upon fertilization.

2. Pollination

- **Pollination mechanism**: Black pepper is primarily an insect-pollinated plant. Insects, especially bees and butterflies, transfer pollen from male to female flowers.
- **Pollination timing**: The flowers are typically open during the day, and the pollination process occurs when the flowers are fully open. Flowers typically open in the morning and are receptive for a short period. Anthesis usually occurs early in the day. The flowers are often fully open by mid-morning and can remain receptive until late afternoon. Pollen dehiscence in black pepper occurs shortly before or during anthesis. The anthers open

to release pollen typically in mid-mourning, which is then available for pollination.

Factors affecting pollination timing

- **Climate and weather**: Temperature and humidity can influence the timing of anthesis and pollen dehiscence. Warmer temperatures and higher humidity often accelerate these processes.
- **Plant variety**: Different varieties of black pepper may have slight variations in anthesis and dehiscence times.

3. Flowering and fruit development

- **Flowering season**: Black pepper plants usually flower in the rainy season. The exact timing can vary based on the geographic location and climatic conditions.
- **Fruit development**: After fertilization, the ovary develops into a peppercorn, which is a drupe containing a single seed. The fruit matures on the spike, turning from green to red when ripe.

4. Reproductive biology

- **Sexual reproduction**: Black pepper plants are dioecious, meaning individual plants are either male or female. Only female plants produce the peppercorns.
- **Pollination requirements**: Successful fruit development requires effective pollination. Inadequate pollination can lead to poor fruit set and lower yields.

5. Breeding and genetics

- **Hybridization**: To improve traits such as yield, disease resistance, and quality, breeders may use controlled pollination and hybridization techniques.
- **Genetic studies**: Understanding the genetics of flowering and fruit development can help in developing new varieties with desirable traits.

6. Cultivation practices

- **Pollinator management**: Ensuring the presence of effective pollinators is crucial for high fruit yield. Farmers may use strategies to attract and maintain pollinators in the field.
- **Flowering management**: Proper management of irrigation, fertilization, and pruning can influence flowering and fruiting patterns.

Crossing techniques in Black pepper

Crossing techniques in black pepper (Piper nigrum) are essential for developing new varieties with improved traits such as higher yield, disease resistance, and enhanced flavor. Black pepper, being a perennial vine, presents unique challenges and opportunities for breeders. Here's an overview of the key crossing techniques used in black pepper breeding:

1. Controlled Pollination

This technique is crucial for creating hybrid varieties with specific characteristics such as higher yield or disease resistance.

- **Selection of Parent Plants**: The first step is to select the parent plants. Typically, one parent (the female) should have desirable traits such as disease resistance or yield, while the other parent (the male) should contribute complementary traits like better quality or vigor.
- **Emasculation**: Since black pepper flowers are bisexual (having both male and female reproductive organs), emasculation is often performed to remove the anthers (the pollen-producing parts) from the female parent before they mature. This prevents self-pollination and ensures that only the pollen from the selected male parent is used.
- **Pollination**: After emasculation, pollen from the selected male parent is collected and manually applied to the stigma of the female flower. This is usually done using a fine brush or directly by gently shaking the male flower over the female.
- **Bagging**: The pollinated flower is then covered with a bag to prevent contamination from unwanted pollen. The bag is typically made of a fine mesh or paper and is kept in place until fruit set is confirmed.

- **Application**: Controlled pollination is used to create hybrids that combine the best traits of both parents, such as crossing a high-yielding variety with one that has strong disease resistance.

2. Mass Selection and open-pollination

- **Selection of superior plants**: In a field of black pepper, plants that show desirable traits such as high yield, large berries, or disease resistance are selected. These plants are allowed to cross-pollinate naturally (open-pollination).
- **Harvest and evaluation**: Seeds from the selected plants are harvested, and the next generation of plants is grown. These plants are then evaluated for their performance.
- **Repetition**: The process is repeated over several generations to gradually improve the population.

- **Application**: This method is useful for improving a population of black pepper plants over time and is often used in conjunction with more controlled breeding methods.

3. Inter-specific and intra-specific hybridization

- **Inter-specific hybridization**
  - Black pepper can be crossed with related species like Piper colubrinum, which is resistant to Phytophthora foot rot, a major disease in black pepper. The hybridization process is similar to controlled pollination but requires careful handling due to potential compatibility issues.
  - **Challenges**: Inter-specific hybridization can result in sterility or other reproductive issues in the offspring, so breeders must carefully select hybrids that are fertile and can be backcrossed to black pepper to stabilize the desirable traits.
  - **Outcome**: Successful inter-specific hybrids may exhibit strong resistance to diseases while retaining the desirable traits of black pepper, such as high-quality berries.
- **Intra-specific hybridization**:
  - This involves crossing different varieties of black pepper within the same species. For example, crossing a variety with high yield potential with another that has a compact growth habit suitable for high-density planting.
  - **Application**: This technique is commonly used to combine traits like yield, quality, and plant architecture within black pepper.

4. Double-cross and multiple-cross breeding

- **Double-cross breeding**:
  - Two initial crosses are made between four parent plants, resulting in two hybrid lines. These two hybrids are then crossed to produce a double-cross hybrid, which combines traits from all four original parents.
  - **Application**: This method is useful when breeders want to combine multiple traits, such as yield, disease resistance, and berry quality, into a single variety.
- **Multiple-cross breeding**:
  - This involves sequentially crossing a plant with several others over multiple generations to introduce a wide range of desirable traits.
  - **Application**: Used to develop complex hybrids that incorporate diverse traits from several parent plants, such as combining drought tolerance, disease resistance, and high yield.

5. Backcrossing

- **Initial Cross**: A black pepper variety with a desirable trait (e.g., disease resistance) is crossed with a high-yielding commercial variety.
- **Backcrossing**: The offspring that inherit the desired trait are then crossed back with the commercial variety repeatedly over several generations. This process is called backcrossing and is done to ensure that the final hybrid retains most of the genetic makeup of the commercial variety while incorporating the new trait.
- **Selection**: After several backcrosses, plants are selected based on the presence of the desired trait and overall performance.

- **Application**: Backcrossing is particularly useful for introducing single traits, such as resistance to Phytophthora, into a popular black pepper variety without altering its other characteristics.

6. Embryo rescue and tissue culture

- **Embryo Rescue**: After hybridization, if the embryo begins to develop but is at risk of aborting due to incompatibility, it is excised from the developing seed and cultured in a controlled environment to complete its development.
- **Tissue Culture**: Once the embryo develops into a plantlet, it is multiplied through tissue culture techniques to produce multiple clones for field planting.

- **Application**: Embryo rescue is especially important in inter-specific hybridization where embryo abortion is common. It allows breeders to recover hybrids that would otherwise not survive.

7. Progeny Testing

- **Planting and observation**: Seeds from a cross are planted, and the resulting plants are observed for the traits of interest, such as yield, disease resistance, and berry quality.
- **Selection**: The best-performing progeny are selected for further breeding or for direct use as new varieties.

- **Application**: Progeny testing is essential for validating the success of a cross and for selecting the best individuals to advance in the breeding program.

8. Hybrid seed production

- **Controlled environment**: Hybrid seed production often takes place in controlled environments such as greenhouses to prevent unwanted cross-pollination.

  - **Isolation techniques**: Physical barriers or spatial isolation are used to ensure that only the desired cross occurs.
  - **Harvesting**: Once pollination is successful, seeds are harvested, cleaned, and stored for future planting.
- **Application**: Hybrid seed production is crucial for commercializing new black pepper varieties with enhanced traits, ensuring that farmers receive uniform, high-quality seeds.

Equipment and resources

1. Controlled pollination equipment

- Pollination brushes: Fine brushes for manual pollination.
- Bagging materials: Bags or covers to protect flowers from external pollen.

2. Grafting tools

- Grafting knives: For making precise cuts.
- Grafting clips: To secure grafts.

3. Tissue culture facilities

- Culture vessels: Sterile containers for growing plant tissues.
- Growth chambers: Controlled environments for tissue culture.

4. Genetic tools

- PCR equipment: For marker-assisted selection and genetic analysis.
- Genetic transformation kits: For introducing new genes.

Summary

- **Controlled pollination** and **hybridization** are traditional methods used to improve and develop new varieties of black pepper.
- **Grafting and budding** are used for propagation and trait integration.
- **Tissue culture** provides a method for rapid propagation and mass production.
- **Genetic improvement** techniques, including marker-assisted selection and genetic transformation, are advancing the development of improved black pepper varieties.

These crossing techniques are fundamental to the development of improved black pepper varieties, enabling breeders to combine desirable traits and address challenges such as disease resistance, yield improvement, and adaptability to different growing conditions. By employing these methods, black pepper breeding programs can create varieties that meet the needs of both farmers and consumers, contributing to the sustainability and profitability of pepper cultivation.

1. Cardamom

Cardamom floral biology

The floral biology of cardamom (Elettaria cardamomum), commonly known as small cardamom, is quite intricate and plays a significant role in its pollination and reproductive processes.

1. Flower structure

- **Inflorescence**: Cardamom flowers are borne on a racemose inflorescence, which emerges from the base of the plant. Each inflorescence carries numerous flowers.
- **Flowers**: The flowers are zygomorphic (bilaterally symmetrical) and hermaphroditic, meaning they contain both male (stamens) and female (pistils) reproductive organs. They are typically white with purple or yellow markings.

2. Anthesis (Flower opening)

- The flowers of cardamom generally open early in the morning, often around 6:00 to 7:00 AM. This timing is crucial as it aligns with the activity patterns of pollinators.

3. Anther dehiscence

- Anther dehiscence, the process by which pollen is released from the anthers, usually occurs soon after the flower opens. The pollen is generally available for a few hours after anthesis, coinciding with the presence of pollinators like bees.

4. Stigma receptivity

- The stigma of the cardamom flower becomes receptive shortly after the flower opens and remains so for a short duration. Stigma receptivity and anther dehiscence are often synchronized to maximize the chances of successful pollination.

5. Pollination behaviour

- **Pollinators**: Cardamom is primarily pollinated by insects, with bees being the most effective pollinators. Insects are attracted to the flowers by their colour and nectar.
- **Self-pollination**: Although cardamom flowers are hermaphroditic, they are

largely cross-pollinated due to the position and structure of the reproductive organs, which limits self-pollination.

6. Fertilization

- After successful pollination, fertilization occurs within the ovary, leading to the development of seeds inside the characteristic cardamom pods.

Understanding the floral biology of cardamom is crucial for cultivation practices, especially in optimizing conditions for effective pollination and maximizing yield. This includes managing the environment to favor the presence and activity of pollinators during the critical hours of anthesis.

Crossing technique in cardamom

Crossing techniques in cardamom (Elettaria cardamomum), commonly known as "true cardamom," are essential for developing new varieties that offer improved yield, disease resistance, and quality. Cardamom is a perennial crop that is often propagated vegetatively, but sexual reproduction through crossing is critical for introducing new genetic traits and creating hybrid varieties. Below is an overview of the key crossing techniques used in cardamom breeding:

1. Controlled pollination

  - **Selection of parent plants**: The first step is selecting the parent plants based on their desirable traits. One plant may be chosen for its high yield, while another might be selected for its resistance to pests or diseases.
  - **Emasculation**: Since cardamom flowers are bisexual, emasculation involves removing the anthers (which produce pollen) from the flower to prevent self-pollination. This ensures that only the pollen from the selected male parent will fertilize the female flower.
  - **Pollination**: Pollen is collected from the male parent and manually transferred to the stigma of the emasculated female flower. This is typically done with a fine brush or by gently shaking the male flower over the female.
  - **Bagging**: The pollinated flower is then covered with a bag to prevent contamination by unwanted pollen. The bag is usually made from paper or fine mesh and is removed once the fruit sets.

- **Application**: Controlled pollination is crucial for developing hybrids with combined traits, such as higher yield and improved resistance to diseases like cardamom mosaic virus (CdMV).

2. Mass selection and open-pollination

  - **Selection of superior plants**: In an open-pollinated cardamom field, the best-performing plants, based on traits like yield, pod size, and disease resistance, are selected.

- **Propagation**: Seeds or vegetative cuttings from these selected plants are collected and used to propagate the next generation.
- **Open-pollination**: The selected plants are allowed to cross-pollinate naturally. This technique is often combined with vegetative propagation to ensure the uniformity of the population.

- **Application**: This method is particularly useful for improving landraces or farmer varieties by gradually enhancing their performance through selection over multiple generations.

3. Hybridization

- **Intra-specific hybridization**
  - **Method**: This involves crossing different varieties within the same species, Elettaria cardamomum. For instance, a variety with large pods may be crossed with one that has high resistance to fungal diseases.
  - **Application**: Intra-specific hybridization is commonly used to develop varieties with improved yield, larger pod size, and better quality.
- **Inter-specific hybridization**:
  - **Method**: Although more challenging, inter-specific hybridization involves crossing cardamom with closely related species to introduce new traits such as disease resistance. This requires careful management as the resulting hybrids may have issues with fertility.
  - **Application**: This technique can be used to introduce traits from wild relatives into cultivated cardamom varieties, such as enhanced resistance to pests or environmental stressors.

4. Recurrent selection

- **Initial selection**: In the first cycle, plants with the desired traits (e.g., high yield, disease resistance) are selected from a diverse population.
- **Crossing and evaluation**: The selected plants are intercrossed, and their progeny are evaluated for the same traits. The best-performing progeny are selected for the next cycle.
- **Repetition**: This process is repeated for several cycles to progressively enhance the target traits within the population.

- **Application**: Recurrent selection is particularly effective for improving quantitative traits like yield or resistance to multiple diseases, which are controlled by many genes.

5. Backcrossing

- **Initial cross**: A superior cardamom variety is crossed with a donor plant that has the desired trait (e.g., disease resistance).
- **Backcrossing**: The resulting hybrid is then crossed back with the superior parent variety. This process is repeated over several generations, with selection for the desired trait at each stage.
- **Final selection**: After several backcrosses, plants that retain the desired trait from the donor while exhibiting the superior characteristics of the original variety are selected.

- **Application**: Backcrossing is ideal for introducing single traits, such as resistance to a specific disease, into a high-yielding commercial variety without altering its other desirable traits.

6. Mutation breeding

- **Mutagenesis**: Seeds or vegetative tissues of cardamom are exposed to physical mutagens (such as radiation) or chemical mutagens to induce mutations.
- **Screening and selection**: The treated plants are grown, and mutants showing desirable traits (e.g., higher yield, better flavor) are selected for further breeding.

- **Application**: Mutation breeding can be used to develop new traits that are not present in the existing gene pool, such as improved resistance to diseases or pests.

7. Tissue culture and embryo rescue

- **Embryo Rescue**: In cases where hybridization results in embryo abortion (common in wide crosses), embryos are excised from the developing seeds and cultured in vitro to complete their development.
- **Micropropagation**: Tissue culture is used to propagate the rescued embryos or selected hybrids rapidly, ensuring the production of genetically uniform plants.

- **Application**: Tissue culture is particularly useful for propagating hybrids or mutants that have desirable traits but are difficult to propagate through traditional methods.

8. Progeny testing and field trials

- **Planting and Evaluation**: Seeds from crosses are planted, and the resulting plants are observed and evaluated for traits such as yield, pod quality, and disease resistance.

- **Field Trials**: These progeny are tested in different environmental conditions to assess their adaptability and stability.
- **Selection**: The best-performing progeny are selected for further breeding or direct release as new varieties.

- **Application**: Progeny testing is essential for validating the success of breeding efforts and for selecting the most promising candidates for further development.

9. Hybrid seed production

- **Controlled environment**: Hybrid seed production often takes place in greenhouses or isolated plots to prevent cross-pollination from unwanted sources.
- **Pollination control**: Strict control is maintained over pollination, often involving emasculation and manual pollination to ensure that only the selected crosses are made.
- **Harvesting**: Once the fruit sets, seeds are harvested, processed, and stored for distribution.

- **Application**: Hybrid seed production is crucial for maintaining the uniformity and quality of cardamom varieties, ensuring that farmers receive reliable planting material.

These crossing techniques are fundamental to the advancement of cardamom breeding, enabling the development of improved varieties that meet the needs of both farmers and consumers. By combining traditional breeding methods with modern technologies, cardamom breeding programs can create varieties that are more productive, disease-resistant, and better suited to diverse growing conditions.

2. Cinnamon

Cinnamon floral biology

The floral biology of cinnamon (*Cinnamomum verum* or *Cinnamomum zeylanicum*) is integral to its reproductive process and cultivation. Here are the key aspects:

1. Flower structure

- **Inflorescence**: Cinnamon flowers are arranged in small panicles or cymes, which are clusters of flowers with a somewhat flat-topped appearance.
- **Flowers**: The individual flowers are small, measuring about 3-5 mm in diameter, and are typically greenish to yellowish-white. They are bisexual (hermaphroditic), containing both male (stamens) and female (pistil) reproductive organs.

2. Anthesis (Flower opening)

- **Timing**: The flowers of cinnamon typically open in the early morning. The exact timing can vary depending on environmental conditions, but it generally occurs around dawn.
- **Duration**: The flowers remain open for a limited period, often closing by late morning or early afternoon.

3. Anther dehiscence

Anther dehiscence in cinnamon occurs shortly after the flowers open. The anthers release pollen early in the morning when the flowers are most receptive to pollination.

4. Stigma receptivity

The stigma becomes receptive soon after the flower opens and generally remains receptive throughout the morning. This period of stigma receptivity overlaps with the time when pollen is available, ensuring that fertilization can occur if pollen reaches the stigma.

5. Pollination behaviour

- **Pollinators**: Cinnamon is primarily pollinated by small insects, particularly bees and flies. These pollinators are attracted to the flowers by their scent and the small amount of nectar they produce.
- **Self-Pollination**: Although the flowers are hermaphroditic, cinnamon plants typically undergo cross-pollination due to the structure of the flowers and the action of pollinators. However, some degree of self-pollination can occur.

6. Fertilization

Following successful pollination, fertilization takes place within the ovary, leading to the development of seeds. These seeds are housed within small, berry-like fruits.

7. Floral Scent

Cinnamon flowers emit a mild, pleasant fragrance that plays a role in attracting pollinators, although the scent is not as strong as in some other flowering plants.

Understanding the floral biology of cinnamon is crucial for successful cultivation, particularly in ensuring that the flowers are effectively pollinated, which can impact the yield of seeds and the overall health of the cinnamon trees. In commercial cultivation, maintaining a healthy population of pollinators and optimizing environmental conditions during the flowering period are key strategies for improving pollination success.

Crossing technique in cinnamon

The "crossing technique" in cinnamon cultivation refers to the process of breeding different varieties or strains of cinnamon plants to produce offspring with desirable traits, such as improved flavor, disease resistance, or growth characteristics. Here's an overview of how this technique generally works:

1. **Selection of parent plants**: The first step involves selecting parent cinnamon plants that have desirable traits. For example, one plant might have superior flavor, while another might be more resistant to disease.
2. **Controlled pollination**: Once the parent plants are selected, controlled pollination is conducted. This involves manually transferring pollen from the male flower of one plant to the female flower of another. This ensures that the offspring will have genetic material from both parent plants.
3. **Growing hybrid offspring**: After pollination, the fertilized flowers produce seeds, which are then grown into new plants. These plants, known as hybrids, will exhibit a mix of traits from both parents.
4. **Selection and testing**: The hybrid plants are then grown and evaluated for their traits. The most promising plants are selected for further breeding or commercial production.
5. **Stabilization (Optional)**: If the goal is to create a new, stable variety, further breeding is done to stabilize the desirable traits over several generations.

This technique is used not only to improve the quality of cinnamon but also to adapt the plants to different growing conditions and to increase yield and resilience.

3. Clove

Nutmeg floral biology

The floral biology of nutmeg (*Myristica fragrans*) is quite distinctive and plays a significant role in its reproductive processes. Nutmeg is a dioecious species, meaning that individual plants are either male or female.

1. Flower structure

- **Inflorescence**: Nutmeg flowers are small and typically found in clusters. On female trees, flowers are solitary or in small groups, while male flowers are more numerous and arranged in clusters.
- **Male flowers**: These are small, pale yellow, and tubular, with a central column of stamens. Male flowers produce pollen but do not bear fruit.
- **Female flowers**: Female flowers are slightly larger, also pale yellow, and have a central ovary. These flowers are responsible for fruit development.

2. Anthesis (Flower opening)

- **Timing**: The flowers of nutmeg generally open in the early morning. The timing can be influenced by environmental conditions, such as temperature and humidity.
- **Duration**: The flowers remain open for a few days, during which pollination can occur.

3. Anther dehiscence (in male flowers)

- **Timing**: In male flowers, anther dehiscence typically occurs shortly after the flowers open. The release of pollen coincides with the receptivity of female flowers to ensure successful pollination.

4. Stigma receptivity (in female flowers)

- **Timing**: The stigma of the female flowers becomes receptive around the same time the male flowers release pollen. This receptivity lasts for a few days, overlapping with the pollen release period.
- **Synchronization**: This synchrony between male pollen release and female stigma receptivity is crucial for effective cross-pollination.

5. Pollination behaviour

- **Pollinators**: Nutmeg is primarily pollinated by small insects, particularly beetles. These pollinators are attracted to the flowers by their scent and nectar.
- **Cross-Pollination**: Due to the dioecious nature of nutmeg, cross-pollination between male and female plants is essential for fruit production. The separation of male and female flowers on different trees ensures that cross-pollination is necessary for fertilization.

6. Fertilization and Fruit development

- After successful pollination, fertilization occurs within the ovary of the female flower. This leads to the development of the nutmeg fruit, which contains the valuable seeds used as a spice (nutmeg) and an aril (mace).
- The fruit is a drupe, which eventually splits open to reveal the seed (nutmeg) covered by a red aril (mace).

7. Floral scent

- Nutmeg flowers produce a subtle fragrance that helps attract pollinators, particularly beetles, which are the primary agents of pollination.

Understanding the floral biology of nutmeg is essential for cultivation, particularly in ensuring the presence of both male and female plants in an orchard to facilitate cross-pollination. The success of pollination directly affects the yield of nutmeg fruits and, consequently, the production of nutmeg and mace

Crossing technique in Clove

The "crossing technique" in clove (Syzygium aromaticum) cultivation involves the intentional breeding of different clove trees to produce offspring with desirable traits, such as improved disease resistance, higher yield, or better essential oil content. Here's an overview of the crossing technique in cloves:

1. Selection of parent trees

- **Identify desirable traits:** Parent trees are selected based on traits like high yield, disease resistance, or high-quality essential oil production.
- **Genetic diversity:** Choosing genetically diverse parent trees helps enhance the chances of combining beneficial traits in the offspring.

2. Controlled pollination

- **Isolation of flowers:** To prevent unintended pollination, flowers from selected parent trees are isolated.
- **Hand pollination:** Pollen is manually collected from the male parent tree and transferred to the female flowers of the selected tree. This is usually done during the flowering season when the trees are most receptive.

3. Seed collection and planting

- **Harvesting seeds:** Once pollination is successful, seeds are harvested from the female parent tree.
- **Germination:** These seeds are then germinated in a controlled environment, ensuring the best possible conditions for growth.

4. Selection of hybrid offspring

- **Evaluation:** The offspring are grown and monitored for desirable traits such as growth rate, resistance to pests and diseases, and quality of the clove buds.
- **Further breeding:** Promising hybrid plants can be selected for further breeding to stabilize the desired traits.

5. Commercial propagation

- Once a desirable hybrid is developed, it can be propagated through vegetative means like grafting or tissue culture to maintain the consistent quality of the new variety.

This crossing technique in clove cultivation helps develop superior clove varieties that are better suited to different environmental conditions, have improved yield, or possess better resistance to pests and diseases.

4. Nutmeg

Floral biology in nutmeg

The floral biology of **nutmeg** (*Myristica fragrans*), an important spice-producing tree, is key to understanding its reproductive processes, pollination, and ultimately its fruit production. Here's an overview of the flower structure, anthesis, pollen dehiscence, and stigma receptivity in nutmeg:

1. Flower structure

Nutmeg is a **dioecious** plant, meaning that male and female flowers are borne on separate plants. This separation of sexes requires cross-pollination between male and female plants for fruit production. The key features of the flower structure are:

a. Male flowers

- **Structure**: Male flowers are small, cream to pale yellow, and borne in clusters. Each male flower contains a short stalk (pedicel) and has 4–6 tepals (undifferentiated petals and sepals).
- **Stamens**: The male flowers have **stamens** that are fused into a column-like structure called the **synandrium**. The synandrium contains numerous pollen-producing anthers.
- **Nectar glands**: Male flowers produce nectar, attracting pollinators.

b. Female flowers

- **Structure**: Female flowers are slightly larger than male flowers and solitary, with 4–6 tepals like male flowers. They are generally creamy or pale yellow in color.
- **Gynoecium**: Female flowers contain a single ovary, which is superior (located above the point where other flower parts are attached). The ovary is unilocular (one-chambered) and contains a single ovule.
- **Stigma**: The stigma, located at the tip of the style, is the receptive part for pollen. It is generally broad and lobed.

2. Anthesis

Anthesis refers to the period during which the flower is fully open and functional. In nutmeg:

- **Timing**: Anthesis in nutmeg typically occurs during the early morning. Both male and female flowers open in the morning, with anthesis lasting for one day in most cases.
- **Male flower anthesis**: Male flowers open to expose the synandrium, where the anthers release pollen. The flower remains open for a short time, often closing later in the day after pollen release.
- **Female flower anthesis**: Female flowers open slightly earlier or simultaneously with male flowers. The stigma becomes receptive during this period, and successful pollination can occur if pollen is transferred from male flowers.

3. Pollen dehiscence

Pollen dehiscence refers to the process of the anthers releasing pollen. In nutmeg:

- **Male flowers**: The anthers in the male flowers dehisce via longitudinal slits, releasing pollen. This typically happens early in the morning, during or just after anthesis. The pollen is fine and needs to be carried to female flowers by insects (primarily beetles or other pollinators).
- **Pollen viability**: Pollen is viable for a short period after release, making timing crucial for successful pollination.

4. Stigma receptivity

Stigma receptivity refers to the period when the stigma of the female flower is capable of receiving and supporting pollen germination. In nutmeg:

- **Timing**: Stigma receptivity in female nutmeg flowers begins early in the morning, around the time of anthesis, and continues for one to two days.
- **Optimal receptivity**: The stigma is most receptive on the first day of anthesis. After this period, its ability to support pollen germination declines rapidly.
- **Signs of receptivity**: A receptive stigma may appear moist and sticky, which helps in trapping pollen grains. It may also undergo slight color changes during this time.

5. Pollination mechanism

- **Cross-pollination**: As nutmeg is dioecious, cross-pollination between male and female plants is necessary for fruit set. Pollinators, mainly small beetles, visit male flowers to collect pollen and then transfer it to the female flowers.
- **Insect-pollinated**: The small, inconspicuous flowers of nutmeg are adapted to insect pollination. Beetles are attracted by the nectar produced in the male flowers, and during their foraging, they inadvertently carry pollen to female flowers.

6. Post-pollination

After successful pollination, the fertilized ovule in the female flower develops into the nutmeg fruit, a **drupe** that contains the nutmeg seed (the spice) and an aril (mace, another spice).

Understanding the floral biology of nutmeg—including flower structure, anthesis, pollen dehiscence, and stigma receptivity—is crucial for managing nutmeg plantations and ensuring successful pollination and fruit set. Since nutmeg is dioecious and dependent on cross-pollination, the synchronization between male pollen release and female stigma receptivity, along with effective pollinator activity, is key to its reproductive success.

Crossing technique in Nutmeg

The "crossing technique" in nutmeg (*Myristica fragrans*) involves the deliberate breeding of different nutmeg trees to produce offspring with improved traits such as higher yield, better quality seeds, enhanced resistance to diseases, and adaptability to varying environmental conditions. Nutmeg trees are dioecious, meaning they have separate male and female trees, which makes controlled breeding a bit more challenging but also crucial for achieving desired outcomes.

1. Selection of parent trees

- **Identifying desirable traits:** Parent trees are selected based on traits such as high-quality nutmeg production (better mace or nutmeg content), resistance to pests and diseases, and overall vigor.
- **Male and female trees:** Both male and female parent trees are selected. The female trees produce the nutmeg seeds, while male trees provide the pollen.

2. Controlled pollination

- **Isolation of flowers:** Flowers from the chosen female tree are isolated to prevent accidental pollination from other sources.
- **Hand pollination:** Pollen from the selected male tree is manually transferred to the female flowers. This is done during the flowering period when the flowers are most receptive.
- **Timing and technique:** Timing is crucial since the pollination window is narrow. Proper technique ensures successful fertilization.

3. Seed collection and germination

- **Harvesting seeds:** Once pollination is successful, the female trees produce fruit that contains the nutmeg seeds. These seeds are harvested when they mature.

- **Germination:** The seeds are then germinated in a controlled environment. The resulting seedlings are monitored for the desired traits.

4. Selection of hybrid offspring

- **Evaluation:** The offspring plants are grown and carefully evaluated for traits such as growth rate, yield quality, resistance to diseases, and overall hardiness.
- **Further breeding:** Promising hybrid plants may be selected for additional breeding cycles to stabilize the desirable traits across generations.

5. Vegetative propagation (Optional)

- **Cloning superior plants:** Once superior plants are identified, they can be propagated vegetatively through methods such as grafting or air layering to ensure that the offspring maintain the desirable traits consistently.

6. Field testing and commercial cultivation

- **Field trials:** The selected hybrids are often subjected to field trials to test their performance under different environmental conditions.
- **Commercial scale-up:** Once a stable and superior hybrid is identified, it can be propagated on a larger scale for commercial cultivation.

The crossing technique in nutmeg helps to develop new varieties that are more productive, resilient, and suited to specific growing conditions, thereby improving the efficiency and profitability of nutmeg cultivation.

5. Allspice

All spices floral biology

The term "allspice" refers to the dried unripe berries of the tree *Pimenta dioica*, a member of the myrtle family (Myrtaceae). Allspice gets its name because its flavour resembles a combination of cinnamon, nutmeg, and cloves. Here's an overview of the floral biology of allspice:

1. Flower structure

- **Inflorescence**: Allspice flowers are arranged in small, branched clusters known as panicles. These panicles are typically located at the ends of branches.
- **Flowers**: The individual flowers are small, typically white or pale green, and have five petals. Each flower contains both male and female reproductive organs, making the plant hermaphroditic (bisexual).

2. Anthesis (Flower opening)

- **Timing**: The flowers of allspice generally open during the early part of the day, often in the morning.
- **Duration**: The flowers remain open for a few days, during which pollination can occur.

3. Anther dehiscence

- **Timing**: Anther dehiscence, or the release of pollen from the anthers, occurs shortly after the flowers open. This coincides with the receptivity of the stigmas to maximize the chances of successful pollination.

4. Stigma receptivity

- **Timing**: The stigma, which is the part of the female reproductive organ that receives pollen, becomes receptive around the time the anthers release pollen. This synchronization ensures that the flowers can self-pollinate or be cross-pollinated effectively.

5. Pollination behaviour

- **Pollinators**: Allspice is primarily pollinated by insects, particularly bees and other small flying insects. These pollinators are attracted to the flowers by their scent and the presence of nectar.
- **Self-pollination and cross-pollination**: Although the flowers are hermaphroditic and can self-pollinate, cross-pollination often occurs with the help of pollinators, leading to potentially more vigorous offspring.

6. Fertilization and fruit development

- After successful pollination and fertilization, the flowers develop into small, green berries that eventually ripen into the dark brown, aromatic berries known as allspice. These berries are harvested before they fully ripen and are dried to produce the spice.

7. Floral scent

- The flowers emit a mild fragrance, which is not as strong as the scent of the dried berries, but it plays a role in attracting pollinators to the flowers.

Understanding the floral biology of allspice is crucial for successful cultivation, particularly in regions where the trees are grown for spice production. Ensuring that pollinators are present and active during the flowering period can significantly impact fruit set and the overall yield of allspice.

Crossing technique in Allspice

The crossing technique in allspice (*Pimenta dioica*), also known as Jamaican pepper, involves the process of breeding different plants to combine desirable traits such as better fruit quality, higher essential oil content, increased resistance to diseases, and adaptability to various environmental conditions. Allspice is typically propagated by seeds, but controlled crossing can be used to enhance the genetic diversity and improve the overall quality of the plants.

1. Selection of parent plants

- **Identify desirable traits:** Parent plants are selected based on traits such as high yield, large fruit size, high essential oil content, and disease resistance.
- **Genetic Diversity:** Selecting genetically diverse parent plants increases the chances of producing offspring with a combination of beneficial traits.

2. Controlled pollination

- **Isolation of flowers:** To ensure that the desired cross-pollination occurs, flowers from the selected parent plants are isolated from other plants.
- **Hand pollination:** Pollen from the male flowers of one plant is manually transferred to the female flowers of the selected plant. This is typically done during the peak flowering season when the plants are most receptive.
- **Monitoring:** After pollination, the flowers are monitored to ensure successful fertilization and fruit set.

3. Seed Collection and germination

- **Harvesting seeds:** Once the fruits mature, the seeds are harvested from the female plants.
- **Germination:** The seeds are then germinated under controlled conditions to produce new plants. Seedlings are monitored for vigor and growth.

4. Selection of hybrid offspring

- **Evaluation:** The offspring plants are grown and evaluated for the presence of desirable traits such as fruit size, yield, essential oil content, and resistance to pests and diseases.
- **Further breeding:** Promising hybrids may be selected for additional rounds of breeding to stabilize the desirable traits across subsequent generations.

5. Vegetative propagation (Optional)

- **Cloning superior plants:** Once superior hybrids are identified, they can be propagated vegetatively (e.g., through cuttings or grafting) to ensure that the desirable traits are consistently passed on.

6. Field testing and commercial cultivation

- **Field trials:** The selected hybrids are tested in different environments to assess their performance under varying conditions.
- **Commercial scale-up:** Successful hybrids that demonstrate superior traits are propagated on a larger scale for commercial cultivation.

Challenges

- **Dioecious nature:** Allspice plants are dioecious, meaning they have separate male and female plants, which can complicate the breeding process. Careful management of male and female plants is required for successful crossing.
- **Long maturity period:** Allspice plants take several years to mature and produce fruit, making the breeding process time-consuming.

The crossing technique in allspice aims to produce new varieties that offer better quality, higher yields, and increased resilience, ultimately contributing to the sustainability and profitability of allspice cultivation.

6. Vanilla

Floral biology in Vanilla

Vanilla (*Vanilla planifolia*), a key species in the production of vanilla flavoring, has a fascinating floral biology that is crucial for its pollination and fruit production. Here's an overview of the flower structure, anthesis, pollen dehiscence, and stigma receptivity in vanilla:

1. Flower structure

Vanilla is an **epiphytic orchid**, and its flowers are relatively large, complex, and short-lived. Key components of the vanilla flower structure are:

a. Basic floral structure

- **Perianth**: The flower has six petal-like structures—three **sepals** and three **petals**. One of the petals is modified into a distinctive structure called the **labellum** or **lip**, which is used to attract pollinators and aid in pollination.
- **Color**: The flowers are typically yellow-green or pale green.

b. Column (Gynostemium)

- **Unique to orchids**, vanilla flowers have a fused structure known as the **gynostemium** or **column**, which combines the reproductive organs (stamens and pistil).
- **Anther**: Located at the top of the column, the anther contains the **pollinia**, a mass of pollen grains. Unlike most flowers where pollen is freely dispersed, orchids like vanilla have **pollinia** that must be transferred as a unit.

- **Stigma**: The **stigma** is located under the anther but is separated from it by a small membrane or flap called the **rostellum**, which prevents self-pollination.

c. Reproductive barrier

- The **rostellum** blocks the pollen from directly contacting the stigma. For successful pollination, the rostellum must be bypassed manually or through specific pollinator behavior.

2. Anthesis

Anthesis refers to the period when the flower is fully open and functional. In vanilla:

- **Timing**: Vanilla flowers open early in the morning, and anthesis lasts for **just one day**. If pollination does not occur within this short window, the flower wilts and drops.
- **Pollinator availability**: In its native habitat (Mesoamerica), **Melipona bees** are the natural pollinators. However, outside this region, vanilla flowers are not naturally pollinated due to the absence of these specific bees.
- **Manual pollination**: In commercial vanilla cultivation, **hand pollination** is often necessary to achieve fruit set. The farmer or worker manually lifts the rostellum and presses the pollinia onto the stigma to ensure fertilization.

3. Pollen dehiscence

Pollen dehiscence in vanilla differs from many other plants because of the unique structure of the pollinia:

- **Pollinia Structure**: The pollen is not released freely from the anthers. Instead, the pollen grains are packed into the pollinia, which must be physically transferred to the stigma.
- **Timing**: The pollinia are available for transfer as soon as the flower opens in the morning during anthesis. The pollinia remain intact and do not naturally release pollen grains unless transferred to the receptive stigma.
- **Manual handling**: In cultivation, hand pollination is performed during the early hours of anthesis when the flowers are fresh, and the stigma is most receptive.

4. Stigma receptivity

The **stigma receptivity** in vanilla flowers is a critical factor for successful pollination and fruit set:

- **Timing**: The stigma is receptive on the same day the flower opens, with peak receptivity occurring in the early morning during anthesis.

- **Short window**: Stigma receptivity lasts for only one day. If the flower is not pollinated within this time frame, the stigma loses its receptivity, and the flower wilts.
- **Signs of receptivity**: The stigma appears sticky and moist during its receptive period, which helps capture and hold the pollinia during pollination.

5. Pollination process

In its natural environment, **Melipona bees** can bypass the rostellum and transfer pollinia from the anther to the stigma. However, in most vanilla-growing regions where these bees are absent, **hand pollination** is the standard practice:

- **Manual pollination**: A small tool (like a thin stick or bamboo sliver) is used to lift the rostellum, allowing the worker to press the pollinia onto the stigma. This mimics the action of the natural pollinator.
- **Success rate**: With proper technique, hand pollination is highly effective, but it must be done within the short window of stigma receptivity.

6. Post-pollination

After successful pollination, the fertilized ovary develops into a **long, green pod**, which is the vanilla bean. The fruit takes several months to mature, and during this time, the plant directs resources to pod growth, which will later be harvested and cured to produce vanilla flavor.

The floral biology of vanilla is characterized by complex structures and processes that require specific interactions between flower components (such as the column and rostellum) to achieve successful pollination. Due to the unique pollination mechanism and the absence of natural pollinators in most regions where vanilla is cultivated, **manual pollination** is essential to ensure fruit set. Understanding the timing of anthesis, pollen dehiscence, and stigma receptivity is crucial for growers to maximize vanilla production.

Crossing technique in Vanilla

The crossing technique in vanilla (Vanilla planifolia), which is the primary species used in vanilla production, involves the intentional breeding of different vanilla plants to produce offspring with desirable traits such as higher yield, better flavor profile, disease resistance, and adaptability to different environmental conditions. Vanilla is a tropical orchid that requires specific conditions for successful pollination and fruiting, and because it is typically propagated vegetatively, controlled breeding is essential for genetic improvement.

1. Selection of parent plants

- **Identifying desirable traits:** Parent plants are selected based on traits like high-quality pod production, strong vanilla aroma, resistance to diseases (such as Fusarium wilt), and vigor.
- **Diverse genetic material:** Selecting genetically diverse parent plants can help combine beneficial traits and improve genetic diversity.

2. Controlled pollination

- **Manual pollination:** Vanilla flowers have a very short lifespan, typically opening for just one day. Therefore, manual pollination is essential. This involves transferring pollen from the male part (anther) of one flower to the female part (stigma) of another flower. This process is delicate and usually performed by hand using a small stick or a needle.
- **Preventing self-pollination:** In vanilla, a natural barrier (the rostellum) prevents self-pollination. During manual pollination, this barrier is bypassed to ensure that cross-pollination occurs between selected plants.

3. Pod development and seed collection

- **Fruit set and maturation:** After successful pollination, the flower develops into a pod, which takes several months to mature (usually 8-9 months). The pods are harvested when fully mature.
- **Seed extraction:** Seeds are extracted from the mature pods. However, vanilla is predominantly propagated vegetatively (from cuttings) due to the variability in traits from seed propagation. Crossing techniques are typically more focused on generating new cultivars for research or genetic improvement.

4. Germination and growth of hybrid seedlings

- **Germination:** Vanilla seeds are tiny and difficult to germinate, requiring a sterile, nutrient-rich medium such as agar. The seeds are germinated under controlled laboratory conditions.
- **Growth and selection:** Once germinated, seedlings are grown in nurseries where they are evaluated for traits like growth rate, disease resistance, and potential yield.

5. Selection of superior hybrids

- **Evaluation of offspring:** The hybrid plants are assessed for their traits, with particular attention paid to the quality of the vanilla pods, disease resistance, and adaptability to different growing conditions.

- **Further breeding:** Promising hybrids can be selected for further breeding to stabilize the desired traits across generations or to introduce new traits.

6. Vegetative propagation of superior hybrids

- **Cloning through Cuttings:** Once a superior hybrid is identified, it is typically propagated vegetatively through cuttings. This ensures that the desirable traits are maintained consistently across all new plants.

7. Field testing and commercial cultivation

- **Field Trials:** The selected hybrids undergo field trials to evaluate their performance in different environmental conditions and to ensure they meet commercial standards.
- **Scaling up production:** Successful hybrids are propagated on a larger scale and introduced into commercial cultivation.

Challenges

- **Labor-intensive pollination:** Vanilla cultivation is labor-intensive, especially during the pollination process, which must be done by hand.
- **Long growth cycle:** Vanilla plants take several years to mature and produce their first harvest, making the breeding process time-consuming.
- **Disease management:** Vanilla is susceptible to several diseases, so breeding for disease resistance is a key focus area in crossing techniques.

The crossing technique in vanilla aims to create new varieties that improve yield, quality, and resilience, addressing the challenges of vanilla cultivation and enhancing the sustainability of this valuable crop.

Development of self compatible vanilla variety

Developing a self-compatible vanilla variety involves overcoming the natural reproductive barrier in *Vanilla planifolia*, which is typically self-incompatible. In self-incompatible plants, pollen from the same plant or genetically similar plants cannot fertilize the ovule, preventing self-pollination and requiring cross-pollination to produce fruit. Here's how a self-compatible vanilla variety might be developed:

1. Understanding self-incompatibility in Vanilla

- **Biological Basis:** Vanilla's self-incompatibility is a genetic mechanism that prevents self-pollination. It is likely controlled by specific genes that recognize and reject self-pollen.
- **Overcoming the barrier:** The goal is to either naturally or artificially overcome this genetic mechanism to allow for self-pollination.

2. Selection and breeding

- **Identifying natural variants:** Researchers can search for natural mutants or rare individuals within vanilla populations that exhibit self-compatibility. These plants would naturally produce fruit when self-pollinated.
- **Hybridization:** Cross-breeding these self-compatible individuals with other vanilla plants might enhance or stabilize the trait.

3. Inducing mutations

- **Mutagenesis:** Chemical or radiation-induced mutagenesis can be used to create genetic variations. The plants are then screened to identify individuals that have lost their self-incompatibility.
- **CRISPR and genetic engineering:** Advanced techniques like CRISPR-Cas9 could be used to directly edit the genes responsible for self-incompatibility, potentially creating self-compatible vanilla plants.

4. Controlled pollination and evaluation

- **Testing Self-Compatibility:** Once potential self-compatible plants are identified or created, controlled pollination experiments are conducted. Flowers are manually pollinated with pollen from the same plant to test if they can set fruit.
- **Assessing Traits:** The resulting plants are evaluated for pod production, seed viability, and overall plant health. It's crucial that the self-compatibility trait does not negatively affect other desirable traits.

5. Breeding and propagation

- **Stabilizing the trait:** If a self-compatible plant is found, further breeding may be necessary to stabilize the trait. This might involve backcrossing with the parent lines or other selected varieties.
- **Vegetative propagation:** Once a stable self-compatible variety is developed, it can be propagated through cuttings or tissue culture to ensure consistency in the trait.

6. Field trials and commercialization

- **Field testing:** The self-compatible varieties undergo extensive field trials to evaluate their performance in various environments. These trials help ensure that the plants are productive, disease-resistant, and produce high-quality vanilla pods.
- **Commercial production:** After successful trials, the self-compatible vanilla variety can be scaled up for commercial cultivation. This variety could significantly reduce the labor costs associated with manual pollination in vanilla production.

7. Benefits and considerations

- **Reduced labor costs:** Self-compatible vanilla plants would not require manual pollination, reducing labor costs and increasing the efficiency of vanilla production.
- **Increased yield:** Self-compatible plants might produce more consistent yields, as every flower has the potential to be pollinated without relying on external factors.
- **Genetic diversity:** Maintaining genetic diversity in cultivated vanilla is crucial to avoid potential vulnerabilities. Even with self-compatible varieties, cross-pollination could be encouraged to preserve genetic health.

Challenges

- **Maintaining quality:** Ensuring that self-compatibility does not negatively impact the quality of vanilla pods is essential.
- **Environmental factors:** Self-compatibility may not be the only factor in successful vanilla production. Environmental conditions, disease management, and proper cultivation practices remain critical.

Developing a self-compatible vanilla variety is a complex process that requires a combination of traditional breeding, genetic research, and advanced biotechnology. Success in this area could revolutionize vanilla cultivation, making it more sustainable and economically viable.

7. Coriander

Floral biology of Coriander

The floral biology of coriander (*Coriandrum sativum*), also known as cilantro, is important for understanding its reproductive processes, which directly impact seed production. Here's an overview of the key aspects:

1. Flower structure

- **Inflorescence**: Coriander flowers are arranged in compound umbels, typical of the Apiaceae family. Each umbel consists of several smaller umbellets, and each umbellet contains multiple small flowers.
- **Flowers**: The individual flowers are small, typically about 2-4 mm in diameter, and can be white or pale pink. The flowers are usually actinomorphic (radially symmetrical) and hermaphroditic, meaning each flower contains both male (stamens) and female (pistil) reproductive organs.

2. Anthesis (Flower opening)

- **Timing**: The flowers of coriander generally open in the morning. The timing of anthesis can vary depending on environmental conditions such as temperature, light, and humidity.
- **Duration**: Each flower remains open for about one to two days, during which time it is receptive to pollination.

3. Anther dehiscence

- **Timing**: Anther dehiscence, the process of pollen release, typically occurs soon after the flowers open. The release of pollen often coincides with the peak activity of pollinators.

4. Stigma receptivity

- **Timing**: The stigma becomes receptive shortly after the flower opens, usually at the same time that the anthers release pollen. Stigma receptivity lasts for a brief period, usually overlapping with the time of anther dehiscence.
- **Self-Pollination and Cross-Pollination**: Coriander can self-pollinate, but cross-pollination is common and beneficial, often leading to better seed set and plant vigor.

5. Pollination behaviour

- **Pollinators**: Coriander is primarily pollinated by insects, including bees, flies, and other small pollinators. These insects are attracted to the flowers by their colour and nectar.
- **Cross-Pollination**: While self-pollination can occur, the structure of coriander flowers and the activity of pollinators promote cross-pollination, which enhances genetic diversity and can lead to more robust seed production.

6. Fruit and seed development

- After successful pollination and fertilization, the flowers develop into small, globular fruits known as schizocarps. Each fruit splits into two mericarps, which are commonly referred to as coriander seeds. These seeds are used both as a spice and for planting.

7. Floral scent

- Coriander flowers emit a mild fragrance, which helps attract pollinators. The scent, along with the small size and colour of the flowers, is part of the plant's strategy to ensure effective pollination.

8. Growth habit

- Coriander is an annual herbaceous plant, completing its life cycle in one growing season. The timing of flowering and seed production is critical for agricultural practices, especially in regions where coriander is grown for both leaves (cilantro) and seeds.

9. Agricultural practices

- In coriander cultivation, managing pollinator presence and ensuring favorable environmental conditions during flowering are important for maximizing seed yield. Since coriander is often grown for both its leaves and seeds, understanding its floral biology helps in optimizing harvest times for both products.

Understanding the floral biology of coriander is vital for effective cultivation, particularly when the goal is to maximize seed production. Proper management of flowering, pollination, and environmental conditions can significantly impact the yield and quality of coriander seeds.

Crossing technique in coriander

The crossing technique in coriander (Coriandrum sativum), also known as cilantro, is used to develop new varieties with improved traits such as enhanced flavor, increased yield, disease resistance, and bolting resistance (delay in flowering). Coriander is an annual herb, and crossing involves careful selection and controlled pollination between different plants to combine desirable traits.

1. Selection of parent plants

- **Identifying Desirable Traits:** Parent plants are selected based on specific traits such as leaf size, aroma, resistance to pests and diseases, and bolting resistance. For example, one parent may have a strong aroma, while another may resist bolting (premature flowering).
- **Genetic diversity:** Selecting genetically diverse parent plants ensures a greater chance of producing offspring with a combination of beneficial traits.

2. Controlled pollination

- **Isolation of flowers:** To prevent unwanted cross-pollination, the flowers of the selected parent plants are isolated. This can be done using physical barriers like mesh bags or growing the plants in isolation.
- **Hand pollination:** In coriander, flowers are typically small and clustered in umbels (flower heads). For controlled crossing, pollen is manually transferred from the male parent's flowers (anthers) to the female parent's

flowers (stigmas). This process ensures that the pollen from the desired male parent fertilizes the female parent.

- **Timely pollination:** Timing is crucial, as coriander flowers have a specific period during which they are receptive to pollination.

3. Seed collection and germination

- **Harvesting seeds:** Once pollination is successful and seeds are set, they are harvested when fully mature. Coriander seeds, which are technically fruits called schizocarps, are collected from the plants.
- **Germination:** The harvested seeds are then germinated under controlled conditions to grow new plants. These plants represent the first generation (F1) hybrids with mixed traits from both parent plants.

4. Selection of hybrid offspring

- **Evaluation:** The F1 hybrid plants are grown and evaluated for the desired traits. This includes checking for enhanced flavor, higher yield, disease resistance, and slower bolting. The best-performing plants are selected for further development.
- **Further breeding:** If necessary, these hybrids can be backcrossed with one of the parent plants or crossed with other hybrids to stabilize the desired traits in subsequent generations (F2, F3, etc.).

5. Field testing and propagation

- **Field trials:** The selected hybrids undergo field trials to assess their performance under different environmental conditions. This step is crucial for ensuring that the new variety is robust and performs well in diverse growing conditions.
- **Propagation:** Once a superior hybrid is identified, it can be propagated for commercial cultivation. Coriander is typically propagated by seeds, so the new variety is grown and multiplied through seed production.

6. Commercialization

- **Scaling up production:** The new coriander variety is produced on a larger scale, with seeds being distributed to farmers and gardeners for cultivation.
- **Variety registration:** The new variety may be registered with appropriate agricultural authorities to ensure its protection and promote its commercial use.

Challenges

- **Maintaining consistency:** Ensuring that the desirable traits are consistently expressed in the offspring is crucial, as coriander can have a lot of genetic variability.

- **Environmental factors:** The performance of the new variety must be stable across different environmental conditions, as coriander is sensitive to factors like temperature and soil conditions.

Benefits

- **Improved varieties:** Through crossing techniques, coriander varieties can be developed with better flavors, higher yields, and greater resistance to environmental stresses and diseases.
- **Sustainability:** Improved varieties can reduce the need for chemical inputs like pesticides, making coriander cultivation more sustainable.

The crossing technique in coriander is a powerful tool in plant breeding, enabling the development of superior varieties that meet the needs of both consumers and farmers.

8. Fennel

Floral biology of Fennel

The floral biology of fennel (*Foeniculum vulgare*) plays a critical role in its reproduction, pollination, and seed production. Understanding the floral structure, anthesis, pollination mechanisms, and seed development is essential for effective cultivation and breeding practices. Below is an overview of the key aspects of fennel's floral biology:

1. Flower structure

- **Inflorescence**: Fennel produces a compound umbel, which is a characteristic feature of the Apiaceae family. Each umbel is composed of smaller umbels, called umbellules, which contain numerous small flowers.
- **Flower morphology**: Each flower in the umbellule is small, yellow, and has five petals, five stamens, and a bicarpellary ovary. The flowers are typically bisexual, meaning they have both male (stamens) and female (pistils) reproductive organs.
- **Ovary**: The ovary is inferior and contains two locules, each of which develops into a seed.

2. Anthesis (Flower opening)

- **Timing**: Anthesis in fennel typically occurs during the morning hours, with flowers opening sequentially within the umbel. The process of anthesis can vary depending on environmental conditions such as temperature and humidity.
- **Duration**: The flowering period of fennel can extend over several weeks, ensuring a prolonged period for pollination.

3. Pollination

- **Pollination mechanism**: Fennel is primarily cross-pollinated, though some self-pollination can occur. The small, yellow flowers attract various pollinators, including bees, wasps, and other insects, which play a significant role in the transfer of pollen.
- **Nectar production**: The flowers produce nectar, which serves as a reward for pollinators, enhancing the efficiency of cross-pollination.
- **Self-Incompatibility**: Although fennel has the capability of self-pollination, it exhibits a degree of self-incompatibility, which promotes genetic diversity through cross-pollination.

4. Anther dehiscence

- **Timing**: Anther dehiscence, the release of pollen from the anthers, usually occurs shortly after the flower opens. This timing ensures that pollen is available when pollinators visit the flowers.
- **Pollen dispersal**: Pollen is dispersed by both wind and insects, although insect pollination is more efficient due to the attraction of pollinators to the nectar and flower colour.

5. Seed setting and development

- **Fertilization**: Following successful pollination and fertilization, the ovary begins to develop into a fruit. In fennel, the fruit is a schizocarp, which eventually splits into two mericarps (seed units).
- **Seed maturity**: Seeds typically take several weeks to mature after pollination. The seeds are small, ridged, and contain essential oils that give fennel its characteristic aroma and flavour.
- **Yield**: Seed yield can vary depending on environmental conditions, plant health, and the effectiveness of pollination. High seed yield is often associated with well-managed pollination and favourable growing conditions.

6. Environmental factors affecting floral biology

- **Temperature**: High temperatures can accelerate flowering and anthesis, while cooler temperatures may prolong the flowering period.
- **Humidity**: Humidity levels can influence the timing of anthesis and anther dehiscence. Optimal humidity levels ensure efficient pollen release and viability.
- **Light**: Fennel requires adequate sunlight for optimal flower and seed development. Light intensity can affect the growth rate and the timing of flowering.

7. Implications for cultivation and breeding

- **Pollinator management**: Encouraging the presence of pollinators, such as bees, is crucial for ensuring good seed set and high yield in fennel cultivation.
- **Breeding**: Understanding the floral biology helps breeders select parent plants for hybridization, focusing on traits like flower structure, timing of anthesis, and pollination efficiency.

Understanding the floral biology of fennel is essential for optimizing cultivation practices, improving yield, and guiding breeding programs aimed at developing superior fennel varieties.

Crossing technique in fennel

The crossing technique in fennel (*Foeniculum vulgare*) is used to breed new varieties with desirable traits such as improved flavor, increased yield, better disease resistance, and enhanced bulb size. Fennel is a versatile plant grown for its seeds, leaves, and bulbs (in the case of Florence fennel). Breeding fennel through crossing involves the careful selection of parent plants and controlled pollination to produce hybrids that express the desired characteristics.

1. Selection of parent plants

- **Identifying desirable traits:** Parent plants are selected based on specific traits such as large and tender bulbs (for Florence fennel), aromatic seeds, disease resistance, and overall plant vigor.
- **Genetic diversity:** Choosing parent plants with diverse genetic backgrounds increases the likelihood of producing offspring with a combination of beneficial traits.

2. Controlled pollination

- **Isolation of flowers:** To ensure controlled pollination, the flowers of selected parent plants are isolated from other plants. This can be done using physical barriers like mesh bags or by growing the plants in isolated plots to prevent cross-pollination with unintended plants.
- **Manual pollination:** Fennel flowers are arranged in umbels, with each umbel containing numerous small flowers. During controlled crossing, pollen is manually transferred from the male parent's anthers to the female parent's stigmas. This is typically done using a fine brush or similar tool to ensure that only the desired pollen fertilizes the female flowers.
- **Monitoring pollination:** After pollination, the flowers are monitored to ensure that fertilization occurs and seeds begin to develop.

3. Seed collection and germination

- **Harvesting seeds:** Once pollination is successful and the seeds have matured, they are harvested from the female plants. Fennel seeds are usually harvested when they turn brown and dry on the plant.
- **Germination:** The harvested seeds are then germinated under controlled conditions. The resulting seedlings represent the first generation (F1) hybrids, which combine traits from both parent plants.

4. Selection of hybrid offspring

- **Evaluation:** The F1 hybrid plants are grown and evaluated for the expression of desired traits, such as bulb size, flavor, resistance to pests and diseases, and overall growth habit. The best-performing plants are selected for further breeding or direct cultivation.
- **Stabilization (If necessary):** If the desired traits are not fully expressed or stable, further breeding may be required. This could involve backcrossing with one of the parent plants or crossing F1 hybrids to produce an F2 generation. The process continues until the desired traits are consistently expressed.

5. Field testing and propagation

- **Field trials:** The selected hybrids undergo field trials to evaluate their performance in different environments and under various growing conditions. This step is crucial for determining the robustness and adaptability of the new variety.
- **Propagation:** Once a superior hybrid variety is identified, it can be propagated for commercial cultivation. Fennel is typically propagated by seeds, so the new variety is grown and multiplied through seed production.

6. Commercialization

- **Scaling up production:** The new fennel variety is produced on a larger scale, with seeds being distributed to farmers and gardeners for cultivation.
- **Variety registration:** The new variety may be registered with relevant agricultural authorities to protect the intellectual property and promote its commercial use.

Challenges

- **Ensuring uniformity:** One of the challenges in breeding fennel is ensuring that the desirable traits, such as bulb size and flavor, are uniform across the crop, especially in commercial production.

- **Environmental sensitivity:** Fennel's growth and flavor can be influenced by environmental factors, so breeding for consistency under various conditions is important.

Benefits

- **Improved varieties:** Breeding through crossing techniques can produce fennel varieties that are more flavorful, easier to grow, and more resistant to pests and diseases.
- **Enhanced yield:** New varieties can potentially offer higher yields, making fennel cultivation more profitable and efficient.

The crossing technique in fennel is a vital aspect of plant breeding, helping to develop new varieties that meet the demands of both consumers and producers, and ensuring that fennel remains a viable and valuable crop in agriculture.

9. Fenugreek

Floral biology of Fenugreek:

The floral biology of fenugreek (*Trigonella foenum-graecum*), a member of the Fabaceae family, is essential for understanding its reproductive mechanisms, pollination, and seed production. Fenugreek is widely cultivated for its seeds and leaves, which are used in cooking and traditional medicine. Here is an overview of the key aspects of fenugreek's floral biology:

1. Flower structure

- **Inflorescence**: Fenugreek typically produces flowers in the axils of the leaves, either singly or in small clusters. The flowers are borne on short pedicels.
- **Flower morphology**: The flowers of fenugreek are typical of the Fabaceae family, with a papilionaceous structure, meaning they have a butterfly-like shape.
  - **Calyx**: The flower has a green, five-lobed calyx.
  - **Corolla**: The corolla consists of five petals— a standard (large upper petal), two wings, and two fused lower petals forming the keel, which encloses the reproductive organs.
  - **Stamens**: Fenugreek flowers have ten stamens, which are diadelphous, meaning nine of the stamens are fused into a bundle, and one is free.
  - **Gynoecium**: The pistil consists of a single carpel with an ovary containing several ovules, a style, and a stigma.

2. Anthesis (Flower opening)

- **Timing**: Anthesis in fenugreek usually occurs in the early morning. The exact timing can vary depending on environmental conditions such as temperature and light.
- **Duration**: Each flower remains open for a short period, usually for a single day, but the overall flowering period can extend over several weeks as different flowers on the plant bloom sequentially.

3. Pollination

- **Pollination mechanism**: Fenugreek is primarily self-pollinating, with flowers often pollinating themselves before opening (cleistogamy). However, some degree of cross-pollination can occur, facilitated by insects such as bees.
- **Self-Pollination**: The structure of the flower promotes self-pollination, as the stamens and pistil are in close proximity, allowing pollen to transfer to the stigma within the same flower.
- **Cross-Pollination**: In cases where the flower is exposed and cross-pollination occurs, insects such as bees play a role in transferring pollen between flowers.

4. Anther dehiscence

- **Timing**: Anther dehiscence, the release of pollen from the anthers, typically occurs just before or during the flower's opening.
- **Pollen Dispersal**: In fenugreek, pollen is often released inside the flower before it opens, ensuring that self-pollination occurs efficiently.

5. Seed setting and development

- **Fertilization**: After pollination and successful fertilization, the ovary develops into a pod (legume) containing multiple seeds.
- **Pod development**: The pods are slender, elongated, and slightly curved, each containing 10-20 seeds. The pods mature over several weeks after pollination.
- **Seed maturity**: Seeds develop and mature within the pods, which dry and turn yellow-brown when ready for harvest. The seeds are small, hard, and have a characteristic bitter taste.

6. Environmental factors affecting floral biology

- **Temperature**: Temperature can influence the timing of flowering and seed set in fenugreek. Optimal temperatures promote better flower development and seed production.

- **Light**: Adequate sunlight is crucial for fenugreek's growth and flowering. Light affects the rate of photosynthesis, which in turn influences flowering and pod development.
- **Water availability**: Consistent moisture is essential during the flowering and pod development stages. Drought stress can lead to reduced flower and seed production.

7. Implications for cultivation and breeding

- **Seed production**: Understanding fenugreek's floral biology is important for maximizing seed yield, especially in self-pollinating crops where maintaining genetic purity is critical.
- **Breeding**: In breeding programs, selecting for traits such as flower number, pod size, and seed quality can lead to the development of improved fenugreek varieties.

This understanding of fenugreek's floral biology is crucial for optimizing agricultural practices, ensuring good pollination, and supporting breeding efforts aimed at improving yield and quality in fenugreek cultivation.

Crossing technique in Fenugreek

The crossing technique in fenugreek (*Trigonella foenum-graecum*), a leguminous plant commonly grown for its seeds and leaves, involves the deliberate breeding of different plants to develop new varieties with improved traits such as higher yield, enhanced nutritional content, disease resistance, and better adaptability to different environmental conditions. Here's an overview of the process:

1. Selection of parent plants

- **Identifying desirable traits:** Parent plants are selected based on traits such as seed yield, leaf production, resistance to diseases (like powdery mildew), drought tolerance, and nutritional content (e.g., higher protein or fiber).
- **Genetic diversity:** Choosing genetically diverse parent plants increases the likelihood of producing hybrids with a combination of beneficial traits.

2. Controlled pollination

- **Flower structure:** Fenugreek flowers are typically self-pollinated, but for controlled crossing, the flowers can be manually pollinated. Fenugreek flowers are small and possess both male (anthers) and female (stigmas) reproductive organs.
- **Emasculation:** To ensure cross-pollination, the anthers (male parts) of the flower from the female parent plant are removed before they mature to prevent self-pollination. This process is known as emasculation.

- **Hand pollination:** After emasculation, pollen from the selected male parent plant is manually transferred to the stigma of the female parent plant. This is done using a fine brush or similar tool to ensure that the pollen from the desired plant fertilizes the female flower.
- **Bagging:** After pollination, the flower may be covered with a small bag to prevent unwanted pollen from other plants from contaminating the cross.

3. Seed collection and germination

- **Harvesting Seeds:** After successful pollination, the fertilized flowers develop pods containing seeds. These seeds are harvested once they mature and dry on the plant.
- **Germination:** The seeds are then germinated under controlled conditions to produce the first generation (F1) hybrid plants. These hybrids will exhibit a mix of traits from both parent plants.

4. Selection of hybrid offspring

- **Evaluation:** The F1 hybrids are grown and carefully evaluated for the desired traits, such as higher yield, better disease resistance, or improved nutritional content. The best-performing plants are selected for further breeding.
- **Stabilization (If Necessary):** If the desired traits are not fully stable, further breeding may be conducted. This can involve backcrossing with one of the parent plants or crossing among the F1 hybrids to produce F2 and subsequent generations, stabilizing the desired traits.

5. Field testing and propagation

- **Field Trials:** The selected hybrids are subjected to field trials in different environments to assess their performance, yield, disease resistance, and overall adaptability to various growing conditions.
- **Propagation:** Once superior hybrids are identified, they are propagated on a larger scale. Fenugreek is typically propagated by seeds, so the new variety is multiplied through seed production.

6. Commercialization

- **Scaling up production:** The new fenugreek variety is produced on a larger scale, with seeds distributed to farmers and growers for commercial cultivation.
- **Variety registration:** The new variety may be registered with relevant agricultural authorities to protect intellectual property and promote its commercial use.

Challenges

- **Self-pollination:** Fenugreek is predominantly self-pollinating, which makes controlled cross-pollination labor-intensive. Ensuring successful crosses requires careful emasculation and pollination techniques.
- **Environmental variability:** Like other crops, fenugreek's growth and yield can be influenced by environmental factors, so the new variety must be tested across various conditions to ensure stability and performance.

Benefits

- **Improved varieties:** Through crossing techniques, breeders can develop fenugreek varieties with higher yields, better disease resistance, and enhanced nutritional content, benefiting both farmers and consumers.
- **Enhanced adaptability:** New varieties can be tailored to perform well in different climates and soil conditions, making fenugreek cultivation more versatile and sustainable.

The crossing technique in fenugreek is an essential tool in plant breeding, allowing the development of improved varieties that meet specific agricultural and market needs, contributing to the crop's economic and nutritional value.

10. Cumin

Floral biology of cumin

The floral biology of cumin (*Cuminum cyminum*), a small but important spice crop, plays a critical role in its reproduction and cultivation. Here's an overview of the key aspects:

1. Flower structure

- **Inflorescence**: Cumin flowers are arranged in compound umbels, which are characteristic of the Apiaceae (Umbelliferae) family. Each umbel consists of multiple small, individual flowers that are grouped together on short stalks, all radiating from a common point.
- **Flowers**: The individual flowers are small, typically about 2-3 mm in diameter, and are white or pinkish in colour. Each flower is hermaphroditic, meaning it contains both male (stamens) and female (pistil) reproductive organs.

2. Anthesis (Flower opening)

- **Timing**: The flowers of cumin usually open in the morning, with anthesis often occurring between dawn and mid-morning. The exact timing can vary depending on environmental conditions such as temperature and humidity.

- **Duration**: Each flower remains open and functional for a short period, usually one day.

3. Anther dehiscence

- **Timing**: Anther dehiscence, which is the process of pollen release, occurs shortly after the flowers open. The release of pollen typically happens during the morning hours when the flowers are most active and pollinators are present.

4. Stigma receptivity

- **Timing**: The stigma of cumin flowers becomes receptive to pollen shortly after the flowers open. Stigma receptivity usually overlaps with the period of anther dehiscence, which enhances the chances of successful pollination.
- **Self-Incompatibility**: Cumin exhibits partial self-incompatibility, meaning that while self-pollination can occur, cross-pollination between different plants often leads to better seed set and higher yields.

5. Pollination behaviour

- **Pollinators**: Cumin is primarily pollinated by insects, including bees, flies, and other small pollinators. These insects are attracted to the flowers by their colour and the presence of nectar.
- **Cross-Pollination**: Cross-pollination is common in cumin due to the activity of insect pollinators moving between flowers and plants. This cross-pollination is important for genetic diversity and can lead to improved plant vigor and seed production.

6. Fruit and seed development

- Following successful pollination and fertilization, the flowers develop into small, elongated fruits known as schizocarps, which split into two mericarps (seed-like structures) upon maturity. These mericarps are what is harvested and used as cumin seeds.

7. Floral scent

- Cumin flowers emit a mild fragrance, which is attractive to pollinators. The scent, combined with the flower's visual appeal, plays a key role in attracting insects necessary for pollination.

8. Growth habit

- Cumin is an annual herbaceous plant, and it completes its life cycle within a single growing season. Understanding the timing of flowering and pollination is crucial for optimizing seed production, which is the primary agricultural output of cumin.

9. Cultivation practices

- In cumin cultivation, ensuring adequate pollination is critical for good seed yield. This can involve managing the crop environment to attract pollinators and ensuring that the plants are healthy and capable of producing viable flowers.

Understanding the floral biology of cumin is essential for effective cultivation, as the reproductive success of the plants directly impacts seed yield, which is the primary product harvested from cumin crops.

Crossing technique in cumin

The crossing technique in cumin (*Cuminum cyminum*), a popular spice known for its aromatic seeds, is employed to develop new varieties with improved traits such as higher yield, better disease resistance, and enhanced flavor. Cumin is typically self-pollinated, but controlled cross-pollination is possible and used in breeding programs to combine desirable traits from different plants. Here's an overview of the process:

1. Selection of parent plants

- **Identifying desirable traits:** Parent plants are selected based on traits such as seed yield, oil content, resistance to pests and diseases (like wilt and blight), and adaptability to environmental conditions (e.g., drought tolerance).
- **Genetic diversity:** Selecting genetically diverse parents increases the chances of producing offspring with a combination of beneficial traits.

2. Controlled pollination

- **Flower structure:** Cumin flowers are small and typically arranged in umbels. They have both male (anthers) and female (stigmas) reproductive parts, making them usually self-pollinating.
- **Emasculation:** To facilitate cross-pollination, the anthers of the selected female parent plants are removed before they release pollen. This prevents self-pollination. This process is delicate and needs to be done with precision to avoid damaging the flower.
- **Hand pollination:** After emasculation, pollen from the selected male parent is manually transferred to the stigma of the female parent. This is done using fine tools, such as a brush or tweezers, ensuring that only the desired pollen fertilizes the flower.
- **Bagging:** The flowers may be covered with a small bag after pollination to prevent contamination by unwanted pollen from other plants.

3. Seed collection and germination

- **Harvesting seeds:** Once fertilization occurs, the flower develops seeds. These seeds are harvested when they are fully mature and dry on the plant.
- **Germination:** The seeds are then germinated under controlled conditions to produce the first generation (F1) hybrid plants. These plants will carry a mix of traits from both parent plants.

4. Selection of hybrid offspring

- **Evaluation:** The F1 hybrid plants are grown and evaluated for desired traits such as yield, disease resistance, flavor, and overall plant vigor. The best-performing plants are selected for further breeding or direct cultivation.
- **Stabilization:** If the desired traits are not stable or fully expressed in the F1 generation, further breeding may be conducted. This could involve backcrossing with one of the parents or crossing F1 hybrids to produce an F2 generation, continuing until the desired traits are stabilized.

5. Field testing and propagation

- **Field Trials:** The selected hybrids are tested in field conditions to evaluate their performance across different environments. This step is crucial for determining how the new variety performs under various climatic and soil conditions.
- **Propagation:** Once a superior hybrid is identified, it is propagated for commercial cultivation. Cumin is typically propagated by seeds, so the new variety is grown and multiplied through seed production.

6. Commercialization

- **Scaling up production:** The new cumin variety is produced on a larger scale, with seeds being distributed to farmers for commercial cultivation.
- **Variety registration:** The new variety may be registered with relevant agricultural authorities, protecting intellectual property and promoting its commercial use.

Challenges

- **Self-pollination:** Cumin is predominantly self-pollinating, which makes controlled cross-pollination labor-intensive and requires careful manual intervention.
- **Environmental sensitivity:** Cumin's growth and yield can be significantly affected by environmental factors, so breeding for stability across different conditions is important.

Benefits

- **Improved varieties:** Through crossing, breeders can develop cumin varieties with higher yields, better flavor, enhanced resistance to pests and diseases, and improved adaptability to different growing conditions.
- **Sustainable cultivation:** New varieties can help reduce the need for chemical inputs like pesticides, making cumin cultivation more sustainable.

The crossing technique in cumin is an essential part of breeding programs aimed at improving the quality and productivity of this valuable spice, benefiting both producers and consumers.

11. Aijwain

Floral biology of Ajwain

The floral biology of **aijwain** (*Trachyspermum ammi*), a medicinal and culinary plant in the Apiaceae family, involves several key aspects:

Flower structure

- **Inflorescence:** Aijwain flowers are small and are organized in compound umbels, typical of the Apiaceae family. Each umbel contains numerous tiny flowers, which are usually white or pinkish.
- **Floral parts:** The flowers are generally bisexual, with both male (stamens) and female (pistil) reproductive organs. The flowers have five petals, five stamens, and a bicarpellary ovary with two locules.
- **Pollination:** Aijwain flowers are primarily pollinated by insects (entomophilous pollination), especially small bees, flies, and other pollinators that are attracted to the nectar and pollen.
- **Flowering season:** Aijwain typically flowers during late spring to early summer. The flowering period may vary slightly depending on the local climate and growing conditions.

Reproductive biology

- **Fertilization:** The flowers are protandrous, meaning the male parts mature first, which promotes cross-pollination. However, self-pollination can also occur if cross-pollination is unsuccessful.
- **Seed formation:** After fertilization, the ovary develops into a schizocarp, which eventually splits into two mericarps, each containing a single seed.

Ecological and Agricultural significance

- **Medicinal properties:** The seeds, which are the primary product, contain essential oils like thymol, known for their medicinal properties, including antimicrobial and digestive benefits.

- **Cultivation:** Understanding the floral biology is crucial for cultivating aijwain effectively, particularly in ensuring adequate pollination and seed set for optimal yield.

This overview provides insights into the floral biology of aijwain, which is important for both its ecological role and its value in traditional medicine and culinary uses.

Crossing technique in Aijwain

Ajwain (Trachyspermum ammi), also known as carom seeds or bishop's weed, is a spice with significant culinary and medicinal uses. Breeding and improving ajwain through crossing techniques involve a careful and scientific approach, as ajwain is typically self-pollinating. Here's how crossing techniques can be applied to ajwain:

1. Understanding Ajwain's reproductive biology

- **Self-Pollination**: Ajwain is primarily a self-pollinating plant, meaning that it tends to fertilize itself without the need for external pollen sources. This makes it challenging to introduce genetic diversity naturally.
- **Controlled cross-pollination**: To introduce new traits or improve existing ones, breeders need to perform controlled cross-pollination, which involves manually transferring pollen from one plant to the stigma of another.

2. Crossing technique in Ajwain

- **Selection of parent plants**: The first step is to select parent plants that exhibit desirable traits, such as disease resistance, higher yield, or improved flavor. The chosen parents should ideally complement each other's strengths and weaknesses.
- **Emasculation**: Since ajwain is self-pollinating, it is essential to prevent self-pollination. This can be done through emasculation, where the anthers (male parts) are removed from the flowers of the plant designated as the female parent before they release pollen.
- **Pollination**: After emasculation, pollen from the male parent plant is carefully collected and manually applied to the stigma of the emasculated flowers of the female parent. This ensures that only the desired pollen fertilizes the ovules.
- **Bagging**: To prevent unwanted pollen from contaminating the cross, the flowers are usually covered with a bag after pollination. This step ensures that the cross is successful and no external pollen interferes with the process.

- **Seed Collection**: Once the cross-pollinated flowers mature and seeds are formed, the seeds are harvested. These seeds will carry the genetic material from both parent plants.
- **Growing and selection**: The seeds are then grown into plants, and the offspring (F1 generation) are evaluated for the desired traits. The best-performing plants can be selected for further breeding or commercial cultivation.

3. Goals of cross-breeding in Ajwain

- **Trait improvement**: The primary goal of cross-breeding in ajwain is to combine desirable traits from two different parent plants. This could include improving yield, enhancing disease resistance, increasing essential oil content, or developing new varieties with better adaptability to different climates.
- **Genetic diversity**: Introducing new genetic material through crossing helps increase the genetic diversity within ajwain populations, which can be beneficial for long-term breeding programs and the overall resilience of the species.

4. Challenges in Ajwain cross-breeding

- **Low outcrossing rate**: Due to its self-pollinating nature, achieving successful cross-pollination in ajwain can be challenging and requires careful management.
- **Time-consuming process**: Breeding new varieties through crossing is a time-consuming process, requiring multiple generations to stabilize and select the desired traits.
- **Limited genetic information**: Unlike more widely studied crops, ajwain may have less genetic information available, making it more challenging to apply advanced breeding techniques.

5. Application of advanced techniques

- **Molecular markers**: In modern breeding programs, molecular markers can be used to identify and select plants with desired traits at the seedling stage, speeding up the breeding process.
- **Tissue culture and genetic engineering**: While traditional crossing techniques are the norm, advanced methods like tissue culture and genetic engineering might be explored to introduce specific traits more effectively.

In conclusion, crossing techniques in ajwain involve careful manual pollination methods to introduce new traits and improve the crop. These techniques are crucial for developing improved ajwain varieties that can meet the needs of farmers and consumers alike.

Floral biology of Ginger

The floral biology of ginger (*Zingiber officinale*) is key to understanding its reproduction, though ginger is primarily propagated vegetatively through rhizomes rather than by seeds. However, the plant does produce flowers under certain conditions. Here's an overview of the floral biology of ginger:

1. Flower structure

- **Inflorescence**: Ginger flowers are borne on a spike-like inflorescence that emerges from the rhizome, separate from the main leafy shoot. The inflorescence is typically 10-20 cm long and is composed of a cluster of small flowers, often with greenish-yellow petals and a distinct labellum (a specialized petal) that may be purple or red.
- **Flowers**: Each flower is hermaphroditic, containing both male (stamens) and female (pistil) reproductive organs. The flowers are tubular, with a prominent labellum that acts as a landing platform for pollinators.

2. Anthesis (Flower opening)

- **Timing**: Ginger flowers usually open in the early morning. The exact timing can vary depending on environmental factors such as temperature and humidity.
- **Duration**: The flowers remain open for one to two days, during which they are receptive to pollination.

3. Anther dehiscence

- **Timing**: Anther dehiscence, which is the release of pollen from the anthers, occurs shortly after the flowers open. This release is typically synchronized with the receptivity of the stigma to facilitate successful pollination.

4. Stigma receptivity

- **Timing**: The stigma becomes receptive around the same time the pollen is released. Stigma receptivity generally lasts for the duration that the flower remains open.

5. Pollination behaviour

- **Pollinators**: In the wild or in cultivation where flowering is encouraged, ginger is primarily pollinated by insects, particularly bees. The flowers' structure and scent attract pollinators, which facilitate the transfer of pollen from the anthers to the stigma.
- **Self-Pollination and Cross-Pollination**: Ginger flowers can self-pollinate because they contain both male and female organs. However, cross-pollination by insects is also possible and may lead to greater genetic diversity.

6. Fruit and seed development

- While ginger can produce fruit and seeds following successful pollination, this is uncommon in cultivated ginger. The fruit is a small capsule, but seed production is rare because ginger is typically propagated vegetatively through its rhizomes rather than by seed.

7. Floral scent

- Ginger flowers have a mild fragrance, which helps attract pollinators. However, the scent is not as strong or distinctive as in other flowering plants.

8. Vegetative propagation

- Although ginger does flower and can reproduce sexually, it is primarily grown through vegetative propagation. Farmers plant sections of the rhizome, which then sprout and grow into new plants. This method ensures consistency in the crop, as all plants are genetically identical to the parent.

Understanding the floral biology of ginger is more of academic interest in cultivated varieties, as commercial propagation relies almost exclusively on the rhizome. However, in wild or native settings, flowering and seed production can play a role in the plant's life cycle.

Floral biology of Turmeric

The floral biology of turmeric (*Curcuma longa*) is an important aspect of its reproductive biology, although like ginger, turmeric is primarily propagated vegetatively through its rhizomes. Here's an overview of turmeric's floral biology:

1. Flower structure

- **Inflorescence**: Turmeric flowers are borne on a spike-like inflorescence that arises directly from the rhizome. The inflorescence is usually 12-20 cm long and is composed of numerous bracts (modified leaves), which are green at the base and gradually turn pink or reddish towards the top.
- **Flowers**: The flowers themselves are small, tubular, and yellow. They are hidden within the bracts and are hermaphroditic, containing both male (stamens) and female (pistil) reproductive organs. Each flower consists of a three-lobed corolla (the petal structure) and a prominent labellum, which is often yellow and serves as a landing platform for pollinators.

2. Anthesis (Flower opening)

- **Timing**: The flowers of turmeric typically open in the early morning. Anthesis occurs when the environmental conditions, such as light and temperature, are favorable.

- **Duration**: The flowers remain open for about one day, during which time they are receptive to pollination.

3. Anther dehiscence

- **Timing**: Anther dehiscence, the process by which pollen is released from the anthers, occurs shortly after the flower opens. The release of pollen is synchronized with the period of stigma receptivity to optimize the chances of fertilization.

4. Stigma receptivity

- **Timing**: The stigma becomes receptive around the same time as pollen release and remains receptive for the duration of the flower's life, which is typically one day.

5. Pollination behaviour

- **Pollinators**: Turmeric flowers are primarily pollinated by insects, particularly bees. The flowers produce a mild fragrance and nectar, which attract pollinators. The structure of the flower encourages pollinators to come into contact with both the anthers and the stigma, facilitating pollination.
- **Self-Pollination and Cross-Pollination**: While turmeric flowers can self-pollinate due to the presence of both male and female organs in the same flower, cross-pollination by insects is also possible and may occur.

6. Fruit and seed development

- Although turmeric can produce seeds following successful pollination, seed production is rare and not typically used for propagation in cultivated varieties. Instead, turmeric is almost exclusively propagated vegetatively using pieces of the rhizome, which ensures uniformity in the crop.

7. Floral scent

- Turmeric flowers have a mild fragrance, which helps to attract pollinators such as bees. The scent, along with the colour of the flowers, plays a role in the pollination process.

8. Vegetative propagation

- In agricultural practice, turmeric is propagated through its rhizomes rather than by seeds. The rhizomes are divided into small pieces, each containing a bud, and planted to grow new plants. This method of propagation is preferred as it maintains the genetic uniformity of the crop.

Understanding the floral biology of turmeric is useful for those interested in the plant's natural reproduction, but for practical agricultural purposes, vegetative propagation through rhizomes is the dominant method used to cultivate turmeric

B) Field trials and experimental designs in spice breeding

Field trials and experimental designs are critical components of spice breeding programs. They help evaluate the performance of new spice varieties in real-world conditions, ensuring that the selected traits (such as yield, disease resistance, and flavor) perform well under various environmental factors. Here's an overview of how field trials are conducted and the experimental designs commonly used in spice breeding:

1. Purpose of field rials in Spice Breeding

- **Performance Evaluation**: Field trials are used to assess the performance of new and existing spice varieties in terms of yield, quality, disease resistance, and adaptability to different environmental conditions.
- **Selection of superior varieties**: The data collected from field trials helps breeders identify the best-performing varieties for further development and potential commercialization.
- **Adaptation to local conditions**: Trials are often conducted in multiple locations to evaluate how a variety performs under different climatic and soil conditions, ensuring the new varieties are suitable for a wide range of environments.

2. Types of field trials

- **Preliminary yield trials (PYT)**: These are initial trials that test a large number of breeding lines to identify those with potential. These trials typically involve small plots and are conducted in one or two locations.
- **Advanced yield trials (AYT)**: After promising lines are identified in PYT, they are tested in more detailed trials with larger plots across multiple locations to evaluate their performance more thoroughly.
- **Multi-Location Trials (MLT)**: These trials involve testing the best-performing varieties across several different locations, representing diverse environmental conditions. MLTs provide data on the stability and adaptability of a variety.
- **On-farm trials**: Conducted in collaboration with farmers, these trials assess the performance of new varieties under actual farming conditions. They help gather farmer feedback and gauge the practical utility of the varieties.

3. Experimental designs in field trials

- **Randomized Complete Block Design (RCBD)**: This is one of the most commonly used designs in spice breeding trials. In RCBD, experimental plots are grouped into blocks, each containing all the treatments (varieties). Within each block, treatments are randomly assigned to plots. This design helps control for variability within the trial site.

- **Advantages**: Controls for environmental variation across the field, making it easier to detect differences between varieties.
- **Application**: Suitable for trials where the field has varying conditions (e.g., soil fertility, moisture levels).

- **Split-Plot Design**: This design is used when there are two levels of experimental factors, such as variety and irrigation level. The main plot might be assigned one factor (e.g., irrigation), and the sub-plots within each main plot are assigned another factor (e.g., variety).
  - **Advantages**: Allows the study of interactions between different factors (e.g., how different irrigation levels affect different spice varieties).
  - **Application**: Useful when testing the effect of multiple factors on spice performance.
- **Lattice design**: This design is used when a large number of varieties need to be tested. Lattice designs help reduce the number of comparisons needed, making the trial more manageable while still controlling for variability.
  - **Advantages**: Efficient for trials with many varieties, reduces the number of plots required compared to RCBD.
  - **Application**: Commonly used in preliminary screening of a large number of breeding lines.
  - **Augmented design**: Used in early stages of breeding when many unreplicated lines are tested alongside a few replicated control varieties. This design allows the evaluation of a large number of lines with fewer resources.
  - **Advantages**: Efficient for initial screening of many lines, saving time and resources.
  - **Application**: Suitable for evaluating breeding lines that are not yet advanced enough to justify full replication.
- **Factorial design**: This design tests the effect of two or more factors (e.g., variety, fertilizer type) and their interactions on the outcome.
  - **Advantages**: Provides detailed information on the interaction between multiple factors affecting spice growth.
  - **Application**: Used when breeders want to understand how different factors combined affect spice yield or quality.

4. Data collection in field trials

- **Phenotypic traits**: Data on plant height, leaf size, flowering time, and other morphological traits are collected regularly.
- **Yield data**: Yield is a critical metric, including total yield per plant, yield per hectare, and the number of marketable units (e.g., peppercorns, turmeric rhizomes).

- **Quality assessments**: Flavor, aroma, color, and the concentration of active compounds (e.g., curcumin in turmeric, piperine in black pepper) are evaluated through laboratory analysis.
- **Disease and pest resistance**: Observations on the incidence and severity of diseases or pest infestations are recorded to assess resistance.
- **Environmental data**: Soil characteristics, weather conditions, and irrigation practices are monitored to correlate environmental factors with plant performance.

5. Statistical analysis of field trial data

- **Analysis of variance (ANOVA)**: ANOVA is used to determine if there are statistically significant differences between the varieties tested. It helps identify whether observed differences in performance are due to the varieties themselves or random variation.
- **Stability analysis**: This involves analyzing data from multi-location trials to assess how consistent a variety's performance is across different environments.
- **Genotype by Environment Interaction (GxE)**: This analysis helps breeders understand how different varieties perform in various environments, which is crucial for selecting varieties that are stable and adaptable.

6. Implementation of results

- **Selection of varieties**: Based on the trial results, breeders select the best-performing varieties for further development or commercialization.
- **Breeding decisions**: Data from field trials guide future breeding efforts, such as identifying desirable traits to focus on or deciding which parent lines to use in crosses.
- **Farmer adoption**: Successful varieties are promoted to farmers, often through extension services or agricultural demonstration plots.

Field trials and experimental designs are fundamental in spice breeding, ensuring that new varieties are rigorously tested and proven to perform well under a range of conditions before they are released to farmers. These trials help breeders make informed decisions and provide farmers with reliable, high-performing spice varieties.

C) Data collection and analysis

Data collection and analysis are critical components of spice breeding programs. Accurate data collection and robust analysis ensure that breeders can identify the best-performing varieties and make informed decisions about which traits to select for future breeding efforts. Here's an overview of the key aspects of data collection and analysis in spice breeding:

1. Types of data collected

Data collection in spice breeding encompasses various traits and environmental factors that influence plant growth, yield, and quality. These can be broadly categorized as:

- **Phenotypic data**
  - **Growth traits**: Includes plant height, leaf area, stem diameter, root length, and overall plant vigor.
  - **Reproductive traits**: Number of flowers, fruit set percentage, number of seeds per fruit, and time to maturity.
  - **Yield data**: Total yield per plant, yield per hectare, weight of individual fruits or seeds, and marketable yield (e.g., usable spice portions like peppercorns or turmeric rhizomes).
  - **Quality traits**: Flavor, aroma, color, and concentration of bioactive compounds (e.g., curcumin in turmeric, piperine in black pepper).
  - **Resistance traits**: Incidence and severity of diseases (e.g., fungal infections, viral diseases) and pest infestations (e.g., nematodes, insect pests).
- **Environmental Data**
  - **Soil characteristics**: pH, nutrient content, texture, and moisture levels.
  - **Climate data**: Temperature, rainfall, humidity, and light intensity.
  - **Irrigation practices**: Amount, frequency, and method of irrigation.
- **Genetic Data**
  - **Molecular markers**: DNA markers linked to desirable traits, used in marker-assisted selection (MAS).
  - **Genomic data**: Information on the genetic makeup of the plants, often used in genomic selection or to understand genotype-environment interactions.

2. Methods of data collection

Data collection methods vary depending on the type of data being gathered and the traits of interest:

- **Manual measurements**
  - Traits like plant height, leaf area, and fruit size are often measured manually using tools like rulers, calipers, and scales.
  - Disease and pest incidence are recorded through visual inspections, often using standardized scoring systems.

- **High-throughput phenotyping**
  - Advanced imaging technologies (e.g., drones equipped with multispectral cameras, ground-based sensors) are used to capture data on plant traits such as canopy cover, leaf chlorophyll content, and biomass, enabling rapid assessment of large breeding populations.
  - Automated systems can analyze hundreds of plants in a short time, increasing the efficiency of data collection.
- **Laboratory analysis**
  - Quality traits such as flavor, aroma, and bioactive compound concentration are measured using laboratory techniques like gas chromatography, mass spectrometry, and high-performance liquid chromatography (HPLC).
  - Nutrient content and soil properties are analyzed using standard soil testing methods.
- **Digital tools and software**
  - Data collection apps and digital tools are increasingly used to record and manage field data, reducing errors and improving data integration.
  - Geographic Information Systems (GIS) and data loggers are used to track environmental variables and correlate them with plant performance.

3. Data management

- **Data entry and storage**: Collected data is entered into databases, often using specialized software for agricultural research. This data must be carefully stored and backed up to prevent loss.
- **Data cleaning**: Before analysis, data is cleaned to remove errors, inconsistencies, and outliers that could skew results.
- **Data integration**: Different types of data (e.g., phenotypic, environmental, genetic) are integrated into a single database for comprehensive analysis. This integration helps in understanding complex interactions between plant traits and environmental factors.

4. Data analysis techniques

- **Descriptive statistics**: Basic statistics like mean, median, standard deviation, and range are calculated to summarize the data and provide an initial overview of the performance of different varieties.
- **Analysis of variance (ANOVA)**: ANOVA is used to determine if there are statistically significant differences between the performance of different varieties or treatments. It helps identify whether observed differences are due to genetic factors or random variation.

- **Genotype by Environment Interaction (GxE) Analysis**: This analysis helps breeders understand how different genotypes perform across various environments. It identifies stable varieties that perform consistently well in different conditions, which is crucial for developing widely adaptable spice varieties.
- **Regression analysis**: Used to model relationships between dependent variables (e.g., yield) and independent variables (e.g., environmental factors, genetic markers). This helps in predicting how changes in one variable might influence another.
- **Multivariate analysis**: Techniques like Principal Component Analysis (PCA) and Cluster Analysis are used to reduce data complexity and identify patterns or groupings in the data. These methods help in identifying key traits that contribute to overall performance.
- **Heritability estimates**: Heritability analysis determines the proportion of phenotypic variation in a trait that is due to genetic factors. High heritability suggests that a trait is largely controlled by genetics and can be effectively selected in breeding programs.
- **BLUP (Best Linear Unbiased Prediction)**: This statistical method is used to predict the genetic value of breeding lines based on both phenotypic and pedigree data, allowing breeders to make more accurate selection decisions.

5. Interpretation and application of results

- **Variety selection**: Data analysis helps breeders select the best-performing varieties for further development, commercialization, or as parents in future breeding programs.
- **Breeding strategy refinement**: Analysis results guide adjustments to breeding strategies, such as focusing on specific traits, using particular genetic lines, or targeting certain environmental conditions.
- **Farmer recommendations**: Data from field trials are used to develop recommendations for farmers, such as which varieties are best suited to specific regions or growing conditions.
- **Publication and reporting**: Results are often published in scientific journals, shared at agricultural conferences, or reported to funding bodies and stakeholders. Clear reporting ensures that the findings can be used to advance spice breeding knowledge and practices.

6. Challenges in data collection and analysis

- **Environmental variability**: Variations in climate, soil, and management practices can complicate the interpretation of results, making it challenging to isolate the effect of genetic factors.

- **Data volume and complexity**: The large volume of data generated in modern breeding programs, especially with the use of high-throughput technologies, requires sophisticated data management and analysis tools.
- **Resource constraints**: Conducting comprehensive data collection and analysis can be resource-intensive, requiring significant time, labor, and financial investment.

7. Future directions

- **Big data and machine learning**: As data collection methods become more sophisticated, there is growing interest in using big data analytics and machine learning to uncover complex patterns and make more accurate predictions in spice breeding.
- **Real-time data collection**: The use of IT devices and mobile technology is enabling real-time data collection and analysis, allowing breeders to make quicker decisions during the growing season.
- **Precision breeding**: Combining detailed data analysis with precision agriculture techniques can lead to more targeted breeding efforts, optimizing the development of spice varieties for specific environments and market needs.

In summary, data collection and analysis in spice breeding are essential for developing superior spice varieties that meet the demands of farmers and consumers. The use of advanced technologies and robust statistical methods ensures that breeding programs are efficient, effective, and capable of producing high-quality, high-yielding spice crops.

D) Case studies of successful breeding programs

Case studies of successful breeding programs in spices provide valuable insights into the strategies, techniques, and outcomes that have led to the development of improved spice varieties. These examples highlight how targeted breeding efforts can address specific challenges and opportunities in spice cultivation. Below are a few notable case studies:

1. High-yielding varieties of Black Pepper (*Piper nigrum*)

- **Background**: Black pepper is one of the most widely traded spices globally. However, traditional varieties were often susceptible to diseases such as *Phytophthora* foot rot and had lower yields.
- **Breeding goals**: The primary objectives were to develop black pepper varieties with higher yield, disease resistance, and improved quality.

- **Breeding strategy**
  - **Selection and hybridization**: Breeding programs focused on crossing high-yielding but disease-susceptible varieties with disease-resistant lines.
  - **Marker-assisted selection (MAS)**: MAS was used to identify and select plants with genetic resistance to *Phytophthora* foot rot.
- **Outcome**
  - **Panniyur series**: The development of the Panniyur series of black pepper varieties in India is a significant success. These varieties, such as Panniyur 1, are known for their high yield, improved disease resistance, and adaptability to different growing conditions.
  - **Impact**: These varieties have been widely adopted by farmers in India and other pepper-growing regions, significantly boosting production and reducing crop losses due to disease.

2. Curcumin-enriched Turmeric (Curcuma longa)

- **Background**: Turmeric is valued not only for its culinary uses but also for its medicinal properties, primarily due to curcumin, a compound with anti-inflammatory and antioxidant properties. Traditional turmeric varieties had varying levels of curcumin.
- **Breeding goals**: The goal was to develop turmeric varieties with higher curcumin content to meet the growing demand for nutraceutical and pharmaceutical applications.
- **Breeding strategy**
  - **Germplasm collection and screening**: A wide range of turmeric germplasm was collected and screened for curcumin content.
  - **Hybridization and selection**: High-curcumin lines were crossed and selected to create varieties with consistently high curcumin levels.
  - **Tissue culture**: Micropropagation techniques were used to rapidly multiply selected high-curcumin varieties and ensure genetic uniformity.
- **Outcome**:
  - **Varieties like ‘Prabha’, ‘Suvarna’, and ‘Krishna’**: These varieties were developed with significantly higher curcumin content (over 5%) compared to traditional varieties.
  - **Impact**: These high-curcumin turmeric varieties have gained popularity in the health food and supplement markets, providing farmers with a premium crop that commands higher prices.

3. Disease-resistant Cardamom (*Elettaria cardamomum*)

- **Background**: Cardamom, known as the "Queen of Spices," is a valuable cash crop. However, it is highly susceptible to diseases such as cardamom mosaic virus (CdMV) and fungal infections, leading to significant yield losses.
- **Breeding goals**: To develop cardamom varieties with resistance to major diseases, while maintaining high yield and quality.
- **Breeding strategy**:
  - **Selection and hybridization**: Breeding involved crossing disease-resistant wild relatives with high-yielding cultivated varieties.
  - **Field screening**: Intensive field trials were conducted in disease-prone areas to identify resistant lines.
- **Outcome**
  - **Variety 'Njallani'**: This variety, also known as 'Green Gold,' is a product of farmer-led breeding in Kerala, India. It is known for its high yield, resistance to CdMV, and adaptability to various agro-climatic conditions.
  - **Impact**: 'Njallani' has become one of the most popular cardamom varieties in India, contributing to a resurgence in cardamom cultivation and improved farmer incomes.

4. Capsaicin-rich chili peppers (*Capsicum* spp.)

- **Background**: Capsaicin is the compound responsible for the heat in chili peppers and has applications in both culinary and medicinal products. Traditional breeding sought to enhance capsaicin content for markets demanding spicier peppers.
- **Breeding goals**: The primary goal was to develop chili pepper varieties with high capsaicin content, improved yield, and disease resistance.
- **Breeding strategy**
  - **Hybridization**: Crosses were made between high-capsaicin wild species and cultivated varieties with desirable agronomic traits.
  - **MAS and genetic analysis**: Marker-assisted selection and genetic analysis were used to identify and propagate lines with the highest capsaicin content.
- **Outcome**
  - **Varieties like 'Bhut Jolokia' (Ghost Pepper) and 'Carolina Reaper'**: These peppers are renowned for their extreme heat, with capsaicin levels far exceeding traditional varieties. The 'Carolina Reaper' holds the Guinness World Record for the hottest chili pepper.

- **Impact**: These varieties have created niche markets, attracting interest from food industries and spice enthusiasts globally. Their success has also highlighted the potential of breeding for specialty markets.

5. Drought-resistant Saffron (*Crocus sativus*)

- **Background**: Saffron is one of the most expensive spices in the world, but its cultivation is water-intensive and typically limited to specific regions. Breeding efforts have aimed at developing drought-resistant varieties to expand cultivation areas.
- **Breeding goals**: To develop saffron varieties that require less water and are adaptable to arid and semi-arid conditions.
- **Breeding strategy**
  - **Selection from existing varieties**: Saffron populations were screened for natural variation in drought tolerance.
  - **Field trials**: Selected lines were tested under reduced irrigation regimes to identify the most drought-resistant plants.
- **Outcome**
  - **Drought-tolerant lines**: Although still under development, some promising lines have shown the ability to produce high-quality saffron with significantly less water.
  - **Impact**: These efforts are paving the way for saffron cultivation in regions that were previously unsuitable, potentially increasing global production and reducing the environmental impact of saffron farming.

6. Improved Vanilla (*Vanilla planifolia*) cultivars

- **Background**: Vanilla is a high-value spice with significant global demand. Traditional vanilla varieties are often prone to diseases like fusarium wilt and require labor-intensive hand-pollination.
- **Breeding goals**: To develop vanilla varieties with improved disease resistance, higher vanillin content, and better adaptability to various growing conditions.
- **Breeding strategy**:
  - **Cross breeding**: Efforts have focused on crossbreeding between disease-resistant wild species and high-vanillin content cultivars.
  - **In vitro techniques**: Tissue culture methods have been employed to propagate disease-resistant and high-yielding vanilla plants on a large scale.

- **Outcome**
  - **Varieties with enhanced disease resistance**: Newer cultivars with improved resistance to fusarium wilt and other pathogens have been introduced, helping to stabilize vanilla production.
  - **Impact**: These developments have helped sustain vanilla production in traditional growing regions while opening up possibilities for cultivation in new areas.

These case studies illustrate how targeted breeding programs can address specific challenges in spice production, leading to the development of improved varieties that benefit both farmers and consumers. Through the application of modern breeding techniques, combined with traditional knowledge, these programs have successfully enhanced the productivity, quality, and sustainability of spice crops.

Index